90 0386512 3
G000128344

Multimedia Communications Networks

Technologies and Services

For a complete listing of the *Artech House Telecommunications Library*,
turn to the back of this book.

Multimedia Communications Networks

Technologies and Services

Mallikarjun Tatipamula
Bhumip Khasnabish
Editors

Artech House
Boston • London

Library of Congress Cataloging-in-Publication Data
Multimedia communications networks : technologies and services /
Mallikarjun Tatipamula, Bhumip Khasnabish, editors
p. cm. — (Artech House telecommunications library)
Includes bibliographical references and index.
ISBN 0-89006-936-0 (alk. paper)
1. Computer networks. 2. Multimedia systems.
3. Telecommunication—Traffic—Management. I. Tatipamula,
Mallikarjun. II. Khasnabish, Bhumip. III. Series.
TK5105.5.M867 1998
006.7—dc21 98-19039
CIP

British Library Cataloguing in Publication Data
Multimedia communications networks : technologies and
services. - (Artech House telecommunications library)
1. Multimedia systems. 2. Broadband communication systems
I. Tatipamula, Mallikarjun II. Khasnabish, Bhumip
621.3'821

ISBN 0-89006-936-0

Cover design by Lynda Fishbourne

685 Canton Street
Norwood, MA 02062

International Standard Book Number: 0-89006-936-0
Library of Congress Catalog Card Number: 98-19039
10 9 8 7 6 5 4 3 2 1

Contents

Preface

Multimedia is a keyword in the evolving information age of the 21st century. Recent proliferation of Internet services and the deregulation of telecommunications, in addition to the multitude of enabling multimedia communication networking (MCN) technologies, are causing an explosion in the number and type of multimedia communication service (MCS) providers.

Traditional plain old telephony service (POTS) providers are now allowed to provide both wireline and wireless delivery of high-speed data services, video services, and Internet access services in addition to plain old telephony and low-speed data services. Similarly, the cable TV providers no longer see themselves as being solely in the business of one-way video delivery. They are now exploring the added services of high-speed data, telephony, and two-way video. Following this trend, all the others—primarily competitive access and high-speed data delivery providers—are also jumping into the MCS fray. The technologies for multimedia communications and networking are available, the globalization of technologies and services is taking place, and many telecommunication equipment and service providers are looking forward to implementing these new technologies in their networks in a global environment. Customers are expected to have many choices of multimedia communication services and MCS providers. Network planners need to implement the necessary security capabilities and features to protect the networks and to provide security to customers using those network services.

As the MCNs evolve, they will encompass most aspects of wireline and wireless networks. This involves the implementation of new technologies (e.g., asynchronous transfer mode or ATM), increased use of the existing technologies (e.g., Internet protocol or IP), planning for more distributed network architectures, and the introduction of new services (e.g., desktop multimedia conferencing). An increase in interconnection of various networks, such as telcos, interexchange carriers, CATV, and Internet service providers (ISPs), will also be visible. Some multimedia networks will have an increased dependency on software to control critical network elements and some will have an increased use of external service logic (separate from switch fabric) and external databases. Additionally there will be increased access by customers to control and manage information, for example, customer network management (CNM) services.

The integration of legacy networks and new networks, existing and emerging technologies, and new architectures and revamped/extended architectures, especially in light of the 1996 telecom act and globalization, provides significant opportunity for discussion.

Book Objectives and Intended Audience

We present an in-depth, one-source study of various technologies, services, performance and security issues, and global standards associated with networked multimedia systems. Prepared by 28 leading experts from industry and academia from around the globe, this reference book helps practicing and future communications engineers understand the impact of fast-moving broadband communications technology on networked multimedia services, and how multimedia applications affect network design. It also provides R&D personnel with the practical information needed to support a wide variety of multimedia services.

This book is intended for a broad range of readers interested in multimedia communication networks:

- Graduate students, practicing specialists (EEs, computer scientists) and engineers, researchers, consultants, and planners of telecommunication networks;
- Technical managers, marketing staff and consultants in the telecommunications field, those seeking to gain a basic understanding and up-to-date overview of advances of global technologies and services for multimedia, and those who need a single reference source for the central topics.

It is comprehensive in that it covers the major areas, but it is not intended to be an encyclopedia.

Also this book could be used for a one- or two-term graduate course and for short courses on multimedia communications. For graduate-level course instructors, short course instructors, and R&D personnel, this book is a good source of information on these areas:

1. Theoretical and practical issues in multimedia networks to support various multimedia services;
2. Transmission, switching, networking and OAM technologies, and software architectures for broadband multimedia networks to support multimedia services;
3. End-to-end practical telecommunications network topologies and architectures, and it provides information on telecom network planning taking economical and other factors into consideration;
4. Network security issues, service provider perspectives of multimedia;
5. Applications, experiences of various researchers and implementors with different backgrounds on multimedia, and testbeds from around the world in addition to comprehensive discussions on standards activities for multimedia.

Organization of the Book

Chapter 1 examines the reasons for combining multimedia systems with communication networks, and discusses both history and trends in transmission technology, switching, and services. It also looks in more detail at several applications of broadband networks for multimedia services and at service trials in various parts of the world, reviewing some of the considerations that go into field trials and system deployment. It concludes with a discussion of the telecommunication aspects of developing nations and the prospects for networked multimedia in those areas.

Chapter 2 presents a survey of high-speed LAN/MAN networks in an integrated services environment. It focuses on networking and transmission technologies for multimedia services by presenting a survey of current and emerging networking technologies in this fast-moving field. Interoperability issues are then considered.

Chapter 3 presents a generalized QoS framework and terminology for distributed multimedia application operating over multimedia networks with QoS guarantees. Following this, Chapter 3 evaluates a number of QoS architectures found in the literature. Finally, this chapter presents a short qualitative comparison and discussion.

A wide range of topics in the field of traffic management and control of multimedia networks is covered in Chapter 4. In particular, this chapter discusses issues ranging from traffic modeling and queueing performance models to actual traffic management issues. In spite of the significant amount of research in this area, the provision of optimal networks from the point of view of reliability and efficiency is still a challenge. Many of the traffic management and control decisions are currently performed under uncertainty due to the burstiness and unpredictability of the traffic. Deep insight is required to meet this challenge and the research results reported in this chapter make the reader better equipped to do so.

It has been very challenging to build a large-capacity ATM switch that is capable of supporting multicast functions and meeting various service quality requirements. In Chapter 5, we first describe the performance requirements of designing an ATM switch system and then list the functions that need to be performed by an ATM switch. Several ATM switch architectures are surveyed. Their strengths and weaknesses in performance and implementation complexity are discussed. Finally, two multicast switch design approaches, copy network and broadcast-and-select, are also presented.

Chapter 6 covers various broadband access and trunk transport systems and how they relate to multimedia. The access systems include fiber-rich architectures like fiber-to-the-home and fiber-to-the-curb, as well as other architectures like hybrid fiber coax and wireless broadband technologies. Areas that are covered in this chapter range from end-to-end requirements to customer premises equipment and architectures. Topologies are discussed in terms of technologies (role of SONET/SDH and ATM), services, evolution, and economics. The technological features of ATM are then described. The important concept of a virtual path, which makes the best use of inherent ATM capabilities, is then elaborated. ATM network hardware systems are illustrated. Future transport network requirements are then identified by reviewing the recent evolution of transmission and transport node technologies. The photonic network technologies are then demonstrated to enhance transport network performance.

Recently, customers have been demanding higher quality services, as a variety of new, advanced services are being introduced. This has led to the need to integrate telecommunication services and OAM. Chapter 7 describes the OAM technology at the network element, network, and service levels in the multimedia era. Also, OAM services, such as QoS management and a multimedia navigation service, are suggested. In addition, the OAM system technology and standards-related activities, such as TMN, ATMF, TINA, and DAVIC, are discussed.

Chapter 8 examines the special challenges associated with multimedia networking in a wireless and mobile environment, namely, providing high-quality real-time services in the face of widely varying channel characteristics as experienced by wireless and mobile users. Toward this end, this chapter reviews the historical development of wireless networking technology and points out the physical properties of the wireless channel that yield difficulties for implementing high-quality network services. This chapter also reviews the latest developments in mobile devices and applications and examines the full range of the network protocol stack and how to implement it effectively on top of unpredictable wireless channels. It examines the key issues in implementing quality of service in such a challenging environment, as well as the difficulties inherent in integrating wireless and wireline networks. A viewpoint on the future of wireless multimedia networks is then presented.

Security is an important part of the discussion of multimedia because multimedia services bring new challenges to the security arena. They also highlight weaknesses of the networks that support them. Increased levels of interactivity, increased interconnection of networks, and use of public network services are some of the areas examined relative to security. Chapter 9 introduces the security concepts at a high level and then looks closely at how a security analysis may be done. The overall goal is to provide needed protection to valuable resources while providing authorized entities access to the resources if they demonstrate appropriate need and privileges. The text moves next to examination of a technology expected to be a key building block to multimedia networks: ATM. A high-level discussion is then presented on the security services currently being examined in the industry for key pieces of the ATM protocol—namely, the user plane and control plane. Examples of some experiments and products being developed are presented. Another key technology widely used in support of multimedia is IP; highlights of some security efforts related to IP are reviewed. Finally the role of the standardization organizations in developing specifications for multimedia security is touched on briefly.

Chapter 10 covers distributed platforms and shows current extensions to support multimedia applications. It introduces some operating system extensions to multimedia, the ODP reference model and CORBA. Moreover, this chapter presents an example of an integrated middleware-groupware architecture and shows how ODP can be used to specify large-scale distributed multimedia applications.

Chapter 11 presents the structure and recent activities of telecommunication and information standardization organizations, forums, and consortia, described from the viewpoint of the current status of standards related to multimedia communications, such as ATM/B-ISDN, UPT, FPLMTS/IMT 2000, TMN, and GII.

Acknowledgments

We would like to extend heartfelt thanks to the authors, who responded to our requests to share their knowledge and experience. This book would not have been possible without their talent, hard work, and dedication. We are honored to work in the company of such outstanding authors.

Our sincere thanks to our colleagues at Northern Telecom, Ottawa, for their support, in particular, to Rodger Williams, Ravi Subramanian, Mike Aalders, and Pierre Cousineau, for their encouragement. Mallik thanks his colleagues, William Schmidt of Motorola for his help, and Chris Brady of Motorola for his support and encouragement. In addition Bhumip is grateful to Bert Basch, Dean Casey, Roman Egorov, Steven Gringeri, Arianne Lewis, and Vijay Samalam of GTE Labs for their help and support.

We are also thankful to Erica Bobone, Traci Beane, Mark Walsh, and other supporting staff at Artech House Publishers for their cooperation, help, and patience in making this book a successful one.

Last, but not least, Mallik would like to thank Latha, Sashank, his parents and other family members for their support. Mallik would also like to thank his teacher, Professor J. P. Raina of IIT Madras. Bhumip sincerely acknowledges the patience, support, and understanding of Ashmita, Inrava, Srijesa, Ringan, Swapan, Munni, Tomaghno, Chayan, Tuhin, Bhula, Kanan, and Komando.

Mallik Tatipamula
Bhumip Khasnabish

1

Introduction

Networked multimedia systems offer many appealing advantages over standalone systems, whereas multimedia networks pose many design challenges not faced in other sorts of telecommunication systems. In the coming chapters, experts in technologies and applications for networked multimedia seek to explain the issues relevant to the hardware/software system engineer, the applications developer, the content provider, and the user.

Before we consider any specifics of application or network design, it is appropriate to consider why networked multimedia is important and what makes a multimedia system different from, for example, an ordinary PC or an unintelligent home entertainment device. We then present some background on networking and switching technology and discuss some considerations relating to user applications that can be provided by means of multimedia communication networks.

1.1 Multimedia Communications Networks

There are perhaps as many definitions of *multimedia*—many of them coined by marketers rather than developers—as there are multimedia systems. Fundamentally, though, a multimedia system is a communication system in which more than one of the set of media forms comprising still and moving images, graphics, text, and sound are reproduced on a device embodying useful computational intelligence. This computational intelligence[1] provides the distinction

1. Here we are discussing computational intelligence that is in some way apparent and accessible to the user. Nowadays, even "dumb" devices such as video monitors incorporate microprocessors.

between a computational multimedia system and a device that acts merely as a media transducer. From creative, communicative, and usability standpoints, a multimedia application becomes more interesting and rewarding in proportion to the usage of the computational intelligence in the system. Specifically, the ever-increasing computational power available at price points appropriate for widespread deployment enables the multimedia device to handle functions beyond data decoding, display formatting, and user input. This desired higher level functionality can in many cases be characterized as *responsiveness,* or the ability of the system to react or adapt in a meaningful way to the user's actions, identity, and viewing circumstances. Although not strictly required in a multimedia system, responsiveness provides a compelling advantage in that it strengthens the link between the content or application creator and the user [1].

In a networked multimedia system, the user's equipment is connected to remote sources of information, and some or all of the content originates with storage devices or real-time input transducers that reside elsewhere. A typical user, of course, expects to see no significant quality difference between locally and remotely stored content. Because of the immediacy of multimedia data, and the audiovisual ways in which network problems manifest themselves to the user, the task of designing network hardware and software architectures appropriate for these applications is both critical and challenging. Concerns that must be addressed include providing the necessary bandwidth, keeping delay within known and manageable bounds, and either minimizing data losses or designing applications and data representations with some insensitivity to lost data. Quality-of-service issues are the topic of Chapter 3, while design consequences of the characteristics of multimedia data are a recurring theme throughout this book. Readers with a particular interest in the latter topic should note Chapters 4 (especially Section 4.2) and 8 (especially Section 8.2).

1.2 Motivation for Multimedia Communications

Networked multimedia communications originated in a confluence of two technological trends. The first involved the development of multimedia computing, and the second involved advances in networking that allowed reliable, widespread delivery of digital data at relatively high bandwidths. Advances continue in both of these areas, but in less of an applications vacuum than existed at their origin. Multimedia is now a driving force rather than just a way of using already-existing bandwidth or processor cycles. Consequently, it is now the case that—to give just one example—PC architecture is designed to meet the demands of real-time, high-quality video, whereas only a few years ago significant effort in the video compression community was devoted to the development of algorithms suitable for the then-existing sub-optimal architectures.

The origin of multimedia computing was straightforward: by the late 1980s, processors, displays, and internal system bandwidths had developed to the point that it became practical to expand the output capabilities of computing devices beyond the text and simple graphics that had sufficed for many years. Recognizing that various messages are best conveyed in different media, multimedia computing developers desired to add the richness of real-world sounds and pictures, thereby expanding the scope of computational applications and also bringing these applications to users for whom traditional text-based interfaces were not acceptable.

At the time, the personal computer seemed poised to become a central communication node replacing a large number of other devices or systems, including television receivers, typewriters, fax machines, telephones, and mail delivery. While some of these applications have been remarkably successful (have you seen many typewriters offered for sale recently?), other questions such as the ultimate relationship of computers and television remain unsettled as of this writing. What has become clear is that processing and memory are now so inexpensive that it is unnecessary to postulate the existence of a single device that does everything; rather than all of the applications migrating to the personal computer, the computation has migrated throughout the household. Nevertheless, the motivation to connect user devices to multimedia networks remains:

1. Networking broadens the amount of material accessible to the user. It is no longer necessary for the physical media to reside at the user's location. As a result, from the user's perspective the storage is essentially infinite and the material is updated without the necessity of action on the user's part. A corollary to this property is that the content creator finds a potentially larger audience than would be available were it necessary to distribute the material on physical media (e.g., CD-ROM).

2. Networking also provides the user with expanded computational power. Processes that require more computing power than is available on-site (a particular concern with low-power portable devices), or that benefit from close coupling between the storage and the processing (e.g., library searches), can execute remotely. Additionally, processing tasks can run in a distributed fashion, with part of an application (graphical user interface, for example) executing locally and the remain-

der executing remotely; this task division can vary from application to application or even over time within a session.

3. Networked storage of multimedia material, whether centralized or distributed, enables several users to access the same data simultaneously. Although offering proper support for this type of access poses some system-design problems, a valuable consequence is the ability to provide cooperative-work applications. For example, several video editors in different locations could edit a television program, and a producer in yet another location could view the results immediately.

Networking enables communication with other people, not just prerecorded content, and indeed generally allows the user to experience live, remote events. Applications for live remote access include conferencing, education, entertainment, or monitoring of remote locations.

1.3 Evolution of Telecommunication Networks

As a result of the introduction of broadband integrated services digital networks (B-ISDN), multimedia communications has become possible in addition to high-speed computer communications and conventional voice and video communication. We can state that these media will be handled flexibly and interactively in the multimedia communication environment.

Figure 1.1 shows the available networks and user requirements such as session time, distance, and transfer speed range applicable for multimedia communications services. Every kind of traffic can be handled using asynchronous transfer mode (ATM) technology, however, at present most user terminals are non-ATM terminals, such as conventional telephones, HDLC/frame relay/ SMDS terminals, and so on. In the existing networks, information such as voice, data, image, and motion video is transmitted via public networks or leased circuits. Although a packet-switching (PS) network is suitable for long-distance, low-density, and middle-speed ranges, it is not suitable for real-time applications. On the other hand, a circuit-switching (CS) network is suitable for real-time applications and high-speed communication services, but it is costly for low-density communication. An ATM network, which can offer various service classes depending on the user requirements, can cover the widest transfer speed range, but is costly over long distances. Leased circuits may cover a wide range of bandwidths, but the number of speed classes is generally restricted compared with the ATM public network services.

Figure 1.2 shows the network evolution and stepwise migration strategy toward B-ISDN including frame relay, cell relay, and SMDS services. Demand

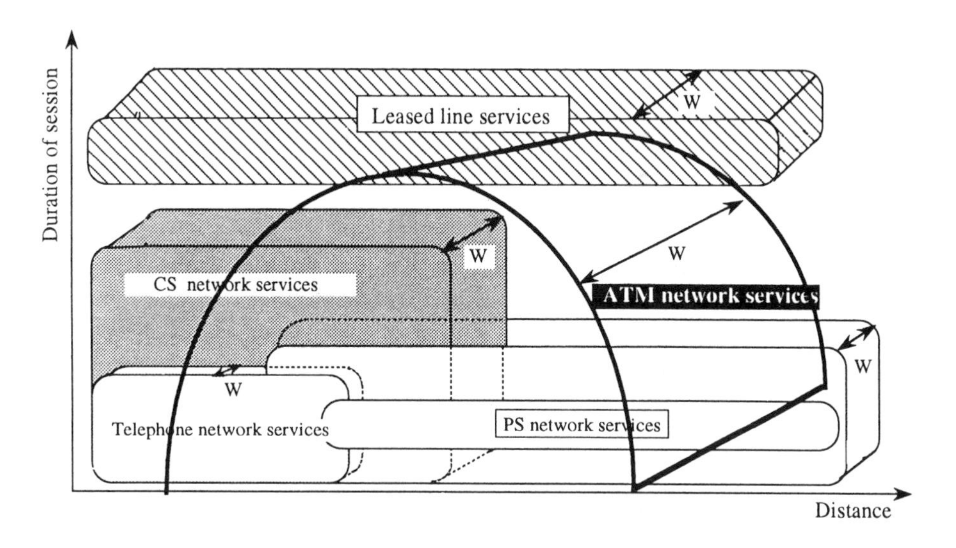

Figure 1.1 Available networks for multimedia communication services.

for multimedia business communications including voice, image, and high-speed data will expand gradually. According to this trend, at least all switching systems in the backbone network will have been replaced by ATM facilities by the year 2010, and optical fibers will link most available subscribers to their central switching offices [2,3].

To support the packet-switching function at very high speed, hardware solutions are mandatory. It was apparent that fixed-size data units were easier to handle and store than variable-size packets, but the biggest advantage of ATM is seen in the fact that it is a universal, service-independent switching and multiplexing technique that is totally independent of the underlying transmission technology and the speed.

In fact, there were some discussions on the best type of telecommunications infrastructure, whether the circuit-switching type or packet-switching type (or frame relay-switching type) should be the common technique adopted by the future integrated communication services networks. Some telecommunication experts recognized that the outstanding flexibility of packet-type switching technology would be more suitable for the public network leading to a future B-ISDN [4,5].

Based on these situations, ATM technology has been actively investigated, though there have been some experiments with hybrid packet and circuit integrated switching systems to satisfy both goals simultaneously [6–8]. The recent evolution toward fixed-length cell-based switching can realize broadband communications in an economical way. Figure 1.3 shows the transitions in ATM switching node requirements in each development stage. The advent of

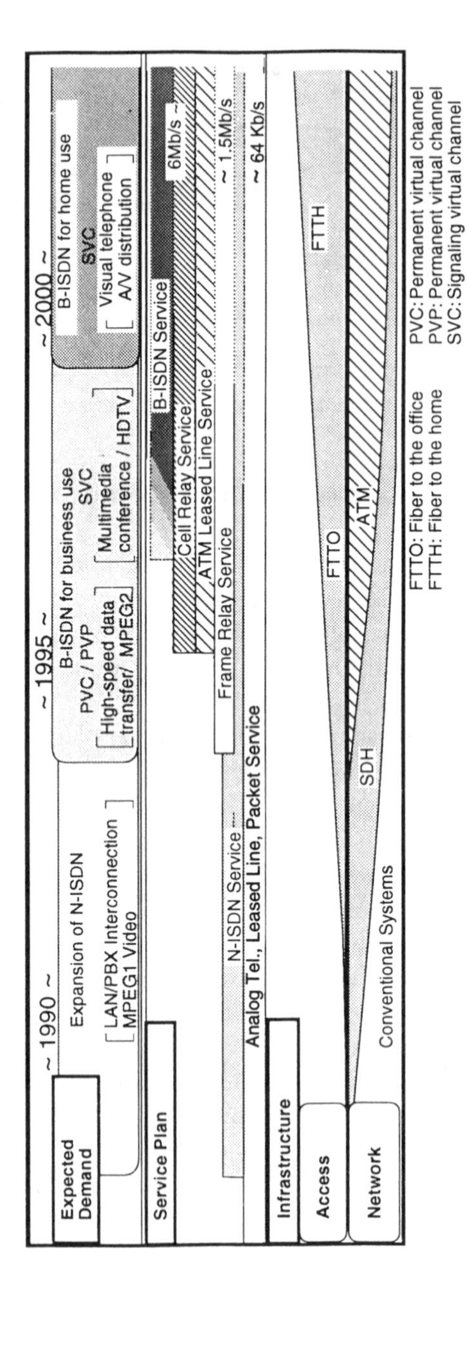

Figure 1.2 Network evolution toward B-ISDN.

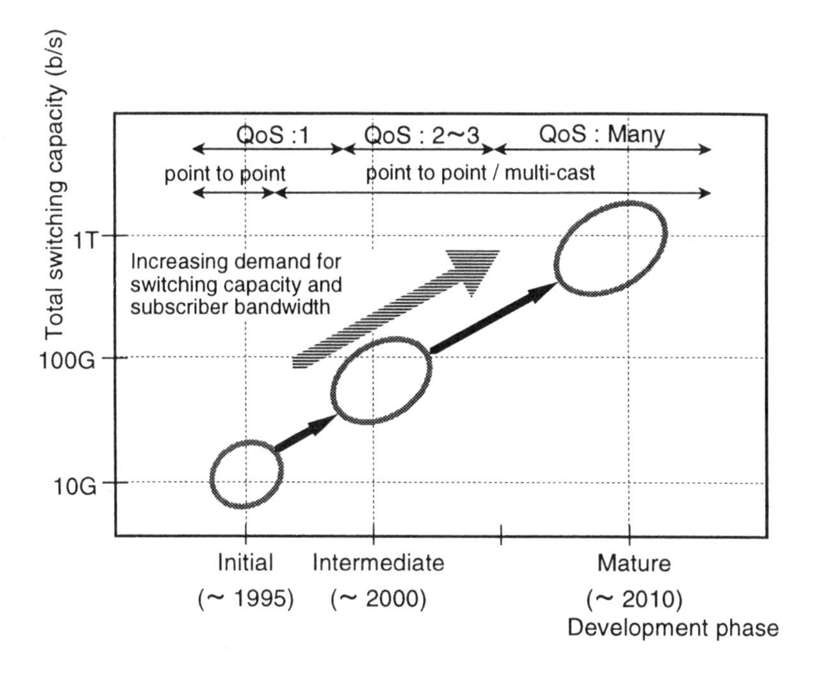

Figure 1.3 Transitions in ATM node requirements.

multimedia communications will create a wide range of quality-of-service (QoS) parameters in addition to conventional network parameters such as call setup delay time, call blocking ratio, and data error rate in packet- and circuit-switching services.

In the initial development stage, switching performance parameters such as delay and cell loss rate and available number of QoS classes are not matters of great urgency in this stage. In this initial stage, the ATM switching system can be used as a transit network facility for existing packet-switching networks, or high-speed data networks for frame relay or switched multimegabit data service (SMDS). It should support high-speed data communications via ATM user network interfaces. The systems should be flexible enough so that capacity and number of virtual channels can be expanded to match the growth in traffic volume during the migration term. Figure 1.4 shows the commercial network configuration for high-speed data communication services in Japan at the initial stage of B-ISDN introduction [9]. During the initial stage of ATM services, users will expect high-speed data communications services at up to 6 Mbps.

The requirements for a practical ATM system are shown in Table 1.1. As for QoS parameters, high-definition full-motion video needs small cell loss ratios, whereas short delay times are needed for high-fidelity services. Providing multiple QoS classes will allow network providers to achieve higher network

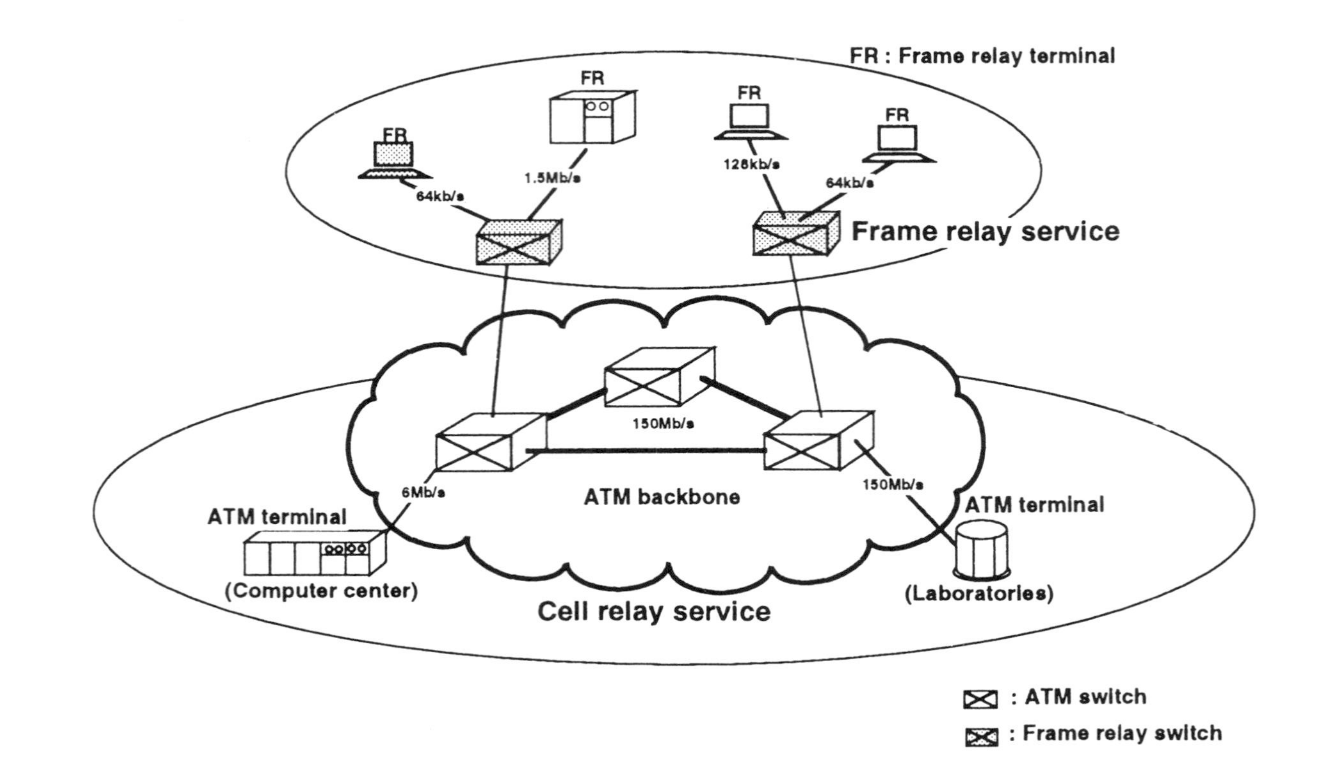

Figure 1.4 High-speed data communication services in Japan.

Table 1.1
Requirements for a Practical ATM System

Item	Requirements	
	Initial stage	**Mature stage**
Transmission speed (a) Subscriber line (b) Trunk line	(a) ~ 1.5M/6Mbps (b) ~ 150M/600Mbps	(a) ~ 150M/600Mbps (b) ~ 2.4G/10Gbps
Switching capacity	~ several tens of 150-Mbps lines	~ several thousands of 150-Mbps lines
Switching performance (a) Delay (b) Cell loss rate	(a) < 1 ms (b) < 10^{-6} ~ 10^{-10}	(a) < several hundred μs (b) < 10^{-11}
Virtual channel/line	several thousands of VCs/line	hundreds of thousands of VCs/line
QoS classes	1 ~ 2 (cell loss rate/cell delay)	many
Application	Point to point CBR/UBR PVC High-speed data communication	Point to point/multicast CBR/UBR/VBR/ABR PVC/PVP/SVC Multimedia communication

utilization efficiencies and also provide users with a satisfactory QoS by executing effective connection admission control [10,11].

Until it reaches the mature development stage, B-ISDN will be mainly used for business services. User terminals that request multicast connections will become much more sophisticated with increased image-handling capacity, and the number of these terminals will increase with the penetration of frame relay, cell relay, and SMDS services. The demand for large-capacity toll switching systems will increase accordingly. However, in the mature stage of ATM services, users will require satisfactory multiple QoS classes for various applications such as CBR (constant bit rate), VBR (variable bit rate), ABR (available bit rate), and UBR (unspecified bit rate). The required transmission speed and the switching capacity for both toll switching systems and local switching systems will increase accordingly in addition to the deployment of the sophisticated QoS control mechanisms.

Communication services are generally specified as guaranteed type services or best effort type services from the viewpoints of communication quality assurance and routing mechanisms [12,13]. The service classification in terms of QoS for circuit switching, packet switching, frame relay and ATM switching is shown in Figure 1.5. Connection-oriented (CO) type packet switching is used to provide reliable communication services when the transmission line's quality cannot be guaranteed. Recently, since the communication infrastructure has matured with the introduction of optical fibers, packet switching is being replaced by frame relay switching, which can eliminate retransmission control within the network and hence realize more efficient packet data transmission.

On the other hand, the connectionless (CL) type packet-switching service is deemed a best effort service since it does not handle retransmission in the network layer when error occurs and error recovery is conducted in the upper transport layer such as the transmission control protocol (TCP). The connectionless function is considered to be useful for accommodating the multimedia database servers expected in the access network. The conventional telephone network service is a circuit-switching type and supports bandwidth guaranteed services. Internet supports best effort services. An ATM network can support both the guaranteed type and the best effort type. More specifically, it can support CBR and VBR for the guaranteed type and can support ABR and UBR for the best effort type. They can be selected to match the user's requirements, for example, VBR for video/high-definition image applications and CBR for voice applications. For the moment it is expected that most ATM users will use UBR or ABR, because it is difficult to specify strictly QoS for VBR taking applications and higher layer protocols into account. Related subjects are discussed further in Chapter 5.

To understand the recent ATM technological background we should also consider the higher layer standard communication protocol for the multimedia

Switching Type	CO/CL	QoS guarantee
Circuit switching	CO	Guaranteed Type
Packet switching	CO	Guaranteed Type (Virtual Circuit)
	CL	Best effort Type (Datagram, IP Packet)
Frame relay	CO	Best effort Type
ATM switching	CO	Guaranteed Type
		Best effort Type

Figure 1.5 Service classification on QoS.

communication network. The network is expected to provide a reliable delivery mechanism (such as TCP) as well as an unreliable delivery mechanism (such as user datagram protocol, UDP) according to the user requirements of cost and reliability. The reliable delivery mechanisms typically use acknowledgment and retransmission to guarantee delivery of packets, whereas the unreliable delivery mechanisms make a best effort to deliver packets; without the overhead and latency of retransmission the communication fee can be reduced. As for connectionless best effort service with QoS requests, for example, IETF (Internet Engineering Task Force) is investigating the concept of the resource reservation protocol (RSVP). It determines whether a router has sufficient resources to support the requested QoS and whether the user has administrative permission to make the reservation. If both checks succeed, RSVP sets parameters in the router packet classifier and scheduler to obtain the desired QoS. RSVP is expected to be applied to small intranets including Internet telephony for the advantage of dynamic resource negotiations. For recent standardization activities, please refer to the corresponding sections of Chapter 11.

Taking the above-mentioned situation into account, the following targets required for ATM technology should be established [14–20]:

1. Wide bandwidth and sufficiently high throughput with QoS control for multimedia applications;
2. Reliability and large switching capability to ensure network expandability;
3. Mutual interconnection with other networks in multivendor environments;
4. Quick response time and user-demanded QoS assurance;
5. Economy;
6. Maximum utilization of trunk circuits by taking user-required QoS levels and ATM traffic characteristics into account;
7. Effective network management and facilitated testing methods.

ATM switching network systems will require more high-speed memory and larger scale logical circuits than conventional systems. Satisfying the above-mentioned basic requirements for ATM systems requires the use of state-of-the-art Large Scale Integration (LSI) technologies, in particular, high-speed and large-capacity gate arrays including embedded RAM technology and short-propagation-delay logical circuits with low power dissipation.

Figure 1.6 shows LSI development trends in communication services considering integration scale and speed. As a general trend, CMOS LSIs are mainly applied to signal (less than several hundreds of megabits per second)

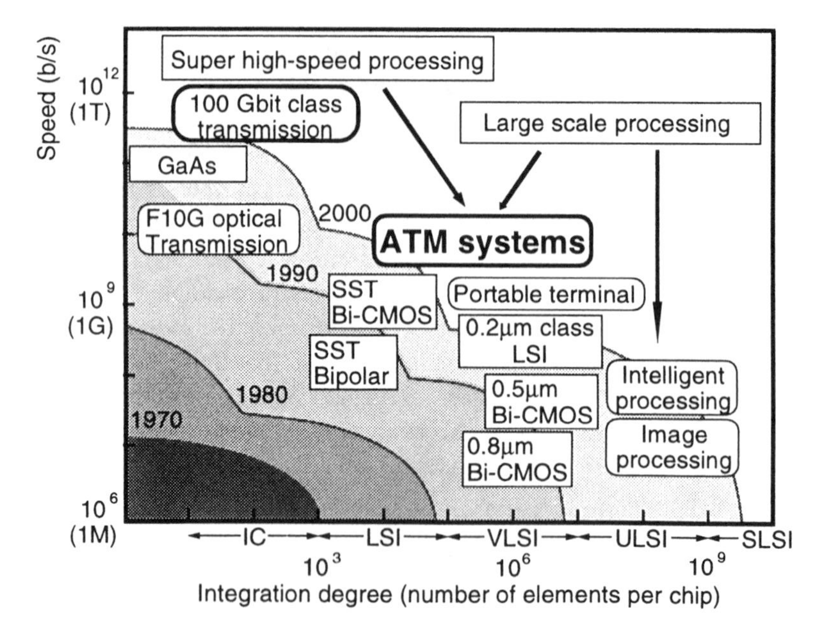

Figure 1.6 LSI development trends.

processing applications that require several hundreds of thousands of logical gates; their benefits include high integration levels and low power dissipation. Bipolar GaAs LSIs, which are more expensive than silicon LSIs, can be applied to ultra-high-speed signal (more than 1 Gbps) processing operations.

1.3.1 Transmission Technology

The portion of the telecommunication network most visible to the user is the *access network* or the part of the system connecting the user to a central office (to use telecommunications terminology) or to a *head-end* (to use cable television terminology). In particular, the user sees the *drop*—the last few meters connecting into the building and ultimately to the subscriber's equipment. In many cases this remains—as it has been for more than a century—a pair of copper conductors carrying a baseband analog signal. As Chapter 6 notes, this portion of the network is now following the earlier shift of other portions of the network from analog to digital. How soon the drop also follows the evolution of the backbone through coaxial cable and eventually to optical fiber is a question whose answer will vary from place to place depending on the physical and population density environment of the installation, the requirements of the services to be provided, and economic constraints. Currently discussed options include asymmetric digital subscriber line (ADSL), which transmits digital signals over existing twisted-pair wiring; fiber-to-the-home; fiber-to-the-curb with coaxial cable for the drop; and wireless systems.

Existing transmission systems utilize the basic pulse transmission technique over optical fiber or twisted wire pair cable and the plesiochronous digital hierarchy (PDH) technique. Due to the advances in LSI technologies mentioned earlier, the newly developed synchronous digital hierarchy (SDH) fiber optic systems or the North American defined Synchronous Optical Networks (SONET) employing network synchronization techniques are currently being widely introduced.

SDH and SONET have two benefits over PDH. The former are superior because they can offer higher speeds and direct multiplexing without intermediate multiplexing stages. This technology is achieved through the use of pointers in the multiplexing overhead that directly identify the position of the payload. If coupled with the SDH/SONET technology, ATM techniques would have the strength of Samson.

In particular, the virtual path (VP) concept exploits the benefits of ATM capabilities and provides the network with a powerful transport mechanism. In an ATM network, VP bandwidth and route can be controlled separately. The independence of VP route and capacity establishment simplifies path bandwidth control. In addition, the direct multiplexing of VPs into transmission links eliminates hierarchical path multiplexing stages and, therefore, eliminates hierarchical digital cross-connect systems [21]. On the other hand, since it is apparent that the cost of access network will increase in fiber-based B-ISDN, cost-effective multiplexed transmission access methods and fiber optic cable distribution techniques should be carefully examined.

The access network technology, the feasibility of SDH-related technology, and the corresponding VP handling techniques are explained in Chapter 6.

As for the LSI technology required for SDH systems, line interface LSI chips for 52M/150M/600M bps SDH transmission have already been developed using 0.8-μm Bi-CMOS and 1-μm bipolar technology. Figure 1.7 shows the developed transmission technology and corresponding random access memory (RAM) technology. The required hardware amount is about several hundred thousand gates and RAM of several thousand bits. For a practical switching system, OAM (operation, administration, and maintenance) cells, which are called monitoring cells (MCs), are employed in both the VP and VC connections for transferring OAM information between ATM nodes. In each line interface, MC drop/insert functions are required for incoming and outgoing directions, respectively. The required hardware amount is proportional to the number of VCs and VPs to be simultaneously processed [9].

1.3.2 Switching System Technology

This section examines ATM switch classification and the corresponding VLSI technologies from the viewpoint of switching architecture.

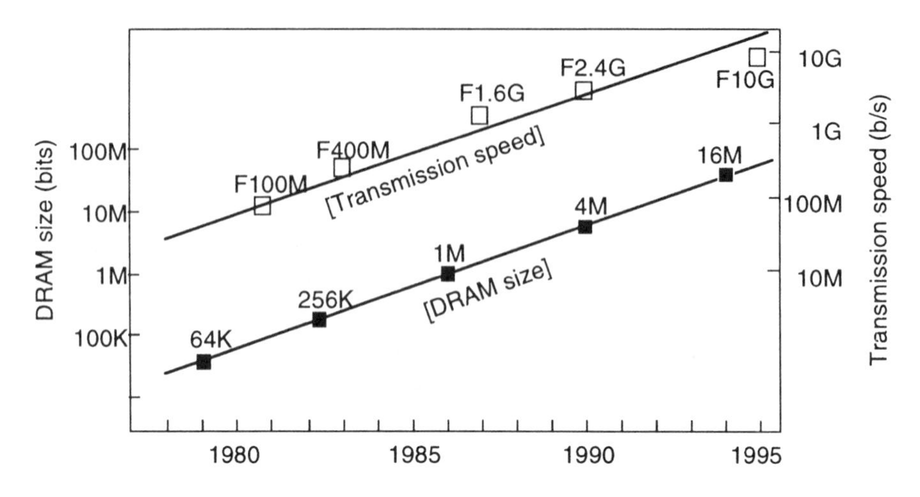

Figure 1.7 DRAM size and transmission speed.

Historically, the Banyan switching element has been a candidate for use in construction of the ATM switch [22]. The principle is that incoming cells are automatically routed according to a self-routing header, which corresponds to the destination address. The sequence of 1's and 0's in the header is arranged so that each switching element is ordered to switch the cell through the fabric appropriately as discussed in Chapter 5. The characteristic of this process is that the actual switching is done in hardware without passing through a processor. This hardware-based switching mechanism is why ATM switches are fast enough to support gigabit-class throughput. However, Banyan switches have several deficiencies as follows:

1. When multiple inputs are active, cell collision occurs easily.
2. If memory buffers are added to every switch element for avoiding cell collision, a complex queuing mechanism is required and, in addition, longer transit delays will occur.
3. In particular, when incoming cells are part of a multicast transmission, the impact of cell loss will be significant, since the corresponding cells are copied many times at the input side and resent through the fabric regardless of the other incoming traffic volume.

To avoid the above-mentioned deficiencies, the Batcher-Banyan switch has been considered. This switch has a separate stage added in front of the Banyan switching fabric for arranging cells into the optimum order to minimize the probability of congestion before actually entering the Banyan switch fabric.

This method, however, is complex and it is difficult to verify fully its performance. For these reasons, the Banyan-type switch fabric has not been successfully adopted.

In this section, the ATM switch architecture is discussed from the viewpoint of practicality. There are four types of ATM switches, all of which are comprised of common buffers, input and output buffers, and cross-point buffers. Keep in mind that an ATM switch utilizes a different paradigm than the circuit switch. In an ATM switch, cell loss can occur, depending on the statistical nature of the virtual connection traffic as loads are handled by the cell switching and buffering strategy method. The ATM switch blocking performance is sensitive to switch architecture as discussed in Chapter 5. The trends in ATM switching systems and ATM switching elements that have been implemented as LSI chips are shown in Figure 1.8.

Existing ATM switch elements offer throughputs of up to 10 Gbps. For example, by using the bit-slice technique shared memory ATM switch, having 16 to 32 ports each offering 600-Mbps throughput, 10- to 20-Gbps throughput can be realized in a switch element LSI; its access time is 20 to 10 ns with the latest LSI circuit technology.

ATM switches can be classified into several categories based on their arrangements of buffer memories and switch matrices. However, there are two

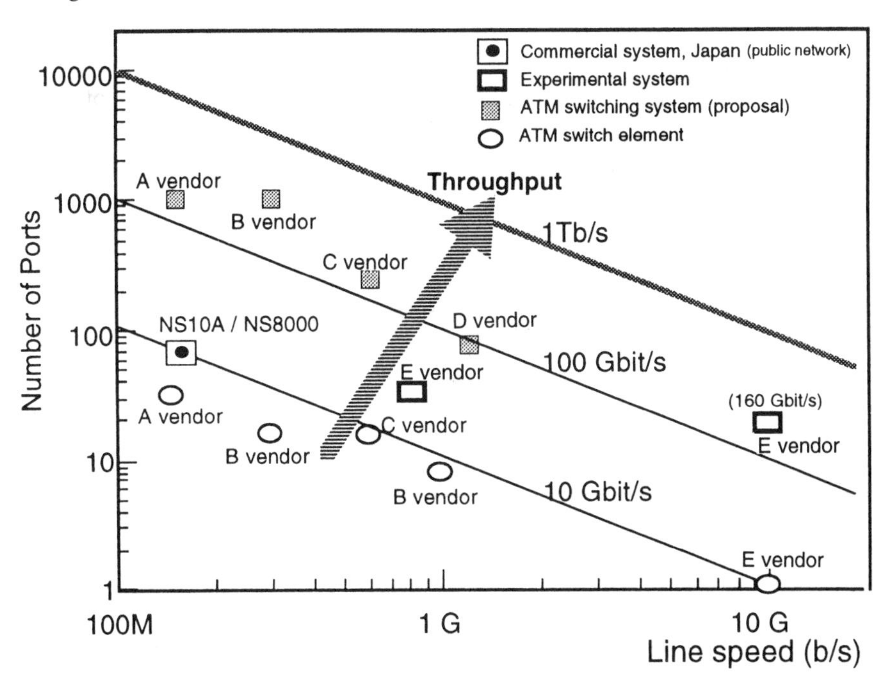

Figure 1.8 ATM switch element and switching system.

basic architectures: the shared resource type and the space-division type as shown in Figure 1.9.

Shared buffer switches need a memory throughput of $2NV$ (where N is the number of input lines and V is the speed of each input line), but output buffering switches need speeds of only $(N + 1)V$. In this sense, output buffering switches are superior to shared buffer switches due to their simple output queuing control and to more relaxed memory access time requirements, which lead to maximum switch throughput with the least cell buffer requirements. The details are discussed in Chapter 5.

One ATM switch LSI chip that is currently available as a shared resource type is the 150-Mbps 32×32 (Bi-CMOS) [23]. Combining several shared buffer-type ATM switches constructs a large-capacity ATM switch. In the input and output buffer-type switch, buffers are placed at the input and output ports.

Internal switching is L times faster than the input or output speed to reduce the effects of head of line (HOL) blocking, which is further discussed in Chapter 5. Therefore, the bandwidth required for the input and output buffers is $(1 + L)V$. By making use of this switch architecture, 160-Gbps switch throughput can be attained [24,25]. Buffers may also be placed at the cross-points of the crossbar switches. ATM switches of this type perform in the same way as output queuing switches with the difference that the queue for each output is distributed to N buffers. The required bandwidth for each buffer is $2V$. Cross-point buffering switches have achieved switch throughputs of up to 40 Gbps with state-of-the-art LSIs that offer 10-Gbps throughput. Considering the current development situation, ultra-high-speed ATM switches whose throughputs exceed 40 Gbps will be realized as input/output buffered crossbar switches or cross-point buffered crossbar switches [9].

Let's consider the future ATM switching system and essential VLSI requirements. Current subscriber and trunk-line transmission speeds range up to 150 Mbps, but these will increase to 600 Mbps and 2.4 Gbps, respectively in the near future. Actually, 1.2-Tbps switch capacity is required in local switches (LSs) given that the normal subscriber line transmission speed in B-ISDN environments is 150 Mbps, the total number of lines is 40,000, and line efficiency is 0.2.

As for ATM line interface cards, the required hardware for each line is approximately 1M gates with several megabytes of high-speed RAM, preferably, embedded RAM with the equivalent number of logic gates. Presently, several 300K to 500K level gate array LSIs (0.8- to 0.5-μm design rule) are used for implementing ATM line interface cards. Figure 1.10 shows the gate array integration effect as realized in the ATM interface LSI. The hardware volume of the line interface cards required will be drastically decreased when 0.2-μm

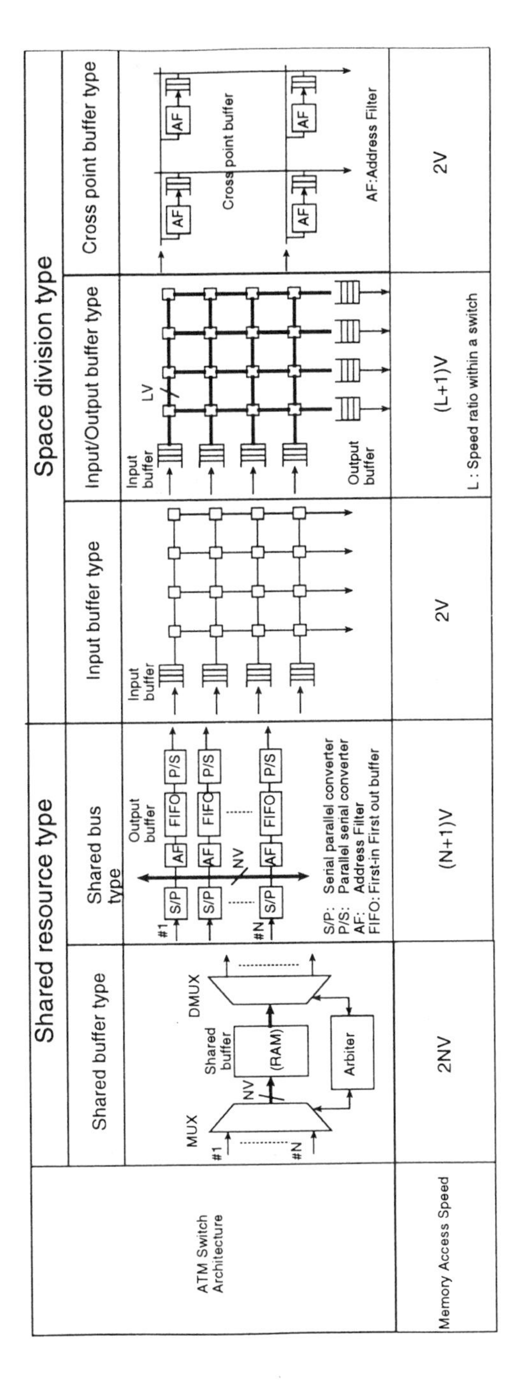

Figure 1.9 Classification of an ATM switch.

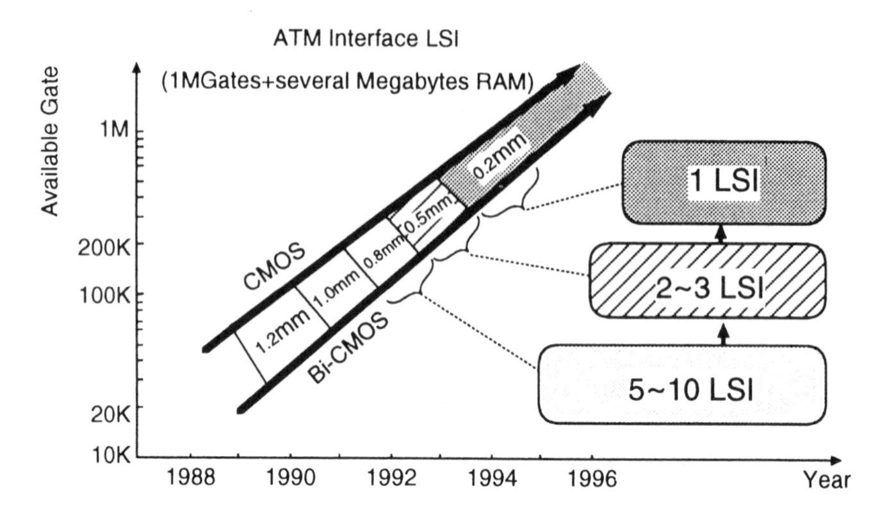

Figure 1.10 Gate array integration effect.

design rule LSIs, in addition to compact E/O conversion modules, are employed to achieve the economical introduction of ATM systems. Fundamentally, ATM is more expensive than Ethernet because it requires much more memory and state information within the line interface cards; however, eventually the required cost will shrink as a result of the above-mentioned VLSI development.

Taking the conditions discussed into account, the following requirements should be effectively achieved for realizing practical B-ISDN service:

1. Ultra-high integration logic LSI technology that offers 1M Gate logic + several megabytes high-speed static RAM;
2. Ultra-high-speed LSI memory technology that can handle gigabit-per-second transmission speed signals.

In addition, power reduction LSI technologies such as 3.3V to 1.5V controlled circuits and corresponding low-switching differential logic circuits technology will be required.

1.3.3 Operation and Management of Broadband Networks

The fundamental standards for ATM and B-ISDN have been well established by the International Telecommunications Union (ITU) and have been adopted by most developed regions throughout the world. Considering the significant advantage the ATM technology can offer in the management and integration of voice, data, and video traffic, there have been strong requirements for network operations and equipment to support multivendor and multioperator environments in a consistent and an interoperable manner.

The telecommunication management network (TMN) aims at the operation, administration, maintenance, and provisioning (OAM&P) of telecommunication networks and services in the present multivendor environment [26–28]. It should be emphasized that TMN interface standardization is indispensable. The main reason is as follows.

The objective of standardization is to enable design flexibility and the interoperability of network elements for use in a global ATM network. It often happens that most public network service providers actually purchase their network elements (NEs) from multiple suppliers. If a common interface is established based on TMN standards, it would help the suppliers to ensure operational compatibility with each corresponding NE type. This would make it easier for them to concentrate on innovative end-system applications to meet diverse customer operational environments.

During the past few years, standards organizations, including the ITU, the American National Standards Institute (ANSI) committee, and the European Telecommunications Standards Institute (ETSI) have worked on developing TMN standards. The TMN standards utilize the Open System Interconnection (OSI) systems management standards, and provide a framework for the management of telecommunications network and services in an open and multivendor environment. The standard bodies have been investigating the TMN principles since 1980, and the focus has mainly been on interface specifications. In the early 1980s, the TMN was envisioned as an integrated management system of telecommunication networks for ensuring interoperability. The ITU standards of the TMN concept were first formally defined in 1988. To promote deeper understanding of TMN, several related ITU-T recommendations such as M3000, M3010, M3100, and M3400 were established [29–32]. The TMN standards define two types of telecommunication resources, which are the key entities in TMN:

1. Managed systems, generally termed network elements (NEs);
2. Managing systems, of which the operating system (OpS) is the most prominent.

TMN standards also specify possible interconnection relationships between these resources in the form of interfaces, such as the OpS interface (Q3) illustrated in Figure 1.11. Protocols specified in the TMN standards should be used by the Q3 interface as shown. In this case, we should also be careful about coexistence with conventional systems; that is, new TMN inter-

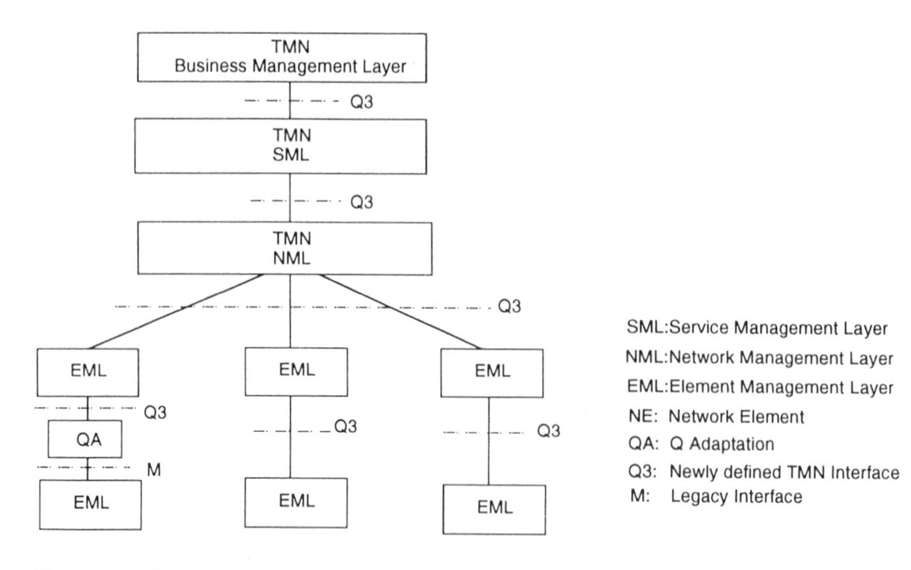

Figure 1.11 TMN logical layer architecture.

faces should coexist with the existing legacy OAM&P interfaces and provide the additional functionality needed.

Q adaptation (QA) can be done between legacy NEs and a TMN operating system to realize information conversion and protocol conversion. Actually, the ATM forum, in addition to the ITU and ETSI, contributed to the definition of the Q3 NE management interface to realize control from a network management system (NMS). It uses common management information protocol (CMIP) as a network management protocol [33] and defines its object classes using Guidelines for the Definition of Managed Objects (GDMO) templates as recommended by ITU-T and the International Organization for Standardization (ISO). In addition, efficient end-to-end management services for a customer network consisting of heterogeneous LANs are also required, considering that networks are continuing to get larger and more complex. The network management function required is discussed in Chapter 7. On the other hand, the performance of communication systems such as a switching system or a transmission system should be verified based on the use of OAM cells, which are standardized in ITU-T (I.610).

The TMN standards are already applied to new telecommunication networks such as TMN-based SDH networks that are being introduced in Japan and the Global System for Mobile Communications (GSM) in European countries [34–37]. Chapter 7 is devoted to TMN principles and reviews the concepts underlying TMN. It also covers the whole topic of TMN applications from a standardization point of view.

1.4 Trends and Driving Forces in Multimedia Communications

Before discussing the motivation driving multimedia communication, let's consider the following application examples:

- *Real-time electronic mail.* Electronic mail (e-mail) is already widely used by millions of people, and some e-mail commercial applications contain audio and low-resolution video as well as text. However, current e-mail services suffer from unpredictable delay owing to the available bandwidth in the network. Recently, users have become very enthusiastic about real time e-mail with high-resolution images, preferably adding the function of seeing and hearing each other as well. If this real-time e-mail technology is established, it would also easily support video teleconferencing.
- *Remote medical diagnosis.* To support further advances in medical science, high-resolution medical images such as x-ray pictures, tomograms, ultrasound images, and electrocardiograms are being used more often to realize more accurate medical examinations. If they are used for remote diagnosis in a telemedicine environment, the required bandwidth for transmitting these types of images is more than twice that required by conventional high-definition TV. In this case, highly reliable transmission should be ensured for more accurate diagnoses. In addition, it will often be necessary for several images to be displayed at the same time for comparison and some may need to be further processed into a more appropriate form. The analyzed results may also be displayed at the same time with the original image. While examining these images, voice communication should be superimposed in addition to other data streams from monitoring devices. This application service is considered to be one of the typical multimedia communication examples.
- *Collaborative computer-aided design.* In most manufacturing industries, including the construction field, the demands on the graphics capabilities of the systems used for computer-aided design (CAD) are increasing. Recently, the number of virtual companies performing design tasks by utilizing remotely located temporary workers is increasing. For this type of situation, large amounts of images and arithmetic data in addition to a voice signal need to be integrated and transmitted at the same time with high speed and reliability for effective collaboration with the members of the companies.

Considering the situations just listed, it is apparent that an explosion in multimedia data traffic will occur. The resulting multimedia traffic will be unpredictable, so complicated end-to-end traffic control is required; in addition, sufficient network facilities and a high-throughput communication mechanism should be provided to withstand unforeseeable traffic loads. As the hardware and software resources become more widespread, it is imperative such multimedia resources be interconnected so as to increase overall performance.

Multimedia communication is being encouraged by the following trends in communication networks:

1. Seamless transition from centralized network control to distributed control;
2. Seamless interconnection of different networks at affordable cost;
3. Seamless integrated transmission of multimedia with multiple QoS levels (including multicast).

Computing and communications originally advanced as separate technologies. However, computing power has been growing exponentially due to advances in semiconductor technology and versatile sophisticated software applications that require faster processing. Based on this technological background, a transition from centralized network control to distributed control is becoming popular. Distributed network control provides two main benefits. One is wide bandwidth at low cost due to more efficient network resource utilization. The other is minimized transmission and processing delay. The latter is especially important given that data sources will become much more widely distributed and advanced multimedia services may require data from several sources to be processed in real time.

The seamless interconnection of different networks is encouraging rapid growth in the number and quality of information suppliers and in the frequency with which different companies communicate with each other. ATM provides the backbone network that effectively connects geographically dispersed LAN sites, and various connectivity levels from dedicated circuits to switched circuits throughout the metropolitan area.

Near-term enhancements to the ATM backbone network will offer bandwidth on demand through call-by-call dynamic networking reconfigurations with minimum capital cost. ATM will realize economical bursty inter-LAN communication.

Currently, basic technologies for implementing ATM LAN and seamless interconnection with WAN have almost been achieved at acceptable cost. Actually, the cost of an ATM switch port is competitive with the cost of a 100-Mbps Ethernet switch port as a result of advances in VLSI technology.

In addition, the optical fiber infrastructure that is being rapidly installed throughout Japan, for example, would support the deployment of multimedia networks.

As explained earlier, new application areas such as collaborative work systems using multimedia facilities will explode given the recent revolutionary improvement in networking capabilities. That is, the seamless integrated transmission of multimedia with multiple QoS levels is now possible because the ATM network can handle a wide range of QoS levels. These applications also require integrated support for point-to-point and multicast connections across wide areas. Higher bandwidth communication is becoming popular in addition to reliable communication with guaranteed quality of service at reasonable costs.

Multicast communication services are expected to become one of the most popular applications, on par with daily publications, software distribution, database replication, and so on in future multimedia communication networks. The multicast communication scheme currently available on the Internet, however, lacks mechanisms for providing service quality assurance to its users. In addition, available bandwidth and host resources differ from each other in most multicast applications. Therefore, to provide multicast communications services in heterogeneous communications environments, it is necessary to develop a QoS guarantee mechanism that can provide an appropriate QoS to each host according to its available bandwidth and resources. To utilize network resources effectively, non-real-time multicasting is also important in addition to real-time multicasting. In this case, multiple QoS class support and the capability for dynamic bandwidth reservation will also be required. Details are discussed in Chapters 3 and 4.

Finally, we should emphasize the importance of the standardization of ATM voice. To introduce and promote the development of the ATM infrastructure effectively, voice communication support such as that required for remote learning, remote medicine, and voice-on-demand applications is indispensable. Since the ratio of voice traffic in private networks is considered to be 60% to 80% in Japan, ATM voice should be flexibly handled by an effective multiplexing method.

Because multimedia communication over the ATM backbone network will enhance the efficiency of communication and really satisfy user demands, the above-mentioned multimedia network services will improve the productivity, work quality, and job satisfaction of knowledgeable workers.

1.5 Services Offered by Multimedia Communications

In the present stage, many kinds of multimedia experiments using the Internet or individual private networks are being conducted in Europe, the United

States, and Japan. Given the increasing bandwidth and PC processing power available, a growing number of new applications that support the real-time transfer of audio and video over the Internet are being developed.

For example in Japan, NTT has been running an experimental multimedia communication network since 1994. At the end of 1995, 127 users/participants were active in the experimental network. Access points offering ATM interfaces with up to 156 Mbps have been utilized for supporting high-speed computer communication for various kinds of multimedia and video on demand, while some private companies have already adopted ATM technology as their backbone LAN technology. The high-speed, broadband backbone network consists of 2.4- and 10-Gbps super-high-speed transmission paths connecting 10 ATM switches located throughout the country.

By making use of this network, versatile multimedia applications such as high-speed inter-LAN communications, large-capacity file transfer, super-high-definition image transfer, remote medical diagnosis, remote education, and so on have been tested. The high-speed computer communication system is shown in Figure 1.12. Supercomputer-generated visualization data are sent over the ATM network and displayed in real time as sequential moving images.

In the experimental network, full utilization of the ATM-based high-speed broadband communication systems is necessary if we are to overcome the obstacles experienced in the present narrowband communication networks. Only ATM can offer the quick response needed by multimedia services such as telepublishing or CAD-related application services with an affordable cost, while offering the reliability demanded by the oil and chemistry fields.

Other popular applications in the experimental network are electronic mail, inter-LAN communication, multimedia information retrieval using

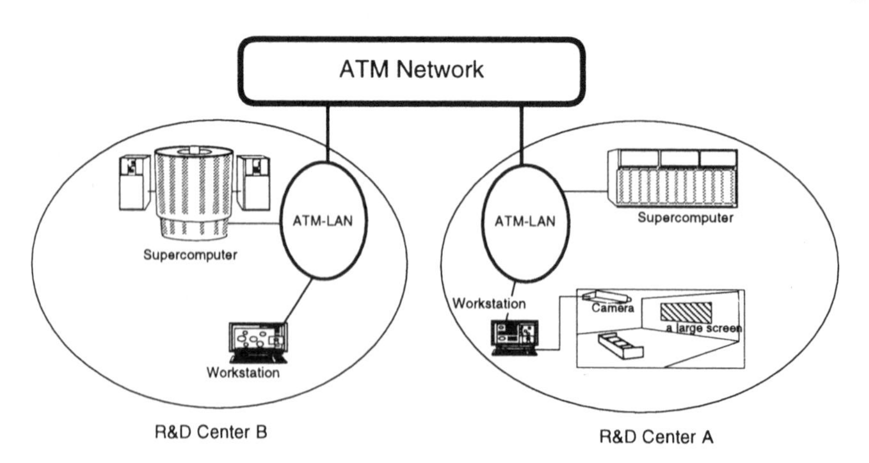

Figure 1.12 High-speed computer communication system.

WWW, on-line transaction processing such as banking, and wideband video conferencing. By the year 2000, wide-bandwidth applications such as video electronic mail, collaboration tools, and desktop video conferencing will become popular. Substantial multimedia systems such as an Internet-based digital network library system, a multimedia teleconferencing system, a video-on-demand (VOD) system, a super-high-definition (SHD) image system, and a collaborative conferencing system are already in the first stage of commercial introduction. Some of the typical multimedia communication systems are described next [38–40].

1.5.1 Digital Network Library System

Figure 1.13 shows a diagram of the digital network library system architecture. The technological aspects of the system are summarized as follows:

1. Simultaneous accessibility to any kind of server through a common high-speed ATM interface;
2. Integration scenario to arrange and lay out the contents captured from the target servers;
3. Background processing technology for translation, speech synthesizing, etc.

By simultaneously accessing an automatic text translation server, text-to-speech synthesizing server, and speech recognition server on demand, the contents captured from the digital library's servers can be processed accordingly and value-added services are provided in addition to the conventional information retrieval service. Thus, by integrating images, video, and speech in an attractive way, this digital network library system is considered to be a characteristic example of multimedia application services.

1.5.2 Multimedia Teleconferencing System

Considerable experimental research has focused on a video conferencing system that gives users a more realistic sense of presence for collaborative work. As shown in Figure 1.14, the system effectively integrates three dispersed locations to create a single, apparently seamless, shared conferencing environment using video codecs, SHD displays, and sound localization technology. The video codes used in the system can transmit HDTV images at a rate of 150 Mbps along with four speech signal channels.

Recently a large-screen high-resolution display system for teleconferencing has been developed, and the appropriate sound system for large-screen telecon-

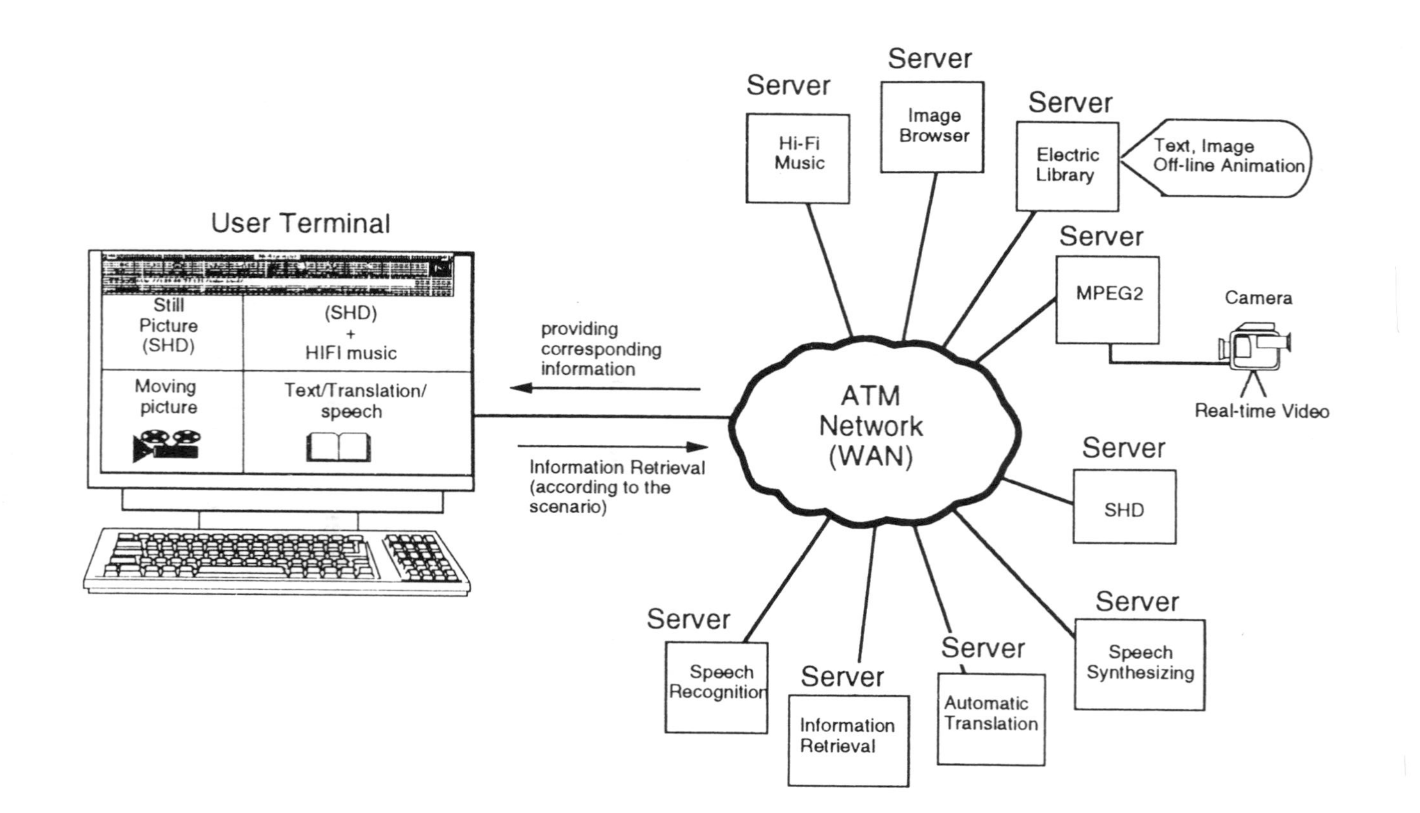

Figure 1.13 Digital network library system.

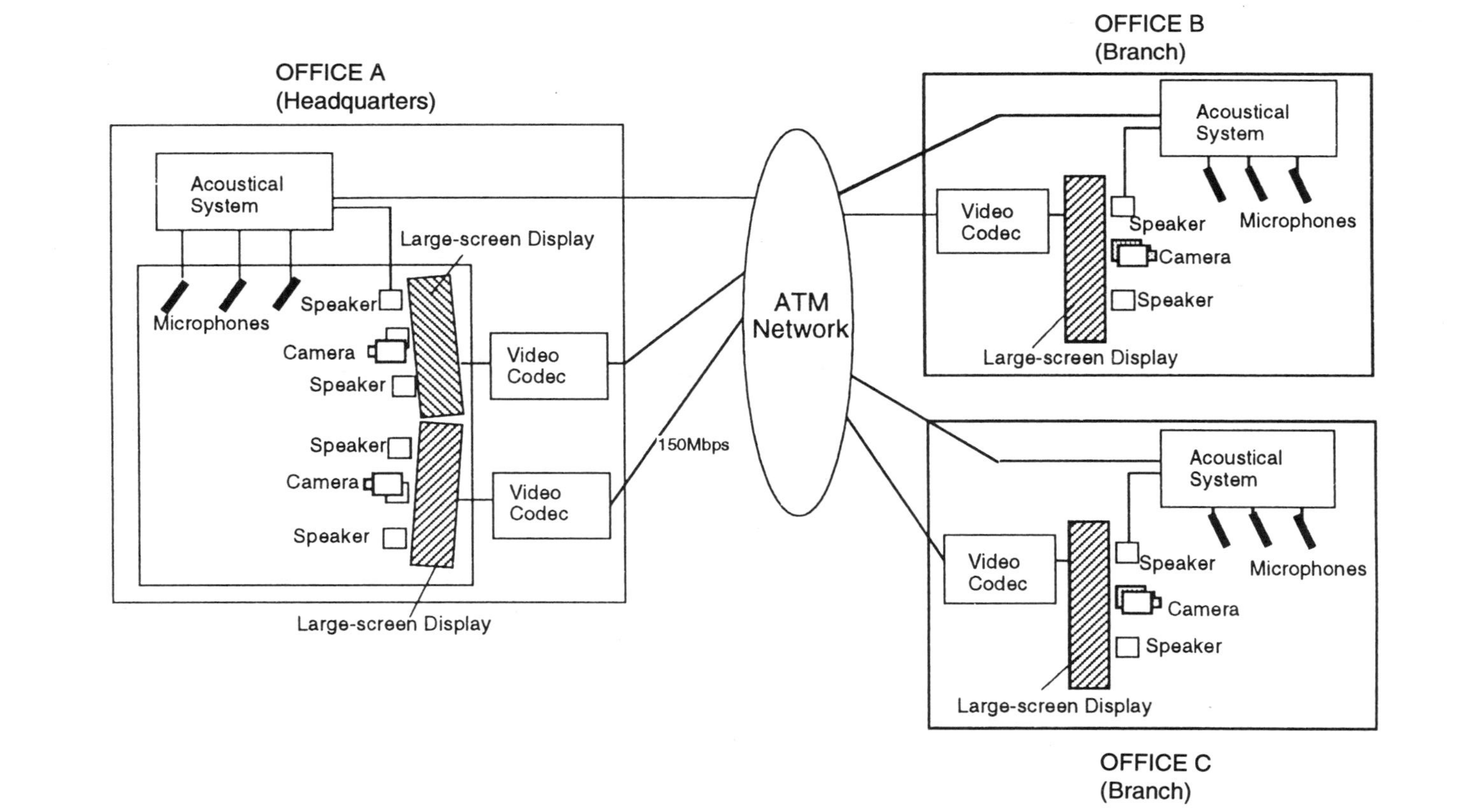

Figure 1.14 Multimedia teleconferencing system.

ferencing is being studied. The key technology is sound localization by accurately reproducing sound to match the visual images from the opposite site.

By effectively introducing the above-mentioned technology, the participants feel that all the participants are in the same room. For realizing the best atmosphere, the optimum placement of a series of microphones and loudspeakers should be carefully examined. Some technological features of the system are as follows:

1. A seamless video image environment is created by employing ATM-based 150-Mbps HDTV encoding, SHD large-screen display technology, and tandem HDTV screens with minimal gap between the screens.
2. A seamless speech environment is created by the acoustical system by exploiting sound-localization technology and through optimal positioning of the microphones and speakers.

For effectively realizing collaborative conferencing, document sharing technology using the WWW systems should be fully utilized.

1.5.3 Super-High-Definition Image System

Because SHD images, having a resolution equal to that of 35-mm film, have a data capacity that is four times greater than that of HDTV, it becomes possible to read stored newspapers with high resolution even when they are displayed at actual size. In fact, since the pixel pitch of the picture tube is so minute and the frames are displayed at a rate of 60 frames per second with noninterlace scanning, viewers can see an extremely high-resolution image with no flicker no matter how close they are to the screen. This SHD image system can be combined with the digital network library system mentioned earlier and can provide versatile application services such as a digital museum or an automatic sightseeing guide. SHD images also support remote medical diagnosis (telepathology), which involves the transmission of photomicrographs of cellular tissue or x-ray pictures to specialists at different locations. As mentioned earlier, utilizing large-bandwidth transmission lines efficiently will widen the range of communication applications.

1.5.4 Video-on-Demand/Multimedia-on-Demand System

VOD systems represent a new type of service that enables people at home or at their offices to receive all sorts of video materials such as movies, on-line shopping, karaoke songs, interactive games, and artwork at any time at their own convenience. Figure 1.15 shows an example of a VOD system based on an ATM network and FDDI LAN. A key feature of the system is that it supports up to several tens of VOD terminals per server for the delivery of MPEG-1 encoded video materials. Besides the regular playback functions available on conventional video systems, VOD systems should soon be able to handle "jump" capability, enabling the reviewer to play back the material from any specified point. The response time for video to be displayed on the screen should be under 1 sec for the entire range of materials. To be more practical, many kinds of VOD servers, which are dispersed through the network, should be flexibly used by each user for his or her own purpose.

In addition to the multimedia communication examples mentioned, there are many other types of applications such as remote education service, remote banking service, electronic newspaper service, electronic museum service, home shopping service, telecommunication karaoke, and so on. A more precise explanation is beyond the scope of this chapter.

Figure 1.16 shows several screen images from Reflection of Presence, an ATM-based multipoint telecollaboration environment developed at the Media Laboratory at MIT [41]. Such systems are generally suggested for enabling productive multimedia "conference calls" among business users seated in their offices, but the same infrastructure can support entertainment-related applica-

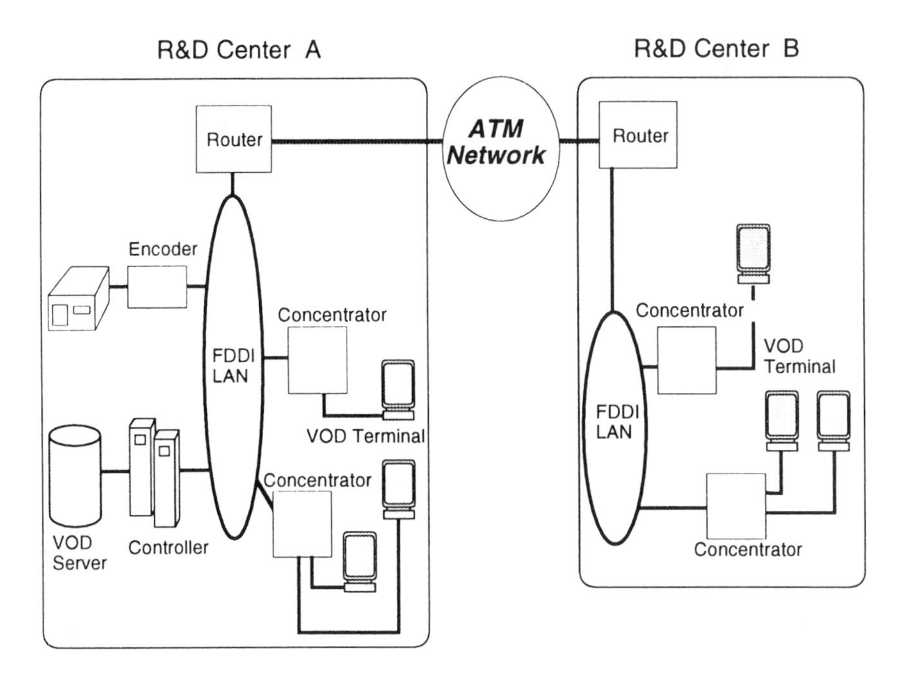

Figure 1.15 VOD system.

Figure 1.16 Reflection of Presence images.

tions such as the linking of cafes or allowing separated groups of people to watch and comment on a sporting event. The Reflection system is implemented not as a central server but rather as cooperating scripts executing on each user's computer. Participants are segmented from their backgrounds and combined into a shared space such that each user sees effectively an electronic mirror occupied by all the participants. Active participation (speaking or gesturing) causes a user to move to the foreground, while passive participants fade toward the background. Each participant's voice is placed at the spatial location corresponding to his or her current position in front of the camera, and name tags or location labels can be attached to users as well. Graphical, photographic, or audiovisual documents accessible via the World Wide Web can be placed into the background and moved or annotated through the use of gesture.

1.6 Initiatives Around the World in Multimedia Networks

What are the purposes of trials? Why not just charge ahead? The primary reason involves economic risk. For example, it was reported in *The Wall Street Journal* on December 6, 1996, that Bell Atlantic, Nynex, and Pacific Telesis Group pulled the plug on Tele-TV, their interactive television joint venture. The three companies supposedly spent around $500 million on the project for such things as investment in high-tech facilities, programming development and personnel, and high-profile talent. Nynex's vice chairman was quoted as saying, "In hindsight, maybe we were a little too aggressive a little too early." Although $500 million may seem like a lot, consider this. It has been estimated that the cost of rewiring America for the delivery of high-quality multimedia content to all homes and businesses will exceed $1 trillion. As can be seen, there is a lot of money involved, a lot of market uncertainty, and therefore a lot of risk. Trials are undertaken primarily to better manage this risk.

In general, trials are meant to look at the following two questions: "Can the technology be made to work effectively and efficiently?" and "Will anybody use it if it's built?" Before giving some examples of trials, let's examine these two questions in a little more detail.

Can it be made to work? Multimedia communications systems are made up of a variety of integrated technologies. There are the client computers and the servers. There is the array of media processing techniques, previously discussed, which must be implemented in hardware and software. There are the communications networks. There is the technology involved with the user interface. And, of course, there are the multimedia applications themselves. Finally, there is the multimedia content that must be produced and formatted. There is no single standardized way of doing any one of these things, much less the entire system. One element of the trial then, is the engineering of a very complex, nonstandardized system. Why not do it in the lab? Lab work is important as a precursor to a trial, but there is no way to deal with issues of scale in the lab. Issues associated with deployment, provisioning, and maintenance must also be studied in a trial situation.

Will anybody use it? There are really two parts to this question. Will anybody use it? And, will anybody pay to use it? In a laboratory, it is possible to study human factor issues around a new service capability. To a large extent, human factors studies consider the questions of ease of use, user friendliness, and required user training. Coupled with focus groups, some insight into customer preferences can be gained. However, until the new service capability is provided in its natural habitat, that is, at the home or business, the willingness to use a capability cannot be fully assessed. Willingness to pay can also be studied as part of a trial, although most trial services are offered at either no cost or greatly discounted cost in order to generate a large number of enthusiastic trial participants. Few bring enough revenue to cover operation of the trial. None bring in enough to cover the up-front investment.

1.6.1 Trials in Developed Countries

In addition to the NTT experiments already discussed, many trials have taken place in developed countries. Some ATM-based systems have been trialed with the cooperation of the ATM Forum,[2] an organization whose mission is to speed the deployment of ATM systems and services by promoting interoperability and cooperation. The ATM Forum has a Residential Broadband Working Group specifically focused on promoting and accelerating the use of ATM to deliver services to the home.

Le Groupe Videotron, a Canadian cable company, began a trial deployment of interactive television in 1989. The trial involved about 1,000 users. Offerings included video games, interactive participation in game shows, and access to a variety of database services. The results published by the company show that users spent more than four times as many hours accessing entertainment as they did the information services. This represents a consistent finding among all trials so far, namely, that entertainment services are more heavily used than information services. This does not directly address the issue of willingness to pay. Obviously, it is possible that customers would be willing to pay more per hour of use for information than for entertainment. Sadly, other studies show that this is usually not the case.

Microsoft and TCI have agreed to be partners in a number of trials. The initial trial will involve only Microsoft employees on or near the Microsoft campus in Redmond, Washington. The trial will expand to include several thousand TCI customers in the Seattle area. The system involves a fiber-to-the-node architecture and uses ATM transport and switching. The initial goals are to test equipment, software, and the end-to-end system. Once satisfied that the system is properly functioning, services will expand to include video-on-demand, pay-per-view movies, home shopping, and database access. Customer preferences will be monitored along with continuing exploration of system integration issues.

Most of the large telephone operating companies have already completed some initial trials. Currently they are expanding their trials with the idea that they will, in fact, be predeployments for the future. *Predeployment* means that services will be offered free or below cost during the trial period, but that eventually the same already deployed technology will be used for the commercial service. This, of course, results in a sharing of the costs between the trial and the initial deployment. An example of this was done by Bell Atlantic. Their previous technology trials have verified acceptable video coding schemes, effective database hardware and software, and the appropriate transport and switching schemes. Now they are planning a series of trials involving several thousand customers in New Jersey and Virginia. As have the other consumer trials, the services provided to the user focus on, but are not limited to, entertainment video.

2. Visit the ATM Forum web site: http://www.atmforum.com

The multicast backbone, or MBone, is a virtual network that lies on top of the physical internet. The MBone might be described as more of an experimental testbed than a trial. Basically, MBone provides the means of allowing point-to-multipoint and multipoint-to-multipoint networks across the Internet, which is itself constrained to point-to-point connections. The original application software consisted of Network Video (NV), Visual Audio Tool (VAT), White Board (WB), and Session Directory (SD). Today the MBone encompasses over 12,500 hosts attached to more than 1,600 routers in about 25 countries. Applications include video conferencing, teleseminars, video and audio broadcasts, and real-time data distribution.

In 1989, Advanced Research Projects Agency (ARPA) and the National Science Foundation (NSF) provided a large grant to the Corporation for National Research Initiatives (CNRI) for research in support of a high-speed national research and engineering network.[3] In turn, CNRI sponsored 5-Gb network testbeds. These were dubbed Blanca, Aurora, Nectar, CASA, and VISTAnet. In total these testbeds involved approximately seven national research laboratories, eight telecommunication service providers, ten universities, and a number of equipment manufacturers. These were technology testbeds with some emphasis on applications. There were no market trials involved. Items investigated included network protocols, computer/network interfaces, congestion control, admission control, error correction schemes, performance guarantees, and supercomputer operating system design. Applications developed and studied included climate modeling, chemical process simulation, three-dimensional medical imaging, and geological imaging.

A current example of a large-scale trial in Europe is the Advanced Multimedia Services for Residential Users (AMUSE) project, partially funded by the European Union [42]. At the time of this writing, AMUSE is setting up residential multimedia service trials in Milan, Italy; Munich, Germany; Mons, Belgium; Aveiro and Lisbon, Portugal; Basel, Switzerland; and Reykjavik, Iceland. Some of these sites will be interconnected by a Pan-European ATM network, and anticipated services include broadcast and on-demand television, shopping, and access to databases.

Trials of wide-area multimedia communications are highly complex affairs, involving many technologies and companies that must all work together. They are also expensive since they include upgrading a segment of the communications infrastructure, developing a critical mass of multimedia content, and developing the applications that cement it together. It is probably safe to say that most of these trials have not fully lived up to the expectations generated by the investment required. Because of this, a lot of industry interest has

3. Visit the CNRI web site: http://www.cnri.reston.va.us/overview.html

developed in multimedia delivery within the intranet, that is networks and applications contained within a single corporation or university. Intranet trials are smaller and less publicized. However, the value proposition of trial and deployment of intranet multimedia communications appears quite strong in many areas.

From all the trials undertaken so far, much has been learned concerning the details of the technology. The most useful information has involved the interworking of equipment. With regard to the higher level lessons learned, we might be able to say the following. Individual consumers appear to be much more interested in entertainment than in accessing information sources. That explains the increasing interest of entertainment providers in these trials. However, nobody has yet been able to prove a value proposition sufficient to either overcome the lack of standards in certain of the involved technologies or to drive the financial investment required to bring high-quality multimedia to the consumer. Thus, although the long-term prospects for high-speed communications to the individual consumer are bright, it is likely that the typical consumer will be faced with degraded quality multimedia for some time yet. If higher quality is desired, they will resort to stand-alone solutions such as CD-ROM.

At the same time, within the smaller, more focused, and more homogeneous environment of a single corporation or university, high-quality multimedia communications is far more feasible in the near term. Most of the uses of multimedia communications surveyed in this chapter, such as in business or medicine, may be first realized through a high-speed intranet.

1.6.2 Multimedia in Developing Countries

The evolution of multimedia computing and its integration into business and personal life is dependent on two primary factors, access to desktop computers or other computing devices, and access to an appropriate communications infrastructure. In the developing world, both of these factors are problematic. To participate fully in the growing globalized economy, the need for rapid development and deployment of these technologies could not be more obvious. For those developing countries already experiencing rapid economic growth, skillful exploitation of these rapidly changing technologies offers great opportunity for improved standing in the new emerging world order. Sadly, for those developing countries with stagnant economic growth, the situation is far more volatile. For some, opportunities for progress will undoubtedly be found through foreign investment. For others, particularly those which are also experiencing political instability, the means for investment in the new technologies may not be found. This will mean falling further behind.

One characteristic of the developed world is that essentially all residents have ready access to a telephone and basic voice service. This is called universal service. In these countries, investment in the communications infrastructure is aimed at delivering higher bandwidth and a rich set of advanced services for the advancement of education, expansion of business productivity, and enriching the options for entertainment. In the developing world, this situation exists only in subsets of the urban population. Also common are situations where only one or two telephones may exist for an entire village. In rural areas, telephones may be unheard of.

Thus, the developing world is faced with a dual challenge: the need to provide universal access and the need to maintain state of the art communications capabilities and services in situations where technological cooperation and/or competition with the developed countries is required. Tanzania is an example of a country where the balance of effort is primarily toward universal service. For the country as a whole, there is about one telephone for every 400 inhabitants. Even in its largest urban city, there is still only one telephone per 60 inhabitants. South America offers a more typical scenario. In Brazil as a whole, there is about one phone per 10 inhabitants. Among medium- to large-sized businesses and high-income families, there is one phone per 1.1 inhabitants. For these, evolution toward digital systems, increased service quality, and introduction of advanced services is necessary. For small businesses and the large majority of the population, increased access to a phone is the primary objective.

Developing countries are following their more developed neighbors in privatizing and deregulating phone services. The two primary goals of this process are to increase the efficiency of the telecommunications service providers and to attract foreign investment. Increasing the efficiency of service providers will lower costs and make rapid expansion of telephone access more probable. Privatization and deregulation of service providers have unique characteristics in developing countries. These include lack of sufficient numbers of appropriately educated personnel, lack of the technology base required to support the higher productivity required, and insufficient capital to fuel expansion.

Few, if any, developing countries have the economic growth necessary to capitalize their desired growth for the telecommunications infrastructure. At least some foreign investment is required. Such investments provide not only capital, but also some of the high-tech talent and experience required. Those countries with projected high economic growth have no difficulty attracting this investment of people and money. Their only problem is how to best manage the process. Countries with little prospect for growth or that have limiting political systems face more difficult times and risk falling further behind.

Multimedia, of course, requires much more than access to a telephone or even to a state-of-the-art communications network. Computers and the ability to both maintain and use them is also necessary. This poses an even more daunting problem for developing countries. The use of computers faces economic, political, and cultural uncertainties in many countries. Personal computers require greater initial investment per user and have more rapid obsolescence than do communications networks. In developing countries, maintenance facilities and technical support are often insufficient, leading to even greater risk and cost. Even among developed countries there are economic barriers to the use of personal computers, and there are apparent wide variations in willingness to embrace new technology and integrate it into business and personal life. In developing countries, there is no evidence of a techno-trophic culture. In addition, in countries with an oversupply of unskilled labor, there is very little incentive to automate.

In the United States, penetration of personal computers is only 32%. In Japan and Germany, other countries with universal telephone service, computer penetration is only 12% and 15%, respectively. In Mexico, which might be considered one of the more advanced developing countries, computer penetration is 2%.

In summary, the developing world when viewed as a whole has neither the communications infrastructure nor the penetration of personal computers to participate vigorously in the emerging world of digital multimedia information systems. However, some countries have the potential to do so, at least in critical business and social segments. For these countries, opportunities are great for exploiting these technologies to further accelerate economic growth toward the benefit of their citizens and their status in the global society. In these countries, a few field experiments are in place. More importantly, particular priority is being placed on educating and training greater numbers of specialists in the relevant technologies. A recently launched organization, the 2B1 Foundation,[4] is engaged in research and philanthropic activities to bring computing and digital connectivity to children throughout the world, particularly in developing countries.

References

[1] Agamanolis, S., and V. M. Bove, Jr., "Multi-Level Scripting for Responsive Multimedia," *IEEE Multimedia Magazine,* Vol. 4, No. 4, Oct.-Dec. 1997, pp. 40–50.

[2] Ishikawa, H., "Evolution from Narrowband Networks," *IEEE Commun. Mag.,* Aug. 1992, pp. 32–36.

4. Visit the 2B1 Foundation web site: http://www.2b1.org

[3] Walters, S. M., "A New Direction for Broadband ISDN," *IEEE Commun. Mag.,* Sept. 1991, pp. 34–42.

[4] Coudreuse, P., and M. Servel, "A Prelude: An Asynchronous Time-Division Switched Network," *Proc. Int. Communications Conf.,* June 1987, pp. 769–773.

[5] Gonet, P., P. Adam, and J. P. Coudreuse, "Asynchronous Time-Division Switching," *Proc. ICC'86,* June 1986, pp. 1720–1724.

[6] Christian, J. J., and K. Karl, "Distributed Processing within an Integrated Circuit Packet Switching Node," *IEEE Trans. Commun.,* Vol. COM-24, No. 10, 1976, pp. 1089–1100.

[7] Ishikawa, H., Y. Kosuge, and N. Miyaho, "A Study on Hybrid Switch Architecture Using Hierarchical Memory System," *Scripta Technica, Electron. Commun. Japan, Part 1,* Vol. 69, No. 1, 1986, pp. 103–111.

[8] Miyaho, N., and A. Miura, "Integrated Switching Architecture and Its Traffic Handling Capacity in Data Communication Networks," *IEICE Trans. Commun.,* Vol. E79-B, No. 12, 1996.

[9] Koinuma, T., and N. Miyaho, "ATM in B-ISDN Communications Systems and VLSI Realization," *IEEE J. Solid State Circuits,* Vol. 30, No. 4, Apr. 1995, pp. 341–347.

[10] Saito, H., "New Dimensioning Concept for ATM Network," *Proc. 7th Int. Teletraffic Congress, Specialist Seminar,* 1990.

[11] Saito, H., and K. Shiomoto, "Dynamic Call Admission Control in ATM Networks," *IEEE J. Select. Areas Commun.,* Vol. 9, No. 7, 1991, pp. 982–989.

[12] Lee, W. C., M. G. Hluchy, and P. A. Humblet, "Routing Subject to Quality of Service Constraints in Integrated Communication Networks," *IEEE Network,* July/Aug. 1995, pp. 46–55.

[13] Mase, K., and H. Kimura, "Quality of Service Issues in Communication Networks," *IEICE B-1,* Vol. J80-B-I, June 1997, pp. 283–295.

[14] Koinuma, T., T. Takahashi, H. Yamada, S. Hino, and M. Hirano, "An ATM Switching System Based on a Distributed Control Architecture," *Proc. ISS'90,* 1990.

[15] Tokizawa, I., K. Kikuchi, and K. Saito, "Transmission Technologies for B-ISDN," *NTT Review,* Vol. 3, No. 3, May 1991, pp. 44–58.

[16] Miyaho, N., M. Hirano, Y. Takagi, K. Shiomoto, and T. Takahashi, "An ATM Switching System Architecture for First Generation of Broadband Services," *Proc. ISS'92,* Oct. 1992.

[17] Miyaho, N., M. Hirano, T. Takenaka, T. Koinuma, and H. Ishikawa, "ATM Switching System Technologies," *IEEE, Denshi Tokyo,* No. 39, 1993.

[18] Miyaho, N., K. Shiomoto, and T. Takenaka, "Development of a Commercial ATM Node as the Backbone Node for High-Speed Data Communication Networks," *Proc. ISS'95,* Apr. 1995.

[19] Miyaho, N., K. Shiomoto, and K. Koyama, "Design Technology of a Commercial ATM Node System," *Trans. Info. Proc. Soc. Japan,* No. 5, May 1996, pp. 741–758.

[20] Miyaho, N., K. Doi, M. Hirano, and Y. Takagi, "ATM Switching System Technologies," *NTT R&D,* No. 3, Vol. 42, 1996, pp. 283–296.

[21] Aoyama, T., I. Tokizawa, and K. Saito, "Introduction Strategy and Technologies for ATM VP-based Broadband Networks," *IEEE J. Sel. Areas Commun.,* Vol. 10, No. 9, Dec. 1992.

[22] Batcher, K. E., "Sorting Networks and Their Applications," *Proc. AFIPS Spring Joint Comp. Conf.*, 1968, pp. 307–314.

[23] Kozaki, T., T. Aiki, M. Mori, M. Mizukami, and K. Asano, "A 156 Mb/s Interface CMOS LSI for ATM Switching Systems," *IEICE Trans. Commun.*, Vol. E76-B, No. 6, June 1994.

[24] Doi, K., K. Endo, H. Yamanaka, T. Takahashi, and M. Suzuki, "A Very High Speed ATM Switch with Input and Output Buffers," *Proc. ISS '92*, Oct. 1992.

[25] Genda, K., Y. Doi, K. Endo, T. Kawamura, and S. Sasaki, "A 160 Gb/s ATM Switching System Using an Internal Speed Up Cross-Bar Switch," *Proc. IEEE Globecom '94*, Nov. 1994, pp. 123–133.

[26] Ejiri, M., and M. Yoshida, "Progress in Telecommunication Services and Management," *IEICE Trans. Commun.*, E-78-B, No. 1, Jan. 1995.

[27] Sidor, D. J., "Managing Telecommunication Networks using TMN Interface Standards," *IEEE Commun. Mag.*, Vol. 33, No. 3, Mar. 1995.

[28] Glitho, R. H., and S. Hayes, "Approaches for Introducing TMN in Legacy Networks," *IEEE Commun. Mag.*, Vol. 34, No. 9, Sept. 1996.

[29] ITU-T Recommendations M3000, "Overview of TMN Recommendations," 1995.

[30] ITU-T Recommendations M3010, "Principles for a Telecommunications Management Network," 1992.

[31] ITU-T Recommendations M3100, "Generic Network Information Model," 1995.

[32] ITU-T Recommendations M3400, "TMN Management Functions," 1992.

[33] ATM Forum af-nm-0027.001, "CMIP Specifications for the M4 Interface," 1995.

[34] Yamagishi, K., N. Sasaki, and K. Morino, "An Implementation of a TMN-Based SDH Management System in Japan," *IEEE Commun. Mag.*, Vol. 33, No. 3, Mar. 1995.

[35] Fujii, N., and T. Yamamura, "ATM Transport Network Operation System in Japan," *IEEE Commun. Mag.*, Vol. 34, No. 9, Sept. 1996.

[36] Towle, T. T., "TMN as Applied to the GSM Network," *IEEE Commun. Mag.*, Vol. 33, No. 3, Mar. 1995.

[37] ITU-T Recommendations G774, "Synchronous Digital Hierarchy (SDH) Management Information Model for the Network Element View," 1992.

[38] Namiki, I., and T. Kimeda, "Broadband Network Applications," *NTT Review*, Special Features (2), Vol. 8, No. 6, Nov. 1996, pp. 62–67.

[39] Asano, S., H. Uose, K. Kitami, and K. Kinoshita, "NACSIS/NTT Joint Research Project on High-speed Multimedia Communications Networks," Vol. 46, No. 3, 1997, pp. 175–180.

[40] Miyawaki, N., "Current R&D Activities at NTT," *NTT Review*, Vol. 8, No. 3, May 1996, pp. 12–19.

[41] Agamanolis, S., and V. M. Bove, Jr., "Reflection of Presence: Toward More Natural and Responsive Telecollaboration," *Proc. SPIE Multimedia Networks: Security, Terminals, Displays and Gateways*, 1997 (in press).

[42] Zahariadis, T., et al., "Interactive Multimedia Services to Residential Users," *IEEE Commun. Mag.*, Vol. 35, June 1997, pp. 61–68.

2

Multimedia Support in Shared Media Local-Area and Metropolitan-Area Networks

2.1 Introduction

Since the early 1980s the use of local-area networks (LANs) has experienced enormous growth. During this period, LANs have become common equipment for many office, manufacturing, and scientific applications [1]. As time goes on, however, these networks are being used for much more traffic-intensive applications than was previously the case. The recent proliferation of high-speed personal computers and workstations has led to the development of applications requiring increasing network bandwidth. In addition, new multimedia applications are being developed at a very rapid pace. Many of these applications are unique in that very tight constraints must be placed on various aspects of network performance in order for them to function properly [2].

Multimedia applications require the concurrent support of voice, video, and data information. Quality of service (QoS) is a measure of the satisfaction experienced by the users of such a system. Requirements for network QoS for multimedia applications are described in [3,4], which are based on ITU-T guidelines (International Telecommunications Union—Telecommunication Standardization Sector). Three classes are defined: QoS class 1 for basic multimedia applications, QoS class 2 for enhanced multimedia applications, and QoS class 3 for premium multimedia applications. Tables 2.1 and 2.2 list some of the typical QoS characteristics expected with these types of applications [3–5].

Table 2.1
QoS Parameter Specifications for the Three Classes of Service

Application	QoS Parameter	Class 1	Class 2	Class 3
Audio	Max delay (ms)	400	400	150
	Frequency range (kHz)	0.3–3.4	0.3–3.4	0.5–6.8
	Level (dBm)	–20	–20	–20
	Min error-free interval (min)	5	15	30
Video	Max delay (sec)	10 (still image)	0.6	0.250
	Frame rate (fps)	—	5	25
	Resolution	—	176 × 144	352 × 288
	Min error-free interval (min)	—	15	30

Table 2.2
Private Network QoS Parameter Specifications for the Three Classes of Service

Qos Parameter	Class 1	Class 2	Class 3
Min. delay (ms)	20	20	10
Error-free ratio	0.995	0.9975	0.999
Severely impaired ratio	0.03	0.01	0.005

Table 2.1 illustrates some of the constraints placed on the network to support certain applications with differing QoS class levels. For example, Table 2.1 shows a 400-ms limit on audio transfer delay with classes 1 and 2. This is reduced to 25 ms with connections to conventional telephones without supplementary echo control. Table 2.2 provides the QoS specifications in terms of delay and transfer rate for private networks. Note the order of magnitude reduction in transfer delay while the data rate has essentially doubled.

Typical multimedia applications vary depending on their real-time and interaction nature. Video conferencing is an example of real-time interactive communication, while video broadcasting is noninteractive. An example of non-real-time, noninteractive communication is sometimes found with multimedia database applications. Multimedia services can be grouped into different types, depending on the underlying requirements placed on the network. Examples of possible services are listed next:

- *Multimedia desktop collaboration.* Real-time multimedia conversational capabilities such as visual conference windows, screen sharing, and real-time document collaboration.
- *Multimedia mail.* Distribution of multimedia files and enhancement of other services such as audio and video mail, message archiving, and paging.
- *Multimedia information services.* Provides a fusion function that may integrate information from multiple sources.
- *Remote learning.* Electronic books, electronic university (virtual classrooms and virtual laboratories), vocational training and retraining.
- *Information on demand.* Electronic news services and custom publishing.
- *Games and entertainment on demand.* Examples of these are interactive television and personal computer games, multiparty games, movies on demand, and real-time video.
- *Home shopping.* On-line catalogs, multivendor competition, video garage sales, and consumer reports.
- *Telecommuting.* Related to multimedia desktop collaboration, and includes such things as interactive workstation and video conferencing.
- *Remote professional services.* Examples are health care, legal, and real-estate services.

Multimedia applications generate traffic with a variety of transmission requirements. For example, so-called *asynchronous traffic* is that for which there are no special real-time constraints [2]. This corresponds to many types of conventional data traffic such as that generated by file transfers and terminal emulation sessions. On the other hand, certain types of traffic require upper bounds on the time taken for packets of that type to be transmitted. This type of traffic is commonly referred to as *synchronous traffic.* This includes such things as packetized voice and video, where there is playout buffering at the destination so that small variations in delay (i.e., jitter) can be smoothed. Finally, traffic can be *isochronous,* which means that the delay required of the network is constant. This is a very strict constraint to impose on a network, but this may be needed when transmitting packetized voice or video directly, when very little buffering is used at the source or destination stations.

In this chapter we discuss the basic design and characteristics of common shared-media LANs and metropolitan-area networks (MANs). We introduce the salient principles of operation of each system, and consider how each one compares from a multimedia standpoint. We start by first reviewing the orginal three IEEE 802 standard LANs, namely, the Ethernet, the token ring, and the token passing bus. In Section 2.3 we discuss the evolution of the Ethernet standard to support Fast Ethernet and the emerging Gigabit Ethernet systems. Following this, in Sections 2.4 and 2.4.1 we consider the FDDI and FDDI-II ring systems. In Section 2.5 we consider the 100-Mbps VG-AnyLAN standard. In Section 2.6 we discuss the IEEE 802.11 wireless LAN. Finally, in Section 2.7 we discuss how shared media networks are being considered for providing multimedia services to the home. In the appendix at the end of the chapter, we provide a brief introduction to asynchronous transfer mode (ATM).

2.2 The Original IEEE 802 Local-Area Networks

In the early 1980s the Institute of Electrical and Electronics Engineers (IEEE) initiated its work on LAN standards. These activities were originally started in response to the proliferation of proprietary LANs that had appeared in the marketplace. At that time it was clear that standards would ensure the product interoperability needed to develop large commercial markets.

The IEEE LAN standards have been developed under the direction of the IEEE 802 committee. Since the first standards were published, many have gone on to adoption by the American National Standards Institute (ANSI) and the International Organization for Standardization (ISO). Rather than remaining fixed, most of the standards are continuously evolving to support new products and to accommodate updated networking technologies.

In this section we start by reviewing the three original IEEE 802 LANs [6–8]. Our objective is to comment on the salient features of each system with a view toward assessing their strengths and weaknesses from a multimedia standpoint.

2.2.1 IEEE 802.3 Ethernet

Networks based on the IEEE 802.3 family of standards are by far the most pervasive of all LANs. These LANs are commonly referred to as *Ethernet.* The actual Ethernet system, however, was developed by Xerox in the early 1980s. This network used coaxial cable as its transmission medium and formed the basis for the orginal baseband versions of the IEEE 802.3 standards [6].

IEEE 802.3 is a shared bus network, where the physical topology provides a broadcast capability for all attached stations. This means that any transmission placed on the medium will eventually propagate past all attached stations. In the original "thickwire" baseband version (10BASE5), stations passively connect to a single shared coaxial cable. This arrangement is illustrated in Figure 2.1

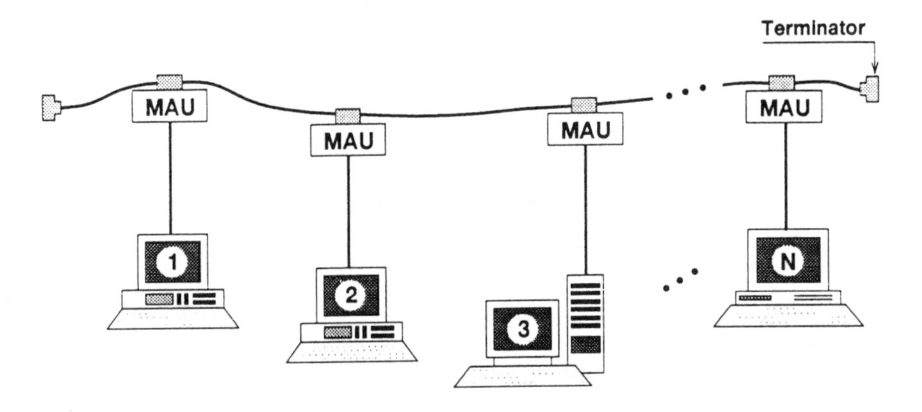

Figure 2.1 Example of an IEEE 802.3 cable segment.

where *N* stations are shown tapping the cable using a medium attachment unit (MAU). The most common configuration in 10BASE5 is shown where a short drop cable connects the MAU to a network adapter card in the station. In this system, the maximum cable segment length is 500m (hence the "5" in 10BASE5) but segments can be extended using repeaters to a maximum of 2.5 km.

A less expensive "thinwire" version of IEEE 802.3 (10BASE2) was also introduced where the coaxial cable connection is made right at the back of the station adapter card using BNC connectors. As a result of the thinner cable and more intrusive coupling, the 10BASE2 system is more restricted in its maximum segment length (185m compared to 500m). However, because of its lower cost, this option has become much more popular.

In a network such as that shown in Figure 2.1, there must clearly be a mechanism to allow the attached stations to share the transmission medium. The rules that the stations use to do this are referred to as a *media access control protocol* (MAC protocol). IEEE 802.3 uses a MAC protocol called *CSMA/CD* (carrier sense multiple access with collision detection). The CSMA/CD protocol is considered to have evolved from less sophisticated protocols such as ALOHA and CSMA.

ALOHA was originally developed at the University of Hawaii for the purposes of connecting various terminal facilities to a main computer using radio links [9]. The ALOHA media access protocol is considered the forerunner of all "random access" protocols. In ALOHA, when a station has a packet to send, it transmits immediately without any regard for other stations that may already be transmitting. As a consequence of this, two or more station transmissions may overlap in time at the intended receiver. This event is referred to as a *collision*, and usually results in the destruction of the packets involved.

Packets that are lost through this process clearly have to be retransmitted. When stations do this they attempt to randomize their retransmission attempts by independently waiting for a pseudo-random "backoff" time before attempting to transmit again. These aspects of ALOHA operation are clearly probabilistic in nature and this is why such protocols are referred to as "random access."

An important innovation that influenced the development of Ethernet was that of the carrier sense multiple access (CSMA) protocol [10,11]. In CSMA, an active station first "senses" the transmission medium to see if it is currently being used. If so, the station defers to the station(s) transmitting and schedules its attempt for a later time. There are a number of protocol variants. In the 1-persistent version, a station that senses the channel is busy continues sensing until it becomes idle, then transmits. It should be clear that even when carrier sensing is used, collisions can still occur. This is caused by the fact that it takes a nonzero time for a transmission at one station to be sensed at another. During this interval, the second station may transmit, leading to a collision.

From an operational standpoint, carrier sensing is only reasonable when the propagation delay times on the medium are small in relation to the transmission time of a packet. When this is *not* the case, the information gained by sensing the channel may be of little use, since the sensed transmissions may have already been completed by the time they are sensed. This performance property is embodied in the so-called *normalized propagation delay parameter* a, where $a = \tau / T$. Here, τ is the propagation delay time from one end of the medium to the other, and T is the average packet transmission time. Generally carrier sensing is efficient when $a \ll 1$ and of little use when $a \gg 1$. For this reason, in products incorporating CSMA it is important that a remain as small as possible. In practice, this is difficult because one would like to continually increase LAN transmission speeds. Doing this will clearly decrease T, resulting in an increase in a. This is an important consideration that has had a major influence on the evolution of Ethernet standards.

In CSMA, when stations collide they continue transmitting for the full duration of the packets being sent. In Ethernet protocols, *collision detection* is added to CSMA to attempt to reduce this wastage. When transmitting stations detect that a collision is occurring, they stop transmitting (after a *collision reinforcement* interval), and then back off for some random time period. In IEEE 802.3 Ethernet, the 1-persistent version of CSMA/CD is used.

As discussed earlier, in IEEE 802.3, colliding stations randomly back off following detection of a collision. The backoff algorithm used is referred to as *truncated binary exponential backoff.* In this scheme, a colliding station randomly selects a backoff time from an integer range of *slot times.* A slot time duration is defined by the standard and is set to be somewhat greater than twice the worst-case propagation delay time between stations on the medium. The intent is that regardless of the physical position of two colliding stations, if each chooses a different slot time backoff interval, they will not recollide on the subsequent attempt. Each time a given packet is collided with, its backoff range is doubled, which leads to an exponential growth in mean backoff times. In this scheme, the backoff range used is $[0, 2^k]$ slot times, where k is the number of collisions that the packet has incurred. This procedure continues until $k = 10$ collisions, at which point the backoff range does not grow any further. If the number of unsuccessful attempts reaches 16, an error is signaled.

Although binary exponential backoff is very simple to implement, it has been widely criticized from a multimedia viewpoint. Because retransmission attempts are scheduled randomly, the algorithm induces larger variations in packet delay than is necessary from strict queueing considerations. In addition, with CSMA/CD there is no means to offer time-sensitive traffic priority over normal data traffic. If a conventional data application starts a large file transfer for example, then there is no way to ensure that bandwidth is allocated properly to more time-sensitive data streams. Because of this, conventional Ethernet systems are generally very poor when it comes to supporting mixtures of traffic with varying delay requirements. Many experiments have been performed that verify this. See [12] for some early attempts at mixing digitized voice and data traffic on Ethernet. However, even though this is the case, many experimental multimedia systems use Ethernet under very tightly controlled situations, for example, when a very small number of multimedia stations are attached.

2.2.1.1 Ethernet Performance

The maximum throughput possible under CSMA/CD is often a strong function of the a parameter. For example, when the loading is uniformly spread across a large number of stations, simple arguments can be made to approximate the maximum throughput by the expression [13]

$$C_{\text{CSMA/CD}} \approx \frac{B}{1 + \gamma \cdot a} = \frac{B}{1 + \gamma \tau B/b} \tag{2.1}$$

where $\gamma \approx 5.4$. Note that $a = \tau B/b$, where B is the transmission bit rate (in bits per second) and b is the mean number of bits per packet. Capacity in this expression is in units of bits per second with a maximum of B, the transmitted data rate. From (2.1) we can see that as B increases for a fixed τ and b, the capacity will eventually saturate. This means that there will come a point where increasing the transmission rate will not significantly improve the traffic-carrying ability of the network. For this reason, CSMA/CD does not scale well with increasing transmission rates. As an example, consider $B = 10$ Mbps,

τ = 10 μs, and b = 1,000 bits, which is typical for Ethernet. In this case, the capacity evaluates to $C_{CSMA/CD}$ = 6.5 Mbps. This means that only 6.5 Mbps of the full 10 Mbps is usable by the stations. Now consider what happens if we were to increase the data rate by 10 Mbps to B = 20 Mbps. In this case $C_{CSMA/CD}$ = 9.6 Mbps. Thus we see that even though we have added 10 Mbps to the channel data rate, our usable capacity has only increased by about 3.1 Mbps (i.e., 6.5 to 9.6 Mbps)! It is clear from this example that conventional Ethernet is already pushing the limits of efficiency for the CSMA/CD protocol. For this reason, Ethernet operation at higher speeds requires other changes so that the a parameter does not decrease substantially. This issue is discussed further in Section 2.3 when Fast Ethernet is discussed.

Various other physical layer standards have evolved for IEEE 802.3. As discussed earlier, options include those for both thick- and thinwire coaxial cable (10BASE5 and 10BASE2). In addition, broadband cable (10BROAD36) and fiber (10BASE-F) options are included. In 1990, IEEE 802.3 approved the 10BASE-T standard, which has become enormously popular.

2.2.1.2 Hub-Based Ethernet, 10BASE-T

10BASE-T is a hub-based Ethernet option that can use unshielded twisted-pair cable. An advantage of this is that suitable telephone cabling already exists in many buildings. This may obviate the need for installing new wiring, which is usually very expensive. In the hub architecture, wiring radiates out from central locations referred to as wiring closets. In 10BASE-T the maximum length of these links is set at 100m. An example of this is shown in Figure 2.2. In the figure a hub is shown with physical connections fanning out to a set of attached stations. Although this arrangement is a "hub-and-spoke" physical layout, the system still functions as a broadcast bus. At the hub, the cables are wired together electrically, so that the system is a broadcast one, just as in the coaxial cable versions.

There are many advantages to using this hub-based approach. Since the hubs are centrally located, no link in the network need be longer than the set limit of 100 m. This strict limit permits the use of less expensive, poorer quality twisted-pair cables at the full Ethernet data rate of 10 Mbps. A degree of centralized management can also be provided at the hubs. This may include such things as traffic monitoring and the ability to disable hub ports that are malfunctioning. In 10BASE-T there are two twisted pairs per station, one for transmission and the other for reception. This setup makes collision detection very simple since the hubs can explicitly indicate collisions on the station reception pair when it sees that more than one station is transmitting.

From the perspective of supporting future multimedia applications, the 10BASE-T standard has much to offer. In its basic mode of operation, the network operates more or less the way it does with the other IEEE 802.3

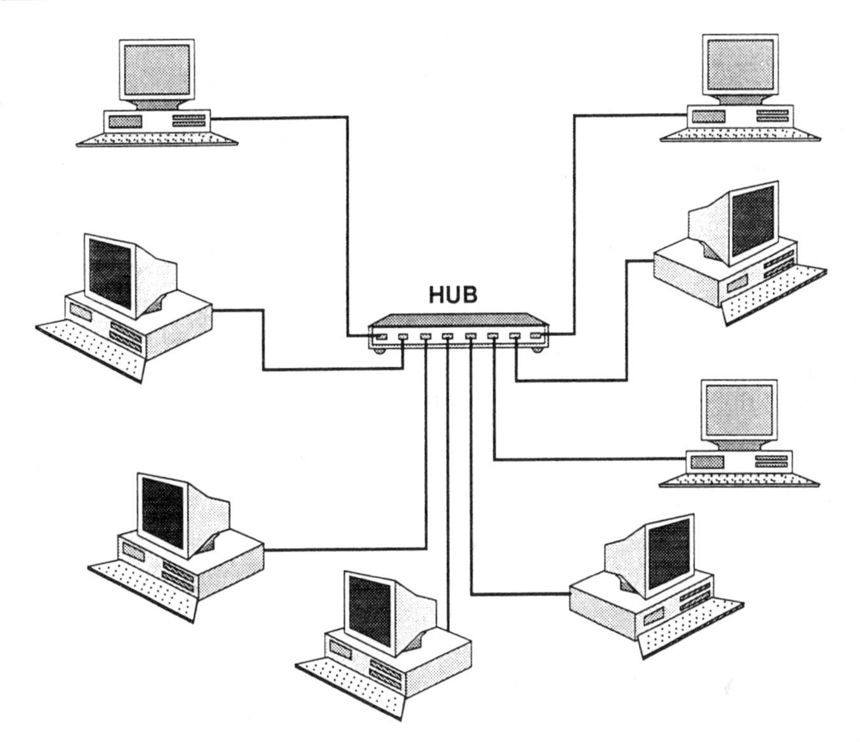

Figure 2.2 Hub-based layout used in 10BASE-T.

physical options. The arguments made against CSMA/CD's ability to support time critical traffic still hold. However, because of the hub-based design, some key evolution paths are possible. For example, rather than using a conventional multiport repeater hub, a "switched hub" can be installed instead. When this is done, the total network throughput that can be supported may be much higher, and sharing of the links does not use the CSMA/CD protocol. Collisions between stations do not occur. As a result, some of the random aspects of system performance associated with CSMA/CD are removed. It is also possible for the hubs to prioritize time-sensitive traffic so that it can coexist with normal traffic in a compatible way. For these reasons, it is expected that future multimedia installations can benefit greatly from evolving 10BASE-T type networks. In addition to all these important factors, the 10BASE-T option allows an evolution path toward Fast Ethernet, which operates at a transmission rate of 100 Mbps. Further discussion of these points is postponed until Section 2.3.

2.2.2 IEEE 802.5 Token Ring

The token ring was one of the original LAN proposals considered by IEEE 802. IBM was a strong advocate of this system and a simple token passing loop option had originally been developed as part of IBM's SDLC protocol definitions [14]. The standardized version of the token ring is defined in IEEE 802.5 and includes provisions for both 4- and 16-Mbps operation over shielded twisted-pair cable [7]. There is also a newer version at 4 Mbps running on unshielded twisted-pair cable.

In a token ring, the physical medium forms a closed loop, which is shared in its entirety by the attached stations. An example of this is shown in Figure 2.3. In the figure, station 1 has possession of the ring and is currently transmitting a packet. All of the other stations are processing the packet and repeating the transmitted bits downstream on the ring. The figure shows that station 1 has "broken" the ring and is simultaneously removing the bits that have returned after propagating completely around it.

Only one station is permitted to transmit new packets onto the ring at a time, and this is enforced by use of a *token*. In IEEE 802.5, the token is a special (24-bit) packet that is passed around the ring from one station to the next. In the basic token passing mechanism, a station wishing to transmit first "captures" the token packet, then transmits its packets, and finally regenerates and releases the token back onto the ring. An important consequence of this mechanism is that access to the ring is highly organized and fair. Stations are serviced in strict round-robin order as the token propagates around the ring. As a result, much of the random and uncontrolled behavior associated with

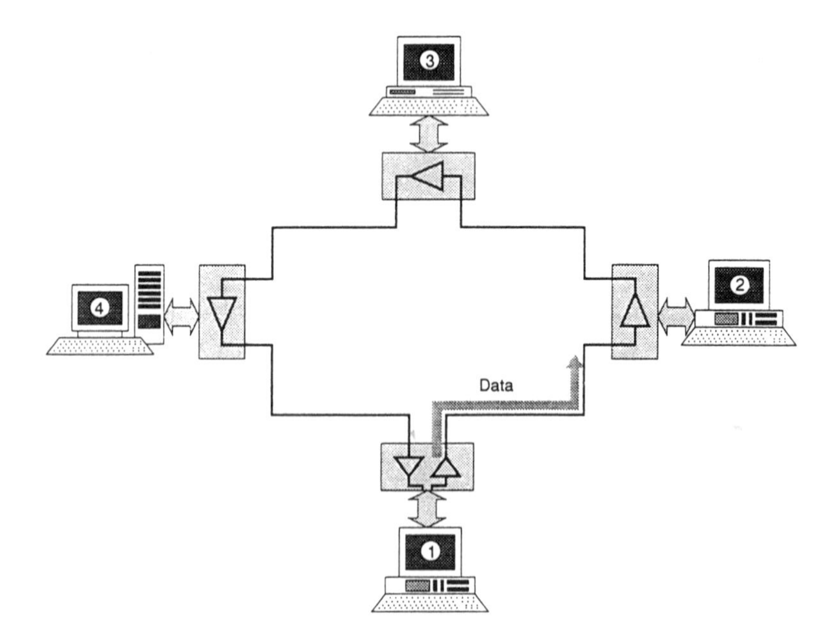

Figure 2.3 Example of an IEEE 802.5 token ring.

CSMA/CD does not occur. In addition, IEEE 802.5 incorporates a priority mechanism that schedules higher priority packets over lower priority ones in a very strict manner. Consequently, this type of media sharing mechanism is much superior from a multimedia viewpoint. Applications with strict real-time constraints can be prioritized over those of less contrained applications. When this is done properly, delay guarantees can be strictly enforced. In addition, the 16-Mbps data rate combined with the more efficient media access protocol offers more bandwidth to the applications than 10-Mbps Ethernet. For these reasons the token ring is a much more viable option for supporting multimedia applications than conventional Ethernet. Simple calculations show that a token ring can support more than one hundred 64-Kbps voice connections while maintaining packetization delays below 10 ms [2].

In the early LAN days, the physical ring architecture was perceived as highly negative. Since attached stations must actively repeat bits along the ring, this was viewed as a possible source of network unreliability. If a station were to fail, then this would inhibit the operation of the ring. These considerations in part led to the development of the IEEE 802.4 token passing bus network discussed in Section 2.3. This negative perception however proved to be inaccurate due to the hub-based design adopted for IEEE 802.5. As in 10BASE-T Ethernet, stations connect to the ring at centralized hubs referred to as MAUs (see Figure 2.2). To attach, a station must provide power to a relay in the MAU, which physically connects it into the ring. If the power is then lost at the station, the relay is released and the ring bypasses the failed station.

A negative aspect of the token ring is that the token itself must be maintained by the stations. It is possible for the token to be lost because of corruption by noise or if a station loses power while in possession of the token. Recovery procedures ensure that the token can be regenerated in these situations. The protocol can also recover if duplicate tokens form on the ring. The controlled access nature of the ring helps simplify these types of procedures. For example, assume that upper limits are set on the number of stations, the ring latency per station, and the lengths of the transmission links. Then a strict upper limit on the total ring latency can be easily calculated. If the token or a valid packet transmission fails to appear within this time, recovery procedures can be invoked immediately. Maintenance procedures of this type are initiated by an active monitor station, which is "elected" from among the connected stations. The active monitor periodically issues a special frame that indicates to the other stations that the monitor is still functional. Otherwise recovery procedures will elect a new one.

In IEEE 802.5, a transmitting station must wait until it has finished transmitting its final packet and the start of that packet has propagated completely around the ring before releasing the token. This is often referred to as *late token release* or *release after reception.* From a performance point of view, this means that each time a station uses the token it must incur an idle waiting time of roughly one complete ring delay before passing the token on. Assuming N stations each transmit for THT seconds per token capture, then the maximum throughput or capacity is given by

$$C_{TR} = \frac{B}{1 + t_{token}/\text{THT} + \max(\tau - T, 0)/\text{THT} + \tau/(N \cdot \text{THT})} \quad (2.2)$$

Here, B is the channel transmission rate in bps and T is the packet transmission time. Note that THT $= b_t/B$ where b_t is the total number of bits transmitted per station token capture. In a large high-data-rate system, $\tau \gg T$. Assuming both $N \gg 1$ and $t_{token} \ll$ THT we can simplify (2.2) to

$$C_{TR} \approx \frac{B}{1 + \tau/\text{THT}} = \frac{B}{1 + \tau B/b_t} \quad (2.3)$$

We can see from this result that when B is increased, the second denominator term will eventually dominate the first and the capacity will become insensitive to B. This was also the case for Ethernet as discussed in Section 2.2.1. For this reason, when token rings are scaled to much higher operating speeds, late token release is not appropriate. We will return to this issue later when discussing early token release in Section 2.4.

It is also instructive to compare (2.3) with (2.1) for CSMA/CD. Assuming each token ring station transmits one packet per token capture, then $b_t = b$. The two expressions are now equal except for γ in the second term of (2.1). Because of this factor, a token ring even using late token release scales better with B than does CSMA/CD.

2.2.2.1 Token Ring Priority Mechanism

IEEE 802.5 has a unique priority mechanism that very tightly controls ring access between different priority levels. Up to eight priority classes are possible. The algorithm used relies on the late token release feature for proper operation. The basic mechanism works as follows. There is a 3-bit priority encoded into the token packet itself. When the token arrives at a station, it can only be captured if the priority of awaiting frames is at least as high as that of the token priority. If this is not the case, then the station will have to wait for a token with lower priority. A mechanism exists for stations to request tokens of a particular priority. This is done using a 3-bit field that is provided in the header of both transmitted data frames and the token frame. These are referred to as *reservation bits* or *R-bits.* When a token or data frame passes by, a station can write these reservation bits provided it is reserving a priority that is higher than the one currently encoded into the R-bits. In this case it may overwrite any lower priority reservation currently encoded by the R-bits. Note that this can only be done to the token if the current priority level of the token is higher than the desired priority. The idea behind this is that when a circulating frame returns to the transmitting station, the contents of the R-bits are examined. When the time comes to release the token, the station may promote that token's priority to a higher requested value. When this happens, the same station is responsible for lowering the token's priority to its previous level after it circulates once around the ring. The previous token state is pushed onto a local stack so that multiple token promotions can be properly managed.

Note that the effect of this mechanism is that requests for higher priority transmissions will very quickly preempt lower priority token passing cycles. When the high-priority packets have been sent, the token will be returned to its original priority by the preempting station. Then the original token passing cycle will continue where it left off.

Since its original inception, *early token passing* has been added as an option to IEEE 802.5. When this is done, however, the priority mechanism just discussed will not operate as described since stations may have to release the token before learning what priority requests have been made. Early token passing is discussed further in Section 2.4 on FDDI.

2.2.3 IEEE 802.4 Token Passing Bus

The main LAN contenders under the original IEEE 802 standards committee were Ethernet and the token ring. However, in addition to the standards that eventually evolved from these two proposals was the 802.4 token passing bus [8]. This system may be viewed as a hybrid of the other two and was initiated by those wanting both strict time bounds on station access times and high network reliability. It was argued that the probabilistic nature of the CSMA/CD protocol made Ethernet undesirable. Although the token ring did not have this shortcoming, the seemingly unreliable physical ring architecture was viewed as being unacceptable. In essence, the token passing bus combines the positive protocol features of token passing with the high reliability of an Ethernet passive bus. This standard became part of the MAP/TOP standards, which defined protocols for use in factory, industrial, and office environments [15].

As mentioned earlier, the IEEE 802.4 token passing bus access method employs token passing on a passive physical bus. In a token ring such as IEEE 802.5, the physical ring arrangement forms a closed loop, which automatically defines the position of each active station on the ring. In IEEE 802.4 this is not the case and stations must therefore form a "logical" or "virtual" ring. The details of this procedure add an enormous degree of complexity to the protocol. In addition to the usual maintenance procedures associated with managing token passing, procedures must also be in place to maintain the logical ring itself.

An example of this system is shown in Figure 2.4. In the figure a set of four stations is shown that have formed a logical ring along which the token is passed. The token passing sequence is illustrated by the dotted lines in the figure. Note that the actual packet transmissions occur on the shared bus and the token packet is addressed to the intended station when a token pass is made.

In the token passing bus, the logical ring is formed in decreasing order of the participating station's MAC addresses. Thus stations pass the token to a station with a lower address than itself. This is true for all stations except the one with the lowest address, which passes the token back to the station with the highest. This is illustrated in the Figure 2.4 where the MAC addresses are shown for each station. A novel aspect of the design is that a station is only aware of the identity of the stations immediately upstream and downstream of itself in token passing order. For each station these are referred to as the station's previous station (PS) and next station (NS). In Figure 2.4, for station 30, NS = 20 and PS = 40. When a station passes the token, it addresses the packet to its NS and checks to make sure that the token pass was successful. If no valid channel activity is seen following the token pass, the station will

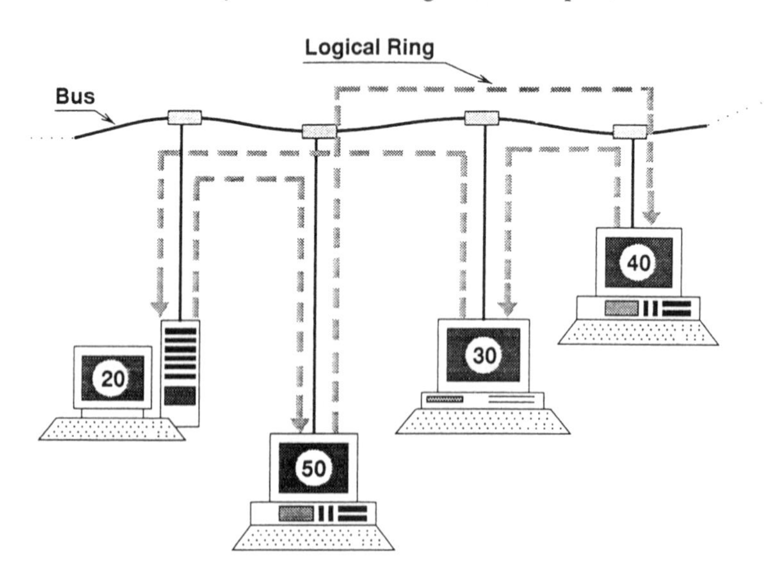

Figure 2.4 Example of an IEEE 802.4 token passing bus.

eventually assume that the next station is not functional. In this case, having each station know the PS address permits simple ring recovery. The token passing station broadcasts a *who_follows* frame containing the address of the nonresponsive NS station. The station whose PS matches the entry in the who_follows frame responds and the logical ring is patched around the defunct station. For example, if the power at station 20 in Figure 2.4 is shut off, the token pass from station 30 will be unsuccessful. Station 30 then issues the who_follows frame containing station 20's address. When station 50 sees this and notes that station 20 is its PS, it responds (by transmitting a set_successor frame) and the logical ring is patched around the failed station. Clearly this mechanism will only work if a single station has dropped out of the ring since the last time the token arrived. In Figure 2.4 if stations 20 and 50 both lose power prior to the arrival of the token at station 30, then there would be no response to the who_follows frame and the logical ring would have to be completely reconstructed.

Clearly there must also be a way of adding stations into the ring. The basic mechanism works as follows. While holding the token, each station will occasionally permit other stations the opportunity to join the ring. Note that the stations to be added must have MAC addresses in the range between the current token holder and its NS. In Figure 2.4, station 40 would solicit entrants over the address range from 31 to 39. This opportunity is given by having the token holding station transmit a *solicit_successor* frame. The basic solicit_successor frame has a response window associated with it. The response window is a period of time equal to roughly twice the worst-case end-to-end propagation delay of the medium. Any station within the indicated address range may transmit into the response window with a *set_successor* frame. If a single station is waiting to be added then it would respond and be added into the ring. Both the current station, NS, and the added station would update their NS and PS values. If more than one station responds, then the set_successor frames will collide and the current station must isolate a single station to be added in the current round. This is done by transmitting successive *resolve-contention* frames. Each of these frames has four response windows following it. The first time, each competing station transmits into the response window corresponding to the first 2 bits of its MAC address. If a competing station sees transmission activity in a response window before the one that it is to transmit into, that station has lost the competition. On each subsequent resolve-contention, the remaining stations use the next 2 bits in their address and so on. Eventually a single station will transmit and the set_successor frame will be successful. This station is then patched into the logical ring. All the other unsuccessful stations must wait for another station to add opportunity. Note that the station with the lowest address uses a *solicit_successor_2* frame, which contains two response windows. One is for stations with addresses lower than it and the other is for addresses higher than its NS.

The procedures involved in creating the logical ring are quite complex. When a set of stations has determined that the ring must be initialized, a competition occurs whose purpose is to select a single station, which starts the ring initialization process. The competition involves having stations repeatedly transmit *claim-token* frames until only a single station remains. Each transmitted claim-token frame has a data field length that is either 0, 2, 4, or 6 times the length of a response window. The value used by a particular station starts with the first 2 bits of the station's MAC address. On each subsequent transmission, the next 2 bits are used and so on. After each claim-token transmission, a station listens to the medium. If it senses that another station is still transmitting then it drops out of the competition. Eventually the station with the highest address will sense the medium idle and know that it has won. This station becomes the only station in the ring and begins to add stations using the mechanism discussed earlier.

In IEEE 802.4, the efficiency of the basic token passing mechanism is generally better than that of the late token passing scheme used in the IEEE 802.5 token ring. The reason for this is that the token is released immediately in 802.4 so that the extra overhead discussed in Section 2.2 is not incurred. Assuming N stations in the logical ring, each of which transmits for THT seconds whenever the token arrives, the maximum throughput is given by

$$C_{\mathrm{TB}} = \frac{B}{1 + t_{\mathrm{tok}}/\mathrm{THT} + \sum_{i=0}^{N-1} \tau_{i,i+1}/(N \cdot \mathrm{THT})} \tag{2.4}$$

In this expression we have numbered the stations from 0 to $N - 1$ and $\tau_{i,i+1}$ is the propagation delay time for a signal to travel from station i to station $i + 1$ in the token passing sequence. Note that the subscripts of $\tau_{i,i+1}$ are taken modulo N. As before, assuming that $t_{\mathrm{tok}} \ll \mathrm{THT}$ and if the average propagation delay between stations is τ_{avg}, then

$$C_{\mathrm{TB}} \approx \frac{B}{1 + \tau_{\mathrm{avg}}/\mathrm{THT}} \tag{2.5}$$

Comparing this result with that of (2.3), we see that the token bus performance will be better provided that $\tau_{\mathrm{avg}} < \tau$ (assuming that B is the same in both cases). This will always be the case since the propagation delay from one token bus station to the next can be at most τ seconds.

Note that in general a token ring using *early* token release can perform with a higher efficiency than the IEEE 802.4 bus. The reason for this is that in a token ring, stations automatically pass the token to the station that is closest to it in the downstream direction. This tends to minimize token passing overhead. This is not true in the token passing bus since the token passing sequence is determined by the MAC addresses of the stations, which generally will be arranged arbitrarily along the bus.

An important optional feature of IEEE 802.4 is its priority access method. Four classes of service are defined. Unfortunately, the strict type of priority scheduling used in the IEEE 802.5 token ring is not possible. Recall that in 802.5 the transmitted packets circulate completely around the ring and return to the transmitting station before the token is released. It is during this circulation that stations set reservation bits that request higher priority tokens. This mechansim is clearly not possible in IEEE 802.4. The priority mechanism used instead is based on a timed token rotation algorithm. Whenever the token arrives, the station holding it can transmit high-priority packets for THT seconds. However, the maximum token holding time for lower priority packets is dependent on the measured token rotation time. In essence, the time that the token takes to return to the station in question is measured. If the token returns quickly, then it may be held longer at the station than if the token returns slowly. This mechanism is very similar to the priority meachanism used for the FDDI ring and a detailed discussion of this is deferred until then. Suffice it to say that the priority mechanism is not as strict in its allotment of capacity as is the case for IEEE 802.5.

From the preceding discussion we can see that one of the drawbacks of the IEEE 802.4 standard is the complexity of the protocol. However, from a multimedia perspective the protocol has much more to offer than does CSMA/CD.

2.3 IEEE 802.3 100BASE-T Fast Ethernet

Since the development of the original Ethernet standard, considerable activity has been devoted to enhancing its performance. This has been done primarily by operating the network at higher data rates. So-called Fast Ethernet operates at 10 times the transmission rate of the original 10-Mbps version. Fast Ethernet was developed by the same standards body that produced the original Ethernet (IEEE 802.3 [6]) and is considered an upgrade path for the 10BASE-T standard discussed previously. The Fast Ethernet additions are a supplement to the IEEE 802.3 standard, referred to as IEEE 802.3u [16]. To reflect its 100-Mbps transmission rate, Fast Ethernet is denoted *100BASE-T.* This standard uses the same hub topology as is used in 10BASE-T (see Figure 2.2) and exploits the same advantages as 10BASE-T in terms of its use of inexpensive unshielded twisted-pair cable options. Because Fast Ethernet was developed under the auspices of IEEE 802.3, a lot of care has gone into issues dealing with the interoperability of the original and newer systems. This includes autonegotiation procedures, wherein 10BASE-T and 100BASE-T components that are interconnected can automatically negotiate a common transmission speed [17].

As previously discussed, the original Ethernet operates over distances that may result in a significantly large normalized propagation delay parameter, a. In going to Fast Ethernet, if the transmission rate were merely scaled from 10 to 100 Mbps, this would further increase a by a factor of 10. (Recall that $a = \tau B/b$, where B is the transmission rate in bits per second.) As a result, if the same type of CSMA/CD protocol is used, the physical span of the network (i.e., τ) would have to be decreased by the same factor in order to maintain the same value of a. (Maintaining the same value for a would result in the same protocol efficiency on the shared channel under the assumptions discussed in Section 2.2.1.) For this reason, in Fast Ethernet the physical size of the *collision domains* is contrained to be roughly an order of magnitude smaller than that of conventional Ethernet. The actual maximum span is 412m compared with 3,000m for conventional Ethernet. Note that this is only one consideration and there are other reasons that Ethernet does not scale well with bit rate.

With Fast Ethernet, the signal timing is decreased by a factor of 10, but the rest of the MAC protocol remains essentially unchanged. Thus the packet formats and MAC protocol are practically identical to that used in conventional Ethernet. The clear advantage here is that upward compatibility from 10BASE-T systems is easily accomplished. However, Fast Ethernet uses the same CSMA/CD MAC protocol as conventional Ethernet. This includes the probabilistic aspects of the backoff algorithm and its lack of packet priorities as discussed in Section 2.2.1. As a result, from a multimedia viewpoint the same criticisms can be made of this system as of conventional Ethernet. However, note that the 10-fold increase in transmission speed will obviously lead to improved performance over that of conventional Ethernet. Thus small multimedia networks may achieve acceptable performance, but such systems will not generally scale to larger sizes. Note that Fast Ethernet also supports switched-hub operation, which can be used to improve performance from a multimedia perspective. This is discussed in greater detail in subsequent sections.

Fast Ethernet supports a variety of different cabling types through three different types of transceiver standards. The 100BASE-T4 transceiver accommodates Category 3, 4, and 5 UTP cable using four pairs of wires. Multiple pairs are used simultaneously so that the signaling rate per pair is lower than the total. This feature permits the use of Fast Ethernet using poorer quality twisted-pair wiring. A novel scheme is employed where two of the pairs are dedicated to each direction (i.e., station to hub and hub to station, respectively), and the other two pairs are shared by the station and hub for transmission in both directions. The data rate per pair is 33.33 Mbps (× 3 pairs = 100 Mbps) using "8B6T" ternary coding so that the symbol clock rate is 25 MHz per pair. Using the dedicated pair makes station collision detection very simple. A disadvantage of this pair sharing scheme is that it is not possible to have simultaneous two-way transmission on each link (i.e., full-duplex operation). This option is available for the other Fast Ethernet options discussed later.

In 100BASE-TX, Category 5 UTP and 150Ω STP cabling can be used with only two wire pairs. Finally, in 100BASE-FX, multimode fiber is used with a transmission wavelength of 1,300 nm. In both these options, transmission occurs on one pair and reception on the other. For this reason, full-duplex operation is possible when a switched hub is used. In the twisted-pair wire systems, link spans are limited to 100m and are implemented using the same hub-based architecture as in 10BASE-T. However, in the fiber transceiver option, 2-km link spans are supported. A novel aspect of the 100BASE-TX and 100BASE-FX options is that they use existing media types standardized for the ANSI FDDI token ring system discussed in Section 2.4.

Note that coaxial cable, which is commonly used in conventional Ethernet, is not supported by Fast Ethernet standards. Also, the standards are specified so that external transceivers of various types can be connected to a 100BASE-T adapter though a media independent interface (MII), so that the different media options can be accommodated using the same computer adapter card, if desired. Full-duplex operation is also supported in all transceiver types except 100BASE-T4. This accommodates the use of Ethernet switches with Fast Ethernet.

2.3.1 Fast Ethernet Hubs

As in conventional 10BASE-T systems, Fast Ethernet hubs are available in both shared media and switched versions. In a *shared media hub*, a single 100-Mbps transmission channel is shared among all connected stations so that only one transmission can proceed at a time. This is accomplished by copying an incoming transmission on any port to all other ports (so that it can be sensed by the CSMA/CD algorithm). An example of this is shown in Figure 2.5, which has eight Fast Ethernet ports connected through a shared 100-Mbps bus. In the figure, station 4 is shown transmitting. When this happens, the hub broadcasts the transmission to all connected stations. In this way, the hub emulates the broadcast nature of a conventional shared cable system. Such a hub is often referred to as having a repeater architecture since received bits are

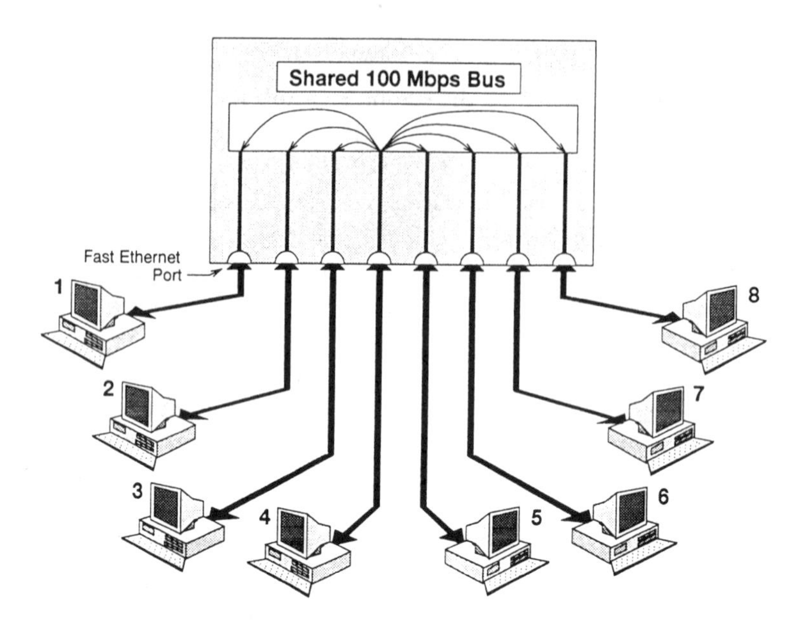

Figure 2.5 Fast Ethernet shared media hub.

broadcast as soon as possible after they arrive. Note that in this hub, the system is simplified by keeping the delay through each port module as small as possible.

In a *switched hub*, packets are buffered at the hub and bridged between the connected stations, so that in general many simultaneous transmissions can occur. This arrangement is shown in Figure 2.6. In this case, a given packet is completely buffered in the hub if necessary, and only sent toward its destination station, rather than broadcasting it to all stations. This obviously leads to much higher total load handling capabilities. In the figure we show two simultaneous transmissions occurring between stations 4 and 1 and stations 5 and 7. Because of their increased complexity, switched hubs are generally much more expensive than their shared media counterparts. Because packets generated on different ports now cannot collide with each other, each port consists of its own collision domain.

Some vendors also provide hybrid-type hubs, which combine the features of shared media and switched hubs. This is done hierarchically by first grouping the ports into a number of separate collision domains. This is accomplished using shared media hub interconnections between each group of ports. These collision domains are then interconnected by a switch, as in a switched hub arrangement. An example of this is shown in Figure 2.7. In this case we have eight ports, which are grouped into two collision domains interconnected by a switch. In each domain there can be at most one transmitted packet at a time. Because the switch isolates the two shared buses, both collision domains

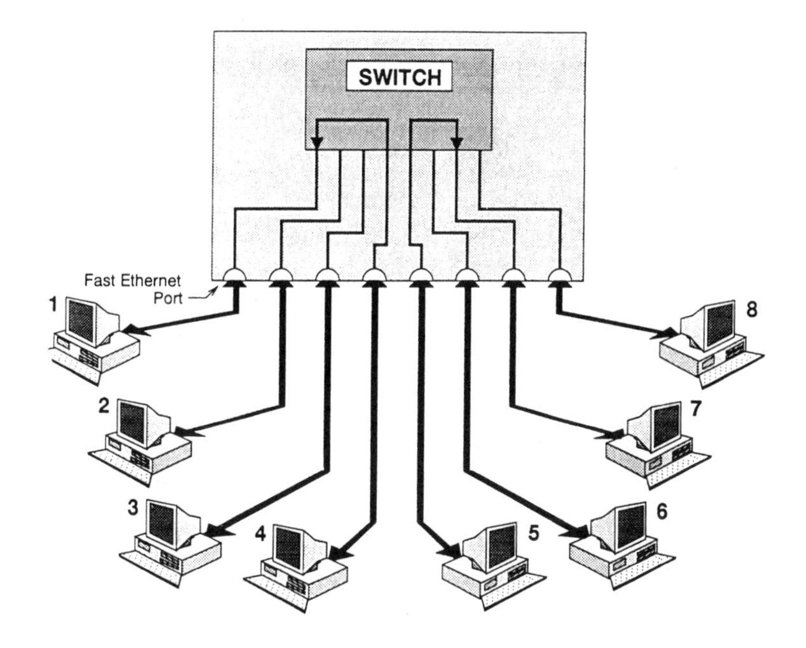

Figure 2.6 Fast Ethernet switched hub.

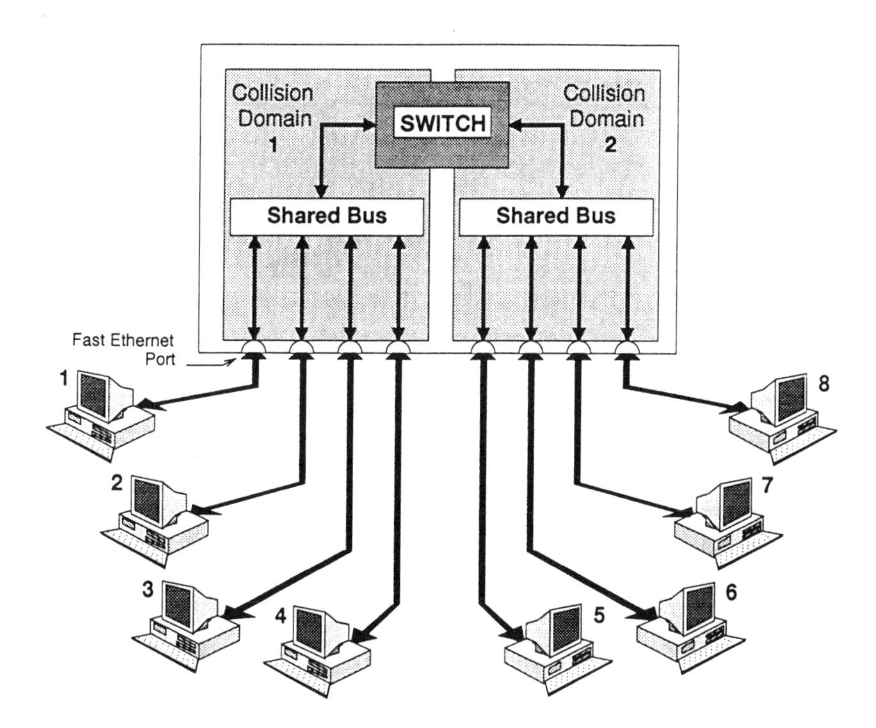

Figure 2.7 Fast Ethernet hybrid switch.

can operate simultaneously thus improving the maximum throughput carried by the system. The advantage of this approach is that it is less expensive than a pure switched hub but offers some of the improved performance capabilities of hub switching.

Some vendors also provide dual-speed operation in their switched hubs so that a 10BASE-T or 100BASE-T station can connect to any given hub port. The type of station connected to the port is detected by the hub when the attachment is made. Similarly, dual-mode adapters are available that can support both 10- and 100-Mbps hub connections. This provides a simple evolution path from 10- to 100-Mbps operation, since replacing the 10BASE-T hub by a 100BASE-T version will upgrade the system to fast operation. It is possible to do this in a transparent way, without making changes to the many stations that may be connected. For this reason, it is often recommended that dual-mode transceivers be installed in newer machines wherever possible. As a further aid for the purposes of evolution, some 100BASE-T hubs have at least one switched 10BASE-T port, which permits bridging between the two systems. The same is true for some 10BASE-T hubs. They may include a switched 100BASE-T port that can be used to interconnect these systems. These types of hubs provide the network designer with a great deal of flexibility in terms of accommodating existing slower speed networks, and upgrading the bandwidth using newer 100-Mbps systems. Switched hub systems are viewed as an essential product in supporting multimedia operation compared to the shared media versions.

As discussed previously, when time-sensitive traffic is mixed with normal data traffic over the Ethernet, there is no mechanism to offer it preferential treatment. As a result, the time-sensitive traffic may suffer intolerable variations in total delay. An obvious way of mitigating this effect is to introduce buffering at the receiving station. In such a scheme, variations in delay are smoothed out by the buffer. This, however, introduces a delay in series with the traffic flow. Some multimedia applications such as one-way video playback are quite tolerant to large amounts of this additional delay. However, other applications are not. In video conferencing, this delay can become noticeable if it exceeds much more than about 20 ms. In a recent study, both 10BASE-T and 100BASE-T Ethernets were subjected to various multimedia sources [18]. In this study it was found that when a 10BASE-T system was supporting as little as 5% background data traffic (consisting of random bursts of 10-Kbyte packets), only 18 one-way training video sources or one video conference could be supported (assuming 20-ms maximum delay with 0.1% probability of loss). When the same background load is used over 100BASE-T, a very large number of training video sources or as many as 37 video conferences could be supported. When the traffic load was increased to 15%, this number fell to 25. In the

100BASE-T experiment, the video streams were sent at a rate of 1.5 Mbps as compared to 384 Kbps for 10BASE-T. It is clear that the factor of 10 increase in transmission rate for Fast Ethernet is significant in terms of supporting multimedia transmission. This is clearly true when the background data traffic is fairly low. When this is not the case, then Fast Ethernet will experience degradation. As discussed before, this is because there is no way to prioritize the time-sensitive packets over the others. However, using switched hubs may dramatically improve the multimedia performance of such a system.

2.3.2 Gigabit Ethernet

Currently a large effort is under way to define new versions of Ethernet that can operate at transmission speeds up to 1 Gbps. So-called Gigabit Ethernet is being promoted by Compaq, Sun Microsystems, 3Com, and many other companies as members of the Gigabit Ethernet Alliance. These activities will eventually be standardized under IEEE 802.3z, which was formed in late 1995. At the time of this writing it is hoped that the standard will be approved early in 1998. A number of current predictions suggest that the market for Gigabit Ethernet will exceed that of ATM[1] for the next 10 years.

Gigabit Ethernet includes support for both switched full-duplex, point-to-point links and for half-duplex shared media operation, both using a hub-based architecture. The initial focus will be for use in backbone interconnection applications using the switched option. It is expected that this will happen in mid- to large-size corporations where aggregation of traffic requires the gigabit-per-second transmission speeds. A major issue is that of cabling. The initial version of Gigabit Ethernet transmits conventional Ethernet frames over Fiber Channel. Fiber Channel includes a physical layer transmission standard that operates over optical fiber. This means that the upgrade path from 10BASE-T and 100BASE-T systems may require the installation of fiber optic cabling. This is a strong negative factor that may dramatically increase the cost of Gigabit Ethernet in certain installations. Of course, when upgrading existing fiber-based backbones, this may not be a concern. The intent for Gigabit Ethernet is to eventually have it operate over unshielded twisted-pair (UTP) cable. The 802.3z committee has plans to operate over 100m runs of Category 5 UTP by 1998. This objective is referred to as 1000BASE-T. Currently a number of companies are working on how to go about this.

Compared with ATM, Gigabit Ethernet proponents hope to provide similar levels of performance at much lower cost. This is partly because the

1. The reader is referred to the appendix at the end of the chapter for a brief discussion of ATM.

technology is based on Ethernet, rather than a new technology, such as ATM. This permits much more seamless integration of the new system. From a multimedia perspective, ATM still has many ultimate advantages. It is much more advanced in incorporating applications with varying qualities of service. This may be a moot point, however, because few LAN links currently carry a significant amount of time-sensitive traffic.

As discussed in Section 2.2.1, the original Ethernet protocol is very sensitive to the normalized propagation delay parameter *a*. It was seen that when going from 10- to 100-Mbps Ethernet, the physical span of the collision domains was divided by about 10, in part, to maintain roughly the same value for *a*. Increasing the speed by yet another factor of 10 makes this problem that much worse. If the physical span was reduced by another factor of 10, the network links would not be long enough to allow for upgrading of existing hub-based Ethernet systems. Although collision operation is a part of the IEEE 802.3z standard, in practice the switched arrangement may become the dominant option. This also permits full-duplex operation of the links, which further increases performance. However, it is expected that the shared channel version could be offered for less cost per port.

To accommodate CSMA/CD collision operation, the standard uses both *carrier extension* and *packet bursting*. In the former, short frames are extended by having the transmitter append special extended carrier symbols to them so that all frames appear to be at least 512 bytes long even if they are shorter. This ensures that collision detection operates properly. The minimum frame length is therefore still 512 bits, but packets shorter than 512 bytes undergo carrier extension.

In both 10- and 100-Mbps versions of Ethernet, a parameter referred to as the *slot time* was defined to be 512 bit-times in duration. This is chosen so that if all packet transmission times exceed the slot time, then collision detection is guaranteed with a very high degree of reliability. When Ethernet data rates went from 10 to 100 Mbps, since the slot time was maintained at 512 bit-times, the slot time actually shrank by a factor of 10. However, this was compensated for by the roughly factor-of-10 decrease in physical network span. In Gigabit Ethernet we have yet another factor of 10 increase in the data rate. However, since the physical span of the system remains the same, it is necessary to increase the number of bits that define the slot time by about a factor of 10. This is accomplished by setting the slot time to 512 *bytes*, which gives an 8-fold increase instead. Of course, extending the packets introduces overhead that amounts to a direct loss in capacity.

In packet bursting, stations can transmit multiple packets back to back thus increasing the effective packet length for each channel acquisition. This has the effect of decreasing the value of *a*. However, the manner in which this is done has been carefully designed so that only the first packet in a burst may suffer a collision and so that the existing MAC interface is preserved.

2.4 ANSI Fiber Distributed Data Interface

The fiber distributed data interface (FDDI) is a set of standards developed by ANSI for a token ring LAN operating at 100 Mbps [19–21]. This activity started in the early 1980s and the standard for the FDDI MAC was first published in 1987. In recent years, FDDI has been extensively used to perform backbone interconnection of first-generation IEEE 802 LANs. However, as PC and workstation capabilities increase, FDDI is being increasingly used to directly interconnect high-performance multimedia systems.

As discussed in Section 2.2.2, ring LANs have a natural performance advantage over bidirectional bus networks such as 10BASE2 or 10BASE5 Ethernet. A novel feature of FDDI that further extends that advantage is that *early token release* is used by the stations. In this scheme, when a station has finished transmitting on the ring, the token is released immediately. This is in contrast to that normally used in the IEEE 802.5 ring, where a station waits until its own packets rotate completely around the ring before transmitting the token. Clearly, waiting for packets to return incurs an additional overhead equal to the total ring latency each time a station captures the token and transmits. This is because the station in question is prevented from transmitting while waiting for its packets to return on the ring. In early token release this is not the case, and higher efficiencies can be obtained using this mechanism. Using the same performance model as that in Section 2.2.2, it can be shown that the capacity of a token ring with early token release is given by

$$C_{\mathrm{ETR}} = \frac{B}{1 + t_{\mathrm{tok}}/\mathrm{THT} + \tau/(N \cdot \mathrm{THT})} \tag{2.6}$$

Assuming as before that $t_{\mathrm{tok}} \ll \mathrm{THT}$, we can write

$$C_{\mathrm{ETR}} \approx \frac{B}{1 + \tau/(N \cdot \mathrm{THT})} \tag{2.7}$$

Comparing this with (2.3), we see that the second denominator term is smaller by a factor of $1/N$. For this reason, token rings using early token release scale to higher bit rates much more readily than those using late token release.

Before leaving this issue, note that (2.6) and (2.7) assume that THT is fixed at all times. As discussed later, FDDI uses a dynamic scheme for determining THT and thus these equations are not directly applicable to FDDI. Also, the performance gains in using early token release are not obtained without a penalty. In IEEE 802.5, the priority mechanism relies on information that is returned in the packets as they propagate around the ring past the other stations. This information must be obtained before the token is released, which is clearly not possible under early token release. For this reason, the priority mechanism in FDDI is not as elegant, as discussed later.

In its most general form, FDDI consists of a "dual ring of trees" physical topology. An example of this is shown in Figure 2.8. The main ring is shown at the top of the figure connected through DAS_1, DAC_1, and DAC_2. This is a dual ring, which is indicated by the solid and dotted lines. The dual ring is most often configured as a single active ring, with the second ring acting as a hot standby. In the figure, the standby ring is shown by the dotted lines.

In FDDI, the two types of nodes are *stations* and *concentrators*. Stations are nodes that generate and receive information. Concentrators are merely nodes that permit several stations to attach physically to the ring. Because FDDI uses a wiring-hub style physical layout, the concentrators normally reside in wiring closets, which radiate cable to the stations and other concentrators.

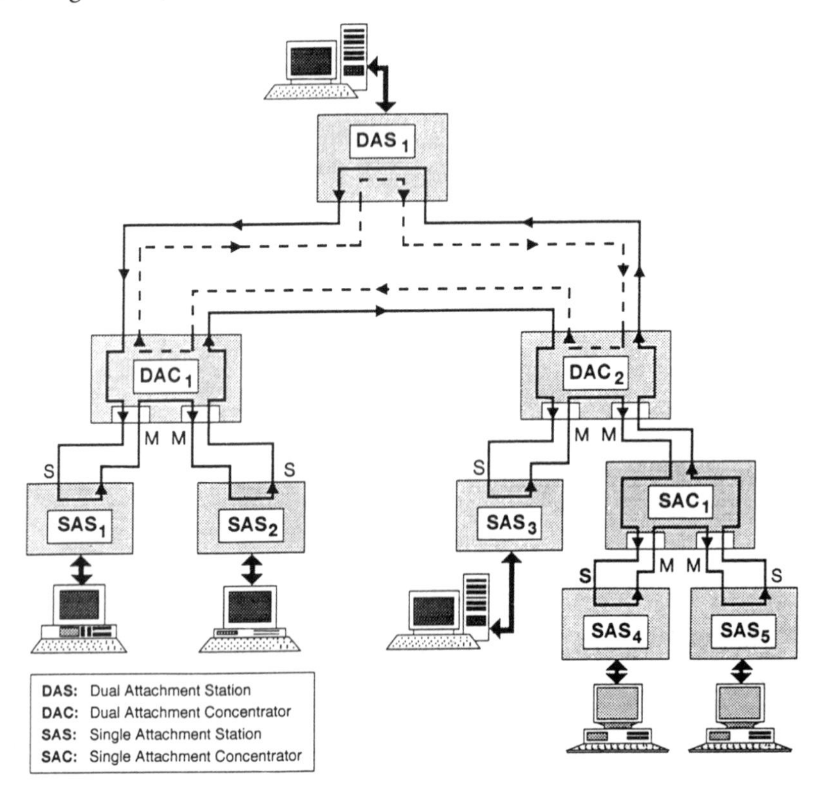

Figure 2.8 FDDI ring components.

In addition to the two node types, there are two types of stations, which are referred to as *single-attachment stations* (SASs), and *dual-attachment stations* (DASs). A SAS connects only to a single ring, whereas a DAS connects to both rings in the dual-ring configuration. These options are shown in Figure 2.8, where station SAS_1 is connected as a SAS and station DAS_1 is a DAS. Depending on the dual-ring mode employed, a DAS can be either a single-MAC or dual-MAC station. In the former case the secondary ring is used in standby mode, and thus only carries traffic if a fault has occurred. This is how we have depicted the network in Figure 2.8 by drawing the secondary ring as a dotted line. In this case, a DAS station only ever sees one active ring and thus a single MAC entity is sufficient. In dual-MAC DASs, both rings are active simultaneously and thus a DAS can transmit on both at the same time. For this reason, two MAC entities are needed, one for each ring.

As mentioned earlier, concentrators are used to provide attachments for a number of stations. These attachment points are referred to as M-ports or spurs. In addition to this, concentrators normally have additional ports that are used to connect the attached stations to the rest of the FDDI network. Several types of concentrators are defined in the FDDI standard. In this regard, concentrators may have no attachments, a single attachment port, or a dual attachment port. These are referred to as null, single, and dual attachment concentrators, abbreviated by NAC, SAC, and DAC, respectively. A NAC only provides a ring connection between its directly attached stations. Similarly, SACs and DACs have single-ring and dual-ring attachment capabilities. In Figure 2.8 examples of these types of concentrators are shown.

The FDDI dual-ring feature permits single-ring operation even in the presence of a cable fault, since the ring can be wrapped around at either side of the break. This capability is illustrated in Figure 2.9. In this figure we only show the dual-ring part of the network from Figure 2.8. Here the link between DAS_1 and DAC_2 has failed. These nodes have wrapped the appropriate ports and parts of the standby ring have been used to keep a single ring operating between the three dual attachment nodes. In the case where both rings are normally used, a cable failure would result in a degraded performance mode with one operating ring.

From its initial inception, the FDDI MAC protocol was designed with multimedia traffic in mind. FDDI uses a timed token access technique similar to that used in the IEEE 802.4 token passing bus network. A salient feature of the protocol is that the worst-case time to access the ring from any station may be strictly upper bounded. This is clearly unlike conventional Ethernet-type protocols, which can at best only provide statistical bounds on station access times.

To accommodate multimedia applications, FDDI defines three types of traffic: synchronous, asynchronous, and restricted asynchronous. Packets that

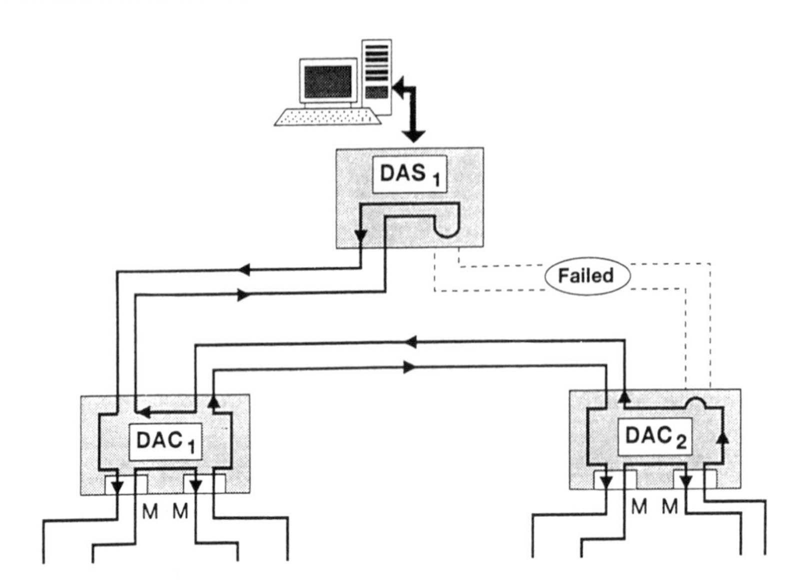

Figure 2.9 FDDI dual-ring failure.

are highly delay sensitive would normally be transmitted as synchronous traffic. Packets that are not time critical would typically be transmitted as asynchronous packets. Multiple asynchronous priority classes are also supported by the MAC protocol, so that different traffic types can be accorded various qualities of service. Note that although the synchronous traffic class is a part of the FDDI standard, many existing implementations do not support it.

2.4.1 Timed Token Rotation Algorithm

In FDDI, when the token arrives at a station, a token holding time is determined that governs how long the station may hold the token. During this holding time, the station is permitted to transmit. Although it would be possible to set the token holding time for all stations when the network is set up, in most cases this would be very inefficient. Recall that one of the objectives of the protocol is to ensure that a given station can access the ring within some known bounded time period. If the token holding times are determined in advance, then in order to meet the desired access time bound, operation under worst-case conditions would have to be assumed. This problem is illustrated as follows.

Consider the worst-case time, t_w, that it takes for the token to pass some point on the ring on two consecutive occasions. Clearly if this time is upper bounded, the time to access the ring for any station will also have a strict upper bound. We will assume that there are n active stations on the ring, that the total ring latency is D seconds (i.e., the total time for a transmitted bit to propagate completely around the ring), and that each active station holds the token for at most THT seconds. Then it is clear that

$$t_w \leq n \cdot \mathrm{THT} + D \tag{2.8}$$

This expression assumes that all active stations hold the token for the full THT seconds. It is obvious that placing upper bounds on the three variables on the right-hand side of this equation will permit us to bound the access time for any station. Thus for a given desired maximum ring rotation time, the THTs that are permitted must satisfy

$$\mathrm{THT} \leq \frac{t_w - D}{n} \tag{2.9}$$

As discussed in Section 2.2.2, token ring efficiency in a given token passing cycle can be calculated by dividing the total packet transmission time in the cycle by the total cycle duration. In the case that we are considering there are n active stations holding the token for THT seconds each. This results in an efficiency of

$$\eta = \frac{n \cdot \mathrm{THT}}{n \cdot \mathrm{THT} + D} \tag{2.10}$$

Now consider the problem of ensuring that there is a bound on station access times. In practice, the number of active stations n is not known beforehand. Therefore, if a THT must be selected in advance, in order to ensure the bound is achieved, a worst-case value of $n = N$ must be used (i.e., we must assume that all stations are active). Using this value in (2.9) and combining it with (2.10), we obtain

$$\eta = \frac{1}{1 + ND/n(t_w - D)} \tag{2.11}$$

In this expression the efficiency is increased (to a maximum of 1) when the second denominator term is reduced. If *all* stations are active, then $n/N = 1$ and the efficiency can be quite large. However, consider the case where only a single station is active. In this situation, $n = 1$ and the N/n factor in (2.11) can be very large, leading to a low overall ring efficiency. It is clear from this discussion that a more reasonable approach would be to decrease the token holding times when n is large (thus maintaining our bounded ring rotation time), and increase the token holding times when n is small (which permits efficient operation in this case). This would obviously involve a dynamic selection of the token holding times. This is the objective of the FDDI timed token algorithm.

Intuitively, the timed token algorithm operates as follows. The standard defines a parameter referred to as the *target token rotation time* (TTRT), which can be considered the "nominal" ring rotation time for the token. TTRT can be viewed as a target objective that the stations collectively try to achieve for the token rotation time on the ring. To achieve this, each station measures the current token rotation time as seen by that station. If a station finds that the current token rotation time is less than TTRT, then the station will permit itself a larger token holding time, thus tending to increase the next token rotation time. Similarly, if a station finds that the token rotation time is currently larger than TTRT, it reduces its token holding time accordingly.

The algorithm used can be described using timers. Each station has a timer that continually measures the elapsed time since that station last received the token. As mentioned before, this time is referred to as the current token rotation time or TRT of that station. Rather than starting from zero and counting up, the TRT timer is initialized to the value of TTRT and counts down. If the token circulates around the ring before the counter reaches zero, then the token is considered "early." On arrival of an early token, the value in the TRT timer is a measure of how early the token is. This value is saved in another counter referred to as the *token hold timer* (THT). This value is used to determine how much additional token hold time the station is permitted. In a similar way, if the TRT timer reaches zero before the token arrives, the token is viewed as being "late." In that case no additional token holding time is permitted. To track these events, each station also has a late counter (LC), which is initially set to 0.

When the station starts, TRT begins to count down from its initial value of TTRT. If it counts to zero before a token is received, then LC is incremented to 1 and TRT is restarted at TTRT. The value LC = 1 simply records the fact that there has been a late token arrival event at the station. When this occurs, TRT is restarted at TTRT and continues to time the current token rotation. If TRT decrements to zero a second time before a token is received, then LC is incremented to 2. This event indicates that a problem has occurred and error recovery procedures have been initiated. This may happen, for example, if the token is lost because a station loses power while holding it.

In FDDI, a station can always transmit a certain number of synchronous packets when the token arrives. However, as discussed earlier, transmission of asynchronous packets may be prohibited if the token arrives late at the station. Accordingly, each station i is given a fixed allocation of time, S_i, which is referred to as the synchronous allocation of station *i*. Whenever a token arrives, a station *i* can always transmit synchronous frames for a time given by S_i.

We now summarize the procedures given. The token is early if it arrives at a station while LC = 0. When this happens the station *i* in question does the following:

THT ← TRT
TRT ← TTRT
Restart the TRT.

Station *i* can then transmit according to the following rules:

1. It can transmit synchronous packets for a maximum time given by its fixed allocation S_i.
2. After it has finished its fixed allocation, the THT timer is enabled and starts counting down. It can transmit asynchronous frames as long as THT > 0.

If a station receives the token and it is late (i.e., LC = 1), then LC is set to 0, and TRT continues to run. The station can transmit synchronous packets for its fixed allocation time, S_i, but cannot transmit any asynchronous packets.

In FDDI, the synchronous allocations and TTRT are chosen so that

$$\sum_{i=1}^{N} S_i + D + TP_{\max} + TT \leq \text{TTRT} \tag{2.12}$$

Here $TP_{\max}$ and TT are defined to be the maximum packet and token transmission times, respectively. Note that when all stations are busy, a simple argument can be made that shows that the token rotation time at any station satisfies the following inequality:

$$\text{TRT} \leq \sum_{i=1}^{N} S_i + D + TP_{\max} + TT + \text{TTRT} \tag{2.13}$$

By combining the preceding two equations, we see that

$$\text{TRT} \leq 2 \cdot \text{TTRT} \tag{2.14}$$

Thus this algorithm ensures a strict upper bound on the access time to the network. This fact also explains the error recovery procedures that are invoked if LC is ever incremented to 2. From (2.14) this should never happen unless the token is lost or is unable to propagate because of a fault on the cable.

In FDDI, multiple priorities for asynchronous traffic are implemented using the same timed token rotation mechanism. Essentially, a time threshold is chosen for each priority class. In the case of priority level j, packets of this priority can only be transmitted if the TRT is less than the selected threshold. If the token is early, the hold time for this class must not exceed the earliness of the token.

2.4.2 FDDI-II

In FDDI, the nominal rotation time of the token is given by the value assigned to the target token rotation time, TTRT. However, we have seen that the actual token rotation time may be as large as twice this value in any particular token rotation cycle. Thus there is some statistical variation in the times at which the token is received at each station, even though its value is upper bounded by $2 \cdot$ TTRT. Recall that regardless of the measured token rotation time, a station can always transmit its quota of synchronous packets. While this is an impressive bandwidth guarantee, the variation in station access times is not very compatible with certain conventional traffic types. For example, in the digital circuit switching used in the telephone network, each voice source is sampled at a rate of 8,000 times per second. Each sample is then encoded as a 1-byte word resulting in a total bit rate of 64 Kbps. When a voice call is established, 8-bit time slots are allocated in both directions along the transmission path from source to destination. These time slots recur and must be used *exactly* every 125 μs (i.e., 1/8000 seconds). This class of traffic, where transmission opportunities are provided periodically, is referred to as *isochronous traffic.*

Isochronous traffic could be carried on an FDDI network using the synchronous traffic class. However because of the statistical fluctuations in token acquisition times, additional buffering would be needed in the stations to smooth out these variations. FDDI-II is an extension to FDDI that permits isochronous traffic to be directly carried over an FDDI ring. Thus in FDDI-II, a station that supports isochronous transmission is given access to the ring at exact multiples of 125 μs for this purpose [20].

An important feature of FDDI-II is that it operates using the same wiring standards as FDDI so that rewiring is not necessary. FDDI-II nodes are capable of operating in conventional FDDI mode, or in hybrid mode which accommodates isochronous functionality. Ring operation is at 100 Mbps as in conventional FDDI, and FDDI-II nodes have backwards compatibility to standard FDDI. However, to operate in hybrid mode, all stations must have FDDI-II functionality. If even one station on the ring is a conventional FDDI station, then all FDDI-II stations will default to this mode and isochronous operation cannot be used. This is a negative aspect of the design, since it does not permit a graceful upgrade path from conventional FDDI.

To ensure periodic access, FDDI-II uses a cycle scheme that permits isochronous stations to transmit every 125 μs. Thus every 125 μs, a new transmission cycle starts. Since the data rate in FDDI-II is 100 Mbps, a total of 1,562.5 bytes can be transmitted in this 125-μs period. Of these bytes, 1,560 are used in the cycle data itself and 2.5 byte-times separate each cycle. The 1,560 bytes are divided into 16 wideband channels each consisting of 96 bytes. Each wideband channel thus provides a bandwidth of 6.144 Mbps. Because a single direction of a telephone call requires only 1 byte every 125 μs, one wideband channel can support as many as 96 such calls. Alternately, it is possible to place a single NTSC television signal into one wideband channel.

Note that of the remaining $1{,}560 - 16 \times 96 = 24$ bytes, 12 are allocated to the cycle header and the remaining 12 are dedicated to packet-switching mode. This is so that even if all wideband channels are allocated to isochronous mode, bandwidth is still available to allot to the packet mode.

The format of a complete 125-μs cycle is shown in Figure 2.10. Rather than transmitting the bytes of each wideband channel contiguously, they are interleaved across all wideband channels. Thus the first bytes of all 16 wideband channels are transmitted first, followed by the second bytes and so on. One of the advantages of this scheme is that when a single source is accessing a full wideband channel, it can transmit into that channel roughly uniformly across the cycle (in 1-byte bursts) rather than having to transmit all 96 bytes as one single continuous burst. This helps to minimize buffering at the stations.

In hybrid ring mode, each of the 16 wideband channels can be allocated for use in either isochronous mode or packet mode. When accessing various wideband channels in packet mode, the bytes are obviously mixed in with those being accessed in isochronous mode. For this reason, there will be gaps in time between consecutive symbols in the packet-mode data packets. However, as far as data-mode protocol operation is concerned, operation is the same as that discussed earlier for conventional FDDI.

When accessing the ring in isochronous mode, FDDI-II allows different transmission channels consisting of different bit rates. This is done by grouping bytes within a given wideband channel. For example, grouping n consecutive bytes would result in a transmission channel of rate $n \times 64$ Kbps. Considerable flexibility is provided for, and submultiplexing is allowed so that transmission channels of lower bit rates can be provided. This can go as low as access to a single bit per cycle, which gives an 8-Kbps channel rate.

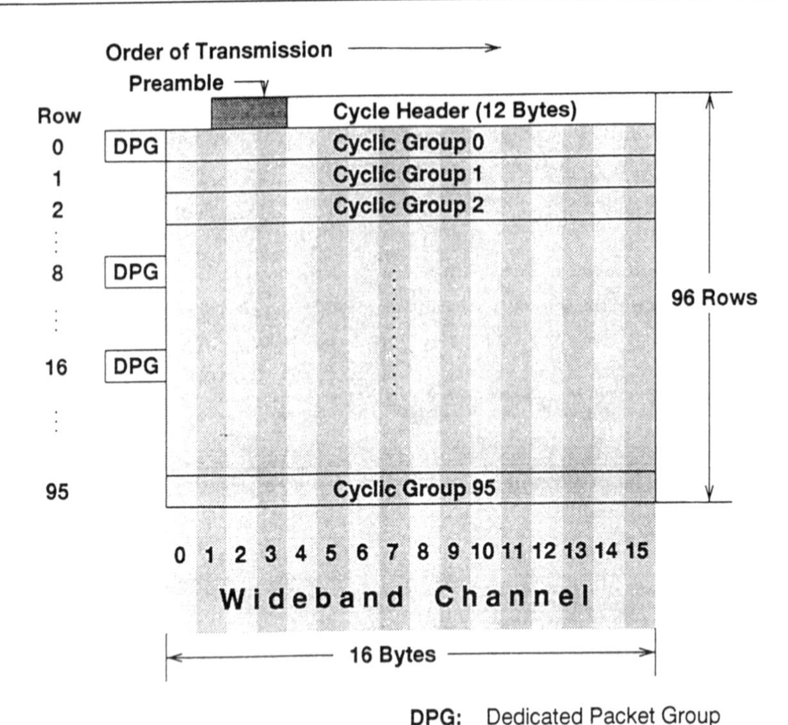

Figure 2.10 FDDI-II frame format.

In FDDI-II, the same node types are used as in conventional FDDI. In addition to this, FDDI-II stations are either monitor stations or nonmonitor stations. It is the monitor stations that have the responsibility for generating the cycles discussed. Each monitor station has an assigned rank, and the one with the highest rank is elected as the cycle master, which is responsible for generating and transmitting the cycles. As part of this activity, it inserts an elastic buffer in series with the ring, so that an integral number of cycles can propagate on the ring at once.

In summary, FDDI-II provides an effective way of transmitting conventional circuit-switched or isochronous traffic which might otherwise be difficult to handle. Packet-switched traffic is also accommodated using the conventional FDDI protocol. To reflect the different design of the system, in 1988 work was initiated that resulted in an updated version of the MAC protocol referred to as MAC-2. Details of this are published in the ANSI X3.239 standard.

2.5 IEEE 802.12 100VG-AnyLAN

Prior to the early 1990s, the FDDI ring was one of the few LAN standards that would accommodate 100 Mbps shared media operation. However, in many installations an inexpensive upgrade of existing networks was needed. In Section 2.3 we discussed the development of the 100BASE-T Fast Ethernet standards which responded, in part, to this need. Another system developed for this purpose was the IEEE 802.12 100VG-AnyLAN [22,23]. Both Fast Ethernet and 100VG-AnyLAN were intended to provide 100-Mbps operation at lower cost than FDDI. The 100VG-AnyLAN system was also designed to operate over Category 3, 4, and 5 UTP cable. There are also plans to support additional cabling systems such as shielded twisted-pair and fiber.

In the 100BASE-T standard, the same CSMA/CD protocol is used as in earlier versions of Ethernet. As discussed in Section 2.3, this constrains the physical design of the network, which inevitably leads to a large degree of network segmentation. In addition, 100BASE-T maintains the undesirable random nature of CSMA/CD and the lack of packet priorities. This latter feature is essential for handling the different types of traffic generated by multimedia applications. In 802.12, the ability to handle larger network spans with less segmentation was an important criterion in the design of the new standard. The new MAC protocol, called demand priority (DP), reduces the propagation constraints imposed by CSMA/CD and allows two priority classes, normal and high priority. In addition, the scheduling performed is similar to that of token ring networks, which makes it much more reasonable for multimedia applications. For ease of upgrading, the protocol uses either the IEEE 802.3 or IEEE 802.5 (token ring) frame formats.

In 1992, the initial DP protocol was proposed to the IEEE 802 project group. This was originally advanced by Hewlett-Packard, AT&T, and IBM as 100BASE-VG. At that time the IEEE 802 committee decided to develop both CSMA/CD and DP as 100-Mbps standards. The "Demand Priority Access Method, Physical Layer and Repeater" specification was published as IEEE 802.12 in 1995.

2.5.1 Architecture and Protocols

100VG-AnyLAN uses a hub-like topology as is the case in 100BASE-T Ethernet. In this case, the hubs are referred to as *repeaters* and are arranged in a hierarchical fashion as shown in Figure 2.11. In the figure, each end station is connected to one port of its repeater. The stations communicate data frames with other stations through the repeater to which they are connected. Note that an 802.12 repeater is more complex than an 802.3 shared media hub, because it explicitly controls the communications of its attached stations. In 802.3, the hub acts mainly as a simple physical-level relaying device.

As shown in Figure 2.11, each repeater has a special *uplink* port through which it can be cascaded to another repeater at a higher level. The repeater at

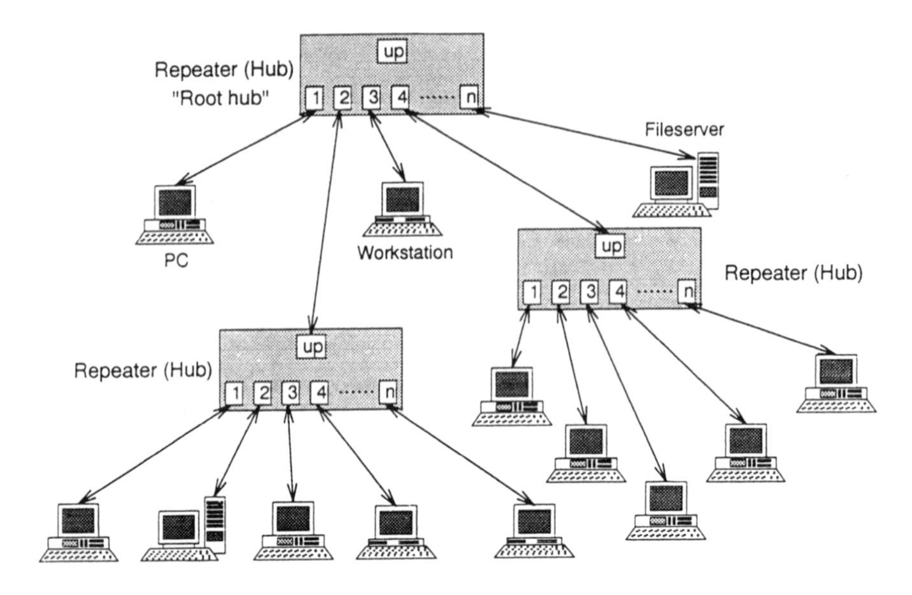

Figure 2.11 A hierarchical demand priority network architecture.

the first level is known as the *root* repeater. Some of the ports of the root repeater may be connected to other repeaters. In the example, we have shown a two-level hierarchy.

The DP protocol is a simple centralized reservation method. First consider how the protocol operates for one repeater handling frames at a single priority level. When a station wants to transmit, it sends a request to its repeater. The repeater continually scans each of its ports in round-robin order looking for requests. When a request is found, the repeater issues a grant to the requesting station. The station can then transmit a single frame toward the repeater. The repeater checks the destination address of the transmitted frame and redirects it to the appropriate output port. Note that under normal conditions the destination station is the only end station who receives the frame. However, in the case of multicasting, the frame is sent to the appropriate group of stations. From this discussion it can be seen that each repeater must know the MAC addresses of all of its attached stations. This information is exchanged when a station is first initialized.

It is instructive to describe the protocol operation in the following way. We can view each repeater port as containing a "register" that can hold a single request. While scanning the port, if the repeater finds the register empty, it continues to scan. Otherwise it removes the request and grants the station permission to transmit a single frame. While the station is transmitting, we consider the scanning to have halted and the repeater sets a pointer, PTR, to the register following the one where it stopped scanning. Once the station is finished transmitting, the scanning starts again from the PTR position. Note that the scheduling achieved by this mechanism is very similar to that of a token ring. However, as discussed later, this system may be more efficient because certain token-passing overheads may be eliminated.

Now consider a single repeater system when there are both normal- and high-priority frames being generated. In this case we view each repeater port as having both a *normal-priority register* and *high-priority register*. When a station has a high-priority frame to send, it places a request in the high-priority register (and likewise for the normal-priority register). The repeater continuously scans the high-priority registers, granting requests in round-robin order as discussed earlier. As before, it sets a pointer (H-PTR, in this case) to the next port whenever scanning stops. Scanning of the normal-priority registers occurs in a similar way. However, normal-priority scanning is only enabled when all of the high-priority registers are empty. Whenever there is high-priority traffic being transmitted, the low-priority scanning is disabled and a pointer (N-PTR) records the point where the scanning was interrupted. Once all high-priority frames have been transmitted, H-PTR records the point where the last packet was transmitted, and the normal-priority scanning continues from N-PTR. Note that while normal-priority scanning and transmission are occurring, the high-priority scanning continues. The arrival of a high-priority request will disable normal-priority scanning.

We can see that this simple mechanism provides priority to high-priority traffic over normal-priority traffic. To prevent higher priority traffic from completely dominating, there is a mechanism whereby a normal-priority frame can be promoted to high priority if it has not been transmitted within some timeout interval.

The protocol just described must be changed slightly to accommodate the hierarchical interconnection of repeaters as shown in Figure 2.11. Essentially control starts with the root repeater and is passed downward to subsequent stations and repeaters in round-robin order. This is done by having the root repeater scan its ports as described earlier. When a request is found, access is granted to that port. If a station is directly connected to the port, then everything works as discussed previously. If a repeater is connected to the port, then the grant passes control of the system to the repeater in question. That repeater then operates as described before. Note that to maintain fairness, a repeater that is passed control in this manner can only issue grants for one complete scan cycle of its ports before returning control to the higher level repeater.

In these modes of operation, when a repeater has outstanding requests from an attached station or repeater, it makes a request on its "up" port to the repeater at the next level. That way requests at one level result in one or more requests at a higher level. A complicating factor is that when control is passed to a repeater below the root level, a preemption mechanism is necessary. Consider the case where a lower level repeater is currently granting or servicing normal-priority requests. If a high-priority request is then generated by a station on another repeater, it is necessary to preempt the current normal-priority transmissions. To accomplish this, signaling is defined that allows a higher level repeater to preempt one at a lower level, for this purpose.

It can easily be seen that the 802.12 repeater is superior from a performance point of view to the 802.5 token ring. This is because the stations have more opportunities over the round-robin cycle to send request signals to the repeater. Furthermore, although the grant signal circulates among the stations like a token, it is only sent to the stations that have registered a request for the cycle. Stations that do not have any data to send are not involved and consequently no time is lost in circulating a grant to those stations.

From a multimedia standpoint, IEEE 802.12 is much superior than standard 100BASE-T with shared media hubs. The scheduling in 802.12 is highly organized giving fair round-robin access, and without the additional random aspects of 100BASE-T. In addition to this, the high-priority class can be used to carry delay-sensitive traffic. Provided that the total number of high-priority multimedia streams is limited, the required access delays can be guaranteed.

As in the 100BASE-T standard, when using Category 3 twisted-pair cable, multiple pairs are used in parallel to achieve the desired 100-Mbps data rate. At the physical level, the MAC data frame is split into four separate channels, which is referred to as *quartet channeling*. This is done by first dividing the MAC frame into 5-bit data quintets. The 5-bit quintets are scrambled and 5B6B encoding is used to encode the 5-bit data quintets into 6-bit sextets. The mapping of 5-bit quintets to 6-bit codes is carefully chosen to closely bound the imbalance between the number of 0's and 1's. The scrambling is mainly used to prevent the energy of the transmitted signal to be focused in particular frequencies that can cause interference with other devices. Finally, the remaining fields of the frame such as the preamble, start-of-frame delimiter, and end-of-frame delimiter are added to the data frame on each data channel. In quartet channeling, each channel represents a transmission pair using twisted-pair cable. Additional multiplexing is used in the case of 2-STP and optical fiber cable to convert the four channels into two or one, respectively. The actual signaling of the data on the physical medium is done by nonreturn to zero (NRZ) encoding. The clock rate for the 4-UTP links is 30 MHz for each wire pair, so that 30 Mbps are transmitted on each pair. Consequently, the total data rate is 120 Mbps. Considering the 5B6B coding, the achieved data rate is 100 Mbps.

The transmission of data is done in half-duplex mode, so that signals are sent in only one direction at a time on the physical link. For control signals, however, full-duplex operation is used. In the quartet channeling, two of the channels are used for transmission of control signals from the repeater to the station and the other two for control signal transmission from station to the repeater. This mode is called *link status control* signaling. The different modes of communication between a repeater and a station are shown in Figure 2.12.

In the link status control signaling, a combination of two low-frequency tones, referred to as tone 1 and tone 2, is used. Tone 1 is generated by transmitting a 30-MHz alternating pattern of sixteen 1's followed by sixteen 0's. This produces a 0.9375-MHz signal. Tone 2 is generated by transmitting a 30-MHz alternating pattern of eight 1's followed by eight 0's, which produces a 1.875-MHz signal. Table 2.3 shows the tone combinations used in link status control.

Table 2.4 shows the cable types allowed in the IEEE 802.12 specification and their maximum distances. As shown, four-pair unshielded twisted-pair (UTP) (Category 3, 4, or 5), two-pair shielded twisted-pair (STP), or optical fiber cable can be used. If 0.8-μm transceivers are used, the maximum optical

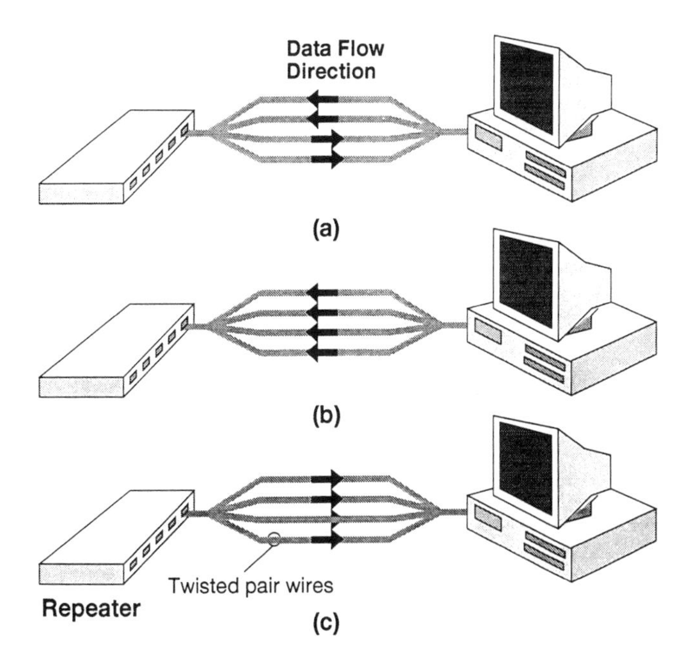

Figure 2.12 The modes of communication in a DP network: (a) full-duplex link status control, (b) half-duplex data transmission from station to repeater, and (c) half-duplex data transmission from repeater to station.

Table 2.3
Link Status Control Signaling

Tone Pattern Transmitted	Meaning: Received by Station	Meaning: Received by Repeater
1-1	Idle	Idle
1-2	Incoming data packet	Normal priority request
2-1	Reserved	High-priority request
2-2	Link training request	Link training request

Table 2.4
Cabling Options for a DP Network

Category	Cable Type	Maximum Distance (m)
3, 4	4 pair UTP	100
	2 pair STP	100
5	4 pair UTP	200
0.8-μm Tx/Rx	Optical fiber	500
1.3-μm Tx/Rx	Optical fiber	2000

fiber link is 500m, while this increases to 2 km when 1.3-μm transceivers are used.

2.6 IEEE 802.11 Wireless LAN

In the past several years, IEEE 802 activities have been extended to include a standard for wireless local-area networks. Wireless LANs have the obvious advantage that cabling such as twisted pair, coax, or fiber is not needed. The cost of providing this is often a significant fraction of the total associated with installing and maintaining a LAN. At this point in time, various vendors are moving to incorporate this standard into their product offerings.

The IEEE 802.11 wireless standards activity has been going on since 1990 [24]. The standard is meant to accommodate both *infrastructure-based* networks and *ad hoc* networks, which may be set up very quickly. The former type of network functions more as a replacement or extension to conventional cabled LANs, but provides for station mobility. Ad hoc networks are those that are formed at will when the need arises. This may happen, for example, when two or more people with laptop computers wish to communicate within a room during the course of a business meeting.

The standard is designed to accommodate both mobile and portable stations. A portable station is one that can be moved from one location to another, such as a laptop computer, but would not normally be used while in motion. In contrast, a mobile station normally communicates while in motion, such as a cellular telephone. Normally a system that "supports mobility" tries to ensure that higher layer functionality is maintained as the portable or mobile station is moved.

An example of an 802.11 wireless LAN is shown in Figure 2.13. Stations that communicate over a single wireless communication channel belong to a *basic service set* (BSS). We can think of a BSS as being a wireless coverage area shared by a set of stations using the same MAC protocol. In the figure, two BSSs are shown. In an infrastructure-based network, basic service sets are interconnected through a distribution system (DS) via an access point (AP). This is also shown in Figure 2.13. The DS would normally be a wired network that performs the functions necessary to interconnect multiple BSSs so that stations connected to different BSSs can communicate. Note that a DS provides special mobility-oriented functions such as managing dynamic associations between stations. These are transparent to the layers above the MAC layer, so that application sessions can be maintained even when a station moves from one BSS to another.

A group of BSSs interconnected in the way described is called an *extended service set* (ESS). BSSs may partially overlap, be physically disjointed, or be

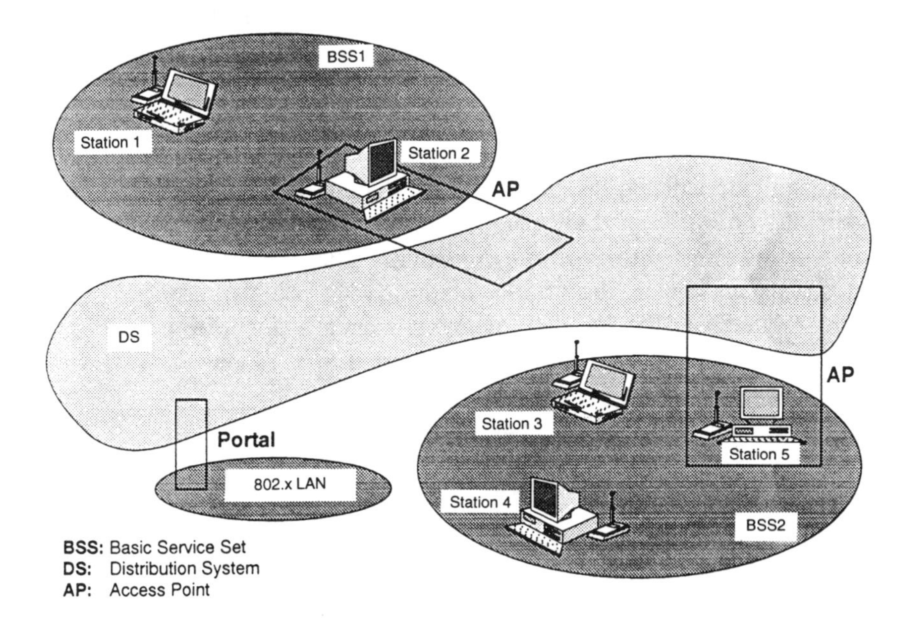

Figure 2.13 Example of an IEEE 802.11 system.

physically colocated for redundancy. One or more independent BSSs or ESS networks may be present in the same space. Note that the standard does not guarantee mobility between BSSs belonging to different ESSs. In Figure 2.13 an attachment into another IEEE 802 LAN is shown. In 802.11 parlance this is performed at a portal.

In 802.11, stations use a media access protocol known as distributed foundation wireless media access control (DFWMAC). This was originally proposed by AT&T Global Information Solutions/NCR Microelectronics Products Division, Symbol Technologies, and Xircom. The 802.11 media access protocol includes options for both centralized and distributed control. This is done by making the protocol hierarchical. At the upper level is the *point coordination function* (PCF). This is a centralized mechanism that can divide time on the channel into a "superframe" consisting of two separate regions. This is shown in Figure 2.14. In the first region, access to the channel is highly organized using a polling scheme. In this case stations would normally be polled in sequence by a central station referred to as a *point coordinator*. Packets that are time sensitive would likely be transmitted during this period. In the second region, stations are free to access the channel using a CSMA protocol. Access to the channel during this interval is less organized and more suited to conventional data traffic. The superframe would normally be repeated at periodic intervals so that time-sensitive traffic can be accommodated.

We first consider how the system operates during the CSMA portion of the superframe. The version of CSMA used in IEEE 802.11 is called CSMA/CA. This is also referred to as the distributed coordination function or DCF. The "CA" in CSMA/CA stands for "collision avoidance."

DCF is similar in many ways to other CSMA protocols. However, there are some important differences. In many versions of CSMA, an active station can transmit immediately on sensing that the channel is idle. Before transmitting under DCF, however, a ready station must sense that the channel is idle for a time period referred to as an *interframe space* (IFS). Different values of IFSs are used depending on the priority of the packet being transmitted. Three different values of IFSs are defined in the standard. In order of decreasing priority (and increasing length) they are referred to as follows.

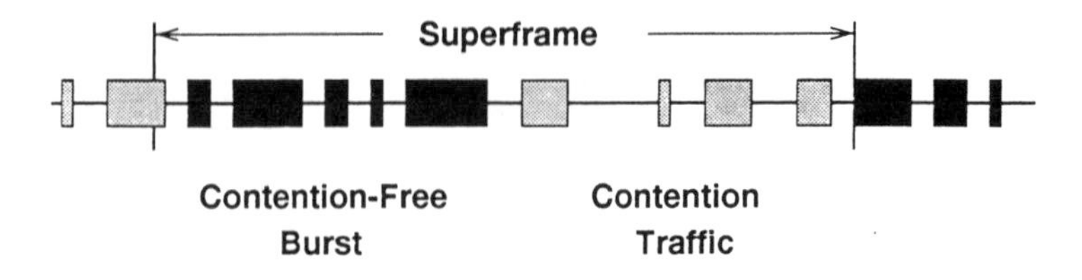

Figure 2.14 IEEE 802.11 superframe.

- *Short IFS (SIFS).* Used for transmitting ACKs, CTS, and multiple-packet segments and by stations responding to polling in the PCF;
- *Point coordination function IFS (PIFS).* Used to send any data packets in the contention-free period;
- *Distributed coordination function IFS (DIFS).* Used by the stations to transmit packets in the contention period.

Note that the relative lengths of the IFSs are such that a single station transmitting using SIFS will always transmit successfully before any station using PIFS. The same is true for PIFS and DIFS. The difference between the assigned values for SIFS and PIFS are such that the PIFS station would sense the channel busy (from the SIFS packet) while still waiting for the channel to be idle for PIFS.

When stations transmit data packets in the contention part of the superframe, they use DIFS. This is illustrated in Figure 2.15. In the figure, the first transmission occurs following an idle period on the channel of DIFS seconds. Once this channel activity is finished, CSMA/CA contention then occurs following the second DIFS interval shown.

Whenever a station receives a packet addressed to it, the station replies immediately with a short *acknowledgment frame* (ACK). This lets the transmitting station know that the packet was successfully received. If an ACK is not heard, then the transmitting station presumes that the packet was lost and it will have to be retransmitted. To prevent collisions with data packets, the ACK is transmitted using SIFS. An example of this activity is shown in Figure 2.16. In the figure, the source transmits after the usual DIFS interval. On receiving the packet, the destination replies with an ACK using SIFS. Using this mechanism, an ACK does not experience collisions.

The backoff mechanism used in 802.11 is different from that used in IEEE 802.3. If a station senses that the channel is busy, it first waits until the

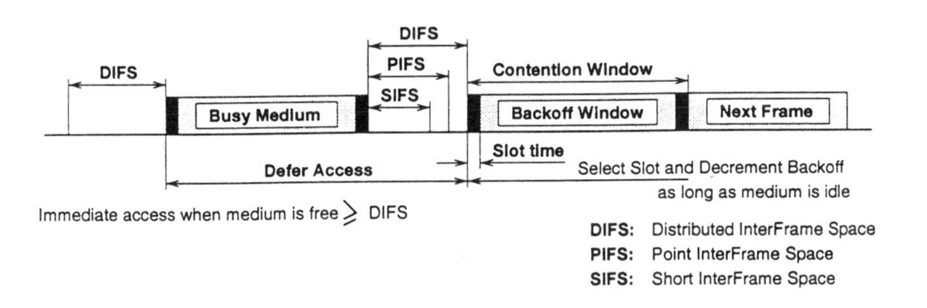

Figure 2.15 IEEE 802.11 IFS.

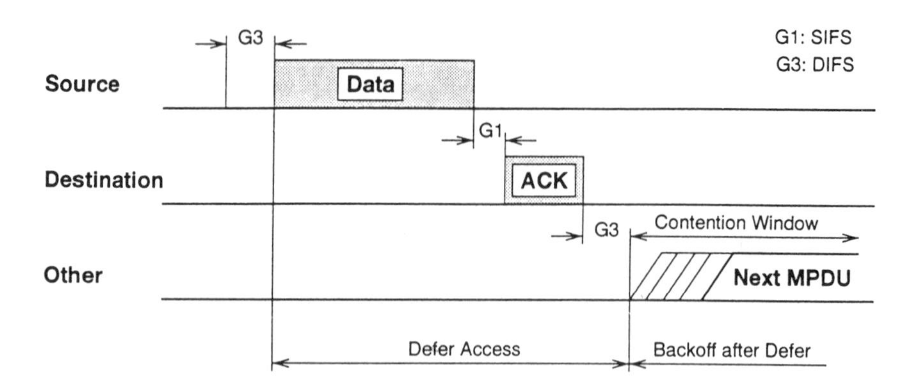

Figure 2.16 IEEE 802.11 data packet and ACK transmission.

channel becomes idle for DIFS seconds. Following this, the station performs a backoff using a randomly chosen pseudo-random backoff interval. The backoff time is given by CW · Random( ) · Slot time, where CW is an integer between CW_{min} and CW_{max}. CW starts at the initial value and increases exponentially after each retransmission attempt (i.e., 1, 2, 4, 8). Random( ) returns a pseudo-random number uniformly distributed over (0,1). The station must then see the channel idle for a total elapsed time equal to the chosen backoff before transmitting.

We now consider the operation of the PCF. The point coordinator seizes control of the medium away from the other stations and controls frame transmissions during the contention-free part of a superframe. It generates the start of the superframe by transmitting a beacon frame using PIFS. This ensures that it will have priority over stations transmitting data frames using DIFS. Following this, the point coordinator polls a set of stations in order. When a station is polled, it can transmit a data packet to another station. It does this by transmitting using SIFS. An example of this is shown in Figure 2.17. The figure shows two polling packets, D1 and D2. In the first case the polled station responds and its destination replies with an ACK. These transmissions both use SIFS. Following this the point coordinator issues the next polling packet using SIFS. Note that if either the polled station or destination station fails to respond, the point coordinator initiates its next transmission using PIFS.

Collision detection is not performed in the IEEE 802.11 protocol. This is because detecting collisions on a shared wireless channel is generally much more difficult than on wired media. This is partly due to the modulation used and because of the larger variations in path loss. A transmitting station may have a very difficult time determining if its transmission is being destroyed due to a collision at a recipient station. If the receiving station is much closer

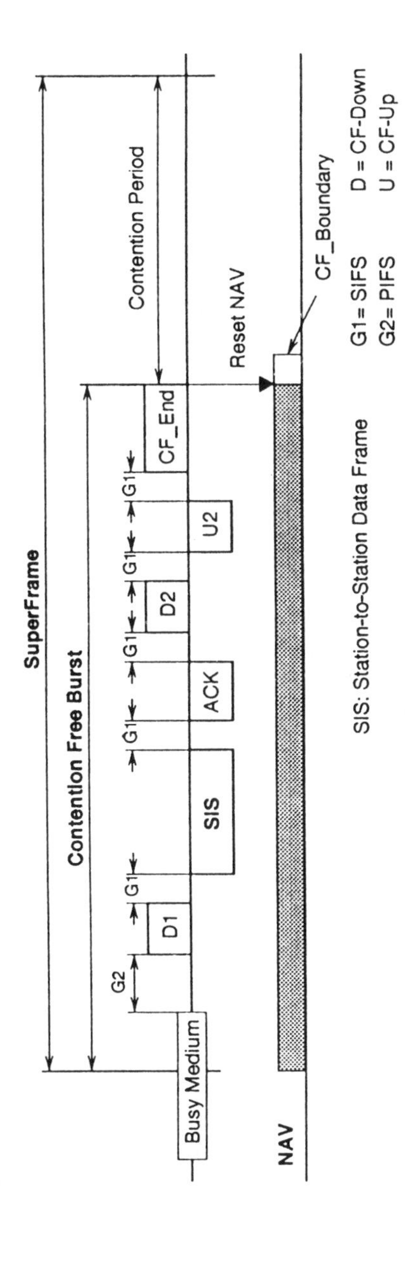

Figure 2.17 PCF polling.

to one of the senders, it may "capture" the packet with the highest receive power. As a result, colliding stations waste the channel for at least a full packet transmission time. IEEE 802.11 includes a novel feature that approximates the actions of collision detection using short RTS/CTS packets. This is done as follows. Instead of transmitting a full-sized data packet, a sender transmits an RTS packet to the same destination instead. If the RTS is collided with, then the channel is only wasted for about the duration of the RTS packet. When a station is sent an RTS, it replies immediately (using SIFS) with a short CTS packet. When the sender sees this CTS, it responds immediately with the full data packet. This is done using SIFS so that other stations cannot collide with it. An example of this type of exchange is shown in Figure 2.18. The figure shows the RTS/CTS handshake and the ACK transmitted by the destination as discussed previously. It can be seen that the net effect of this is that if a collision occurs, the wastage is much lower than would be the case without RTS/CTS. If no collision occurs, then there is added wastage due to the RTS/CTS exchange. However, since RTS/CTS packets are very small, there may be a gain in overall performance.

The IEEE 802.11 protocol also includes a "virtual carrier sense" feature using *network allocation vectors* (NAVs). Whenever certain packets are transmitted, they include a field that indicates the expected time that the channel will be busy due to subsequent transmissions. In Figure 2.18, for example, when the RTS is sent, the sender anticipates the full duration of the exchange and indicates this value in the RTS. The destination station includes the same information in its CTS reply. Stations that hear this value update their NAV and are not permitted to contend for the channel until the time specified. The advantage of doing this is that sometimes all stations may not have full connectivity within a single BSS. Consider what would happen in Figure 2.18

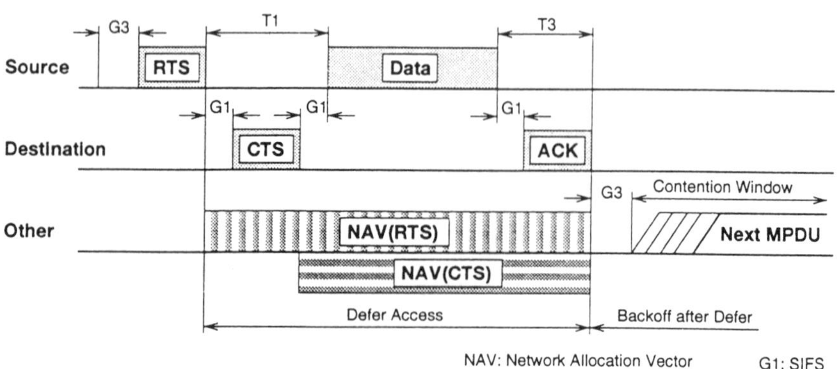

Figure 2.18 RTS/CTS operation.

if a third station cannot hear the destination station. Since it would not hear the CTS, it may transmit, colliding with the destination. However, provided it can hear the sender, its NAV would be set so that this collision would not occur. In most of the figures presented in this section, the NAV set by various stations is shown.

The IEEE 802.11 standard includes options for frequency hopping spread spectrum, direct sequence spread spectrum, and infrared physical layers. The first two options operate in the 2.4-GHz ISM frequency band. Both 1- and 2-Mbps operation are defined.

2.7 Support for Multimedia to the Home

In previous sections we have considered shared media networks that are used primarily in local-area business, scientific, and manufacturing scenarios. This section considers the problems of supporting multimedia applications to the home. In this case, low-cost solutions are particularly essential.

Two different approaches are discussed, Isochronous Ethernet (IEEE 802.9) and the HDSL/ADSL/VDSL family of schemes. Essentially, the two approaches were selected because they provide a realistic method of delivery of high-performance and low-cost connectivity by leveraging the currently installed networking infrastructure available to homes and businesses (via UTP telephone wires) and 10BASE-T Ethernet widely deployed in offices. As discussed in Section 2.2.1, 10BASE-T (IEEE 802.3) is very popular and has become the dominant LAN technology. It is a connectionless shared media 100BASE-T, and does not have the quality of service characteristic of ATM[2]. Note, however, that there is much activity in the networking research community to support some limited sense of quality of service in IP-based networks. This section describes a connection-oriented method that preserves the legacy interest in delivery of ATM to the office. There is concern that interest in ATM to the desktop is waning due to its high cost in comparison with other network technologies such as higher speed Ethernet. As discussed later, approaches are emerging that preserve legacy networks and provide a more general framework for low-cost support of ATM.

2.7.1 High-Data-Rate Digital Subscriber Line

High-data-rate digital subscriber line (HDSL) is very attractive because it uses the standard telephone wiring found in businesses and homes. This enables

2. See the appendix a at the end of the chapter for a brief description of ATM.

telephone companies to quickly deploy standard T1 carriers to customer premises. This deployment is achieved through the transport of a standard DS1 signal over simple 26-gauge wiring found in most houses and businesses.

A full-duplex 1.544-Mbps link over two pairs of copper wire at distances up to about 12,000 ft from the central office is possible. Alternatively, a full-duplex 768-Kbps signal on a single wire pair is possible. This represents a significant increase in capabilities beyond basic narrowband ISDN service and in fact can be used to obtain primary-rate ISDN service, or other services such as frame relay for IP networks, over already installed wiring.

Four HDSL loop architectures are under consideration:

1. *Dual duplex.* Two pairs of subscriber loops (four wires) are used for full-duplex communication. The data rate will be at 1.544 Mbps and carry a DS1 signal format. Unconditioned loops without repeaters have been proposed.
2. *Dual simplex.* This approach uses two subscriber loops but carries a full DS1 signal in one direction only.
3. *Single pair.* A single subscriber loop is used to support a full-duplex DS1 signal. This option still has unresolved technical issues and is not being considered for initial deployment.
4. *Provisioned single-loop.* This approach creates a full-duplex 768-Kbps link in which a full T1 frame is used but 12 of the 24 channels are disabled.

A number of possible approaches were considered for the standard, but the two-binary one-quaternary (2B1Q) line code was selected as the most promising. This is the same line code used in basic rate ISDN. This coding scheme is used with echo cancellation and adaptive filtering.

Discrete multitone (DMT) is a multicarrier system using discrete Fourier transforms to create and demodulate individual carriers. DMT would use FDM for upstream multiplexing. It is likely that DMT will be used in place of 2B1Q in future systems.

2.7.2 Asymmetrical Digital Subscriber Line

Consider two common applications, namely, browsing on the World Wide Web (WWW) and video distribution. Such applications can essentially use as much bandwidth as is available. Note, however, that the data rate required by the customer in both directions is not equal. Net browsing and video distribution have asymmetric traffic requirements. Network endpoints operating in this context receive much more (orders of magnitude) traffic than they generate. Such applications require a high-bandwidth channel from the service provider to the customer endpoint but only low bandwidth in the reverse direction for control. Asymmetrical digital subscriber line (ADSL) was developed to exploit this characteristic.

2.7.2.1 Asymmetric Flows

The potential data rates available through ADSL depend on the distance between the customer and the central office. Define the upstream and downstream channels as the low- and high-capacity channels, respectively. Downstream typically flows from the service provider (server) to the customer endpoint, whereas upstream flows from the endpoint to the server. ADSL transmits analog voice, signaling, control information, and data (from 1.5 to 6 Mbps) downstream from the network to the user. The reverse upstream path supports analog voice, signaling, and control information up to 576 Kbps.

By allowing the higher rate in only one direction, ADSL operates on a single pair of copper wires. A 6-Mbps downstream channel might be possible at distances of up to 12,000 ft. The total rate may increase to 9 Mbps and beyond due to advances in digital signal processing and more improved modulation and coding schemes. The approach is sensitive to distance since it is impacted by channel noise. Forward error correction (FEC) is used to reduce impulse noise errors, and correction on a symbol-by-symbol basis reduces errors caused by continuous noise coupled into the line. FEC is used since it is more effective with video and other situations when link-level or network-level protocols such as LAPB are not appropriate.

2.7.2.2 ADSL Structure

ADSL can greatly exceed the original characteristics of narrowband ISDN. It can transform the current voice-based network to support voice, graphics, and perhaps full-motion video to the home. Higher rates are possible with more sophisticated cabling schemes. However, it takes a significant investment of capital and time to deploy new cable widely, while ADSL is based on a technology that would allow multimedia to the home today.

An ADSL modem is used on each end of a twisted-pair telephone line, creating the following three channels (Figure 2.19):

1. A high-speed downstream digital channel (1.5–6.1 Mbps);
2. A medium-speed upstream digital duplex channel (16–640 Kbps);
3. An analog voice channel for traditional telephone service.

Note that both the high-speed channel and the full-duplex channel can be further multiplexed. Filtering the voice before the modem ensures a reliable

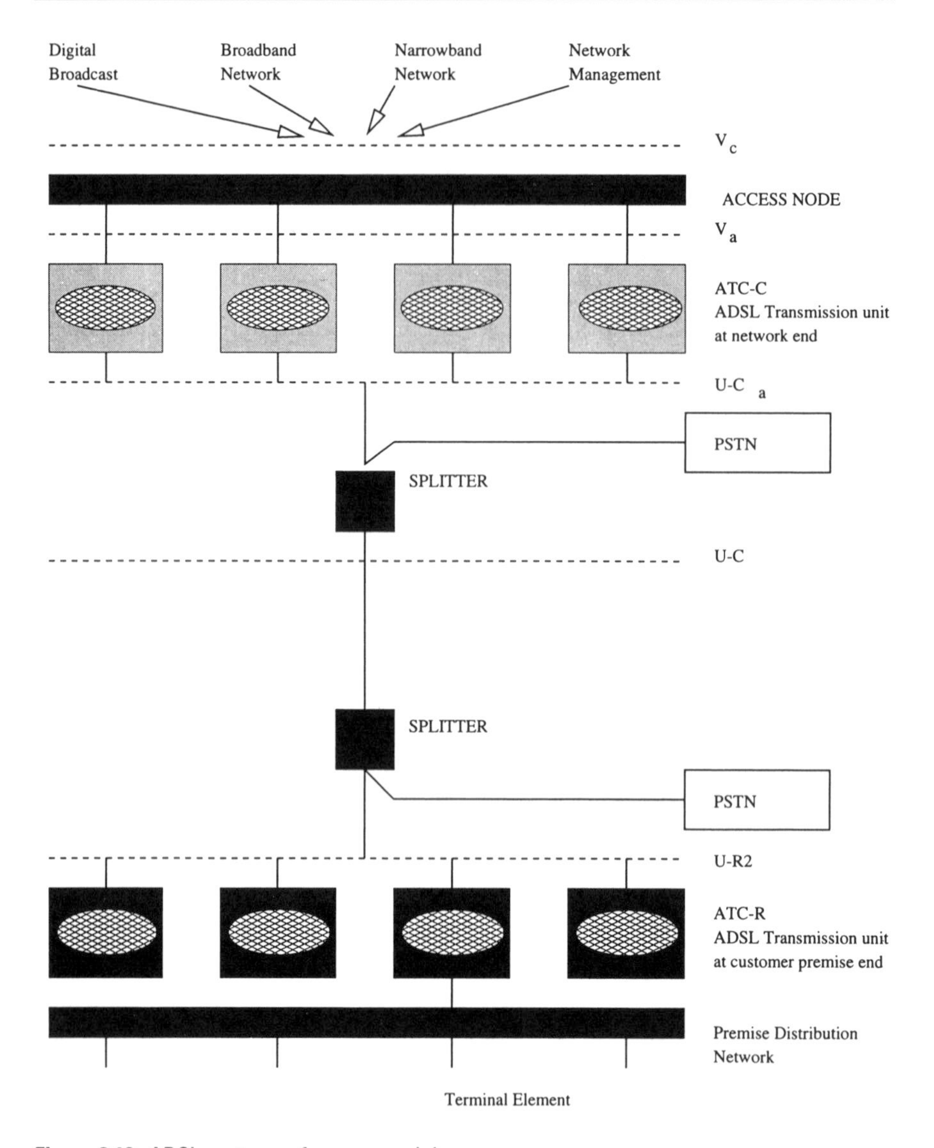

Figure 2.19 ADSL system reference model.

channel. This allows the voice channel to function even if the ADSL modem loses power or is damaged.

Downstream data rates depend on a number of factors, including the line length, wire gauge, bridged taps and other impairments, and EMI interference. Longer lines, higher frequencies, and smaller wire gauges increase signal attenuation and decrease the potential speed. For example, a data rate between 1.5 and 2 Mbps is possible with 24-gauge wire at a distance of about 5.5 km.

However, with 26-gauge wire the maximum distance is reduced by about a kilometer. At shorter distances, say, 3.5 km, a data rate of 6 Mbps can be achieved with 24-gauge wire. Customers located beyond these distances can be reached with optical fiber to the local loop.

The analog voice channel resides at the DC end of the band within the baseband 4-kHz region. This is split off prior to the ADSL processing required to generate and separate other channels. Frequency-division multiplexing (FDM) or echo cancellation is used to obtain the high speeds. The channel bandwidth is partitioned through FDM into the upstream and downstream channels. The downstream channel can be further divided through TDM into a combination of high- and low-speed channels.

The upstream channel is similarly multiplexed into low-speed channels. As with V.32 and V.34 modems, echo cancellation is used to separate the upstream and downstream channels that are overlapped. Echo cancellation is very effective but has increased complexity and cost.

The downstream channels are multiplexed with the duplex channel and the control channels together into blocks. An error correction code is appended to each block creating the aggregate flow. Transmission errors are corrected, depending on the FEC and the block length.

Interleaving data within sub-blocks creates superblocks. This provides the receiver with the ability to correct any combination of errors within a specific span of bits. For example, an error burst of 500 μs can usually be corrected if the data is interleaved at 20-ms lengths.

2.7.2.3 ATM and ADSL

ATM can natively support the requirements needed by multimedia traffic. ADSL and VDSL can be configured to carry ATM traffic. This is very attractive since it would speed wide-scale deployment of ATM. However, the actual configuration, achievable data rates, and quality of service depend on the physical conditions of the specific phone wire and surrounding conditions. As ATM becomes more widespread, ADSL modems can accommodate it with variable rates. The ATM Forum has recognized ADSL as a physical-layer transmission protocol for unshielded twisted-pair media [25].

2.7.3 Very-High-Speed Digital Subscriber Line

Very-high-speed digital subscriber line (VDSL) transmits high-speed data over short distances. The distance limitation is on the order of 300m with twisted-pair copper telephone lines. The maximum downstream rate is between 50 and 55 Mbps over lines up to 300m in length. The potential downstream speed is reduced to about 13 Mbps when the length is beyond 1500m. The upstream rates operate at a speed between 1.6 and 2.3 Mbps. This creates an asymmetric configuration in a fashion similar to that of ADSL.

The data channels will be separated in frequency from the baseband so VDSL can be overlaid on existing analog voice service. The two high-speed channels will be separated via FDM and may use echo cancellation in the future for higher rates or symmetric service.

VDSL technology will resemble ADSL, although ADSL must support a larger dynamic range and is considerably more complex. VDSL units may have to implement physical-layer media access control for multiplexing upstream data.

Four line codes have been proposed for VDSL. The following description is based on a passive configuration. Carrierless AM/PM (CAP) is related to suppressed carrier QAM. The idea is that QPSK would be used upstream with a form of TDMA for multiplexing. Discrete Fourier transforms, with FDM for upstream multiplexing, are used to create and demodulate individual carriers with a second multicarrier option called discrete multitone (DMT). Wavelet transforms are used to create individual carriers for the multicarrier system of discrete wavelet multitone (DWMT). DWMT uses FDM for upstream multiplexing and also allows TDMA. Four-level baseband signaling that filters the baseband signal, with TDMA for upstream multiplexing, is used with the forth line code candidate, simple line code (SLC).

VDSL is still in the definition stage and although some preliminary products exist, it is not yet fully available commercially. There are some unresolved issues with VDSL such as the maximum distance VDSL can reliably operate for a given data rate. VDSL uses the frequency ranges of amateur radio, which is a problem since telephone wiring acts as an antenna in that band. This increase in noise and the signal degradation impacts the maximum characteristics achievable in VDSL.

2.7.3.1 ADSL and VDSL

VDSL has evolved from ADSL. The data rate of VDSL is approximately an order of magnitude greater than ADSL. However, VDSL has lower complexity than ADSL since ADSL must contend with much larger dynamic ranges and longer distances (noise) than VDSL.

An extremely attractive aspect of both ADSL and VDSL is that it uses existing wires to carry high-speed information from the central telephone office to the home. However, service degrades the further the home is from the central office. An ADSL modem, as specified in the ANSI T1.413 ADSL standard, transmits at 1.5 to 8 Mbps downstream and 160 to 800 Kbps upstream, at distances of up to 5 miles. VDSL can support a data rate of 13 to 55 Mbps over unshielded twisted-pairs but over a shorter distance of approximately 1 mile. This rate depends on the length and gauge of the wire. The capacity is reduced with increasing distance because of increased attenuation in a signal transmitted over a wire as the wire lengthens.

In the case where the physical distance is long, ADSL will offer a single channel. As the distance is reduced, either through fiber being deployed to the central office or by just being close to a central office, the performance and number of channels possible through ADSL increases and VDSL becomes possible.

Improved performance is possible if there is optical fiber from the central office to the *neighborhood*. This is not fiber-to-the-home, but to neighborhood access points. This allows the length of copper wire to the home to be shorter and VDSL to be used. The distance extends typically to about 1 mile.

2.7.3.2 Impairments Limiting Performance

ADSL and VDSL transmission is done over existing copper wires. A number of channel impairments, however, degrade performance such as crosstalk, RFI, and impulse noise:

- *Crosstalk.* This is a problem for frequencies higher than voice channels. Crosstalk is very common since the wires are bundled with a large group of other wires between the central office and the home.
- *RFI.* Wires act as antennas. The longer the cable, the longer the antenna and RF signals coupled onto the wires introduce noise.
- *Impulse noise.* These are short bursts of noise that can be caused by switching on nearby lights or surges in power. Impulse noise typically has a duration of hundreds of microseconds. However, a short pulse in time has a wide frequency spread and this causes a greater problem for VDSL. Forward error correction in connection with interleaving can correct most of the data affected by the noise impulse.

The U.S. ANSI standards group T1E1.4 has just begun a project for VDSL. System requirements development is the first objective, which will evolve into the system and protocol definitions. The ATM Forum has defined a 51.84-Mbps interface for private network UNIs with UTP wiring.

2.7.4 IEEE 802.9 Isochronous Ethernet

Multimedia technology requires the integration of voice, video, and data. IEEE 802.9a [26,4] is an extension to the ISO/IEC 8802-3 (IEEE 802.3) 10BASE-T standard that permits this integration.

The Narrowband Integrated Services Digital Network (N-ISDN) provides a switched environment of up to 1.536 and 1.920 Mbps for North America/Europe and Japan, respectively. This provides a connection-oriented network that is useful for multimedia applications. N-ISDN has been available for more than 10 years although the roll-out has been very slow in North America due to technical limitations. Deployment continues today but at a slow pace. ATM and SONET/SDH are more recent additions, where SONET and SDH are widely deployed in place of PDH as the networking "infrastructure" of telecommunications networks. ATM and SONET have helped the move from narrowband ISDN to broadband ISDN. As discussed in a later chapter of this book, ATM and SONET are unique in a number of ways that allow multimedia communication to proceed. The networks are switched, connection oriented, provide quality of service, which allows guaranteed bandwidth, and low-latency communication. This section describes a different method, in place of SONET, to support a broadband environment. It is useful in place of legacy networks and can be used to transport ATM.

IEEE 802.9 provides the functionality of packet-based 10BASE-T and four primary rate ISDN lines on a single interface. Isochronous Ethernet extends the existing 10 Mbps of packet-switched Ethernet (P-channel) services with 96 isochronous connection-oriented 64-Kbps bearer channels (B-channels). Furthermore, interoperability is preserved in that if an Isochronous Ethernet network interface card (NIC) is used in a 10BASE-T environment, the NIC reverts back to standard 10BASE-T by using the appropriate coding scheme.

The B-channels are provisioned in the same manner as N-ISDN B-channels. Furthermore, the B-channels may be aggregated together, designated as a C-channel, and this aggregate channel can be used as a high-data-rate channel to carry ATM cells. This represents a low-cost method of providing ATM benefits to the desktop. Signaling is achieved through a separate 64-Kbps channel, denoted as the D-channel. The D-channel is used for connection establishment, teardown, maintenance, and provisioning of basic and supplementary services.

Isochronous Ethernet utilizes two pairs of Category 3 or better UTP wiring up to distances of 100m. This is the same wiring required for Ethernet 10BASE-T. This is very important since many office buildings have such wiring already installed. The same hub-based network configuration is used as in 10BASE-T.

Isochronous Ethernet, which has been standardized by the IEEE 802.9 Integrated Services LAN committee, has been specified within ITU-T as the LAN that provides the required QoS for multimedia applications utilizing the H.320 recommendation [27]. IEEE 802.9a specifies the architecture, physical-layer components, user-network interface (UNI) signaling, and management for Isochronous Ethernet networks.

As discussed previously, LANs are typically packet based, shared media, and connectionless. Extensions to this connectionless approach are switched 10BASE-T (10 Mbps) and 100BASE-T (100 Mbps). Note that both the broadcast and switched versions of Ethernet do not support end-to-end connection-oriented communication. Edge-level connection-oriented communication can take place in 802.2 but not at the application-to-application level because IP is stateless. The advantage of connection-oriented communication is that it supports minimal latency and consistent performance behavior, which is important for voice and video communication.

2.7.4.1 Isochronous Ethernet Framing

TDM framing is used to create channels to support the P-, C-, B-, and M-channels on a pair of UTP wires. The TDM frame is 125 μs with up to 256 bytes of data per frame, and a 1-byte (64-Kbps channel) start-of-frame delimiter is employed to allow the remote end to detect the beginning of the TDM frame. This allows the system to synchronize to the incoming frame and extract the 8-kHz clock.

There is a total of 16.384 Mbps of TDM framed data, 16.448 Mbps is the total data rate including start-of-frame (SOF). This is achieved through a different encoding from the traditional Manchester coding used with Ethernet. A 4B/5B encoding scheme is used with Isochronous Ethernet and transformed to the NRZI format for transmission over UTP wiring. This implies a line speed of about 20 MHz, which is the same line speed as traditional Ethernet and only requires a different coding scheme.

Within each TDM frame, the different channels are created by associated bytes within the frame. Each frame is guaranteed to have available the bandwidth required for each of the service channels (P-, C-, D-, and M-channels).

Each TDM frame consists of up to 256 bytes of data, a sequence every 125 μs, for a frame rate of 8 kHz. Bytes are reserved within each frame to create the multiple channels. Since TDM is a static allocation of bandwidth, the bandwidth required for each of the service channels (P-, C-, D-, and M-channels) can be guaranteed.

TDM systems require frame synchronization and so the start-of-frame delimiter, actually a 64-Kbps channel since a full byte per frame is dedicated to it, is employed to allow the remote end to detect the beginning of the TDM frame. Achieving frame synchronization allows the phase lock loops (PLLs) at the remote end to synchronize to the incoming frame and retrieve the 8-kHz clock.

The following channels are created and transported over this interface:

- *P-channel.* This is the contention-based channel that essentially supports the CSMA/CD MAC protocol and creates a half- or full-duplex

10-Mbps 10BASE-T transport channel. This channel can be deployed within networks that offer repeated or switched 10BASE-T operation.

- *C-channel.* This forms a full-duplex circuit-switched channel whose rate at a minimum is 64 Kbps and can be increased in multiples of 64 Kbps. A bearer channel (B-Channel) is a channel created by assigning 1 byte per frame forming a rate of 64 Kbps. A C-channel is a ganged together collection of isochronous B-channels. The maximum rate of the channel depends on the mode of operation. It will be 15.872 Mbps in the all-isochronous mode of operation or 6.144 Mbps in multiservice mode.
- *D-channel.* This is the signaling channel used to create connections between source and destination. It is a full-duplex 64-Kbps packet channel. The signaling protocol is defined in the IEEE 802.9a specification and is based on the original ITU Q.93x family of protocols. This family of signaling protocols has a broad range of functionality including session connection and has been employed in ISDN and ATM. This allows for easier interoperability between ISDN, ATM, and Isochronous Ethernet-based networks. The signaling stack used is the link access protocol on the D-channel (LAPD) at layer 2, while the IEEE 802.9a Q.93x UNI procedures are used for layer 3. In addition to call setup and teardown, supplementary service procedures for the provisioning of the C-channel can also be transmitted on this channel. This channel can also be used to transfer packet mode data as with X.25-type networks.
- *M-channel.* This is the maintenance channel, a full-duplex channel formed at a rate of 96 Kbps. This channel is used to exchange control and status information. For example, the channel is used to detect the network clock that is used for frame synchronization and other network management functions such as loopback, error detection within the frame, and backup power notification.
- *Start-of-frame channel.* This channel carries the start of TDM frame synchronization pattern. This channel is important for two reasons: to be able to identify the boundaries of adjacent TDM frames and also to ensure that the 8-kHz system clock remains in synchronization.

2.7.4.2 Operational Modes

The modes of operation are multiservice mode, all-isochronous mode, and 10BASE-T mode. Multiservice mode uses the channel partitioning described earlier. A powerful enhancement over the original Ethernet, in addition to the flexibility of the variable set of possible channels, is the autonegotiation link pulse signaling scheme. This facility can be used to configure the connection automatically. This negotiation facility is executed at power-up and common services between endpoints are identified. Isochronous Ethernet recognizes the capabilities of 10BASE-T and 100BASE-TX systems and dynamically switches between modes.

These modes provide the following services:

- *Multiservice mode.* The TDM structure is used and all channels are provisioned into 256 bytes per frame which repeats on an 8 kHz basis. This creates the P-channel (a 10-Mbps channel operating with CSMA/CD for transmission of ISO/IEC 8802-3 MAC frames), and 96 B-channels (each operating at 64 Kbps), which can be ganged together to form the C-channel (operating at up to 6.144 Mbps with all 96 B-channels), a 64-Kbps D-channel, a 96-Kbps M-channel, and a 64-Kbps SOF delimiter channel. The 16.384 Mbps of data are encoded using the 4B/5B encoding scheme.
- *All-isochronous mode.* This mode places the complete bandwidth of the system for use as B-channels and so a high-capacity C-channel of 15.872 Mbps can be achieved when all of the 248 B-channels are used. This mode is useful for circuit-switched isochronous communication. This mode also provides a 64-Kbps D-channel, a 96-Kbps M-channel, and a 64-Kbps SOF channel. The all-isochronous mode eliminates the P-channel, and the available TDM slots are made available to the B-channels. The total data rate of 16.384 Mbps is transmitted in a TDM frame, using 4B/5B encoding with an NRZI format for transmission on unshielded twisted-pair wiring.
- *10BASE-T mode.* Isochronous Ethernet will automatically configure itself for 10BASE-T operation if the isoEthernet NIC detects that it has been placed in a 10BASE-T environment. In this case, the TDM frame is not used, Manchester coding will be used in place of 4B/5B, and the C, D, M, and SOF channels are not supported. A 10-Mbps P channel will be formed that will support CSMA/CD MAC frames as in a 10BASE-T environment.

Autonegotiation link pulse signaling is a new protocol within the IEEE 802 standard suite. The goal of this scheme is to allow endpoints to configure themselves automatically. The idea is that a device will advertise its capabilities and modes of operation during system power-up.

A major advantage of this approach is that interoperability is preserved since this scheme is employed by both IEEE 802.9a, Isochronous Ethernet, 10BASE-T, and 100BASE-TX networks.

There are some similarities between ATM and Isochronous Ethernet since the signaling protocols are both based on Q.93x [28]. Isochronous Ethernet, operating in either all-isochronous or multiservice mode, can support ATM at a very low cost. This is particularly useful in the context of ATM to the desktop. In all-isochronous mode, ATM cells can be transported by the C-channel. B-channels, or perhaps ganged B-channels, would be used to transport ATM in multiservice mode. Because ATM is a wide-area solution, this may allow users end-to-end QoS support in the wide area at data rates approaching the OC-3c rates but doing so over unshielded twisted pair.

The IEEE 802.9a specification defines the Isochronous Ethernet physical layer. This is further subdivided as follows:

1. *Physical medium dependent (PMD).* This sublayer performs the medium-dependent encoding (NRZI) and provides the electrical interface to Category 3 or better UTP wiring.
2. *Physical signaling (PS).* PS receives the HMUX byte stream and inserts the TDM information. It then encodes the data using the 4B/5B encoding scheme and passes the data to the PMD sublayer.
3. *Hybrid multiplexing (HMUX).* HMUX receives the TDM byte stream from the PS sublayer and demultiplexes the stream into the individual bytes that are associated with the individual and aggregated channels. It is the interface to the P, C, and D channel media access control layers.

2.7.5 Broadband Network Security

The preceding section defined methods where ATM can be delivered to the home at low cost through the existing wiring infrastructure. HDSL/ADSL/VDSL and Isochronous Ethernet were used as examples of some of the emerging network technologies that can support a wide variety of traffic modes. For example, Isochronous Ethernet can concurrently support both connectionless and connection-oriented flows. Furthermore, the connection-oriented flows can be aggregated to create a large connection-oriented pipe.

The significance of this is that ATM can be transported through the physical interfaces described in this section, thereby achieving high-speed communication. With this huge increase in performance and connectivity also comes a downside: network security. In the same way in which a user can reach out from his home or office into the network, someone outside can reach inside through the network. This is good when the user is away from home and wishes to access files located there but bad due to the obvious security concerns.

One potential solution is to power off the systems and not leave them running when not in use. However, this does not eliminate the threat since a hacker can reach into your enclave while you are accessing your machine. Alternative solutions include installing an ATM firewall between a protected enclave (home or office) and the outside network. Unfortunately, the current ATM firewalls that are commercially available are positioned for the high-performance high-end market and therefore extremely expensive. Alternative approaches for low-cost security are being developed. In particular, Project CellBlock is a low-cost FPGA-based firewall being developed at the Univeristy of Maryland at College Park [29]. Techniques are currently being developed to provide protection of IP and non-IP traffic over ATM at a cost that will not inhibit deployment and use.

2.8 Summary

Enormous growth is occurring in the development of multimedia desktop applications. This growth is expected to expand rapidly in the next few years. This expansion will have a huge impact on the type of traffic that future computer networks will have to carry. In particular, networks will have to accommodate various QoS constraints that have not been required in the past.

Multimedia communications requires the integration of multiple service streams, such as video, audio, and data, in a consistent fashion. In this chapter we reviewed many of the popular shared media local-area and metropolitan-area networks. Many of the technologies used in these systems were developed some time ago and yet they will be used to carry multimedia traffic in the future. The basic principles of operation for each system were introduced. Also, we discussed these systems from a performance perspective, commenting on some of the salient features that affect their use. In each case we have focused on the capabilities these networks have for supporting future multimedia applications.

References

[1] Stallings, W., *Local and Metropolitan Area Networks*, 5th ed., Upper Saddle River, NJ: Prentice-Hall, 1997.

[2] Stuttgen, H. K., "Network Evolution and Multimedia Communication," *IEEE Multimedia Magazine*, Fall 1995.

[3] Draft ITU-T Recommendation H.322, "Visual Telephony Systems and Terminal Equipment for Local Area Networks Which Provide a Guaranteed Quality of Service," May 11, 1995.

[4] Popli, S., "Isochronous Ethernet—The Multimedia LAN for Real Time Desktop Connectivity," National Semiconductor Corporation, 1996.

[5] ITU-T Recommendation H.320, "Narrowband Visual Telephone Systems and Terminal Equipment," 1990.

[6] ANSI/IEEE Std. 802.3, "Carrier Sense Multiple Access with Collision Detection," New York: *Institute of Electrical and Electronics Engineers*, 1985.

[7] ANSI/IEEE Standard 802.5, "Token Ring Access Method and Physical Layer Specifications," 1985.

[8] ANSI/IEEE Std 802.4, "Token-Passing Bus Access Method and Physical Layer Specifications," 1990.

[9] Abramson, N. M., "The Aloha System, Another Alternative for Computer Communications," *Proc. 1970 AFIPS Joint Computer Conf.*, 1970, pp. 281–285.

[10] Kleinrock, L., and F. A. Tobagi, "Packet Switching in Radio Channels: Part I—Carrier Sense Multiple Access Modes and Their Throughput-Delay Characteristics," *IEEE Trans. Commun.* Vol. COM-23, Dec. 1975, pp. 1400–1426.

[11] Tobagi, F. A., "Multiaccess Protocols in Packet Communication Systems," *IEEE Trans. Commun.* Vol. COM-28, No. 4, Apr. 1980, pp. 468–488.

[12] Gonsalves, T. A., "Packet-Voice Communications on an Ethernet Local Computer Network: An Experimental Study," *Proc. SIGCOMM'83, ACM SIGCOMM*, Mar. 1983, pp. 178–185.

[13] Tanenbaum, A. S., *Computer Networks*, 2nd ed., Upper Saddle River, NJ: Prentice-Hall, 1988.

[14] "IBM Synchronous Data Link Control: General Information," Research Triangle Park, NC: IBM Corporation, 1975.

[15] Valenzano, A., C. DeMartini, and L. Ciminiera, *MAP and TOP Communications: Standards and Applications*, Reading, MA: Addison-Wesley, 1992.

[16] IEEE Std 802.3u-1995, Supplement to ISO/IEC 8802-3:1993, "Local and Metropolitan Area Networks: Media Access Control (MAC) Parameters, Physical Layer, Medium Attachment Units, and Repeater for 100 Mb/s Operation, Type 100BASE-T (Clauses 21-30)."

[17] Johnson, H. W., *Fast Ethernet: Dawn of a New Network*, Upper Saddle River, NJ: Prentice-Hall PTR, 1996.

[18] Dalgic, I., W. Chien, and F. A. Tobagi, "Evaluation of 10BASE-T and 100BASE-T Ethernets Carrying Video, Audio and Data Graffic," *Proc. of IEEE INFOCOM'94*, Toronto, Canada, June 1994.

[19] ANSI X3.139, "FDDI Token ring Media Access Control (MAC)," 1987.

[20] Jain, R., *FDDI Handbook: High Speed Networking Using Fiber and Other Media*, Reading, MA: Addison-Wesley, 1994.

[21] Ross, F. E., "An Overview of FDDI: The Fiber Distributed Data Interface," *IEEE J. Sel. Areas Commun.* Vol. 7, No. 7, Sept. 1989, pp. 1043–1051.

[22] IEEE Std 802.12-1995, "Demand Priority Access Method Physical Layer and Repeater Specification for 100 Mb/s Operation," 1995.

[23] Watson, G., A. Albrecht, J. Curcio, D. Dove, S. Goody, J. Grinham, M. Spratt, and P. Thalery, "The Demand Priority MAC Protocol," *IEEE Network Magazine*, Vol. 9, No. 1, Jan./Feb. 1995, pp. 28–35.

[24] IEEE Std 802.11, "Standard for Wireless LAN Medium Access Control (MAC) and Physical Layer (PHY) Specification—Draft."

[25] ANSI T1.413-1995, "Network and Customer Installation Interfaces—Asymmetric Digital Subscriber Line (ADSL) Metallic Interface."

[26] IEEE 802.9a Draft D10, "Integrated Services (IS) LAN: IEEE 802.9 Isochronous Services With Carrier Sense Multiple Access With Collison Detection (CSMA/CD) Media Access Control (MAC) Service," Mar. 30, 1995.

[27] "ADSL Forum System Reference Model," Technical Report ADSL Forum TR-001.

[28] Modlin, C. S., "ATM Over ADSL and VDSL: PMD Specific Attributes that Affect the Network," ATM Forum/97-0156, Feb. 1997.

[29] Dowd, P. W., T. McHenry, F. A. Pellegrino, T. M. Carrozzi, and W. B. Cocks, "A Method for High Performance Low Cost Network Security for IP and Non-IP Traffic in an ATM Environment," *Proc. IEEE Symp. FPGAs for Computing Machines (FCCM'97)*, Napa, CA, Apr. 1997.

Appendix 2A: Asynchronous Transfer Mode

ATM is an emerging broadband technology that was first envisioned as spanning from wide-area networks to the desktop, including the wireless domain. This section provides a very brief overview of ATM and is not in any way complete. Compared with the other link-level protocols discussed in this chapter, ATM is very complex and has a much broader scope. Later chapters provide additional detail related to its usefulness for integrated services and, in particular, multimedia networking. One attractive aspect of ATM is its ability to support quality of service. This is the capability for an application to specify the type of service it requires. The network essentially guarantees that level of performance and service to the application by using call admission control and congestion control to ensure that performance is not degraded.

The actual use and deployment of ATM is currently evolving. ATM originally was popular in the local area but is now viewed as being particularly attractive in the wide area. There are strong economic factors that encourage ATM deployment in the wide area. ATM to the desktop is still a desire for a large consumer base that requires end-to-end quality of service. Through the advances in residential broadband techniques (xDSL and HFC), low-cost broadband connectivity is possible that may carry ATM or perhaps IP traffic. This brief appendix provides an introduction into ATM and some of the issues related to it. Refer to later chapters of this book for more specific details on adaptation layers, routing, congestion control, and multicast support.

2.A.1 Cells and Encapsulation

The first thing that is usually discussed regarding ATM is its odd size for the basic unit of transfer. The basic unit of transfer is called a *cell.* A cell is 53 bytes in length—5 bytes of header followed by 48 bytes of payload. The header contains an edge-level address composed of two parts: the virtual path identifier (VPI) and the virtual channel identifier (VCI). The VCI is 16 bits in length while the VPI is either 12 or 8 bits for a user-network interface (UNI) or for a network-network interface (NNI), respectively. The (VPI, VCI) used across a particular link does not have end-to-end significance. Both NNI and UNI have a 4-bit wide payload-type indicator control field but the UNI also has a generic flow control field (GFC) of 4 bits in length. The last byte of the header is a checksum that can detect double-bit errors and correct single-bit header errors.

Data to be transmitted, passed either by IP or a native interface, is first encapsulated in an ATM adaptation layer (AAL) protocol data unit. The adaptation layer is composed of a convergence sublayer, which is service dependent, and the segmentation and reassembly layer (SAR), which supports the segmentation/reassembly into cells/PDUs and PDUs/cells. The data that have been encapsulated will then be segmented into appropriately sized pieces to be included into the cells. A number of different AALs have been developed during the past few years, each positioned to support a particular class of traffic. Some adaptation protocols also have additional cell-level overhead. The simplest adaptation layer is AAL5. It does not include any cell-level overhead and only adds an 8-byte trailer to the end-of-data segment. For example, an LLC/SNAP encapsulated IP datagram would have the AAL5 trailer appended and then be segmented into 48-byte payloads for inclusion in ATM cells. At the receiving side, the cells are reassembled into the AAL5 PDU. The CRC is computed and compared to the CRC field within the AAL5 trailer. The PDU is accepted if correct and discarded if it has been corrupted. Details of AAL encapsulation and the formats for the different adaptation layers have not been included in this chapter. Additional details can be found in later chapters.

2.A.2 Signaling and Routing

ATM is a connection-oriented protocol—a separate signaling channel is used to create a path for data to flow. Signaling is transmitted on a well-known (VPI, VCI) channel. For example, UNI is transmitted on (0,5).

An ATM switch has the advantage that once the connection has been created, data can flow through the ATM switch with only hardware intervention. Routing is not performed hop by hop as with traditional IP networks. The signaling sets up a connection in the switch fabric. This maps an incoming cell on a particular (VPI, VCI) pair and a particular port to a particular output port and a new (VPI, VCI). The (VPI, VCI) only has edge-level significance and is changed at each ATM switch between source and destination. No software intervention is required as the cells are being switched so the entire operation is essentially a table lookup and an update to the VPI, VCI and cell header fields. Note, however, that there is recent interest in multiprotocol label switching (MPLS), which offers some of these same advantages to IP networks although the technique is not specific to IP.

The nominal signaling transaction uses a three-message sequence that flows between the two communicating endpoints through intervening ATM switches. Within the signaling message, QoS parameters are specified. This specifies the requirements of the flow, in terms of average and peak data rate among many other parameters. As the signal message is passed between ATM switches in its path from the source and destination, the switches examine the QoS request and allocate the resource to the flow or block the connection if the level of resource request cannot be supported. This is known as *call admission control* (CAC). A routing protocol is needed so the ATM switches know how to route a connection request. A number of routing protocols have been developed, differing mainly in level of complexity and the size of the target network. One example is PNNI, which divides the network into a hierarchy. Nodes within the hierarchy that share a common parent form a peer group. A connection request is transmitted with a hierarchically complete source address specified. When the request enters a peer group, the ingress node specifies the precise source route through the peer group. This enables a certain amount of information hiding, thus avoiding the situation in which a node would have to know the complete structure of the network.

The endpoints are identified within the signaling message via one of two addressing schemes. Two popular addressing schemes are E.164, which is very popular in the wide area in the traditional telecommunication environment, and NSAP, which is popular for local private networks. The ATM Forum is trying to resolve the interoperability problems.

2.A.3 Congestion Control

One of the major advantages of ATM is that it can support edge-level flow and congestion control. Consider the situation with a TCP-based connection. Congestion is detected at the endpoints through packet loss. Typically, once congestion is detected, the protocol will throttle itself and enter a mode to reduce the rate at which packets are injected into the network. The trouble is that the latency to detect the congestion can be long, delaying the time for the system to react, and that one flow can monopolize a link and adversely affect the performance of a different flow. Only recently has the IRTF, the research arm of the IETF, recommended that individual IP routers begin to interact in an edge-to-edge manner (discard strategies and transmission scheduling) to improve performance and network stability. ATM has the potential to improve this situation because it can exert more control over individual flows on an edge-to-edge basis. A rate-based congestion control scheme has been adopted where backpressure can be applied to congested links to reduce the traffic and maintain stable network operation. Call admission control and the rate-based congestion control are both used to help maintain stable network performance and be able to provide a guaranteed level of service to a flow.

3

QoS Architectures

3.1 Introduction

Meeting quality-of-service (QoS) guarantees in distributed multimedia systems is fundamentally an end-to-end issue, that is, from application to application. Consider, for example, the remote playout of a sequence of audio and video: In the distributed system platform, QoS assurances should apply to the complete flow of media from the remote server across the network to the point(s) of delivery. As illustrated by Figure 3.1, this generally requires end-to-end admission testing and resource reservation in the first instance, followed by careful coordination of disk and thread scheduling in the end system, packet/cell scheduling and flow control in the network, and, finally, active monitoring and maintenance of the delivered quality of service. A key observation is that for applications relying on the transfer of multimedia and, in particular, continuous media flows, it is essential that quality of service be configurable, predictable, and maintainable system-wide, including the end-system devices,

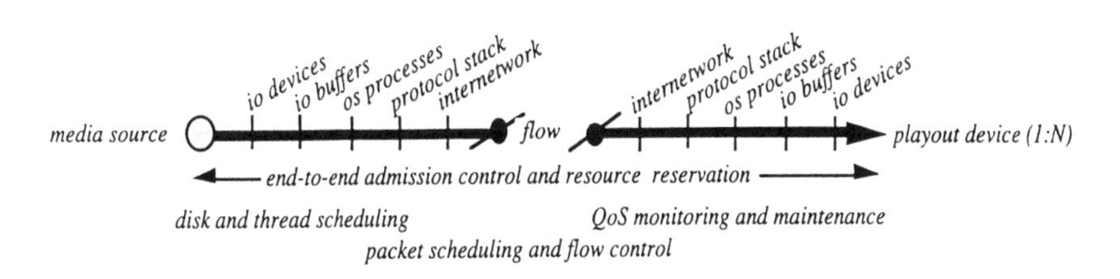

Figure 3.1 End-to-end QoS scenario for a continuous media flow.

The author would like to thank Cristina Aurrecoechea and Linda Hauw who provided considerable input to this chapter.

communications subsystem, and networks. Furthermore, it is also important that all end-to-end elements of distributed systems architecture work in unison to achieve the desired application-level behavior.

To date, most of the developments in the area of QoS support have occurred in the context of individual architectural components [1]. Much less progress has been made in addressing the issue of an overall QoS architecture for multimedia communications. Considerable progress, however, has been made in the separate areas of distributed systems platforms [1–9], operating systems [10–16], transport systems [17–26], and multimedia networking [27–47] support for quality of service. In end systems, most of the progress has been made in the areas of scheduling [12,48,49], flow synchronization [50,51], and transport support [17–26]. In networks, research has focused on providing suitable traffic models [52] and service disciplines [33], as well as appropriate admission control and resource reservation protocols [29,32,34]. Many current network architectures, however, address quality of service from a provider's point of view and analyze network performance, failing to address comprehensively the quality needs of applications. Until recently there has been little work on QoS support in distributed systems platforms. What work there is has been carried out mainly in the context of the open distributed processing [8].

The current state of QoS support in architectural frameworks can be summarized as follows [1]:

1. *Incompleteness.* Current interfaces (e.g., application programming interfaces such as Berkeley Sockets) are generally not QoS configurable and provide only a small subset of the facilities needed for control and management of multimedia flows.
2. *Lack of mechanisms to support QoS guarantees.* Research is needed in distributed control, monitoring, and maintenance of QoS mechanisms so that contracted levels of service can be predictable and assured.
3. *Lack of an overall framework.* It is necessary to develop an overall architectural framework to build on and reconcile the existing notion of quality of service at different system levels and among different network architectures.

In recognition of these limitations, a number of research teams have proposed systems architectural approaches to QoS support. In this chapter these are referred to as *QoS architectures* [53–76]. The intention of QoS architecture research is to define a set of QoS configurable interfaces that formalizes quality of service in the end system and network, providing a framework for the integration of QoS control and management mechanisms.

In this chapter we present, in Section 3.2, a *generalized QoS framework* and terminology for distributed multimedia applications operating over multimedia networks with QoS guarantees. The generalized QoS framework is based on a set of principles that governs the behavior of QoS architectures. Following this, we evaluate a number of QoS architectures found in the literature that have been developed by the telecommunications, computer communications, and standards communities. We then present a short qualitative comparison and discussion in Sections 3.4 and 3.5, respectively. Finally, in Section 3.6 we offer some concluding remarks.

3.2 Generalized QoS Framework

This section describes a set of elements used in building QoS into distributed multimedia systems. These elements include QoS principles that govern the construction of a generalized QoS framework, QoS specifications that capture application-level *flow* requirements, and QoS mechanisms that realize the desired application end-to-end QoS behavior.

The notion of a flow is an important abstraction that underpins the development of QoS frameworks. Flows characterize the production, transmission, and eventual consumption of a single media source (namely, audio, video, data) as integrated activities governed by single statements of end-to-end QoS. Flows are simplex in nature and can be either unicast or multicast. Flows generally require end-to-end admission control and resource reservation, and support heterogeneous QoS demand.

3.2.1 QoS Principles

A number of QoS principles motivate the design of a generalized QoS framework:

- The *transparency principle* states that applications should be shielded from the complexity of underlying QoS specification and QoS management. An important aspect of transparency is the QoS-based API [60,77] at which desired QoS levels are stated (see QoS management policy in Section 3.2.2). The benefits of transparency are that it reduces the need to embed functionality in the application, hides the detail of underlying service specification from the application, and delegates the complexity of handling QoS management activities to the underlying framework.
- The *integration principle* states that QoS must be configurable, predictable, and maintainable over all architectural layers to meet end-to-

end QoS [54]. Flows traverse resource modules (e.g., CPU, memory, multimedia devices, network, etc.) at each layer from source media devices, down through the source protocol stack, across the network, up through the receiver protocol stack to the playout devices. Each resource module traversed must provide QoS configurability (based on a QoS specification), resource guarantees (provided by QoS control mechanisms), and maintenance of ongoing flows.

- The *separation principle* states that media transfer, control, and management are functionally distinct architectural activities [55]. The principle states that these tasks should be separated in architectural QoS frameworks. One aspect of this separation is the distinction between signaling and media transfer. Flows (which are isochronous in nature) generally require a wide variety of high-bandwidth, low-latency, nonassured services with some form of jitter correction. On the other hand, signaling (which is full duplex and asynchronous in nature) generally requires low-bandwidth, assured-type services.
- The *multiple timescales principle* [55] guides the division of functionality between architectural modules and pertains to the modeling of control and management mechanisms. It is necessitated by, and is a direct consequence of, fundamental time constraints that operate in parallel between resource management activities (e.g., scheduling, flow control, routing, and QoS management) in distributed communications environments.
- The *performance principle* subsumes a number of widely agreed rules for the implementation of QoS-driven communications systems, which guide the division of functionality in structuring communication protocols for high performance in accordance with systems design principles [78], avoidance of multiplexing [79], recommendations for structuring communications protocols [80], and the use of hardware assists for efficient protocol processing [21,36].

3.2.2 QoS Specification

QoS specification is concerned with capturing application-level QoS requirements and management policies. QoS specification is generally different at each system layer and is used to configure and maintain QoS mechanisms resident in the end system and network. For example, at the distributed system platform level, QoS specification is primarily application oriented rather than system oriented. Lower level considerations such as tightness of synchronization of multiple related audio and video flows, the rate and burst size of flows, or the details of thread scheduling in the end system should all be hidden at this level. QoS specification is therefore declarative in nature; applications specify what is required rather than how this is to be achieved by underlying QoS mechanisms. QoS specification encompasses, but is not limited to, the following:

- *Flow performance specification,* which characterizes the user's flow performance requirements [81]. The ability to guarantee traffic throughput rates, delay, jitter, and loss rates is particularly important for multimedia communications. These performance-based metrics are likely to vary from one application to another. To be able to commit necessary end system and network resources, QoS frameworks must have prior knowledge of the expected traffic characteristics associated with each flow before resource guarantees can be met.
- *Level of service,* which specifies the degree of end-to-end resource commitment required (e.g., deterministic [30], predictive [28], and best effort [80]). While the flow performance specification permits the user to express the required performance metrics in a quantitative manner, level of service allows these requirements to be refined in a qualitative way to allow a distinction to be made between hard and soft performance guarantees. Level of service expresses a degree of certainty that the QoS levels requested at the time of flow establishment or renegotiation will be honored.
- *QoS management policy,* which captures the degree of QoS adaptation [60] that the flow can tolerate and the scaling actions to be taken in the event of violations in the contracted QoS [72]. By trading off temporal and spatial quality to available bandwidth, or manipulating the playout time of continuous media in response to variation in delay, audio and video flows can be presented at the playout device with minimal perceptual distortion. The QoS management policy also includes application-level selection for QoS indications (aka QoS alerts [53]) in the case of violations in the requested quality of service and periodic QoS availability notifications for bandwidth, delay, jitter, and loss.
- *Cost of service,* which specifies the price the user is willing to incur for the level of service [82]. Cost of service is a very important factor when considering QoS specification. If there is no notion of cost of service involved in QoS specification, there is no reason for the user to select anything other than the maximum level of service, for example, guaranteed service.

- *Flow synchronization specification,* which characterizes the degree of synchronization (i.e., tightness) between multiple related flows [50]. For example, simultaneously recorded video persepectives must be played in precise frame-by-frame synchrony so that relevant features may be simultaneously observed. On the other hand, lip synchronization in multimedia flows does not need to be absolutely precise [51] when the main information channel is auditory and video is only used to enhance the sense of presence.

3.2.3 QoS Mechanisms

QoS mechanisms are selected and configured according to user-supplied QoS specifications, resource availability, and resource management policy. In resource management, QoS mechanisms can be categorized as either static or dynamic in nature. *Static resource management* deals with flow establishment and end-to-end QoS renegotiation phases (which we describe as QoS provision), and *dynamic resource management* deals with the media-transfer phase (which we describe as QoS control and management). The distinction between QoS control and QoS management is characterized by the different timescales over which they operate. QoS control operates on a faster timescale than QoS managment.

3.2.3.1 QoS Provision Mechanisms

QoS provision is comprised of the following components:

- *QoS mapping,* which performs the function of automatic translation between representations of QoS at different levels (i.e., operating system, transport layer, and network) and thus relieves the user of the necessity of thinking in terms of lower level specification. For example, the transport-level QoS specification may express flow requirements in terms of level of service, average and peak bandwidth, jitter, loss, and delay constraints. For admission testing and resource allocation purposes, this representation must be translated to something more meaningful to the end system. As illustrated in Figure 3.2, QoS mapping derives the scheduler QoS parameters (namely, period, quantum and deadline times of the threads) from the transport level QoS specification parameters [15].
- *Admission testing,* which is responsible for comparing the resource requirement arising from the requested QoS against the available resources in the system. The decision about whether a new request can be accommodated generally depends on system-wide resource

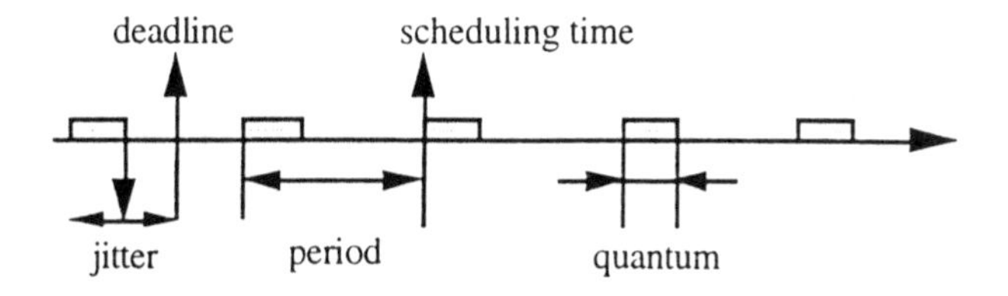

Figure 3.2 QoS parameters derived during QoS mapping.

management policies and resource availability. Once admission testing has been successfully completed on a particular resource module, local resources are reserved immediately and then committed later if the end-to-end admission control test (i.e., accumulation of hop by hop tests) is successful.

- *Resource reservation protocols,* which arrange for the allocation of suitable end-system and network resources according to the user QoS specification. In doing so, the resource reservation protocol interacts with QoS-based routing to establish a path through the network in the first instance; then, based on QoS mapping and admission control at each local resource module traversed (e.g., CPU, memory, I/O devices, switches, and routers), end-to-end resources are allocated. The end result is that QoS control mechanisms such as network-level cell/packet schedulers and end-system thread schedulers are configured accordingly.

3.2.3.2 QoS Control Mechanisms

QoS control mechanisms operate on timescales at or close to media transfer speeds. They provide real-time traffic control of flows based on requested levels of QoS established during the QoS provision phase. The fundamental QoS control mechanisms include the following:

- *Flow scheduling,* which manages the forwarding of flows (chunks of media based on application-layer framing) in the end system [11–16] and network (packets and/or cells) in an integrated manner [33]. Flows are generally scheduled independently in the end systems but may be aggregated and scheduled in unison in the network. This is dependent on the level of service and the scheduling discipline [52] adopted.
- *Flow shaping,* which regulates flows based on user-supplied flow performance specifications. Flow shaping can be based on a fixed-rate throughput (i.e., peak rate) or some form of statistical representation (i.e., sustainable rate and burstiness) of the required bandwidth [30].

The benefit of shaping traffic is that it allows the QoS frameworks to commit sufficient end-to-end resources and to configure flow schedulers to regulate traffic through the end systems and network. It has been mathematically proven [39] that the combination of traffic shaping at the edge of the network and scheduling in the network can provide hard performance guarantees.

- *Flow policing,* which can be viewed as the dual of monitoring. Monitoring, which is usually associated with QoS management, observes whether the QoS contracted by a provider is being maintained, whereas policing observes whether the QoS contracted by a user is being adhered to. Policing is often only appropriate where administrative and charging boundaries are being crossed, for example, at a user-to-network interface [34]. Flow shaping schemes at the source allow the policing mechanism to detect misbehaving flows.
- *Flow control,* which includes both open-loop and closed-loop schemes. Open-loop flow control is used widely in telephony and allows the sender to inject data into the network at the agreed levels given that resources have been allocated in advance. Closed-loop flow control requires the sender to adjust its rate based on feedback from the receiver [22] or network [45]. Applications using closed-loop flow control based protocols must be able to adapt to fluctuations in the available resources. On the other hand, applications that cannot adjust to changes in the delivered QoS are more suited to open-loop schemes where bandwidth, delay, and loss can be deterministically guaranteed for the duration of the session.
- *Flow synchronization,* which is required to control the event ordering and precise timings of multimedia interactions. Lip-sync is the most commonly cited form of multimedia synchronization (i.e., synchronization of video and audio flows at a playout device). Other synchronization scenarios reported include event synchronization with and without user interaction, continuous synchronization other than lip-sync, and continuous synchronization for disparate sources and sinks. All scenarios place fundamental QoS requirements on flow synchronization protocols [25].

3.2.3.3 QoS Management Mechanisms

To maintain agreed levels of QoS it is often insufficient to just commit resources. Rather, QoS management is frequently required to ensure that the contracted QoS is sustained. QoS management of flows is functionally similar to QoS control. However, it operates on a slower timescale, that is, over longer monitoring and control intervals [83]. The fundamental QoS management mechanisms include the following:

- *QoS monitoring,* which allows each level of the system to track the ongoing QoS levels achieved by the lower layer. QoS monitoring often plays an integral part in a QoS maintenance feedback loop, which maintains the QoS achieved by resource modules. Monitoring algorithms operate over different timescales. For example, they can run as part of a scheduler (as a QoS control mechanism) to measure individual performance of ongoing flows. In this case measured statistics can be used to control packet scheduling and admission control [28]. Alternatively, QoS monitoring can operate on an end-to-end basis as part of a transport-level feedback mechanism [25] or as part of the application itself [84].
- *QoS availability,* which allows the application to specify the interval over which one or more QoS parameters (e.g., delay, jitter, bandwidth, loss, synchronization) can be monitored and the application informed of the delivered performance via a QoS signal [60]. Both single and multiple QoS signals can be selected based on the user-supplied QoS management policy (see Section 3.2.2).
- *QoS degradation,* which issues a QoS indication to the user when it determines that the lower layers have failed to maintain the QoS of the flow and nothing further can be done by the QoS maintenance mechanism. In response to such an indication the user can choose either to adapt to the available level of QoS or scale back [71] to a reduced level of service (i.e., end-to-end renegotiation).
- *QoS maintenance,* which compares the monitored QoS against the expected performance and then exerts tuning operations (i.e., fine- or coarse-grain resource adjustments) on resource modules to sustain the delivered QoS. Fine-grain resource adjustment counters QoS degradation by adjusting local resource modules (e.g., loss via the buffer management, throughput via the flow regulation, and queueing delays and continuous media playout calculation via the flow scheduling [72]).
- *QoS scalability,* which comprises QoS filtering (which manipulates flows as they progress through the communications system) and QoS adaptation (which scales flows at the end systems only) mechanisms. Many continuous media applications exhibit robustness in adapting to fluctuations in end-to-end QoS. Based on the user-supplied QoS management policy, QoS adaptation in the end systems can take

remedial actions to scale flows appropriately. Resolving heterogeneous QoS issues is a particularly acute problem in the case of multicast flows. Here, individual receivers may have differing QoS capabilities to consume audiovisual flows; QoS filtering helps to bridge this heterogeneity gap while simultaneously meeting individual receivers' QoS requirements [76].

3.3 QoS Architectures

Until recently, research in providing QoS guarantees has focused primarily on network-oriented traffic models and service scheduling disciplines. These guarantees are not, however, end-to-end in nature. Rather they preserve QoS guarantees only between network access points to which end systems are attached [67]. Work on QoS-driven end-system architecture needs to be integrated with network configurable QoS services and protocols to meet application-to-application QoS requirements. In recognition of this, researchers have recently proposed new communication architectures that are broader in scope and cover both network and end-system domains. In this section, we review a number of QoS architectures that have recently emerged in the literature [53–76]. Each architecture tends to use its own distinctive QoS terminology. We do not attempt to resolve that here. We present, rather, the pertinent and novel features of each architecture and then, in Section 3.4, compare them with the generalized QoS framework introduced in the preceding section.

3.3.1 Heidelberg QoS Model

The HeiProject at IBM's European Networking Center in Heidelberg has developed a comprehensive QoS model that provides guarantees in the end systems and network [57]. The communications architecture includes a continuous media transport system (HeiTS/TP) [23], which provides QoS mapping and *media scaling* [71] as illustrated in Figure 3.3. Underlying the transport is an internetworking layer based on ST-II [27], which supports both guaranteed and statistical levels of service. In addition, the network supports QoS-based routing and QoS filtering. Key to providing end-to-end guarantees is *HeiRAT* (resource administrative technique) [57]. HeiRAT is comprised of a comprehensive QoS management scheme that includes QoS negotiation, QoS calculation, admission control, QoS enforcement, and resource scheduling. The HeiRAT operating system scheduling policy is a rate-monotonic scheme whereby the priority of a system thread performing protocol processing is proportional to the message rate requested.

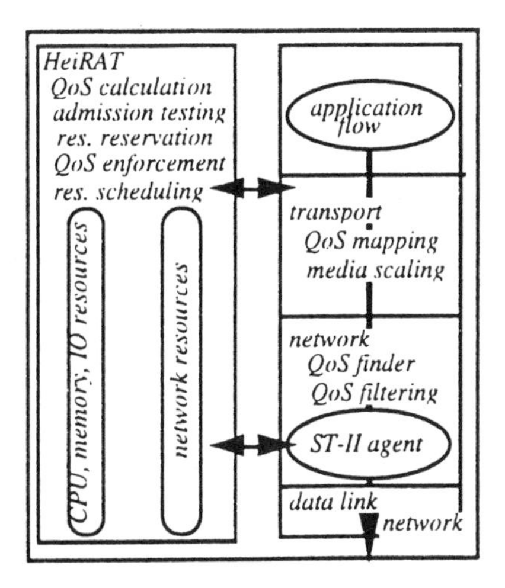

Figure 3.3 Heidelberg QoS model.

The Heidelberg QoS model has been designed to handle heterogeneous QoS demands from individual receivers in a multicast group and to support QoS adaptivity via flow filtering and media scaling techniques. Media scaling [71] and codec translation at the end systems and flow filtering and resource sharing in the network are fundamental to meeting heterogeneous QoS demands. Media scaling matches the source with the receivers' QoS capability by manipulating flows at the network edges. In contrast, filtering accommodates the receivers' QoS capability by manipulating flows at the core of the network as flows traverse bridges, switches, and routers.

3.3.2 XRM

The COMET group at Columbia University is developing an *Extended Integrated Reference Model* (XRM) [9] as a modeling framework for control and management of multimedia telecommunications networks (which comprise multimedia computing platforms and broadband networks). The COMET groups argues that the foundations for operability (i.e., control and management) of multimedia computing and networking devices are equivalent; that is, both classes of devices can be modeled as producers, consumers, and processors of media. The only difference between computing and network devices is the overall goal that a group of devices has set to achieve in the network or end system. The XRM is divided into five distinct planes [55] as illustrated in Figure 3.4:

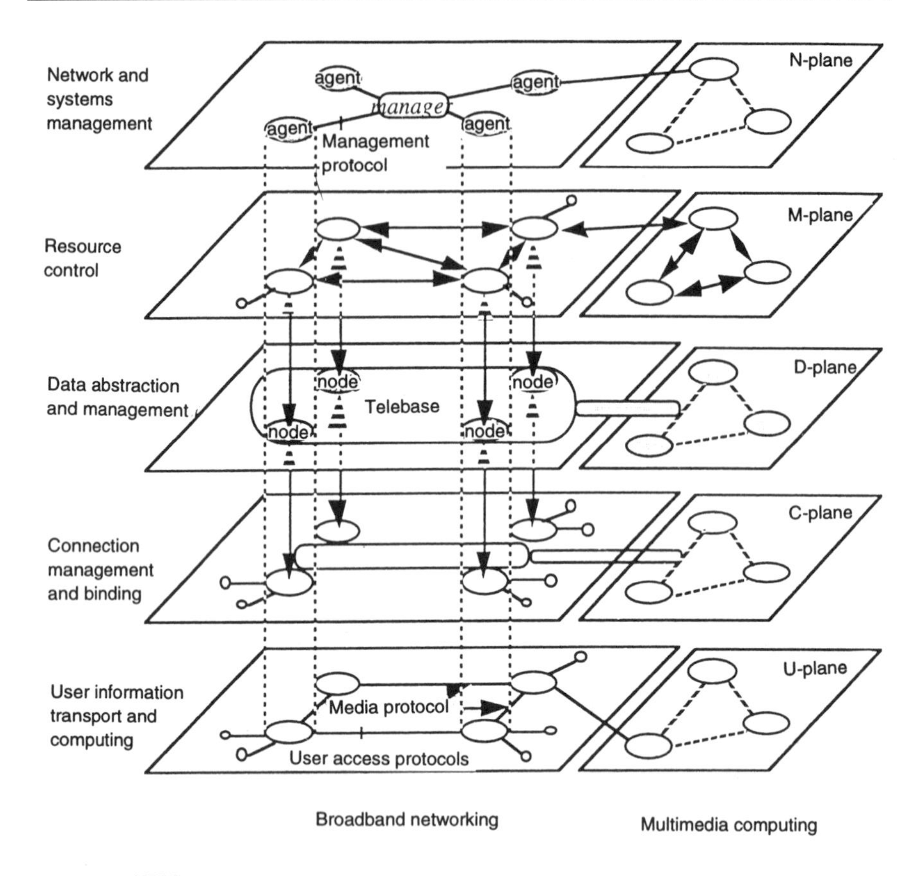

Figure 3.4 XRM.

- The *management function* resides in the network management plane (N-plane) and covers the OSI functional areas of network and system management.
- The *traffic control function* comprises the resource control (M-plane) and connection management and control (C-plane) planes. Resource control constitutes cell scheduling, call admission, call routing in the network, process scheduling, memory management, routing, admission control, and flow control in the end systems.
- The *information transport function* is located in the user transport plane (U-plane), models the media protocols and entities for the transport of user information in both the network and the end systems.
- The *telebase* resides in the data abstraction and management plane (D-plane) and collectively represents the information, data abstractions

existing in the network and end systems. The telebase implements data sharing among all other XRM planes.

The XRM is built on the theoretical work of guaranteeing QoS requirements in ATM networks and end systems populated with multimedia devices. General concepts for characterizing the capacity of network [68] and end-system [59] devices (e.g., disks and switches) have been developed. At the network layer, XRM characterizes the capacity region of an ATM multiplexer with QoS guarantees as a *schedulable region.* Network resources such as switching bandwidth and link capacity are allocated based on four cell-level traffic classes (Classes I, II, III, and C) for circuit emulation, voice and video, data, and network management, respectively. A traffic class is characterized by its statistical properties and QoS requirements. Typically, QoS requirements reflect cell loss and delay constraints. To satisfy efficiently the QoS requirements of the cell level, scheduling and buffer management algorithms dynamically allocate communication bandwidth and buffer space appropriately.

In the end system, flow requirements are modeled through service class specifications with QoS constraints. For example, in the audiovideo unit the service class specification is in terms of JPEG, MPEG-I, MPGE-II video, and CD audio quality flows with QoS guarantees. Quality of service for these classes is specified by a set of frame delay and loss constraints. The methodology of characterizing network resources is extended to the end system to represent the capacity of multimedia devices. Using the concept of a *multimedia capacity region,* the problem of scheduling flows in the end system becomes identical to the real-time bin packing exercise of the network layer. The implementation of XRM, including key resource abstractions (namely, schedulable and multimedia capacity region), is currently being realized as part of a *binding architecture* [9] for open signaling, control, and management of multimedia networks.

3.3.3 OMEGA

During the past 3 years the University of Pennsylvania has been developing an endpoint architecture called the OMEGA architecture [56]. OMEGA is the result of an interdisciplinary research effort that is examining the relationship between application of QoS requirements (which make stringent resource demands) and the ability of local (the operating system) and global resource management (combining communication and remotely managed resources) to satisfy these demands. The OMEGA architecture illustrated in Figure 3.5 assumes a network subsystem that provides bounds on delay errors and can meet bandwidth demands, and an operating system that is capable of providing run-time QoS guarantees. The essence of the OMEGA architecture is resource

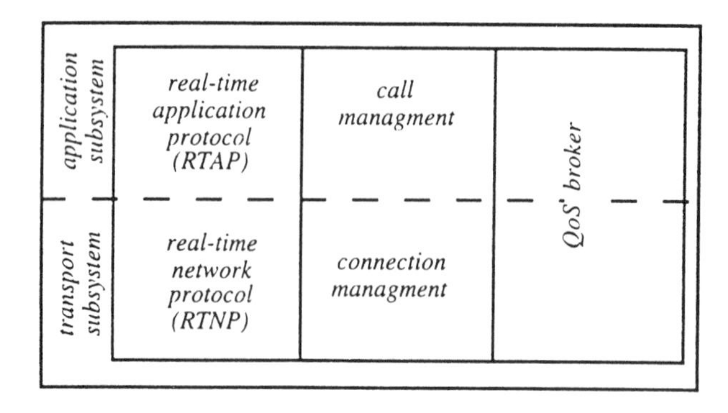

Figure 3.5 OMEGA.

reservation and management of end-to-end resources. Communication is preceded by a call setup phase where application requirements, expressed in terms of QoS parameters, are negotiated, and guarantees are made at several logical levels, such as between applications and the network subsystem, applications and the operating system, and the network subsystem and the operating system. This establishes customized connections and results in the allocation of resources appropriate to meet application requirements and operating system/network capabilities. To facilitate this resource management process, the University of Pennsylvania has also developed a *QoS brokerage model* [74], which incorporates QoS translation, and QoS negotiation and renegotiation (see [75] for full details on similar work on QoS negotiation protocols at the University of Montreal).

3.3.4 Int-Serv Architecture

The work by the Integrated Services (int-serv) Group [43] of the Internet Engineering Task Force (IETF) is a significant contribution to providing controlled QoS for multimedia applications over an integrated services internetwork. The group has defined a comprehensive int-serv architecture [43] and a QoS framework [65] used to specify the functionality of internetwork system elements (known as elements), which make multiple, dynamically selectable QoS available to applications. The behavior of elements, which constitute routers, subnetworks, and endpoint operating systems, is captured as a set of services, of which some or all are offered by each element. Each element is QoS-aware and supports interfaces required by the service definition [43]. The concatenation of service elements along an end-to-end data path provides an overall statement of end-to-end QoS. The following int-serv services are offered in addition to best effort: (1) *controlled delay,* which attempts to provide several levels of delay from which the application can choose; (2) *predicated delay,* which provides a statistical delay bound similar to the Tenet Group's statistical service [30] and the COMET group's guaranteed service [42]; and (3) *guaranteed delay,* which provides an absolute guaranteed delay bound.

Flows in an int-serv architecture are characterized by two specifications: a *traffic specification,* which is a specification of the traffic pattern that a flow expects to exhibit; and a *service request specification,* which is a specification of the QoS a flow desires from a service element. The int-serv architecture, which is restricted to the network but also applicable in the end system, is comprised of four components [43]:

1. A *packet scheduler,* which forwards packet streams using a set of queues and timers;
2. A *classifier,* which maps each incoming packet into a set of QoS classes;
3. An *admission controller,* which implements the admission control algorithm to determine whether a new flow can be admitted or denied; and
4. A *reservation setup protocol* (e.g., RSVP [29]), which is necessary to create and maintain the flow-specific state in the routers along the path of the flow.

In [66] Clark introduces some early work on a *quality of service manager* (QM) as part of the end-system int-serv architecture. As illustrated in Figure 3.6, the QM, which constitutes a user interface, service agents, and dispatcher, presents an abstract management layer designed to isolate applications from underlying details of the specific services provided by a QoS-driven Internet [43]. One motivating factor behind the introduction of a QM is that applications can negotiate desired QoS without needing to know the details of a specific network service, as described earlier. In this case, the QM provides a degree of transparency whereby applications express desired levels of QoS in application-oriented language rather than using communication QoS specifics. The QM is responsible for determining what QoS management capabilities are available on the application's communication path and chooses the path best suited to the application.

3.3.5 QoS-A

The *quality of service architecture* (QoS-A) [54] is a layered architecture of services and mechanisms for quality-of-service management and control of

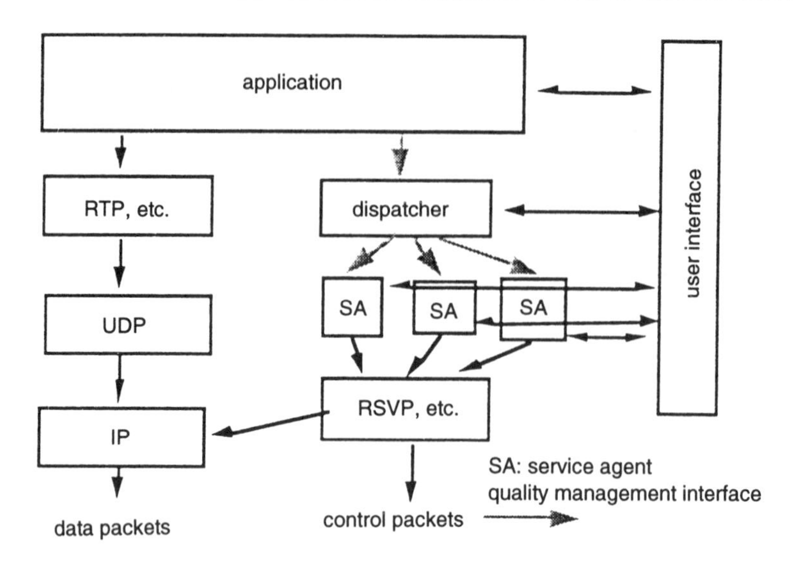

Figure 3.6 Int-serv architecture QoS manager.

continuous media flows in multiservice networks. The architecture incorporates the following key notions: *flows,* which characterize the production, transmission, and eventual consumption of single media streams (both unicast and multicast) with associated QoS; *service contracts,* which are binding agreements of QoS levels between users and providers; and *flow management,* which provides for the monitoring and maintenance of the contracted QoS levels. The realization of the flow concept demands active QoS management and tight integration between device management, end-system thread scheduling, communications protocols, and networks.

In functional terms, the QoS-A (as illustrated in Figure 3.7) is composed of a number of layers and planes. The upper layer consists of a distributed applications platform augmented with services to provide multimedia communications and QoS specification in an object-based environment [5]. Below the platform level is an orchestration layer that provides jitter correction and multimedia synchronization services across multiple related application flows [25]. Supporting this is a transport layer that contains a range of QoS configurable services and mechanisms [15]. Below this, an internetworking layer and lower layers form the basis for end-to-end QoS support.

QoS management is realized in three vertical planes in the QoS-A. The protocol plane, which consists of distinct user and control subplanes, is motivated by the principle of separation. QoS-A uses separate protocol profiles for the control and media components of flows because of the different QoS requirements of control and data. The QoS maintenance plane contains a

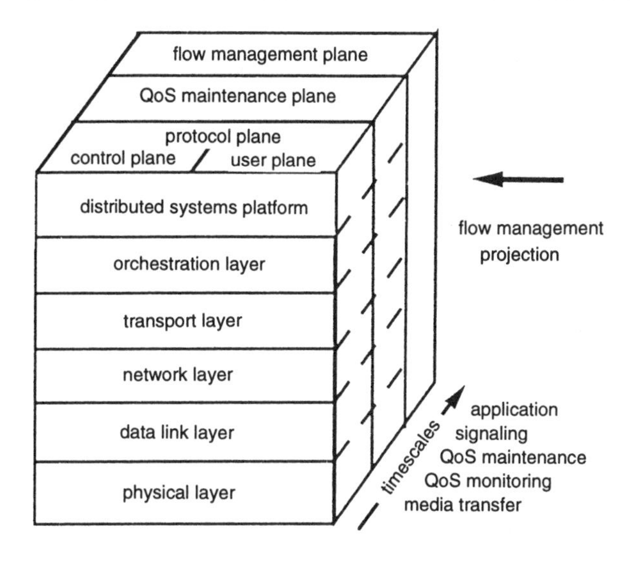

Figure 3.7 QoS-A.

number of layer-specific QoS managers. These are each responsible for the fine-grained monitoring and maintenance of their associated protocol entities. For example, at the orchestration layer [25], the QoS manager is interested in the tightness of synchronization between multiple related flows. In contrast, the transport QoS manager is concerned with intraflow QoS such as bandwidth, loss, jitter, and delay. Based on flow monitoring information and a user-supplied service contract, QoS managers maintain the level of QoS in the managed flow by means of fine-grained resource tuning strategies. The final QoS-A plane pertains to flow management, which is responsible for flow establishment (including end-to-end admission control, QoS-based routing, and resource reservation), QoS mapping (which translates QoS representations between layers), and QoS scaling (which constitutes QoS filtering and QoS adaptation for coarse-grained QoS maintenance control).

3.3.6 OSI QoS Framework

One early contribution to the field of QoS-driven architecture is the *OSI QoS framework* [53], which concentrates primarily on quality-of-service support for OSI communications. The OSI framework broadly defines terminology and concepts for QoS and provides a model which identifies objects of interest to QoS in open system standards. The QoS associated with objects and their interactions is described through the definition of a set of QoS characteristics. The key OSI QoS framework concepts include the following:

- *QoS requirements,* which are realized through QoS management and maintenance entities;
- *QoS characteristics,* which are a description of the fundamental measures of QoS that have to be managed;
- *QoS categories,* which represent a policy governing a group of QoS requirements specific to a particular environment such as time-critical communications; and
- *QoS management functions,* which can be combined in various ways and applied to various QoS characteristics in order to meet QoS requirements.

The OSI QoS framework (as illustrated in Figure 3.8) is made up of two types of management entities, *layer-specific* and *system-wide entities,* that attempt to meet the QoS monitoring requirements by monitoring, maintaining, and controlling end-to-end QoS. The task of the policy control function is to determine the policy that applies at a specific layer of an open system. The policy control function models any priority actions that must be performed to control the operation of a layer. The definition of a particular policy is layer specific and therefore cannot be generalized. Policy may, however, include

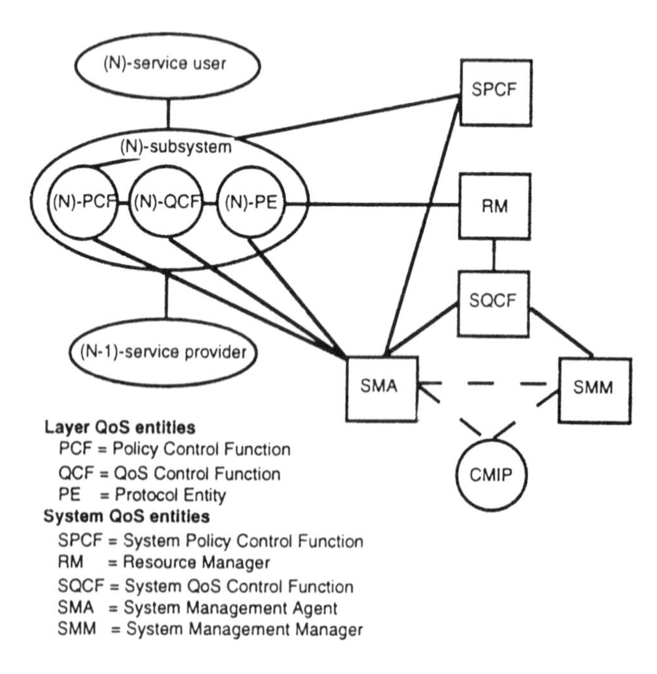

Figure 3.8 OSI QoS framework.

aspects of security, time-critical communications, and resource control. The role of the QoS control function is to determine, select, and configure the appropriate protocol entities to meet layer-specific QoS goals. The system management agent is used in conjunction with OSI systems management protocols to enable system resources to be remotely managed. The local resource manager represents end-system control of resources. The system QoS control function combines two system-wide capabilities: to tune performance of protocol entities and to modify the capability of remote systems via OSI systems management. The OSI systems management interface is supported by the systems management manager, which provides a standard interface to monitor, control, and manage end systems. The system policy control function interacts with each layer-specific policy control function to provide an overall selection of QoS functions and facilities.

3.3.7 Tenet Architecture

The Tenet Group at the University of California at Berkeley has developed a family of protocols [18,30] that runs over an experimental wide-area ATM network. As illustrated in Figure 3.9, the *Tenent architecture* [70] includes a real-time channel administration protocol (RCAP) [32] in addition to real-time Internet protocol (RTIP) and continuous media transport protocol (CMTP) [18]. The former provides generic connection establishment, resource reservation, and signaling functions for the rest of the protocol family. RCAP spans the transport and network layers for overall resource reservation and flow setup. CMTP is explicitly designed for continuous media support. It is a lightweight protocol that runs on top of RTIP and provides sequenced and periodic delivery of continuous media samples with QoS control over throughput, delays, and error bounds. The Tenet Group makes a distinction between deterministic and statistical guarantees for hard real-time and continu-

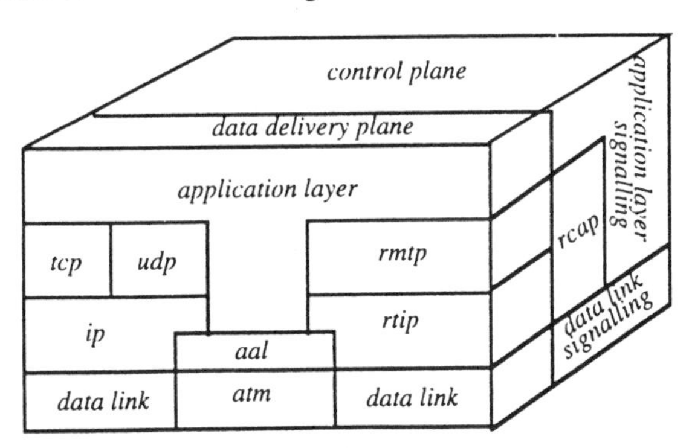

Figure 3.9 Tenet architecture.

ous media flows [31], respectively. In the deterministic case, guarantees provide a hard bound on the performance of all cells within a session. Statistical guarantees promise that no more than x% of packets would experience a delay greater than specified, or no more than x% of cells in a session might be lost.

3.3.8 TINA QoS Framework

The *TINA Architecture* is governed by the separation between telecommunication applications and the TINA distributed processing environment (TDPE). The TDPE software can be visualized as a distributed operating system layer that supports the execution of telecommunication applications. Multimedia services offered by a provider utilize the TDPE and underlying computing and communications capabilities. The TINA QoS framework [62] addresses the specification and realization of QoS support for telecommunications applications. The framework is partly based on the ANSA [8] and CNET QoS frameworks [7]. QoS specification is stated declaratively using the notion of *service attributes* as part of the computational specification. This has been integrated into the TINA-ODL specification language, which provides extensions to OMG-IDL [61]. QoS mechanisms have been specified as part of the TDPE specification for QoS provision and QoS negotiation. These mechanisms consider QoS mapping from the application level to the QoS offered by the TDPE kernel and QoS degradation reports in the case that the contracted QoS fails to meet its agreed targets.

3.3.9 MASI End-to-End Model

The CESAME Project [63] at Laboratoire MASI, Université Pierre et Marie Curie, is developing an architecture for multimedia communications that takes end-to-end QoS support as its primary objective. As with the QoS-A, the MASI architecture (shown in Figure 3.10) offers a generic QoS framework to specify and implement the required QoS requirements of distributed multimedia applications operating over ATM-based networks. The CESAME Project considers end-to-end resource management, which spans the host operating system, host communications subsystem, and ATM networks. The research is motivated by (1) the need to map QoS requirements from the ODP layer to specific resource modules in a clean and efficient manner, (2) the need to resolve multimedia synchronization needs of multiple related ODP streams [4], and (3) the need to provide suitable communication protocol support for multimedia services being developed at Université Pierre et Marie Curie.

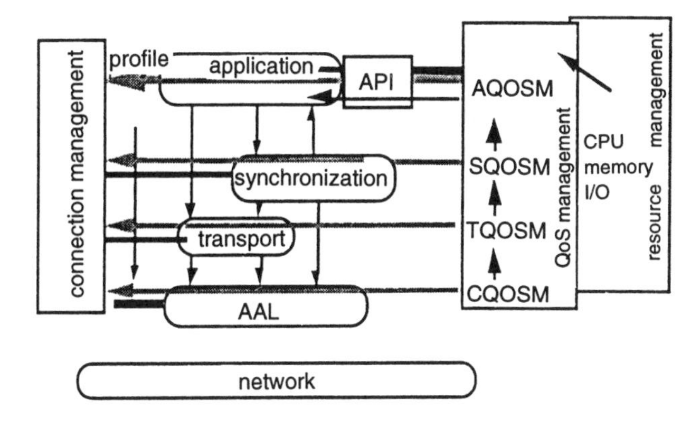

Figure 3.10 MASI schematic.

3.3.10 End-System QoS Framework

At Washington University, Gopalakrishna and Purulkar [58] have developed a QoS framework for providing QoS guarantees within the end system for networked multimedia applications. There are four components of the Washington University end-system QoS framework as illustrated in Figure 3.11: QoS specification, QoS mapping, QoS enforcement, and protocol implementation. QoS specification is at a high level and uses a small number of parameters to allow applications greater ease in specifying their flow requirements. Based on the QoS specification, QoS mapping operations derive resource requirements for each end-to-end application session. Important system resources considered in [58] include the CPU, memory, and network. The third component of the framework is QoS enforcement. QoS enforcement is mainly concerned with providing real-time processing guarantees for media transfer. A real-time upcall

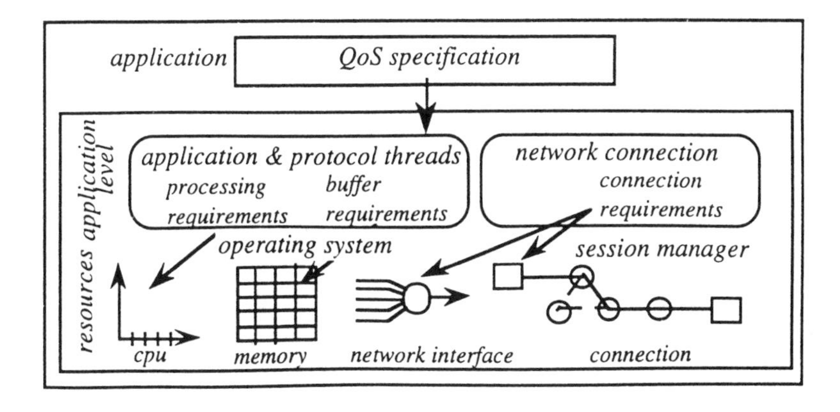

Figure 3.11 End-system QoS framework.

(RTU) facility [67] has been developed for structuring protocols. RTUs are scheduled using a rate monotonic policy [49] with delayed preemption that takes advantage of the iterative nature of protocol processing to reduce context switching overhead and increase end-system scheduling efficiency. The final component of the framework is an application-level protocol implementation model. Protocol code is structured as RTUs with attributes that are derived from high-level specifications by QoS mapping operations.

3.4 Comparison

In this section we present a simple qualitative comparison of the QoS architectures surveyed in Section 3.3. We use the elements of the generalized QoS framework (described in Section 3.2) as a basis for the comparison summarized in Table 3.1.

The legend for the comparison table is as follows:

—	"not addressed"
E/N	"addressed in detail in the end system/network"
(E)/(N)	"mentioned only in the end system/network"
R	"QoS renegotiation addressed in detail"
(R)	"QoS renegotiation mentioned only"
S	"QoS scaling addressed in detail"
D	"QoS degradation addressed in detail"
(D)	"QoS degradation mentioned only"
A	"QoS availability in detail"

3.5 Discussion

All QoS architectures surveyed in Section 3.3 consider the QoS specification (e.g., services contracts, flow specs, and service and traffic classes) to be fundamental in capturing application-level QoS requirements. Although there is a broad consensus on the need for a *flow spec* that captures quantitative performance requirements, there exist two schools of thought on what it should be. On the one hand, XRM and ATM [34] solutions are based on a flow spec that is made up of one or two QoS parameters that identify a traffic class and an average bandwidth. On the other hand, the Tenet, QoS-A, and OMEGA architectures adopt a multivalued flow spec (cf., RFC1633, ST-II, RSVP, HeiTS). Although both of these proposals seem similar, philosophically they are rather different in practice. The COMET group [9] argues that by limiting

Table 3.1
Comparison of QoS Architectures

QoS Framework Elements	QoS Provision			QoS Control					QoS Management	
QoS Models	**Flow Spec and QoS Mapping**	**Adm. Control/ Resource Allocation**	**E2E* Coordination**	**Flow Scheduling**	**Flow Shaping**	**Flow Control**	**QoS Filtering**	**Flow Synchro-nization**	**Monitoring/ Alerts**	**QoS Maintenance**
XRM [9]	E N	E N	(E) N	(E) N	—	N	—	—	N	—
QoS-A [54]	E N	E (N)	E N	E (N)	E	(E)	(E) N	E	E A D	E N R S
ISO [53]	(E) (N)	E N	E N	—	—	—	—	—	E N	E N
Heidelberg [57]	(E) N	E N	E N	E (N)	(E)	(N)	N	—	E D	E R S
TINA [62]	(E)	(N)	N	—	—	—	—	(N)	(N)	—
IETF [43]	E N	—	E	—	—	—	—	—	E N	E N R
Tenet [70]	E N	N	N	N	N	(E)	N	—	E D	E R S
MASI [63]	E (N)	E (N)	E	E	—	—	—	E	E	E
OMEGA [56]	E (N)	E (N)	E (N)	E (N)	E	E	—	—	E	E R
WashU [58]	E	E		E	E	—	—	—	—	E R

*The term *E2E coordination* refers to the coordination of end-system and network resources for flows. This could be provided by a resource reservation protocol (e.g., RSVP [29]), a connection setup protocol (e.g., RCAP [32]), or a signaling protocol (e.g., UNI 4.0 [34]).

a flow spec to a set of well-defined services in the end system and traffic classes in the network, complexity in the end system and network is more manageable. In contrast, the Tenet, QoS-A, and OMEGA architectures consider such an approach unnecessarily limiting. These groups argue that by defining a set of discrete QoS classes, applications may be unduly constrained to conform to a QoS class that may not meet the desired application-level QoS requirements.

Level of service (Section 3.2.2) expresses the degree of certainty at which the QoS levels specified in a flow spec will be honored. Each architecture offers a different set of services to applications. For example, the Washington University QoS framework supports three application classes to which it maps applications-level flows. These include (1) an *isochronous class,* which is suitable for continuous media flows; (2) a *burst class,* which is appropriate for bulk data transfer; and (3) a *low delay class,* which is suitable for applications that require a small response time such as an RPC request. The Washington QoS framework assumes that all applications fall into one of these three general application classes. While all architectures provide services based on both hard (i.e., guaranteed service) and soft (i.e., best effort), QoS guarantees it is difficult to determine which set optimally covers the application base. Additional services found in the literature include the predicted service (IETF), statistical service (Tenet, XRM, and Heidelberg), and the available bit rate service (ATM Forum).

With the exception of the IETF work (which uses an RSVP maintained state), all architectures advocate connection-oriented or "hard state" solutions to network-level QoS provision; that is, hard state couples path establishment and resources reservation. Work in the IETF on an integrated services architecture (using RSVP and IPv6 flows) described in Section 3.3.4 assumes that network-level QoS guarantees can be built using a "soft state" approach; that is, no explicit connection is established but flows traverse intermediate routers on paths that are temporarily (i.e., network state is timed out and periodically refreshed) established. In this instance, path establishment and resource reservation are decoupled. It is argued that a soft state approach provides better scalability, robustness, and eradicates the round-trip call setup time found in connection-oriented approaches. In [47] Turner suggests a hybrid approach called *ATM-soft,* which benefits from the use of the soft state in a native ATM environment. It is still too early to determine which approach is more suitable for future QoS architectures given the need to support both high-end (e.g., telesurgery and time critical applications) and low-end (e.g., video conferencing and audio tools) multimedia applications.

Commonalities exist between QoS control and management strategies found in the end system and network; for example, admission control, resource management, and scheduling mechanisms. The extent to which network-level QoS mechanisms are applicable in the end systems (or vice versa) remains an open issue. End-system and network devices can be modeled in a similar way: The only real difference is the overall goal that end system or network devices are set to achieve. For example, the XRM models the end system as a virtual switch [9] and a set of configurable multimedia devices based on a desk area network (DAN) architecture [85]. It is evident that commonalities exist between scheduling strategies found in switches/routers and end-system operating systems (e.g., fair share techniques can be found in the end-system and network switches/routers). This seems encouraging in the first instance. A counter argument, however, is that end systems have fundamentally different scheduling goals than routers and switches. End systems schedule a wide variety of both isochronous (e.g., continuous media flows) and asynchronous (e.g., RPCs) work, whereas switches and routers are mainly involved with switching/routing of cells/packets. This means that in the end-system application execution times (i.e., a quantum [15] of work as illustrated in Figure 3.2) can vary widely (e.g., uncompressing a video flow is computationally more intensive than displaying video to a screen). In contrast, switch and router schedulers are generally moving packets/cells from queues to ports or vice versa and are optimized for that task. Therefore, techniques resident in switches (such as HRR [39]) may be inappropriate in host operating systems.

3.6 Summary

In this chapter we have argued that multimedia systems designers should adopt an end-to-end approach to meet application-level QoS requirements. To meet this challenge, we have proposed a generalized QoS framework that is motivated by five design principles; that is, the principles of transparency, integration, separation, multiple timescales, and performance. Elements of our generalized framework include QoS specification and static and dynamic QoS management. We have summarized and evaluated key research in QoS architectures for distributed systems and discussed some of the issues that emerged during a comparison of the existing QoS architectures. The work presented in this survey represents a growing body of research that is laying the foundations for future QoS programmable multimedia platforms [9,86]. While the area of QoS research in multimedia networking is mature [87], work on QoS architectures remains in early stages of development with no substantial implementation results having been published to validate the approach. Given that, the work presented in this chapter contributes toward a qualitative understanding of the key principles, services, and mechanisms needed to build QoS into distributed multimedia systems.

References

[1] Hutchison, D., et al., "Quality of Service Management in Distributed Systems," Chap. 11 in *Network and Distributed Systems Management,* M. Sloman, ed., Reading, MA: Addison-Wesley, 1994.

[2] "ANSAware 3.0 Implementation Manual," Castle Park, Cambridge, APM Ltd, Poseidon House, CB3 0RD, UK, 1991.

[3] "Distributed Computing Environment," Open Software Foundation, Cambridge, MA, 1992.

[4] ISO/IEC JTC1/SC21/WG7, "Draft Recommendations X.903: Basic Reference Model of Open Distributed Processing," International Standards Organization, 1992.

[5] Coulson, G., et al., "Extensions to ANSA for Multimedia Computing," *Computer Networks & ISDN Syst.,* Vol. 25, No. 11, 1992, pp. 305–323.

[6] Nicolaou, C., "An Architecture for Real-Time Multimedia Communications Systems," *IEEE J. Sel. Areas Commun.,* Vol. 8, No. 3, Apr. 1990.

[7] Hazard, L., F. Horn, and J. B. Stefani, "Towards the Integration of Real-Time and QoS Handling in ANSA," CNET Report CNET.RC.ARCADE.01, June 1993.

[8] Guangxing, A., "A Model of Real-Time QoS for ANSA," Technical Report APM.1151.00.04, Cambridge, UK: APM Ltd., Mar. 1994.

[9] Lazar, A. A., S. Bhonsle, and K. S. Lim, "A Binding Architecture for Multimedia Networks," *Proc. COST-237 Conf. Multimedia Transport and Teleservices,* Vienna, Austria, 1994.

[10] Bulterman, D. C., and R. van Liere, "Multimedia Synchronisation and UNIX," *Proc. Second Int. Workshop on Network and Operating Systems Support for Digital Audio and Video,* Heidelberg, Springer Verlag, 1991.

[11] Leslie, I. M, D. McAuely, and S. J. Mullender, "Pegasus—Operating Systems Support for Distributed Multimedia Systems," *Operating Syst. Rev.,* Vol. 27, No. 1, 1993.

[12] Govindan, R., and D. P. Anderson, "Scheduling and IPC Mechanisms for Continuous Media," *Proc. Thirteenth ACM Symp. Operating Systems Principles, Asilomar Conf. Center,* Pacific Grove, CA, 1991, SIGOPS, Vol. 25, pp. 68–80.

[13] Jeffay, K., "The Real-Time Producer/Consumer Paradigm: A Paradigm for Construction of Efficient, Predictable Real-Time Systems," *Proc. 1993 ACM/SIGAPP Symp. Applied Computing,* Indianapolis, IN, Feb. 1993.

[14] Tokuda, H., and T. Kitayama, "Dynamic QoS Control Based on Real-Time Threads," *Proc. Fourth Int. Workshop on Network and Operating Systems Support for Digital Audio and Video,* Lancaster University, Lancaster, UK, 1993.

[15] Coulson, G., A. Campbell, and P. Robin, "Design of a QoS Controlled ATM Based Communication System in Chorus," *IEEE J. Sel. Areas Commun.,* Special Issue on ATM LANs: Implementation and Experiences with Emerging Technology, Vol. 13, No. 4, May 1995, pp. 686–699.

[16] Jeffay, K., and D. Bennett, "A Rate-Based Execution Abstraction for Multimedia Computing," *Proc. Fifth Int. Workshop on Network and Operating Systems Support for Digital Audio and Video,* Durham, NH, Apr. 1995.

[17] Danthine, A., et al., "The OSI 95 Connection-Mode Transport Service—Enhanced QoS," *Proc. 4th IFIP Conf. High Performance Networking,* University of Liege, Belgium, December 1992.

[18] Wolfinger, B., and M. Moran, "A Continuous Media Data Transport Service and Protocol for Real-time Communication in High Speed Networks," paper presented at Second International Workshop on Network and Operating System Support for Digital Audio and Video, IBM ENC, Heidelberg, Germany, Nov. 1991.

[19] Feldmeier, D., "A Framework of Architectural Concepts for High Speed Communication Systems," Morristown, NJ: Computer Communication Research Group, Bellcore, May 1993.

[20] Doeringer, W., et al., "A Survey of Light-weight Transport Protocols for High-speed Networks," *IEEE Trans. Commun.,* Vol. 38, No. 11, Nov. 1990, pp. 2025–2039.

[21] Chesson, G., "XTP/PE Overview," *Proc. 13th Conf. Local Computer Networks,* Minneapolis, MN, 1988, pp. 292–296.

[22] Clark, D. D., M. L. Lambert, and L. Zhang, "NETBLT: A High Throughput Transport Protocol," *Computer Commun. Rev.,* Vol. 17, No. 5, 1987.

[23] Hehmann, D. B., et al., "Implementing HeiTS: Architecture and Implementation Strategy of the Heidelberg High Speed Transport System," paper presented at Second International Workshop on Network and Operating System Support for Digital Audio and Video, IBM ENC, Heidelberg, Germany, Nov. 1991.

[24] Schulzrinne, H., and S. Casner, "RTP: A Transport Protocol for Real-Time Applications," Work in Progress, Internet Draft, <draft-ietf-avt-rtp-05.ps>, 1995.

[25] Campbell, A. T., et al., "A Continuous Media Transport and Orchestration Service," *Proc. ACM SIGCOMM '92,* Baltimore, MD, 1992, pp. 99–110.

[26] Keshav and Saran, "Semantics and Implementation of a Native-Mode ATM Protocol Stack," Bell Labs Technical Memorandum, http://www.cs.att.com/csrc/keshav/papers.html, 1994.

[27] Topolcic, C., "Experimental Internet Stream Protocol, Version 2 (ST-II)," Internet Request for Comments No. 1190 RFC-1190, Oct. 1990.

[28] Clark, D. D., S. Shenker, and L. Zhang, "Supporting Real-Time Applications in an Integrated Services Packet Network: Architecture and Mechanism," *Proc. ACM SIGCOMM '92,* Baltimore, MD, Aug. 1992, pp. 14–26.

[29] Zhang, L., et al., "RSVP Functional Specification," Working Draft, draft-ietf-rsvp-spec-10.ps, 1996.

[30] Ferrari, D., and D. C. Verma, "A Scheme for Real-Time Channel Establishment in Wide-Area Networks," *IEEE J. Sel. Areas Commun.,* Vol. 8, No. 3, 1990, pp. 368–377.

[31] Ferrari, D., J. Ramaekers, and G. Ventre, "Client-Network Interactions in Quality of Service Communication Environments," *Proc. 4th IFIP Conf. High Performance Networking,* University of Liege, Belgium, Dec. 1992.

[32] Benerjea, A., and B. Mah, "The Real-Time Channel Administration Protocol," paper presented at Second International Workshop on Network and Operating System Support for Digital Audio and Video, IBM ENC, Heidelberg, Germany, November 1991.

[33] Zhang, H., and S. Keshav, "Comparison of Rate-Based Service Disciplines," *Proc. ACM SIGCOMM,* 1991.

[34] "ATM User-Network Interface Specifications," Version 4.0, ATM Forum, 1996.

[35] Shenker, S., D. Clark, and L. Zhang, "A Scheduling Service Model and a Scheduling Architecture for an Integrated Service Packet Network," Working Draft available via anonymous ftp from parcftp.xerox.com:/transient/service-model.ps.Z, 1993.

[36] Zitterbart, M., B. Stiller, and A. Tantawy, "A Model for Flexible High-Performance Communication Subsystems," *IEEE J. Sel. Areas Commun.*, Vol. 11, No. 4, May 1992, pp. 507–518.

[37] Parekh, A., and R. G. Gallagher, "A Generalised Processor Sharing Approach to Flow Control in Integrated Services Networks—The Multiple Node Case," *Proc. IEEE INFOCOM'93,* San Francisco, Apr. 1993, pp. 521–530.

[38] Golestani, S. J., "A Stop and Go Queueing Framework for Congestion Management," *Proc. ACM SIGCOMM'90,* San Francisco, June 1990.

[39] Keshav, S., "On the Efficient Implementation of Fair Queueing," *Internetworking Res. Exper.,* Vol. 2, 1991, pp. 157–173.

[40] Guerin, R., H. Ahmadi, and M. Naghshineh, "Equivalent Capacity and Its Application to Bandwidth Allocation in High Speed Networks," *IEEE J. Sel. Areas Commun.,* Vol. 9, No. 7, Sept. 1991.

[41] Cruz, R., "A Calculus for Network Delay: Part I: Network Elements in Isolation," *IEEE Trans. Info. Theory,* Vol. 37, No. 1, Jan. 1991.

[42] Hyman, J., A. A. Lazar, and G. Pacifici, "Real-Time Scheduling with Quality of Service Constraints," *IEEE J. Sel. Commun.,* Vol. 9, No. 7, Apr. 1990.

[43] Braden, R., D. Clark, and S. Shenker, "Integrated Services in the Internet Architecture: An Overview," Request for Comments, RFC-1633, 1994.

[44] Floyd, S., "Link-Sharing and Resource Management Models for Packet Networks," Draft available via anonymous ftp from ftp.ee.lbl.gov: link.ps.Z, Sept. 1993.

[45] Jain, R., "Congestion Control and Traffic Management in ATM Networks: Recent Advances and a Survey," *Computer Networks & ISDN Syst.*, Vol. 28, No. 13, 1996, pp. 1723–1738.

[46] Dabbous, W., and C. Diot, "High Performance Protocol Architectures," Technical Report, Sophia Antipolis, France: INRIA, 1995.

[47] Turner, J., "ATM-Soft: A Mini-Proposal for ATM Network Control Using Soft State," Technical Note, Washington University, 1995.

[48] Liu, C., and J. Layland, "Scheduling Algorithms for Multiprogramming in Hard Real Time Environment," JACM, 1973.

[49] Stankovic, J. A., et al., "Implications of Classical Scheduling Results for Real-Time Systems," *IEEE Computer,* Special Issue on Scheduling and Real-Time Systems, Vol. 28, No. 6, June 1995, pp. 16–25.

[50] Little, T. D. C., and A. Ghafoor, "Synchronisation Properties and Storage Models for Multimedia Objects," *IEEE J. Sel. Areas Commun.,* Vol. 8, No. 3, Apr. 1990, pp. 229–238.

[51] Escobar, J., D. Deutsch, and C. Partridge, "Flow Synchronisation Protocol," *IEEE GLOBECOM'92,* Orlando, FL, Dec. 1992.

[52] Kurose, J. F., "Open Issues and Challenges in Providing Quality of Service Guarantees in High Speed Networks," *ACM Computer Commun. Rev.,* Vol. 23, No. 1, Jan. 1993, pp. 6–15.

[53] ISO/IEC JTC1/SC21/WG1 N9680, "Quality of Service Framework," International Standards Organization, UK, 1995.

[54] Campbell, A. T., et al., "Integrated Quality of Service for Multimedia Communications," *Proc. IEEE INFOCOM'93,* San Francisco, Apr. 1993, pp. 732–739.

[55] Lazar, A. A., "A Real-Time Control, Management, and Information Transport Architecture for Broadband Networks," *Proc. Int. Zurich Sem. Digital Communications,* 1992, pp. 281–295.

[56] Nahrstedt, K., and J. Smith, "Design, Implementation and Experiences of the OMEGA End-Point Architecture," Technical Report (MS-CIS-95-22), University of Pennsylvania, 1996.

[57] Volg, C., et al., "HeiRAT—Quality of Service Management for Distributed Multimedia Systems," *Multimedia Syst. J.,* 1996.

[58] Gopalakrishna, G., and G. Parulkar, "Efficient Quality of Service in Multimedia Computer Operating Systems," Report WUCS-TM-94-04, Department of Computer Science, Washington University, Aug. 1994.

[59] Lazar, A. A., "Challenges in Multimedia Networking," *Proc. Int. Hi-Tech Forum,* Osaka, Japan, Feb. 1994.

[60] Campbell, A. T., G. Coulson, and D. Hutchison, "A Quality of Service Architecture," *ACM Computer Commun. Rev.,* April 1994.

[61] OMG, "The Common Object Request Broker: Architecture & Specification," Revision 1.3., OMG, Dec. 1993.

[62] TINA-C, "The QoS Framework," Internal Technical Report, 1995.

[63] Besse, L., et al., "Towards an Architecture for Distributed Multimedia Application Support," *Proc. Int. Conf. Multimedia Computing and Systems,* Boston, May 1994.

[64] Sluman, C., "Quality of Service in Distributed Systems," BSI/IST21/-/1/5:33, British Standards Institution, Oct. 1991.

[65] Shenker, S., and J. Wroclawski, "Network Element Specification Template," Internet Draft, June 1995.

[66] Slides from IETF meeting 31, Integrated Service Working Group, ftp://mercury.-lcs.mit.edu/pub/intserv, 1995.

[67] Gopalakrishna, G., and G. Parulkar, "A Real-Time Upcall Facility for Protocol Processing with QoS Guarantees," paper presented at 15th ACM Symposium on Operating Systems Principles, Dec. 1995.

[68] Hyman, J., A. A. Lazar, and G. Pacifici, "Joint Scheduling and Admission Control for ATS-based Switching Node," *Proc. ACM SIGCOMM'92,* Baltimore, MD, Aug. 1992.

[69] Lazar, A. A., L. H. Ngoh, and A. Sahai, "Multimedia Networking Abstraction with Quality of Services Guarantees," *Proc. SPIE Conf. Multimedia Computing and Networking,* San Jose, CA, Feb. 1995.

[70] Ferrari, D., "The Tenet Experience and the Design of Protocols for Integrated Services Internetworks," *Multimedia Syst. J.,* 1996.

[71] Delgrossi, L., et al., "Media Scaling for Audiovisual Communication with the Heidelberg Transport System," *Proc. ACM Multimedia '93,* Anaheim, CA, Aug. 1993.

[72] Campbell, A. T., and G. Coulson, "Transporting QoS Adaptive Flows," *Multimedia Syst. J.,* 1996.

[73] Anderson, D. P., R. G. Herrtwich, and C. Schaefer. "SRP: A Resource Reservation Protocol for Information Center, Menlo Park, CA: SRI International, Sept. 1990.

[74] Nahrstedt, K., and J. Smith, "The QoS Broker," *IEEE Multimedia*, Vol. 2, No. 1, Oct. 1995, pp. 53–67.

[75] Vogel, A., et al., "Distributed Multimedia and Quality of Service—A Survey," *IEEE Multimedia*, Vol. 2, No. 2, 1995, pp. 10–19.

[76] Miloucheva, I., "Quality of Service Research for Distributed Multimedia Applications," ACM Pacific Workshop on Distributed Multimedia Systems, 1995.

[77] Bansal, V., et al., "Adaptive QoS-based API for Networking," paper presented at Fifth International Workshop on Network and Operating System Support for Digital Audio and Video, Durham, NH, Apr. 1995.

[78] Saltzer, J., D. Reed, and D. Clark, "End-to-end Arguments in Systems Design," *ACM Trans. Computer Syst.*, Vol. 2, No. 4, 1984.

[79] Tennenhouse, D. L., "Layered Multiplexing Considered Harmful," in *Protocols for High-Speed Networks*, New York: Elsevier Science Publishers (North-Holland), 1990.

[80] Clark, D., and D. L. Tennenhouse, "Architectural Consideration for a New Generation of Protocols," *Proc. ACM SIGCOMM '90*, Philadelphia, 1984.

[81] Partridge, C., "A Proposed Flow Specification," Internet Request for Comments, No. 1363, Network Information Center, Menlo Park, CA: SRI International, Sept. 1990.

[82] Kelly, F. P., "On Tariffs, Policing and Admission Control for Multiservice Networks," *Proc. Multiservice Networks '93*, Cosener's House, Abingdon, July 1993, and Internal Report, Statistical Laboratory, University of Cambridge, England, 1993.

[83] Pacifici, G., and R. Stadler, "An Architecture for Performance Management of Multimedia Networks," *Proc. IFIP/IEEE Int. Symp. Integrated Network Management*, Santa Barbara, CA, May 1995.

[84] Jacobson, V., "VAT: Visual Audio Tool," VAT manual pages, Feb. 1993.

[85] Hayter, M., and D. McAurely, "The Desk Area Network," *ACM Operating Syst. Rev.*, October 1991.

[86] Campbell, A. T., "Making Multimedia Networks Programmable," Workshop on Gigabit Network Technology Distribution, Washington, St. Louis, 1996.

[87] Keshav, S., "Report on the Workshop on Quality of Service Issues in High Speed Networks," *ACM Computer Commun. Rev.*, Vol. 22, No. 1, Jan. 1993, pp. 6–15.

Further Reading

Kanakia, H., P. Mishra, and A. Reibman, "An Adaptive Congestion Control Scheme for Real Time Packet Video Transport," *Proc. ACM SIGCOMM '93*, San Francisco, Oct. 1993.

Miloucheva, I., "Quality of Service Research for Distributed Multimedia Applications," paper presented at ACM Pacific Workshop on Distributed Multimedia Systems, 1995.

Tennenhouse, D. L., et al., "A Software Oriented Approach to the Design of Media Processing Environments," *Proc. IEEE Int. Conf. Multimedia Computing and Systems*, Boston, 1994.

Vogel, A., et al., "Distributed Multimedia Applications and Quality of Service—A Survey," *IEEE Multimedia*, 1994.

4

Traffic Management and Control

4.1 Introduction

Efficient utilization of network resources is essential in the provision of cost-effective multimedia services where quality-of-service (QoS) requirements for the different network applications are met and are appropriately differentiated. Efficient traffic management and efficient network design are strongly related. During the network planning phase, it is important to consider the benefit provided by traffic management (e.g., routing and congestion control) in the efficient use of network resources. When a decision is made regarding which traffic management functions are required in the network, the network topology and design need to be considered. For example, if the network is overdimensioned, the network provider may not need complex and sophisticated traffic management features. In some cases, when the new sophisticated features are not available, or are available at a high cost, a decision may be made to use only simple traffic management schemes and to invest more in transmission and switching. Research and development of both traffic management and network dimensioning technologies rely heavily on understanding of QoS requirements of the different services, understanding of the statistical characteristics of the traffic, and knowledge of performance evaluation methods. This chapter provides the reader with background on the research and standardization efforts that have taken place in these areas as well as on current research issues and open problems.

4.2 Source Traffic Modeling

The goal in source traffic modeling is to find a traffic model, defined by a small set of statistics, that can still capture the significant statistical information of the real (modeled) traffic, so that an accurate queueing performance evaluation (e.g., evaluation of loss probability or delay) can be obtained if the model is used as input into a queueing system instead of the real traffic. Before we review some of the traffic models that have been proposed, analyzed, and tested in the state of the art, let us begin with a qualitative discussion, including several simple examples, of the significant statistical characteristics of a traffic stream and their effect on queueing behavior.

For simplicity, as is commonly done, we consider a single server queue with a finite buffer. Let time be divided into consecutive fixed-length intervals. Let A_n be the amount of work measured in units of ATM cells, arriving in the nth interval. Hence, the process $\{A_n, n = 1, 2, 3, \ldots\}$ is the input process. It is assumed to be stationary. For convenience let the service rate be equal to one; and let the buffer size be equal to b. This queueing model resembles an output queue in an ATM switch, where the fixed-length intervals correspond to cell-times (each cell-time is the time it takes to transmit an ATM cell), and the service rate is equal to one cell per cell-time. Henceforth, we use the word *cell-time* to refer to this time interval, and the word *cells* to designate the units of the amount of work arriving for service.

It is clear that the arrival rate, defined as the average amount of work that arrives in an interval (the mean of A_n), is a significant statistic—the higher the arrival rate, the higher the loss probability. We also know that the variance of A_n is significant. This is evident if you examine the following example. Let us consider two alternatives of the $\{A_n\}$ process. In the first, $A_n = 1$ for n odd and $A_n = 0$ for n even. In the second, $A_n = 1{,}000$ for $n = 1, 2{,}001, 4{,}001, \ldots$, and $A_n = 0$ otherwise. In both cases the arrival rate is equal to 0.5, but the second alternative will lead to many more losses than the first. In fact, there are no losses in the first alternative but 900 out of each batch of 1,000 are lost in the second alternative, because only 100 (out of a batch of 1,000) can be admitted into the buffer. Clearly, the variance of A_n is much higher in the second alternative. We can see that the variance (the variability in the traffic) adversely affects queueing performance.

Consider two cases where both the mean and the variance of $\{A_n\}$ are equal, but we can still observe significant differences in queueing performance. In the first case, let $A_n = 100$ for $n = 1, 101, 201, \ldots$, and $A_n = 0$ otherwise; in the second case, let $A_n = 100$ for $n = 1, 2, \ldots, 100, 10{,}001, 10{,}002, \ldots, 10{,}100, 20{,}001, 20{,}002, \ldots, 20{,}100, \ldots$, and $A_n = 0$ otherwise. In the first case all the traffic will be stored in the buffer and served with no loss, while in the second case again most of the traffic will be lost. The question is this: What is the difference between the two cases that causes the devastating losses in the second case with no loss at all in the first? The answer is that the correlation is much higher in the second case. We can conclude that heavy correlation may have a devastating effect on queueing performance. (The exact calculation of the variance and the autocorrelation function in both cases is left as an exercise for the reader.) Moreover, notice that although the variance of the amount of work that arrives in one time interval is the same in both cases, the variance of the amount of work that arrives in larger time intervals is much greater in the second case. If we consider a time interval of 100 cell-times, no more than 100 cells can arrive in such an interval in the first case, while up to 10,000 cells may arrive in a 100 cell-time interval in the second case. Well, input of 10,000 and output of 100 during such an interval must create devastating losses.

Now that we appreciate the effect of correlation or equivalently the variance of the amount or work (number of cells) arriving in large time intervals, it becomes apparent that traffic statistics that can capture this effect would be useful. Let S_n be a random variable representing the amount of work that arrives in n consecutive intervals; so, $\mathrm{VAR}\{S_n\}$ denotes the variance of the amount of work that arrives in n consecutive intervals. As discussed, of particular interest is the value of $\mathrm{VAR}\{S_n\}$ for large n. That is, the variance of the amount or work arriving in large time intervals. However, the limit of $\mathrm{VAR}\{S_n\}$ as n approaches infinity is usually infinite, and the statistic of interest is $\mathrm{VAR}\{S_n\}/n$ as n approaches infinity. This is called the *asymptotic variance rate* (AVR) [1]. The AVR is equivalent to the index of dispersion for counts (IDC) [2], which is very well known. There is a very simple linear relationship between the AVR and the IDC: AVR = $\lambda \times$ IDC, where λ is the mean of A_n.

From measurements of real traffic streams, it has been observed that in many cases both Ethernet and VBR video traffic streams exhibit the phenomenon referred to as *long-range dependence* or *self-similarity* (see [3, 4] and references therein), and traffic which behaves in that way is usually referred to as *fractal* traffic. Such traffic will be bursty over all timescales, and the variance $\mathrm{VAR}\{S_n\}$ as a function of n, due to the long-range dependence, will experience a faster than linear growth. This means that, in this case, the AVR or the IDC as defined is infinite. Therefore, researchers have considered a time-dependent version of the IDC instead of its limit value (see [5,6]). That is,

$$\mathrm{IDC}(n) = \mathrm{VAR}\{S_n\}/\lambda_n \tag{4.1}$$

Another important parameter that is very much related to fractal traffic is the *Hurst parameter*. It can be defined as the value of H ($0 \le H \le 1$) that satisfies the following limit as n approaches infinity for some constant α:

$$\mathrm{VAR}(S_n)/\alpha n^{2H} \to 1 \tag{4.2}$$

By observing (4.2), we can see that if $H > 0.5$ we have long-range dependence, and the variance as a function of n experiences a faster than linear growth. The Hurst parameter can be used as a measure of burstiness. If it is closer to 1, it represents strong long-range correlation and burstiness; if closer to 1/2, the process does not exhibit long-range correlation, but may exhibit short (finite) range correlation, which may still adversely affect queueing performance. The case $H < 1/2$ is related to negative correlation [7].

The art of modeling, of any system or object, usually involves an understanding of the different components of that system or object. When we analyze a traffic stream generated by a source, we observe that three different components are related to the hierarchy of timescales: (1) call scale, (2) burst scale, and (3) cell scale [8]. In addition to these three lower timescales, there are higher layer timescales related, for example, to work habits, holiday periods, etc.

An interesting question is how self-similar traffic occurs. The answer is probably related to the hierarchy of timescales and to the different levels of activities that characterize multimedia traffic. For example, the variance of the amount of work arriving in a timescale of seconds can be high in a VBR video connection due to the fact that some videoframes carry much more information than others (e.g., due to scene changes). Some Web browsers generate more traffic than others transmitting complex images. Some FTPs generate more traffic than others simply because some data files are much larger than others. On a higher timescale, burstiness is created by busy hours, time variability in work pressures and requirements, banks transmitting their files and updating their records at night, disasters, and special occasions that generate unusual amounts of traffic.

In this section we begin with a general description of several mathematical (fractal and nonfractal) source models in Section 4.2.1, and then in Section 4.2.2 we consider real applications and discuss their statistical characteristics and suitable models.

4.2.1 Mathematical Models

A large number of traffic models have been proposed over the years that attempt to capture the different statistical characteristics of real traffic for the purpose of performance evaluation and dimensioning. We discuss here several models that have been popular during the last 15 years. The models we discussed can be divided into two categories: (1) self-similar, which exhibits long-range dependence; and (2) non-self-similar, which typically has an exponentially decaying correlation function. We begin with several models of the second category: the on/off model, the Markov modulated process, and the autoregressive Gaussian process, before we move on to the first category and present several self-similar traffic models.

4.2.1.1 On/Off Model

According to the *on/off model*, traffic is alternating between two states: *on* and *off*. During an *on* period the traffic is generated (transmitted) at a constant rate r, and during the *off* period no traffic is generated. The lengths of the *on* and *off* periods are independent. The *on* periods have a common distribution, and the *off* periods have another common distribution.

Let $E(on)$ and $E(off)$ be the means of the *on* and *off* periods, respectively. Then the probability that the process is in the *on* state is given by

$$P_{on} = \frac{E(on)}{E(on) + E(off)} \tag{4.3}$$

Let X_t be a process representing the state of the on/off process at time t with $X_t = 1$ and $X_t = 0$ denoting the process being *on* and *off* at time t, respectively. Let R_t be the process representing the rate of the on/off process at time t. Clearly, $R_t = rX_t$. Because the random variable X_t is a Bernoulli random variable with parameter P_{on}, the mean of R_t is given by $E(R_t) = rP_{on}$ and its variance is given by $\text{var}(R_t) = r^2(P_{on})(1 - P_{on})$.

The latter two simple equations for the mean and the variance of an on/off process are very useful in traffic modeling. In many cases, we are given the peak cell rate (PCR) (which is r in our case) and the mean cell rate. Using these two equations, we can estimate the variance, which can be further used when we consider many independent sources multiplexed together. In this case, we can benefit from the fact that the variances of independent random variables are additive and calculate the variance of the total multiplexed traffic. Such a variance could be an important parameter for evaluation of CLR. Notice that so far, we have not made any assumption concerning the distribution of the *on* and the *off* periods, so that the result may also be applicable to a wide range of processes including those with long-range dependence.

If we further assume that the *on* and *off* periods are exponentially distributed, then results for the densities of the amount of work arriving in any time interval t can also be obtained [9].

4.2.1.2 Markov Modulated Model

A Markov modulated process is a process in which the arrival rate is modulated by an embedded Markov chain. The advantage of Markov modulated processes is that they are flexible and allow for the incorporation of different correlation patterns. Markov modulated processes are defined at both continuous and discrete time intervals. They can also be classified based on the possibility of group arrivals. The most used Markov modulated processes are:

- *Batch Markovian arrival process (BMAP)*: Continuous time with batch arrivals;
- *Markov modulated Poisson process (MMPP)*: Continuous time with single arrivals;
- *Discrete-time batch Markovian arrival process (D-BMAP)*: Discrete time with batch arrivals;
- *Markov modulated Bernoulli process (MMBP)*: Discrete time with single arrivals.

4.2.1.3 Gaussian Autoregressive Model

A traffic model based on a Gaussian process can be described as a traffic process where the amount of traffic generated within any time interval has a Gaussian distribution. There are several ways to represent a Gaussian process. The Gaussian autoregressive is one of them. Also, in many engineering applications, the Gaussian process is described as a continuous time process. In this section, we define the process as a discrete time one. A discrete time representation may be justified by the nature of ATM where time is usually considered discrete with the smallest time unit being the cell-time.

Let time be divided into fixed-length intervals. Let A_n be a continuous random variable representing the amount of work entering the system during the *n*th interval. The variable A_n may represent the number of bits or ATM cells entering the system during the *n*th interval.

According to the Gaussian autoregressive model we assume that A_n, $n = 1, 2, 3, \ldots$, is a *k*th-order autoregressive process, that is,

$$A_n = a_1 A_{n-1} + a_2 A_{n-2} + \cdots + a_k A_{n-k} b \tilde{U}_n \tag{4.4}$$

where $\tilde{U}_n$ is Gaussian with mean η and variance $\tilde{\sigma}^2$, and a and b are real numbers with $|a| < 1$.

To characterize real traffic, we need to find the best fit for the parameters $a_1, \ldots, a_k, b$, as well as η and $\tilde{\sigma}$. On the other hand, it has been shown in [1,10,11] that in any Gaussian process only three parameters are sufficient to estimate queueing performance. It is therefore sufficient to reduce the complexity involved in fitting all the parameters required in a case of the *k*th-order autoregressive model and use only the first-order autoregressive process. In this case we assume that the A_n process is given by

$$A_n = aA_{n-1} + b\tilde{U}_n \tag{4.5}$$

where $\tilde{U}_n$ is Gaussian with mean η and variance $\tilde{\sigma}^2$, and a and b are real numbers with $|a| < 1$. This model was proposed in [12] for a VBR traffic stream generated by a single source of video telephony. (In [12] $\tilde{\sigma}^2 = 1$.) Let $\lambda = E\{A_n\}$ and $\sigma^2 = \text{VAR}\{A_n\}$.

The A_n values can be negative with positive probability. This may seem to hinder the application of this model to real traffic processes. However, in modeling traffic we are not necessarily interested in a process that is similar in every detail to the real traffic. What we are interested in is a process which has the property that when it is fed into a queue, the queueing performance is sufficiently close to that of the queue fed by the real traffic.

4.2.1.4 Self-Similar Models

There are many ways to generate self-similar traffic. We present three models, each of which has its own justification to be a "realistic" traffic model. The first will be based on the on/off model, the second on the M/Pareto/∞ model, and the third is the fractional Brownian motion (FBM) process, which is in fact a Gaussian process with long-range dependence characteristics.

An on/off model with the *off* (silence) period and/or the *on* (activity) period having a distribution with a heavy tail is self-similar. An example of such a distribution is the Pareto distribution. A random number X has a Pareto distribution with parameters $\theta > 0$ and $1 < \gamma < 2$ if

$$P(X > t) = [\theta/(t + \theta)]^{\gamma} \qquad \text{for } t > 0 \tag{4.6}$$

A heavy-tailed distribution of the *on* period can be justified by transmission of extremely long files (e.g., complex images) occasionally by a source. A heavy-tailed *off* period is related to human/business behavior such as people going on holiday, conferences, etc. To reduce the number of parameters it is convenient to use the Pareto distribution with $\theta = 1$, and we talk about Pareto distribution with parameter $1 < \gamma < 2$. In this case the complementary distribution function is given by

$$P(X > t) = [(t + 1)]^{-\gamma} \qquad \text{for } t > 0 \tag{4.7}$$

The mean of the Pareto distribution is given by

$$E\{X\} = \int_0^{\infty} x(x + 1)^{-\gamma}\, dx = 1/(\gamma - 1) \tag{4.8}$$

The M/Pareto/∞ model has been used to model traffic generated by a large number of sources acccessing a certain buffer in a switch. According to this model, chunks of work (e.g., files) are generated at points in time in accordance with a Poisson process with parameter λ. The size of these chunks of work has a Pareto distribution, and each of them is transmitted at fixed rate r. At any point in time, we may have any number of sources transmitting at rate r simultaneously because according to the model, new sources may start transmission while others are active. If m sources are simultaneously active, the total rate equals mr. The symbol ∞ in the notation of the model is a representation of the unlimited number of active resources transmitting simultaneously.

Again, let time be divided into fixed-length intervals, and let A_n be a continuous random variable representing the amount of work entering the system during the nth interval. For convenience, we assume that the rate r is the amount transmitted by a single source within one time interval if the source were active during the entire interval. We also assume that the Poisson rate λ is per time interval. That is, the total number of transmissions to start in one time interval is λ.

To find the mean of A_n for the M/Pareto/∞ process, we consider the total amount of work generated in one time interval. The reader may notice that the mean of the total amount of work generated in one time interval is equal to the mean of the amount of work transmitted in one time interval. Hence,

$$E\{A_n\} = \lambda r/(\gamma - 1) \tag{4.9}$$

Also, another important relationship for this model, which is provided here without proof, is

$$\gamma = 3 - 2H \tag{4.10}$$

where H is the Hurst parameter.

Using the last two equations, we are able to fit the overall mean of the process $E\{A_n\}$ and the Hurst parameter of the process with those measured in a real-life process, and generate traffic based on the M/Pareto/∞ model.

The FBM [13] is a Gaussian process. That is, the amount of work generated in any time interval has the Gaussian distribution. It is also stationary, and exhibits long-range dependence. If we let Z_t be the amount of work generated within $(0, t)$, then Z_t is Gaussian, $Z_0 = 0$, $E\{Z_t\} = 0$, for all t, $E\{Z_t^2\} = t^{2H}$, and Z_t has a continuous sample path.

To use the FBM for traffic modeling of a process A_n, we recall the process S_n defined earlier as the sum of the A_n values from 1 to n, and we set

$$S_n = nE\{A_n\} + \sigma Z_n \tag{4.11}$$

where σ is the standard deviation of the amount of work arriving in one interval. Notice that $VAR\{Z_1\} = 1$, so by (4.11), $VAR\{S_1\} = \sigma^2$. In other words, σ^2 is the variance of A_n. This way, the traffic model is defined by three parameters: (1) the means, (2) the short-term variance (σ^2), and (3) the Hurst parameter. Notice that by definition, H (the Hurst parameter) of Z_t is also the Hurst parameter of the A_n process. The reader is referred to [14] for a method of generation of an FBM sequence which is beyond the scope of this book.

Although self-similar traffic models do capture important statistics of the real traffic, there is still no agreement on a traffic model that applies to all multimedia traffic streams. The study in [15] demonstrates that a traffic model based on FBM tends to overestimate (lower loss and delay) the performance of traffic while the M/Pareto/∞ tends to underestimate it.

4.2.1.5 Pseudo-Self-Similar Models

Models that are not self-similar, such as those based on Markov chains and autoregressive Gaussian, are more amenable to queueing analysis. On the other hand, self-similar models better capture statistical characteristics of real traffic. It is therefore useful to use models that experience short-range dependence in the mathematical sense, with an exponentially decaying autocorrelation function, and are amenable to queueing analysis but do exhibit enough long-range dependence to capture the heavy autocorrelation for the range of interest. Notice that it is not very important to find a traffic model that has the same autocorrelation function as that of the real traffic for its entire range. For example, if we consider the case of the zero buffer approximation, correlation does not have any effect on queueing performance. The larger the buffer, the longer the range of correlation that should be matched. Now, since no buffer is of infinite size, the significant range of the autocorrelation function is also limited.

In [16] a superposition of two AR(1) processes is used to generate pseudo-self-similar traffic. One of these two processes has its a coefficient [see (4.5)] very close to one (e.g., $a = 0.9999$). The other AR(1) process is stable (e.g., $a = 0.75$). This way the slope of the log–log variance time curve in the superposition traffic can be set at any number between 1 and 2 (recall that the Hurst parameter is half the value of that slope at the limit) for an arbitrarily set period of time. This means that we can set the range within which the correlation is heavy for as long as we need it to be. It has been demonstrated by queueing simulations [15] that if such a process is fed into a single server queue, it leads to performance equal to that obtained by feeding the same queue with a traffic model based on FBM with the same mean, short-term variance, and Hurst parameter. Similar ideas can be applied to generate a pseudo-self-similar process based on Markov modulated processes [17].

4.2.2 Multimedia Traffic

In this section, we describe the different types of multimedia traffic, their statistical characteristics, and their modeling implications. According to Cost 242 [9], multimedia traffic may be divided into three categories according to the timeliness requirements: (1) interactive audio or video communications, (2) transmission of stored information to another temporary storage, and (3) transmission of stored information for immediate use.

The interactive video or audio communication category includes bidirectional, multiparty, and conference traffic. Due to the requirement for real-time responsiveness, these services impose the most stringent delay and jitter requirements. Speech traffic alternates between spurts and periods of silence and hence may be modeled by an on/off process. Because the *on* (or *off*) periods do not have a heavy-tailed distribution, a nonfractal version of the on/off model is suitable. On the other hand for interactive VBR video, a self-similar model may be required. We can learn from [9] that, based on statistical measurements, an MPEG video sequence of a student sitting in front of a workstation had a Hurst parameter of 0.53, but on the other hand the Hurst parameter of a TV sequence of a political discussion was 0.85. Although both sequences were not taken from an interactive application, they could provide some guidelines regarding the level of burstiness we could expect in video conferencing.

The second category includes transfer of text (e-mail messages, files, documents), movies (off-line) and images. It also includes transmissions of data to update databases. By comparison to the first category, services of the second category do not have strict delay and jitter requirements. Hence they may be transferred at a lower delay priority than the real-time services of the first and third categories. The size of files to be transferred varies significantly from a few bytes in an e-mail message to many gigabytes in a movie to be transferred off line. Images may require storage of order of megabytes (for example, an x-ray may require 50-Mbyte storage). This huge variety in size in the different applications indicates that the distribution of a "chunk of work" introduced by a user may have a heavy tail, which in turn leads to long-range dependent traffic. This conjecture is consistent with statistical studies performed on WWW traffic [18,19]. It was demonstrated there that although most documents retrieved over the Web are small, a significant number of large documents are also retrieved, so that the distribution of a size of a document is heavy tailed (Pareto type).

The third category includes audio and video broadcasting, video on demand, and electronic mail (which may belong either to the second or third category). Because they do not have to cater to interactive communication needs, these services have less stringent delay requirements than those of the first two categories. A large enough buffer can be used to absorb the jitter and to present the traffic stream to the end user without delay fluctuation. The size of such a buffer will dictate the delay requirement of these services, which usually have more stringent delay requirements than those of the second category. The services under this category, which involve VBR video, are both bandwidth hungry and bursty. Before we discuss their statistical characteristics in more detail, we provide some relevant information on statistical characteristics of MPEG video sequences.

The MPEG sequence consists of a series of frames. Each frame contains a two-dimensional array of picture elements (pels). There are several types of frames (based on different coding mode): intra (I), predictive (P), and bidirectional (B). An important concept in MPEG traffic is the *group of pictures* (GOP). For example, a 12-frame GOP pattern used in [9] is: IBBPBBPBBPBB. The MPEG sequence is based on sequential transmission of one GOP after the other. Typically I frames will have more bits than P and B frames, and B frames will have the least number of bits. As traffic varies from one frame to the next, important statistical characteristics are the peak to mean ratio of the number of bits in a frame as well as the number of bits in a GOP. Values for the former could be within the range of 7 to 20 while for the latter they lie within 2 to 6. The difference between the two is expected because the latter ratio is related to variability in level of activities in the video while the former also includes the variability related to the coding.

The traffic models used for VBR video sequences (see [12,20–22] and references therein) can be divided into three categories: (1) Markov chains, (2) autoregressive processes (not necessarily Gaussian), and (3) self-similar models.

4.3 Traffic Management

The challenges of traffic management in telephone networks, data networks, and high-speed multimedia networks that have evolved during the 20th century have led to a vast amount of academic and industrial research in this area. Notice that in telephony all connections require the same capacity, while in ATM and multimedia networks different connections require different bandwidth, which may also vary. The fact that ATM and multimedia networks support a multiservice environment, bursty and unpredictable traffic streams, and nonhomogenous demands for bandwidth has introduced significant additional complexity in traffic management, which in turn has resulted in vast amounts of research being done. In this section, we report on many recent important developments.

The set traffic management functions can be divided into two groups: (1) those that take place during connection setup and (2) those that take place during the connection. Examples of those belonging to the first group are connection admission control and routing. Functions belonging to the second group are aimed at protecting the network from congestion and increasing efficiency; for example, policing to confirm that there are no violations of negotiated traffic parameters by the user, traffic shaping to achieve desired modification of traffic characteristics, selective discard, ABR flow control, and cell scheduling. This section covers all of these mechanisms.

4.3.1 Connection Admission Control

Connection admission control (CAC) is the set of actions taken by the network during call setup to determine whether a connection request is to be admitted or rejected. If a call requires more than one connection, CAC should be performed for each connection. Based on the CAC scheme, a connection request is admitted only when sufficient capacity is available end to end so that required QoS can be guaranteed for that connection as well as for all other existing connections.

In this section we discuss several general principles related to CAC, such as how to describe a connection, the criteria for admitting a connection, the different conceptual principles and approaches in CAC design, and how cell delay variation affects CAC. The aim here is to provide the reader with a broad understanding of CAC principles; for more detail, the reader is referred to [9] and references therein.

4.3.1.1 Connection Descriptors

At connection setup, the user and the network negotiate the traffic parameters such as peak cell rate (PCR), sustainable cell rate (SCR), minimum cell rate (MCR), intrinsic burst tolerance (IBT), the cell delay variation tolerance (CDVT), the requested QoS requirements (CLR, CTD, and CDV), and the ATM transfer capability (ATC). The result of this negotiation is a *traffic contract.* The role of the traffic contract is to enable an efficient network operation where QoS requirements of each connection are met. Different ATCs require different traffic parameters, hence the set of connection descriptors may change from service to service. For example, a service requiring a CBR transfer capability does not need to specify the SCR like a VBR service. Also, only an ABR connection can specify minimum cell rate. The CAC can be regarded as a system whose input is the traffic parameters, the CDVT, the QoS, and the ATC of the proposed connection, and the output is a decision whether or not to admit the connection. The CAC input parameters are henceforth called *connection descriptors*.

Before we explain in detail each of the connection descriptors, we should mention that ATM is a moving target, and as it evolves new descriptors may be proposed. Their selection has been and will be based on the following criteria: (1) They should be easily provided by the user, (2) they should have significant contribution to network efficiency in terms of resource utilization, and (3) the network should be able to monitor and verify them. These criteria should always be kept in mind when developing equipment and networks. Notice that it is not easy to meet all criteria at once and the designer may have to consider trade-offs among them. Sacrificing efficiency for the sake of simplicity is one example. It may be interesting to note here that although the ITU-T and ATM Forum recommend setting of the SCR and IBT in the case of a VBR connection, many experts believe that the benefit in efficiency of network resources does not justify the burden imposed on the user in setting these parameters. We now provide definitions of the different descriptors with an intuitive discussion on the need for the CDVT, followed by a description of the generic cell rate algorithm (GCRA).

4.3.1.2 CAC Approaches

Although many proposals for CAC have been studied and analyzed in the literature, there is no clear agreement on which is the best one for a given set of services. In this section we discuss three fundamentally different approaches for implementing CAC. The first is the simplest and the most conservative. It is based on peak rate allocation. The second is based on the zero buffer approximation or rate envelope multiplexing (REM). The third is based on traffic and queueing models and takes into consideration the buffering capacity on the network to share the excess traffic load during periods of heavy traffic.

CAC Based on Peak Rate Allocation

A simplistic description of CAC based on peak rate allocation is as follows. Consider the peak rate of every new connection. If for every network link, on the route of the new connection, the sum of the peak rates of all connections already in progress, plus the peak rate of the new connection, is lower than the total link capacity, admit the new connection, otherwise reject it. This simplistic description does not take into consideration the effect of CDV. Due to CDV, cell loss may occur even if the sum of the peak rates is lower than the link capacity.

CAC Based on Rate Envelope Multiplexing (REM)

While CAC based on peak rate allocation may be efficient for CBR streams, it is usually inefficient for VBR. In a typical VBR connection, the peak rate is several times the mean rate, hence no more than 40% to 50% utilization can be achieved under peak rate allocation. Therefore, to obtain higher efficiency, the CAC should admit connections such that the total peak rates may exceed the available capacity, and to rely on the fact that, most of the time, there is enough bandwidth to satisfy connections transmitting at their peak because others transmit at a lower rate than their peak. In particular, REM is based on the conservative assumption that there is no buffering, or that the amount of buffering is very small. Accordingly, this assumption is also called the *zero buffer approximation.* This assumption is appealing for the following reasons. First, in many real-time applications the buffers are too small to absorb burst scale fluctuations. It is therefore convenient to assume that the buffers will avoid all cell loss that may be due to cell scale fluctuations, and cannot avoid any cell loss due to burst scale fluctuations. In this case we can use burst scale traffic models such as on/off models together with the zero buffer approximation to obtain an accurate evaluation of the cell loss. Second, the zero buffer approximation makes the derivation of CLR much easier because there is no need for queueing analysis. Third, under the zero buffer approximation there is no need to consider the correlation in the arrival process, hence processes with independent arrivals could be considered. This makes the process of traffic modeling and characterization much simpler. Fourth, the zero buffer approximation is a conservative assumption, but not too conservative in certain important cases of real-time traffic as discussed earlier.

Mathematically, let S be a random variable representing the total amount of work arriving in a small time interval (which includes the amount of work arriving by existing connections plus that of the new connection). Let C be the available link capacity. The CLR is given by

$$P_{\text{loss}} = \frac{E\{(S - C)^+\}}{E\{(S)\}} \tag{4.12}$$

In this equation, the CLR is simply the ratio between the amount of work lost and the amount of work arrived. This is the way CLR is defined by the different standards bodies. It is important to recognize the difference between loss probability in terms of amount of work, or number of cells, and the proportion of time that the system is losing cells. The latter is defined by

$P(S > C)$. Many authors use the latter to approximate P_{loss} as defined by (4.12), and it is important to realize that the difference between the two may be very significant. Consider the very bursty process in which a very huge burst arrives during the first cell-time, but there are no arrivals afterwards. In other words, the probability $P(S > 0)$ is almost zero. Hence, $P(S > C)$ is almost zero. However, most of the cells that arrive are lost, that is, P_{loss}, as given in by (4.12), is very close to one.

To implement a CAC based on REM, we need to have the distribution of S at every point in time when a new connection request arrives. This can be estimated either by traffic measurements, or we can rely on traffic models. For example, if the traffic of each of the sources is modeled as a discrete time on/off source with geometric *on* and *off* times, then the distribution of S is binomial. If the total capacity is much larger than the peak rate of each of the active sources, S may be assumed to have a Gaussian distribution. Having the distribution of S, for a newly arriving connection request, we compute the CLR using (4.12). If the predicted CLR is lower than the required CLR for all existing connections as well as for the new connections (as specified in their traffic contract), the connection is accepted, otherwise it is rejected.

CAC Based on Rate Sharing

REM-based CAC will not be very efficient in cases for which the switches use large buffers capable of absorbing large bursts. There is therefore a need for a CAC that takes into consideration the buffering capabilities of the switches. One of the important functions of such a CAC is a CLR predictor. For every newly arriving connection request, such a predictor will estimate what the CLR will be if the connection is admitted, and compare it with the required CLR. One way to predict the CLR is to use a traffic model as discussed in Section 4.2. However, given the doubt over the "effectiveness of different modeling approaches and their applicability to the range of connection types" [9], we discuss here a method that does not rely on modeling the traffic, but on measurements and estimation of the distribution of the amount of cell arrivals during different time intervals [23]. The method is based on the use of (4.12) for the case of finite buffer.

Let $S(t)$ be a random variable that represents the amount of work arriving in an interval time t. We know that if during time t, the amount of work arriving is more than what can be served plus what can be buffered some work must be lost. We also know that the loss must be at least the excess of work over what can be served and buffered. Let CLR(t) be a lower bound for the cell loss ratio based on the random variable $S(t)$, given by

$$\text{CLR}(t) \geq \frac{E\{(S(t) - C(t) - b)^{+}\}}{E\{(S(t))\}} \tag{4.13}$$

where $C(t)$ is the amount of work that can be served during time t and b is the buffer size. Because CLR(t) for all t is a lower bound of CLR, we have that

$$\text{CLR} \geq \max_{t} \text{CLR}(t) \tag{4.14}$$

If we consider the inequality in (4.14) as an equality, we obtain an estimate for the CLR [24]. It is demonstrated in [24] that the right-hand side of (4.14) is an accurate estimator for the CLR.

The key problem here is to have accurate estimation of the distributions of $S(t)$ for different t values. These can be obtained using traffic measurements by which moments or histograms are produced and the required distributions can be estimated. Another important factor in estimating the distributions of $S(t)$ for different t is the traffic contract. If the PCR and its CDVT are specified, we have the point where required distributions should be truncated for small t while the SCR and the IBT tell us where the distributions are truncated for large t values.

4.3.1.3 Effective Bandwidth

Effective bandwidth is a number w_i associated with connection i such that the grade of service requirement is met if

$$\sum_{i} w_i \leq c \tag{4.15}$$

For effective bandwidth to be useful it should have the following characteristics:

1. *Independence.* Effective bandwidth of a given traffic stream is only related to the statistical characteristics of that traffic stream and the network equipment (e.g., buffering capacity). It should not be dependent on the traffic characteristics of any other streams.
2. *Additivity.* The effective bandwidth of the superposition of k streams is equal to the sum of the effective bandwidth of each of the streams.
3. *Efficiency.* It should lead to an efficient operation. For example, if we mix k VBR streams where for each of the streams the CDVT of the PCR is equal to zero, we could set the effective bandwidth to be equal to the PCR for each of the streams. This will be consistent with the definition of effective bandwidth and the first two requirements listed here, but may not necessarily lead to efficient network operation.

If we have an effective bandwidth value for each connection, the CAC function will be simply to compare the sum of these effective bandwidth values (including the one for the connection yet to be admitted), in all network links on the route of the new connection, with the total available capacity. If they exceed the total capacity on at least one of these links, the connection is rejected, otherwise it is admitted.

For the case of REM, we can obtain effective bandwidth by using (4.12). Simply, for a given distribution of S to find the smallest value for C such that P_{loss} is not higher than the required CLR.

For the case of rate sharing CAC, the derivation of effective bandwidth requires a formula for the loss probability that takes into consideration the buffer as well as realistic models of the traffic. This is a very difficult problem. A practical approach would be to use traffic traces generated by the different services and to keep a record of CLR versus buffer size for a large range of traffic streams.

When such simulation results are not available, the following analytical approximate approach (which has its limitation) could be used. Consider a single server queue with infinite buffer fed by traffic stream i, assume that the probability that there are more than b cells in the buffer is given by

$$P\{X > b\} = e^{-f_i(w_i)b} \tag{4.16}$$

where w_i is the service rate (or the effective bandwidth) and $f_i(w_i)$ is some function of w_i. Now if we approximate $P\{X > b\}$ by the CLR for the case where the buffer size is b, we obtain

$$w_i = f_i^{-1}(-\log \text{CLR}/b) \tag{4.17}$$

The w_i meet the requirements of effective bandwidth of stream i. The benefit of this approach is that in many cases, it is relatively easy to compute the function f_i. The drawbacks of this approach are as follows:

1. As demonstrated in Section 4.3.1.2, the left-hand side of (4.16) is not a good approximation for cell loss in many cases.
2. In (4.16) the weight of the negative exponential tail equals one. This is not the case in most models. Note that the weight of the tail is evaluated only for few models (e.g., Gaussian—see Section 4.2.1.3).
3. The basic assumption that the right-hand side of (4.16) is exponential does not apply to self-similar traffic. Nevertheless, equivalent results for some queues with self-similar traffic have been obtained, as discussed in Section 4.2.1.4.

4.3.1.4 Adaptive CAC Based on Measurements

An efficient and safe way to implement CAC is to combine the approaches described earlier allowing us to benefit from the safety of peak rate allocation while at the same time enjoy the efficiency of rate sharing or REM [24]. The loss prediction is based on (4.12) for REM and (4.13) for rate sharing. To estimate the distributions of $S(t)$ for different t values, the CAC relies on real-time traffic measurements. The random variable $S(t)$ is the sum of the contribution of the different connection to the total traffic. For connections that have been in use for a period longer than a set *warming up* period, histograms of the measured traffic are used to estimate their contributions. On the other hand, for connections that have been in use for less than the warming up period their worst case traffic is considered. In particular, it is assumed conservatively that for short t values, $S(t)$ is deterministic and is equal to their specified PCR; and for long t values (longer than their IBT), their $S(t)$ is equal to their specified SCR. Now, if during the operation of the CAC, the measured CLR is higher (lower) than its specified value, the warming up period is increasing (decreasing). Notice that if sources specify only their PCR and the warming up period is long, the CAC behaves as peak rate allocation. Accordingly, the CACs fluctuate between peak rate allocation and rate sharing. A simpler version of this CAC is one where the REM is used instead of rate sharing; in this case, the CACs fluctuate between peak rate allocation and REM. It is demonstrated in [24], based on real traffic traces, that very high utilization levels (up to 90%) can be achieved using this scheme for VBR traffic.

4.3.2 QoS-Based Routing

Routing is a collection of algorithms that determines the routes that data packets will traverse until reaching their destination nodes. Virtual circuit networks, like ATM, use connection-oriented routing. This means that a route is determined once, upon VC setup, and is then used by all the cells of the VC. By contrast, datagram networks, like the Internet, use connectionless routing. This means that a routing decision is made for each individual packet, independently of the previous packets of the same session.

4.3.2.1 Shortest Path Routing

Most of the routing algorithms used in traditional networks, which are not supposed to support QoS, try to find the shortest path from the source to the destination. These algorithms assign to each link a positive length, and associate with each path a length equal to the sum of its links. If each link is assigned the same length 1, the shortest path is the one with the minimum number of hops, and therefore the one that consumes minimum bandwidth from the network. However, many other metrics are possible, like distance, bandwidth, load, or cost. When the length represents load, the shortest path is the one over which packets will suffer lower average queueing delay. In such a case the length assigned to each link will change over time, depending on the prevailing congestion level of the link. Some algorithms use a composite metric that represents several parameters like propagation delay, bandwidth, reliability, and load.

Several algorithms for computing the shortest path between two network nodes in polynomial time are known. The most popular algorithms in computer networks are Bellman–Ford and Dijkstra. The following is the Bellman–Ford algorithm, as executed by a network node s in order to find the distance $\delta_s[v]$ of the shortest path to every network node $v \in V$, and to find the next node $\lambda_s[v]$ on that shortest path. The algorithm assumes that s knows the network graph $G(V, E)$, and the non-negative length $l(u, v)$ associated with every link $(u, v) \in E$:

initialization: for each node $v \in V - \{s\}$ do $\{\delta_s[v] \leftarrow \infty, \lambda_s[v] \leftarrow \text{NIL}\}$
$\delta_s[s] \leftarrow 0$
for $i = 1$ to $|V| - 1$
for each link $(u, v) \in E$
if $\delta_s[v] > \delta_s[u] + l(u, v)$ then $\{\delta_s[v] \leftarrow \delta_s[u] + l(u, v), \lambda_s[v] \leftarrow u\}$

The running time of the Bellman-Ford algorithm is worse than the running time of the Dijkstra's algorithm: $O(|V| \cdot |E|)$ versus $O(|V|^2)$, where $|V|$ is the number of nodes and $|E|$ is the number of edges in the network graph. (See [25] for a description of Dijkstra's algorithm and for complexity analysis of both algorithms.) Nevertheless, it is employed by distance-vector routing protocols. Such protocols avoid the need to broadcast complete topology information to all nodes by having the network nodes run the Bellman–Ford algorithm distributedly and asynchronously in the following way. During initialization, node s sets $\delta_s[v] \leftarrow l(s, v)$ and $\lambda_s[v] \leftarrow v$ if it has an outgoing link (s, v) to node v, or $\delta_s[v] \leftarrow \infty$ and $\lambda_s[v] \leftarrow \text{NIL}$ if it does not have such a link. Then, node s periodically exchanges the vector $\delta_s[]$ with all its neighbors. Upon receiving the vector $\delta_u[]$ from neighbor u, node s performs the following for every node $v \in V - \{s\}$: If $\delta_s[v] > \delta_u[v] + l(s, u)$ then $\delta_s[v] \leftarrow \delta_u[v] + l(s, u)$ and $\lambda_s[v] \leftarrow u$. This algorithm is demonstrated in Section 4.3.2.3, on distance-vector routing. It will be shown that its main weakness is that it takes a long time to reconverge to alternate paths when a failure occurs in the network.

Unlike distance-vector routing protocols, link-state protocols distribute to all network nodes the status of every link in the network. Thus, every node can create a local copy of the network topology and execute any shortest path algorithm. Hence, such protocols use Dijkstra's algorithm, whose running time is better than that of the Bellman–Ford algorithm.

4.3.2.2 Multiple Metrics

To support a wide range of QoS requirements, a routing protocol needs to consider multiple metrics, such as bandwidth, average delay, maximum delay, loss probability, etc. A typical route for a multimedia session should have sufficient bandwidth, a bounded delay, a bounded delay jitter, and a bounded loss rate. Moreover, from all the routes that fulfill each of these requirements the one with the minimum number of hops is preferred, because it wastes resources on a minimum number of links.

The various metrics can be classified into two categories as follows. Let μ be a QoS routing metric, and $\mu(i, j)$ be the value of μ over the link (i, j). Consider a path P established over the links (i_1, i_2) (i_2, i_3) $\cdots$ (i_{N-1}, i_N). Let $\mu(P)$ be the value of μ over path P. Then:

- Metric μ is said to be an *additive* (or *cumulative*) *metric* if $\mu(P) = \sum_{j=1}^{N-1} \mu(i_j, i_{j+1})$. Examples of such a metric are delay, delay jitter, and cost.
- Metric μ is said to be a *concave metric* if $\mu(P) = \mathrm{MIN}_{j=1}^{N-1} \mu(i_j, i_{j+1})$. An example for such a metric is bandwidth.

To satisfy the constraint imposed by an additive metric μ, a route P should fulfill $\mu(P) \le I$ where I is some integer or real number. One exception is when μ represents cost, in which case it is usually required that $\mu(P) \le \mu(P')$ for every route P' between the same source/destination pair.

When μ is an additive metric, finding a route P that satisfies the constraint $\mu(P) \le I$ is a simple problem that can be solved using a shortest path algorithm like Bellman–Ford or Dijkstra. However, finding a route P satisfying $N \ge 2$ additive metric constraints (for instance, the shortest path with delay not larger than I, or any path with delay not larger than I_1 and delay jitter not larger than I_2) is an NP-complete problem because this is a generalization of an NP-complete problem called *shortest weight-constrained path.*

Finding a route P that satisfies any number of concave metric constraints is a simple problem, which can be easily solved by a polynomial-time algorithm. Consider one concave metric μ, and let the constraint be $\mu(P) \ge I$. The algorithm should simply find any route from the source to the destination on the network graph after eliminating from that graph every link (i, j) for which $\mu(i, j) < I$. If there are $N \ge 2$ concave metric constraints $\mu_1 \le I_1 \cdots \mu_N \le$

I_N the algorithm should eliminate from the network graph every link (i, j) for which there exists l, $1 \le l \le N$, such that $\mu_l(i, j) \le I_l$, before trying to find a route from the source to the destination.

Finally, suppose there are $N \ge 1$ concave metric constraints and one additive metric constraint $\mu(P) \le I$. As an example for $N = 1$, consider the problem of finding a route with enough bandwidth and a bounded delay. In such a case the algorithm should first generate the network graph based on the concave metric constraints as in the previous case. Then, the algorithm should assign to each remaining link (i, j) a length equal to $\mu(i, j)$, and find the *shortest* path from the source to the destination. If the length of the shortest path P is not larger than I, then P is a solution. Otherwise, there exists no route that fulfills all the $N + 1$ constraints.

Since the problem of finding a path that satisfies $N \ge 2$ additive metric constraints and $M \ge 0$ concave metric constraints is NP-complete, polynomial-time heuristics that find a suitable path, though not necessarily the optimal one, are sought. The following is a representative algorithm:

1. Let $\mu_1 \le I_1 \cdots \mu_N \le I_N$, $N \ge 2$, be the additive metric constraints and $\mu'_1 \le I'_1 \cdots \mu'_M \le I'_M$, $M \ge 0$, be the concave metric constraints.
2. Remove from the network graph each link (i, j) for which there exists l, $1 \le l \le M$, such that $\mu'_l(i, j) < I'_l$.
3. Select the most important additive metric μ_l, $1 \le l \le N$. Assign to every remaining link e a "length" of $\mu_l(e)$, and find K different "shortest" paths $P_1, P_2, \ldots, P_K$ from the source to the destination, where $K \ge 1$. This step can be performed in several ways. For instance, by running the shortest path algorithm K times, and eliminating from the network graph before the ith execution all (or some of) the links used by every P_j, $j < i$.
4. If one of the K paths fulfills all the other $N - 1$ additive metric constraints, return this path as an output. Otherwise, return "not found."

Of course this algorithm may return "not found" even if a suitable path between the source and the destination does exist. This will be the case if the suitable path is not one of the K paths found in step 3.

4.3.2.3 Distance-Vector Versus Link-State Routing

To make routing decisions, the network nodes should get topology information and maintain routing tables. Two approaches are used to perform this task distributedly. In the first approach, called *distance-vector routing*, each node sends to its *neighboring nodes* only its *entire* routing table. The receiving nodes use the received information in order to update their own routing tables, which they then send to their own neighbors. In the second approach, called *link-state routing*, each node broadcasts to *all the network nodes* information regarding the status of its *local links only*. The network nodes use the received information in order to create and maintain an up-to-date network map, from which they deduce their routing tables.

As already indicated, distance-vector protocols use a distributed and asynchronous version of Bellman–Ford in order to enable each node to compute its routing tables based on the routing tables recently received from its neighboring nodes. The main drawback of the distance-vector algorithm is that it takes a long time to reconverge to alternate paths when a failure occurs in the network. During that time, the routing tables may define loops that may cause congestion in the network.

A distance-vector routing protocol called routing information protocol (RIP) was used in the early ARPANET network. Due to the long convergence time following topological changes, this protocol was replaced by a link-state protocol. Link-state protocols are based on a distributed protocol, called *reliable flooding*, which enables every node to send periodically a copy of a link-state message with up-to-date information regarding the state of its links to all the other nodes. The time interval between update broadcasts varies, depending on whether there has been a substantial change in any link since the last broadcast. Each link-state message specifies the nodes directly connected to the sender, along with the cost of the link to each of them. The network nodes keep the link-state information in a link database, from which they create a complete up-to-date copy of the network map. After creating the network map, a node can use any shortest path algorithm for computing the best route to any other node.

The most important advantage of a link-state protocol over a distance-vector protocol is that it responds much faster to topology changes. Its main disadvantage, however, is scalability. Every node needs to broadcast periodically a link-state message to the whole network and needs to process all the link-state messages it receives. For a network with a large number of nodes, the communication and processing burden is excessive.

Each of these two routing concepts has been enhanced to overcome its main drawback. In the Internet, distance-vector routing is replaced with path-vector routing (the border gateway protocol, used for routing between autonomous systems). The main idea behind path-vector routing is that each routing update, sent by a node to its neighbors, contains not only the distance from the sending node to each destination in the network, but also the list of transit nodes. This increases the size of routing messages and the amount of memory needed for performing routing, but it helps to avoid routing loops and to reconverge much faster to alternate paths when a failure occurs in the network.

Link-state routing is enhanced using the concept of hierarchical routing. The OSPF protocol, used for autonomous system routing in the Internet, has two levels of hierarchy whereas the ATM P-NNI routing protocol supports a maximum of 105 levels. OSPF splits the network into a set of independent areas. Each area behaves as an independent network that runs the link-state algorithm. Hence, the database of every node contains only the states of the links in the area, the flooding protocol delivers a copy of each link-state message only to nodes in the same area, and routes are computed only within the area. To enable routing between nodes in different areas, a collection of nodes, which contains at least one node from each area, forms the network backbone. Each backbone node maintains two routing databases: one for its area and another one for the backbone. Data packets from one area to another reach the backbone node of the origin area, continue across the backbone to the backbone node of the destination area, and then reach their destination nodes. Hierarchical link-state routing is discussed in the following, in the context of ATM P-NNI routing.

4.3.2.4 ATM P-NNI Routing

The following is a brief description of the ATM private network-to-network interface (P-NNI) routing as defined by the ATM Forum. More details can be found in [26,27]. As already indicated, the ATM P-NNI uses hierarchical link-state routing, with a maximum of 105 levels. It is expected, however, that even in a large global ATM network fewer than 10 levels will be used. To support the hierarchy, the P-NNI model defines a uniform network model at each level, with a set of logical nodes connected by logical links. Each lowest level node represents a physical ATM switch with a unique ATM address. Nodes within a given level are grouped into sets of peer groups. A peer group (PG) is a collection of logical nodes that exchange link-state messages, called P-NNI topology state packets (PTSPs), with other members of the group, such that all members maintain an identical view of the group. Each PG is assigned a unique peer group identifier (PGID) and is represented in its parent PG as a single node, called a logical group node (LGN).

Figure 4.1 shows an example for ATM P-NNI hierarchy. All of the logical nodes whose identity is of the form $x \,.\, y \,.\, z$ represent in this figure physical nodes (ATM switches). Nodes 1.1.1 to 1.1.4 are grouped into a PG called PG(1.1), and in the same way five additional PGs are created. Hence the network has 22 logical nodes in level 1 and 6 logical nodes in level 2. In level 3 there exist only two logical nodes: PG(1), which consists of LGNs (1.1) to (1.3), and PG(2), which consists of LGNs (2.1) to (2.2). The upper level

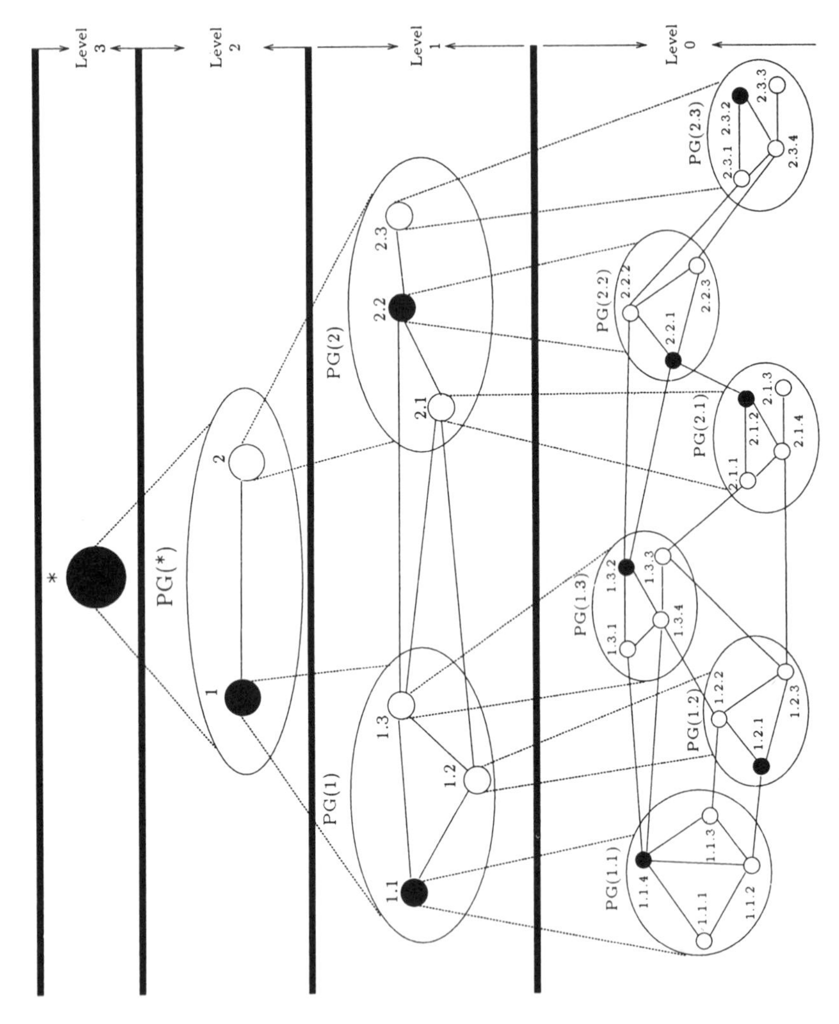

Figure 4.1 An example of the ATM P-NNI hierarchy.

always has only one logical node (LGN * Figure 4.1). The links between the physical (level 1) nodes, like (1.1.1,1.1.4) or (1.3.2,2.2.1), represent physical links or virtual path (VP) connections. By contrast, the links between logical nodes in level 2 or above, like (1.1,1.3) or (1,2), represent a *set* of physical links and VP connections.

As already indicated, each PG is represented in the next hierarchical level by a single LGN. The functions needed to be performed by an LGN are implemented by the peer group leader (PGL). The PGL is determined by means of a distributed election process, executed by all the LGNs in the PG. In Figure 4.1, the PGLs are the LGNs represented by black circles. Note that physical node 1.1.4 functions as a PGL of (1.1). However, since LGN 1.1 is the PGL of PG(1) and LGN 1 is the PGL of PG(*), the functions LGNs 1.1 and 1 are supposed to perform as PGLs are actually implemented by physical node 1.1.4.

Topology information flows horizontally through a peer group and downward into and through child peer groups. A PGL within each group has the responsibility of creating PTSPs that represent the status of the links connecting its PG with other PGs, and broadcasting these PTSPs to the other LGNs in its parent PG. In Figure 4.1, LGN 1.1.4 is the PGL of PG(1.1), LGN 1.2.1 is the PGL of PG(1.2), and LGN 1.3.2 is the PGL of PG(1.3). Hence, these three LGNs exchange PTSPs in PG(1). The PTSPs created by 1.1.4 indicates the status of the links connecting PG(1.1) with other PGs: (1.1.4,1.3.1), (1.1.4,1.3.4), (1.1.3,1.2.2), and (1.1.2,1.2.1). This PTSP exchange may be thought of as the PGL feeding information up the hierarchy. It is necessary for creating the hierarchy and for distributing routing information about child peer groups. Conversely, feeding information down the hierarchy is necessary to allow nodes in the lower level PGs to obtain knowledge about the full network hierarchy in order to select routes to destinations. When information is fed down from one level to a lower level, it is aggregated (summarized). Hence, at the lowest level each node has full information about its peer group, aggregated information about its parent group, more aggregated information about its parent's parent group and so on.

The ATM PNNI is supposed to provide sophisticated QoS routing while still allowing flexibility in the choice of route computation. Hence, each implementation is free to use its own algorithm. This gives rise to source routing, which does not require different switches to agree on the same computation. The source node (the switch of the source host) creates a hierarchical route consisting of a detailed path within the source node PG, a less detailed path within the source node's parent PG and so on until reaching the lowest level PG, which is an ancestor of both the source and destination nodes. When the lowest level common ancestor is reached, a new "source" route is computed to descend to the final destination. A "source" route is also computed when necessary by border nodes that determine the best way to cross their PG.

The path is encoded as a set of destination transit lists (DTLs), which is explicitly included in a stack within the PNNI signaling call setup request. Each DTL contains the description of a path for one level in the hierarchy. It explicitly specifies every node, and optionally every link, used to cross the PG. Each DTL is associated with a pointer that indicates the next element in the list to be processed.

4.3.3 Policing

A policing mechanism monitors a connection during its lifetime in order to enforce the delcared parameters at the connection setup time. Causes for contract violation may vary from inaccurate parameter estimation to malicious user behavior. Different actions may be taken by a policing mechanism whenever a connection violates its contract. A violating cell may be dropped or simply marked for later discard in case it visits a congested node.

The "leaky bucket" (LB) is a fundamental algorithm but assists the sources to comply with the contract and is also used by the policing function to identify violating cells. In a leaky bucket mechanism, a cell needs to consume a token before entering the network. Tokens are generated at a constant rate, and there is a maximum number of tokens (bucket size) allowed in the token pool (bucket). If a cell arrives at the network and there is no token in the pool, it may be discarded or marked. An LB can also be implemented by a counter, a leak rate, and a threshold. The counter is incremented up to its threshold value at every cell arrival and it is decremented at the leak rate. Whenever the counter reaches the threshold value, the LB starts marking (discarding) cells. Several variations of the LB have been defined so far. One of these variations has a buffer for smoothing the traffic instead of dropping or just marking the cell (buffered LB). The policing mechanism adopted by the ATM Forum, called the generic cell rate algorithm, is equivalent to the leaky bucket [28].

In addition to the leaky bucket, several window-based mechanisms have been defined [29]. In a jumping window (JW) mechanism, time is divided into fixed-length intervals called *windows.* The number of cells during any of these windows is limited. For a JW to have a rate equivalent to that of an LB mechanism, the ratio between the upper bound of the number of cells during an interval and the duration of the interval should be equal to the leaky rate. In the triggered jumping window (TJW) mechanism, the start of a window is triggered by the arrival of a cell. Thus, consecutive windows are not necessarily consecutive in time as in the JW. In a moving window (MW) mechanism, a cell is "reminded" by a constant interval. The difference between MW and the JW/TJW is that in the latter, a cell is always "reminded" by a constant interval, whereas in the former it may be "forgotten" much earlier. Finally, in the exponentially weighted moving average (EWMA), the number of cells accepted during a window is determined by an exponentially weighted sum of the number of cells accepted in the previous windows. In the seminal work of Rathbeg [29], a comparison showed that LB and EWMA are much more promising than MW and TJW regarding reaction time, worst-case traffic, and sensitivity to load variation.

One of the main issues in analyzing policing mechanisms is to determine the burstiness of the output process of a policing mechanism (regulator). It was shown in [30] that the first and second moment of the output process of a buffered LB depends on both the cell buffer size and on the bucket size. In [31], an expression for the queueing delay of the output process of a leaky bucket regulator was derived. The worst-case traffic was characterized by an arrival of a bucket size batch followed by no arrivals while the bucket is full. It was demonstrated that the mean queueing delay varies linearly with the bucket size. It was also noticed that asymptotic growth of the average delay as a function of the bucket size is independent of the number of multiplexed sources.

Neither the leaky bucket mechanisms nor window-based mechanisms proved to be effective in policing VBR sources [32,33]. In fact, even sources described by their average and peak rate are difficult to monitor by using these regulators. Actually, the LB and window-based mechanisms lack flexibility due to the small number of parameters available for tuning the policing process. For instance, for bursty sources, if we choose a small threshold, well-behaved cells may be marked and the violation probability may be high. On the contrary, if a high threshold is chosen, the time until overload is detected (reaction time) is large and congestion may occur. Moreover, the burstiness of the output flow of a queueing system increases with the threshold. Another option is to choose a higher leak rate, but then it is not possible to police the average bit rate of the source. The usual solution is to choose a leak rate close to the source's peak rate, which may waste bandwidth since bursty sources usually do not transmit at full speed all the time. In a window-based mechanism the reaction time to a violating behavior is highly dependent on the duration of the window for a certain enforced rate.

In [34], an experiment was presented in which a connection carrying misbehaving VBR traffic was able to pass through a leaky bucket without being tagged. It was also shown that average bursts that are larger and average bursts that are higher than the negotiated ones may cause performance degradation at a downstream queue even with leaky bucket enforcement. Moreover, allocating bandwidth according to the allowed rate is not enough to bound the average delay at a downstream queue.

To overcome the limitation of traditional policing mechanisms, multilevel policing mechanisms were defined. In each state of a multistate policing mechanism, there is a single-state mechanism [35]. For instance, each state of the multilevel leaky bucket (MLLB) mechanism has a leak rate, a counter, and a threshold [36]. Whenever the state *i* threshold is reached, the mechanism jumps to the next state, which has a higher leak rate. Whenever the counter goes back to zero, the mechanism jumps to the previous state. Besides the common parameters, in each state there is a maximum residence time within a fixed duration interval (cycle). If the residence time in a certain state exceeds its maximum allowed value, cells are considered to be violating. It was demonstrated in [35] that a multistate mechanism is more effective in monitoring VBR sources than its single-state counterpart because a multistate mechanism reacts much faster to a source state change. The filtered rate drops sharply when the peak rate is violated in a multistate mechanism, whereas it may even increase in a single-state mechanism. Furthermore, multistate mechanisms are more appropriate for detecting burst size variation than single-state mechanisms. Single-state mechanisms may ineffectively monitor even conforming sources.

4.3.4 Selective Discard Mechanism

Due to the bursty nature of multimedia sources, cell loss always occurs in networks based on statistical multiplexing, such as ATM networks. The only way to avoid loss is to allocate bandwidth based on a source peak rate, which, obviously, makes statistical multiplexing ineffective. The QoS requirements of an application are usually translated into two main performance metrics: the loss rate (the ratio between the lost cells and the total number of transmitted cells) and the length of the gap loss (the number of cells consecutively lost). Coping with different requirements is a challenging task. *Selective discard* is a mechanism aimed at enabling the network to deal with diverse loss requirements.

Before we explain selective discard, it is important to understand the meaning of loss rate and the length of the loss gap. It is common to use the term loss *probability* when loss *rate* should be used.

When we say loss probability, it is implied that the probability of losing cells is the same for all cells. However, since the cell stream is correlated, so is the process of lost cells, in which case some cells are more likely to be lost than others according to their position in the process. In other words, a loss of a particular cell may be influenced by the loss of previous cells. In this way, we cannot apply the term *loss probability* to nonrenewal processes. Although loss rate is a meaningful and measureable parameter, it is an average value and it does not describe the cell loss process. The number of cells consecutively lost (loss gap) gives a more detailed description of the loss process. For a certain value of loss rate, cells may be lost in several different ways. For instance, for a loss rate of 0.25, we may lose one cell out of every four cells or we may lose a quarter of the total number of cells in a role. Depending on the signal recovery procedure at the receiver side, the length of the loss gap may have a different impact on user's perceived QoS.

Selective discard is related only to the management of the buffer space and not to the transmission order. A selective discard mechanism is completely specified by a buffer organization policy and by a push-out policy. A buffer organization policy defines which buffer slot may be occupied by which cell, a push-out policy chooses a cell to be discarded among the cells with the lowest priority. A list of the most common buffer organization policies follows:

1. *Complete partitioning.* Each class has its own buffer space. Actually, there is no buffer sharing. Each buffer space can be optimized to handle its expected traffic stream. Although it is attractive for not having control overhead, losses occur even if there is plenty of free space.
2. *Complete sharing.* Any buffer slot can be occupied by any cell irrespective of its class. Cells are dropped if and only if the buffer is full. It minimizes the aggregated cell loss.
3. *Complete sharing with push-out.* Same as complete sharing but if a cell finds the buffer full and there is an enqueued cell with a lower priority, then this cell is dropped. (Complete sharing is sometimes used to designate complete sharing with push-out.)
4. *Partial sharing.* Each class has a threshold position up to which cells can be accepted into the buffer space. If a cell arrives and there are more enqueued cells than its class threshold, then the cell is lost.
5. *Complete sharing subject to maximums.* Each class has a bound for the number of enqueued cells.
6. *Complete sharing subject to minimums.* Each class has a minimum inviolable number of buffer slots. The rest of the buffer space is shared in a complete sharing fashion.
7. *Complete sharing subject to guaranteed queue minimums.* Similar to complete sharing but each class has a guaranteed number of buffer slots in overflow situations. Notice that the guaranteed queue length is enforced only in overflow situations [37].

Figure 4.2 illustrates loss scenarios for different buffer policies.

A buffer organization policy is called *work-conserving* for an input process with fixed-size packets if it loses cells only in overflow situations. From the above-mentioned choices, complete sharing (with and without push-out) and complete sharing subject to guaranteed queue minimums are work-conserving. Work-conserving queues are of special interest because they minimize the overall cell loss and consequently maximize the throughput.

A push-out policy selects a cell to be dropped among the low-priority cells. The most common policies are last-in-first-drop (LIFD), first-in-first-drop (FIFD), and randomly selected (RAND). Dropping low-priority cells at different positions defines a different queue distribution over a certain period of time and consequently may differently impact the QoS parameters of an application. Figure 4.3 illustrates a scenario that exemplifies the difference between the LIFD and the FIFD policies.

To assess the influence of how the choice of a specific push-out policy impacts the loss rate and the loss gap in a queue, Fonseca and Silvester [38] compared via simulation LIFD, FIFD, and RAND. They simulated a queue with two priority levels fed by one high- and one low-priority two-state MMPP. The arrival rate was set in order to cover a wide range of values of the offered load and of the burstiness (peak to mean rate). The buffer size was varied between 30 and 200. They determined that the FIFD policy gives the lowest

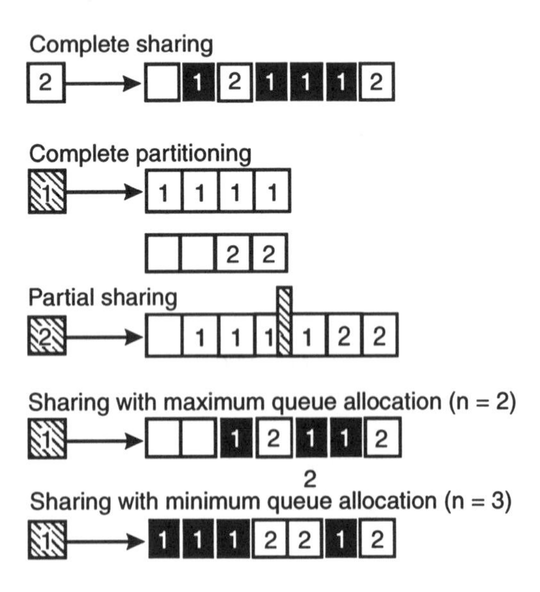

Figure 4.2 Examples of buffer sharing policies.

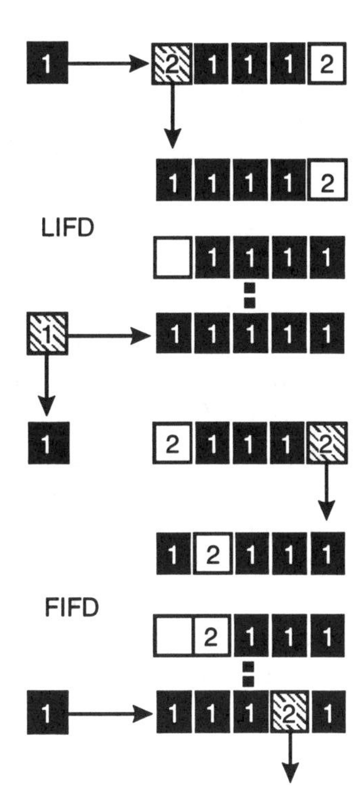

Figure 4.3 An example of the impact of different push-out policies.

high-priority loss rate, whereas the LIFD policy gives the highest values. FIFD always gives the lowest value because it maximizes the waiting time of every low-priority cell before dropping it and consequently maximizes the chances that an arriving high-priority cell will find an enqueued low-priority cell. The percentage difference between these two policies was of the order of 70% for a loss rate of the order of 10^{-7}. The adoption of a specific policy does not have a significant impact on the low-priority loss rate. Regarding the average length of the loss gap, LIFD gives the smallest average for the low-priority class. This relationship can be understood by the following: In the LIFD policy a low-priority cell that would be dropped by the FIFD policy is eventually transmitted. This transmission breaks the loss burst into two smaller bursts. As the low-priority probability increases, the frequency of long bursts also increases and an eventual LIFD transmission brings a huge difference in the average. The maximum burst size for the LIFD is half the value for FIFD, and small loss burst is much more frequent in LIFD than it is in FIFD. The impact of a push-out policy on the high-priority loss gap is not as significant as it is for the low-priority class.

Selective discard has been extensively investigated in the literature [39, 40]. The first studies considered queues with different buffer organization policies but under Markovian input [39]. Li et al. [41] used spectral decomposition to study a system fed by on/off sources. Petr and Frost [42] distinguished several buffer policies based on the time at which actions are taken. In [43] Petr and Frost studied the problem of minimizing the offered load over all multithreshold-type policies. Fonseca and Silvester [44] showed the solution for a complete sharing with push-out queue and with MMPP input. Garcia and Casals [45] provided a solution for complete sharing with push-out and for partial sharing queues with MMPP input [44,46]. Elwalid and Mitra [47] solved a partial sharing queue via fluid-flow approximation. Bae et al. [48] also analyzed partial sharing queues with a Markovian arrival process at discrete and continuous times via the stochastic integral approach. Fourneau et al. [49] studied a push-out queue via stochastic ordering, an approach that derives a simpler Markov chain to compute the bound on the loss rate. Wu and Mark [50] solved a complete sharing queue subject guaranteed queue minimum via a fluid-flow approach. Fonseca and Silvester [44] studied multipriority selective discard. Finally, Gravey and Hebuterne [51] studied the mixing of service and buffer priority in a queue.

4.3.5 Scheduling

In real-time multimedia communication, a cell (packet) can be characterized by the time at which it is delivered. Typically, the maximum allowed transfer delay is bounded by an application-specific value. If a cell arrives after its deadline, the received information is discarded and considered lost. Besides a maximum transfer delay, some applications require a bounded delay jitter.

Delay and jitter guarantees can be provided by different mechanisms. Nonetheless, not all of them are satisfactory for multimedia networks. For instance, current voice networks are based on circuit switching, which guarantees bounded delays. However, circuit switching underutilizes bandwidth when multiplexing multimedia (bursty) traffic. Buffers at the receiver can be allocated to control jitter but a large buffer space may be required. For example, a video source transmitting thirty 2-Mbyte frames per second and experiencing a transmission jitter of 1 sec requires 60 Mbytes of buffer space in a network with no delay control [52]. Alternatively, scheduling can provide an end-to-end jitter bound by delaying cells within the network. In addition, delay bounds can be achieved by prioritizing cells according to their deadline.

Real-time applications can be either soft real-time or hard real-time depending on their ability to tolerate loss. Service disciplines (policies) for hard real time can be classified as rate based or scheduler based. In rate-based disciplines, there is a set of allowed rates. For a certain rate, the allocated bandwidth implies specific bounded delay. Rate-based policies can be grouped into a rate-allocating discipline or rate-controlled discipline. In a rate-allocating discipline, transmitting at a rate higher than the predefined one is permitted whenever it does not interfere with other connections, whereas in rate-controlled disciplines, it is not allowed. In scheduler-based disciplines, priorities are assigned according to the closeness to a cell deadline expiration. Disciplines can also be classified as either work-conserving or non-work-conserving. In work-conserving disciplines, an output link never stands idle if there is a packet waiting for transmission. Conversely, in non-work-conserving disciplines, a packet may be held even if the server is idle. We proceed by explaining some scheduling disciplines.

The fair queueing (FQ) discipline tries to share an output line equally among N connections by giving them $1/N$ of the bandwidth. If a channel uses less than its share, the slack is equally divided among the other competing connections. Fair queueing aims at emulating the (bit-by-bit) round-robin discipline, which is commonly used in operating systems. Since it is impractical to implement exactly bit-by-bit round-robin, each packet is given a finish number, which is the round number the packet would receive had service been furnished in a bit-by-bit fashion. In fair queueing, packets are transmitted according to their finishing number, and a priority scheme can be implemented by giving channels different weights.

The virtual clock discipline aims at emulating TDM while preserving the flexibility of statistical multiplexing [53]. A statistical multiplexer may use a virtual clock in the same way TDM uses a real-time one. One difference between the virtual clock discipline and TDM is that the virtual clock orders packet transmission without changing the statistical nature of packet switching. At admission setup time, every connection flow negotiates its average transmission time. The virtual clock ticks with a frequency equivalent to the inverse of a connection declared average rate. Each received packet is stamped with the current virtual clock value, and an eventual difference between the virtual clock value and the real-time value shows how closely a connection is transmitting according to its claimed rate. Proper action may be taken in a case where a packet is received before it is expected. Since the transmission rate is enforced within an interval, a source may transmit a packet burst within this interval, leading to undesirable queueing behavior. To avoid credit accumulation, an auxiliary clock is also used. The auxiliary clock is updated with the real-time value, and if its current value is lower than the real-time value, packets are stamped with the auxiliary virtual clock value. Packets are then transmitted according to their increasing order of virtual clock stamp. Whenever overflow occurs, a packet with the highest stamp value is dropped. It is also possible to implement a priority mechanism by decreasing a connection virtual clock by a certain amount. Given that the virtual clock discipline does not impose any restriction on a minimum spacing between consecutive packets, it can provide only bandwidth guarantee, but not delay guarantee.

In the generalized processor sharing (GPS) service discipline, the allocated rate to a connection is proportional to its weight. GPS is an idealized discipline: It assumes that all the sessions can be simultaneously served and that the traffic is infinitely divisible. In a realistic system only one session can be serviced at a certain time and an entire packet must be fully served before another packet can be transmitted. Several disciplines try to simulate the idealized GPS scheduling. One of these disciplines is packet-by-packet generalized processor sharing (PGPS), which is equivalent to weighted round-robin (WRR). In PGPS, a packet can be transmitted only after it is fully received (gated queue) [54]. A PGPS implementation is accomplished by adopting a virtual time reference. An arriving packet is stamped with a virtual time value and packets are served according to an increasing order of virtual time. The virtual time computation takes into consideration the time a packet would depart in a GPS system. A packet selected at time t for transmission in a PGPS system is the same one that would be selected in a GPS queue if no other packet arrives after t. In [55] it was shown that a leaky bucket controlled PGPS queue lags behind at most one packet transmission time compared to a GPS queue. A packet in WRR may need to wait N packet transmission times where N is the number of connections in the system, whereas in PGPS it does not need to wait since PGPS approximates GPS within a packet transmission time. Moreover, PGPS handles variable-size packets in a more systematic way than does WRR. PGPS does not punish a connection for credit accumulation as in the virtual clock discipline. A process is said to have exponentially bounded burstiness if the amount of generated traffic within a finite interval is bounded by a exponentially decaying function. In [56], it was shown that the output of a server with the GPS discipline whose input process is exponentially bounded burstiness (EBB) is also EBB. It was also shown that it is possible to derive an end-to-end delay bound per session in a PGPS network with arbitrary topology.

Although a PGPS (WRR) queue may fall behind a GPS queue by at most one packet transmission time, in a PGPS system, a packet may leave the system much earlier than in GPS. Consequently, PGPS may be ahead of GPS in terms of number of bits served for a session. This scenario may occur over and over again, and the difference between a packet system and a GPS system may be quite large. Since in the long run the service received under GPS and under a packet system is the same, there will be alternating states during which GPS is ahead of a packet system and other states in which the packet system is ahead of GPS. The worst-case fair queueing (WF^2Q) method was proposed in order to minimize this discrepancy between an idealized fluid GPS and a packet queue [57]. In WF^2Q, rather than selecting a packet at time *t* among all the packets in the server, a packet is selected among the ones who would have already started service at time *t* in an associated GPS queue. Additionally, the selected packet should complete service first in the GPS system. It was shown that WF^2Q cannot be ahead of the associated GPS by more than a fraction of the maximum packet size, and that WF^2Q does not fall behind the GPS system. Since the maximum discrepancy between WF^2Q and the associated GPS system is less than a packet size, WF^2Q could be considered to be an optimal approximation of GPS. Furthermore, if the same assumptions used to guarantee a bounded end-to-end delay in a PGPS system are adopted [54] then WF^2Q also provides bounded end-to-end delay.

4.3.6 ABR Flow Control

4.3.6.1 The ABR Service

The ABR service is defined for the support of applications that may require minimum bandwidth. It guarantees a low or zero cell loss ratio and a fair share of the available bandwidth to an end system that adapts its traffic in accordance with the feedback received from the network. Such a service is usually required by data applications that cannot predict their own bandwidth requirements, and expect to get some share of the available bandwidth, as in a legacy Ethernet/token-ring LAN. By contrast, the UBR service category offers best-effort service with no QoS guarantees. It does not include the notion of a per-connection negotiated bandwidth and does not make numerical commitments with respect to the cell loss ratio or the cell transfer delay experienced on the connection. It is designed for data applications that want to use any leftover bandwidth and are not sensitive to cell loss and delay. A UBR connection is not rejected on the basis of bandwidth shortage and is not policed by the network.

The ABR service allows applications to utilize the bandwidth not used by the CBR and VBR services, by continually adjusting their instantaneous transmission rates to the time-varying capacity available for ABR. A congestion control scheme is essential for the support of ABR traffic to utilize the available bandwidth without causing congestion, and to divide that bandwidth among the ABR connections in a fair manner. Computing the fair share of an ABR connection on an outgoing link of a switch requires an algorithm for fair allocation. Another algorithm is required at the end system to adjust its rate to the feedback received from the network. The proper source behavior is important to guarantee the low cell loss rate. If a source does not behave as expected, it could increase the loss rate to all ABR users.

4.3.6.2 Max-Min Allocation

The ABR service is supposed to divide the bandwidth left by the VBR and CBR connections among the ABR connections while giving each ABR source a fair share. Many different definitions of *fairness* exist. The ATM Forum has converged toward a particular definition, called *max-min fairness* [58], which maximizes the total throughput. Max-min allocation allows all the VC connections using a common link to share the available bandwidth equally *as long as they are not bottlenecked elsewhere in the network.* The VC connections competing for bandwidth on a link l are divided into two categories: constrained connections and unconstrained connections. Constrained connections are those that cannot use their fair share on l because of limited bandwidth on another link. Unconstrained connections are those that cannot use their fair share on another link because of limited bandwidth on l. For these connections, the link l is referred to as the *bottleneck link.* The basic idea behind max-min fairness is to allocate to the unconstrained connections of l an equal share of the bandwidth left by the constrained connections.

The following example illustrates the concept of max-min fairness. Figure 4.4(a) shows four VC connections, marked *A–D*, crossing a network with four switches connected by three 1-Mbps links. Link (s_3, s_4) is the bottleneck link for VC *A, C,* and *D*. Hence, each of these VCs can be allocated only 1/3 Mbps, whereas VC *B* can be allocated the remaining 2/3 Mbps of link (s_1, s_2). Note that this max-min allocation wastes 1/3 Mbps on (s_2, s_3), because this link is not a bottleneck link for any VC connection.

A centralized algorithm for computing max-min allocations works as follows [59]. Suppose the network has L links, denoted by $1 \cdots L$. For every link l, denote by b_l the capacity (bandwidth) of l, by C_l the set of ABR VC connections crossing l, and by $|C_l|$ their number. A sequence of allocation of

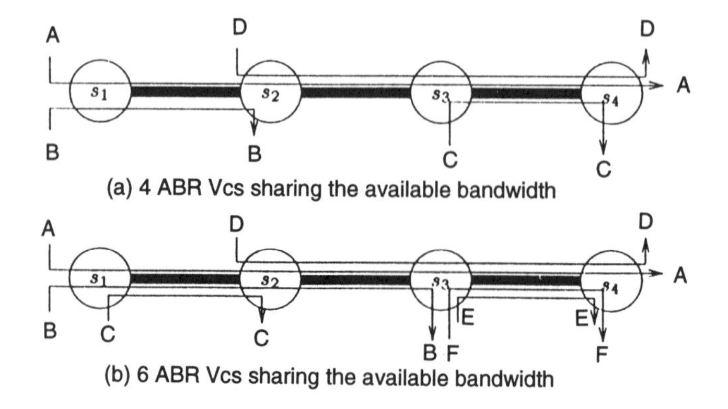

Figure 4.4 Max-min allocation example.

rates can be represented by a vector β whose lth component $\beta[l]$ is the rate allocated by β on link l. Such an allocation is said to be max-min fair if the following holds:

(c1) The rate $\beta(c)$ allocated by β to an ABR VC c is not negative.
(c2) For every link l, $1 \leq l \leq L$, $\beta[l] \leq b_l$.
(c3) $\beta(c)$ cannot be increased without decreasing the rate allocated to another connection c' for which $\beta(c) \geq \beta(c')$.

In [59] it is proven that conditions (c1) to (c3) are fulfilled for β if and only if each connection c has a bottleneck link, namely, a link l over which the rate assigned to c cannot be increased because all the capacity of l has been allocated (i.e., $\beta[l] = b_l$) and because none of the VCs crossing l has been assigned more bandwidth than c (i.e., for every $c' \in C_l$, $\beta(c) \geq \beta(c')$).

The algorithm for finding a max-min fair vector works in iterations until it finds the bottleneck link for each VC connection. It starts with a vector β^0 of all-zero rate, which obviously fulfills conditions (c1) and (c2). It then increases some of the components of β^0 such that (c2) continues holding and a bottleneck link is found, in which case β^1 is defined. To guarantee that (c2) is not violated, the algorithm looks for the link l for which the available bandwidth divided by the number of not-yet-bottlenecked VCs is *minimum*. It then allocates an equal share of the bandwidth of the selected link to every VC crossing that link. This allocation affects the bandwidth available on the other, not yet bottlenecked, links. Then, the algorithm disregards the bottleneck link and all the VCs crossing this link, and continues to the next iteration where a new link is selected. This process continues until all the VCs are bottlenecked.

The algorithm just discussed is not useful in a real dynamic ATM network, because it assumes centralized control. However, its main concepts are used for the design of distributed algorithms, to be executed by the ATM switches and source nodes. These algorithms adjust the ABR bandwidth allocation to changes in the layout of ABR VCs, which affect the number of ABR connections sharing the available bandwidth on every link, and to changes in the layout of CBR and VBR VCs, which affect the amount of bandwidth available for ABR connections on every link. Some of these algorithms are presented and discussed later, in the context of rate-based flow control. A thorough discussion and references to various related works can be found in [59, 60].

4.3.6.3 Credit-Based Versus Rate-Based Control

The congestion control schemes that have been proposed to the ATM Forum fall into two categories: credit-based schemes and rate-based schemes [61]. Credit-based schemes adopt a link-by-link window flow control [62]. A certain number of buffers are reserved at the receiving end of each link for every VC sharing that link. At various times, the receiving end of a link monitors the queue length of each VC and sends credits to the sending end, indicating that there is a certain amount of buffer space available for receiving data cells of the VC. The sending end maintains a credit balance for each VC. Upon forwarding a data cell on the link, the sending end decrements its current credit balance for the VC by one. When the sending end runs out of credit, it stops forwarding cells of that VC.

The main advantage of this per-VC link-by-link flow control mechanism adopted by credit-based schemes is that it can guarantee zero cell loss under ideal conditions. In addition, it can utilize the available bandwidth efficiently because of the short control loop, which consists of a single link. Its main disadvantage, however, is that it requires per-VC buffering at every switch, which must be equal to the delay-bandwidth product. This is because in the worst case there is only one active VC on a link, and the credit must be larger than or equal to the link cell rate times the link round-trip propagation delay. An adaptive version of the scheme solves this problem by giving each VC only a fraction of the round-trip delay worth of buffer allocation. The fraction depends on the rate at which the VC uses the credit: For highly active VCs the fraction is larger than for less active VCs. This adaptive scheme reduces the overall buffer requirements at the switch, but also introduces a ramp-up time: If a less active VC becomes active, it may take some time before it can get the share it deserves.

The rate-based approach is the one that was eventually adopted as the standard by the ATM Forum. The idea is that the source node, and optionally some network switches as well, shapes the connection rate as directed by feedback from the network. The following is a brief history of the evolution of the rate-based approach [61, 63].

The original proposal was a version of the DEC-bit scheme, according to which congestion is signaled by the network in the forward direction through a single-bit congestion indicator in the cell header (EFCI). The destination monitors this bit for a periodic interval, and sends an RM (resource management) cell back to the source in order to decrease the rate when necessary. This *negative polarity of feedback*, where RM cells are sent only to decrease the rate, whereas the absence of such cells indicates the source that the rate can be increased, has the problem that under heavy congestion in both directions of a VC connection, the RM cells are lost and the source increases its rate instead of decreasing it. This problem was fixed in the PRCA scheme by using positive polarity of feedback as follows.

The source keeps decreasing its rate by a constant factor (i.e., exponentially) and increases it only on the receipt of an RM cell from the destination. Such a cell is sent by the destination whenever a data cell with a clear EFCI field is received. To constrain the amount of bandwidth consumed by the RM cells to a fixed proportion of the total ABR bandwidth, the source sets the EFCI bit on every cell except the N_{rm}th one. Hence, an RM cell can be sent by the destination only in response to the N_{rm}th cell, provided that this cell does not encounter a congested switch. The amount of rate increase when an RM cell is received compensates for the rate decrease over the last N_{rm} cells plus an additional fixed increment.

4.3.6.4 Rate-Based Control With Explicit Feedback

The single-bit binary feedback, which can only tell the source whether to increase or decrease its rate, has been shown to adapt too slowly to the dynamics of high-speed ATM networks and in some cases to allocate the ABR bandwidth in an unfair manner. Simulations show that VCs that have relatively long paths (e.g., VC *A* in Figure 4.4(a)) are driven down much below their fair share. This happens due to the following three phenomena [64]: First, such connections are less likely to get back an RM cell because they must encounter no congestion at *every* node along their longer path. Second, such connections have longer feedback delay and can therefore respond less quickly to acquire new available bandwidth. Finally, when their rate drops due to the above-mentioned reasons, the sources of such connections cannot easily climb back up. This is because sources of lower rate connections have fewer opportunities for rate increases. To understand why, recall that a rate increase happens in cycles of N_{rm} cells, after the source has continually decreased its rate, and that each increase compensates for the rate decrease over the last N_{rm} data cells plus a fixed additive increment. This means that lower rate sources, which are allowed to send N_{rm} cells slower than other sources, acquire their fair share more slowly.

Due to these reasons, several researchers advocated rate-based schemes with explicit rate indication, where the network can provide the source with its exact allowed rate rather than with a feedback bit. These schemes not only bring the sources to the optimal rates much more rapidly (with a few round trips), but also can naturally support ABR traffic policing by the network. The idea behind these schemes is that a source generates a steady stream of RM cells that contain a field for the explicit rate. Each switch along the path can reduce the rate if it is higher than what can be locally supported. When an RM cell is received by the destination, it is looped back to the sending host.

A scheme for rate-based control with explicit feedback has been proposed in [65]. In this scheme, the source sends an RM cell after every N_{rm} data cell. The RM cell contains the current rate and the desired rate of the VC. Each network switch computes the fair share on each outgoing link for every VC crossing that link iteratively, using the following formula:

$$\text{fair\_share} = \frac{\text{available\_bandwidth} - \text{bandwidth\_consumed\_by\_underloading\_VCs}}{\text{number\_of\_non\_underloading\_VCs}}$$

where a *non-underloading VC* is a VC whose desired rate is not smaller than the fair share computed during the previous iteration for the considered link, because it is not bottlenecked in some other link. The initial value of fair_share is the available bandwidth divided by the number of active connections.

This scheme has been shown to achieve the max-min fairness in $4N$ round-trips, where N is the number of iterations performed by the centralized procedure described earlier, namely, the number of bottleneck links. However, it has the disadvantage that the switch needs to keep and update information regarding every crossing VC, and that the number of operations required for the computation of the fair share is proportional to the number of the crossing VCs.

The synthesis of the explicit rate concept with PRCA led to a new scheme called the enhanced proportional rate control algorithm (EPRCA). This scheme allows both binary-feedback switches and explicit-feedback switches on the path of each VC. The network switches can make an estimate of the proper rate for each VC without per-VC accounting. The explicit-rate feedback is used as an upper bound on the rate computed by PRCA. As in PRCA, all data cells are generated with EFCI=0, which can be changed to 1 by a congested node in order to inform the destination node of congestion. To allow for explicit rate feedback as an option, an RM cell is sent by the source after every N_{rm} data cell, and then looped back by the destination.

An RM cell contains a DIR bit that indicates the forward or backward direction of the cell, and a CI bit that indicates to the source the presence (1) or absence (0) of congestion. It is generated by the source with CI=0 and carried in the forward direction unchanged by the network. It is looped back by the destination host with CI=1 if the preceding data cell contained EFCI=1, and with CI=0 otherwise. During the backward direction, switches are allowed to change the CI bit from 0 to 1 in order to indicate the presence of congestion in the forward direction.

The optional "enhanced" mechanism assumes the network nodes are able to calculate a fair share for each VC. To this end, an RM cell has two additional fields, called explicit rate (ER) and current cell rate (CCR). The ER field carries the explicit rate. It is initially set by the source to the source peak cell rate, and can then be lowered by intermediate switches. The CCR field indicates the value of the allowed cell rate (ACR) effective at the time of origination of the RM cell, and it is not modified by the network switches.

The switches compute a fair share and reduce the ER field in the returning RM cells if necessary. To simplify the computation of the fair share and to avoid per-VC accounting, each switch calculates the fair share mean ACR (MACR) on receiving an RM(CCR, ER) cell of any crossing VC using the following simple first-order filter:

$$\text{MACR} \leftarrow (1 - \beta)\text{MACR} + \beta \cdot \text{CCR}$$

where $\beta < 1$ is an exponential averaging factor. The switch then reduces the value in the ER field to the new computed MACR multiplied by a *switch-down pressure factor*, which is set close to but below 1. The source decreases its rate continuously after every cell by a reduction factor called RDF: ACR $\leftarrow$ ACR · RDF. When the source receives the returned RM cell it checks the CI bit. If CI=0, ACR is increased to min(ACR + N_{rm} · AIR,ER,PCR), where N_{rm} · AIR is supposed to compensate for the rate decrease over the last N_{rm} data cells plus a fixed additive increment.

4.4 Summary

We have covered in this chapter many of the current research topics in the field of traffic management and control of multimedia networks: from source traffic modeling to actual traffic management issues such as policing, scheduling, flow control, connection admission control and routing. In spite of a significant amount of research in this area, the provision of optimal networks from the point of view of reliability and efficiency is still a challenge. The reader will appreciate that many of the traffic management and control decisions are performed under uncertainty due to the burstiness and unpredictability of the traffic. The question of whether to rely completely on traffic descriptors specified by the user, or also to use real-time traffic measurements, is still debatable among vendors and network providers. In telephony, we could rely for many years on Erlang theory to provide support for congestion prediction and efficient network design and traffic management. The challenge of finding an equivalent theory for the far more complex multimedia networks is still with us.

References

[1] Addie, R. G., and M. Zukerman, "Queues with Total Recall—Application to the B-ISDN," *Proc. ITC 14*, North Holland, 1994.

[2] Heffes, H., and D. Lucantoni, "A Markov Modulated Characterization of Voice and Data Traffic and Related Statistical Multiplexer Performance," *IEEE J. Sel. Areas Comm.* Vol. SAC-4, Sept. 1986, pp. 856–867.

[3] Leland, W. E., et al., "On the Self-Similar Nature of Ethernet Traffic (Extended Version)," *IEEE/ACM Trans. Networking*, Vol. 2, No. 1, Feb. 1994, pp. 1–15.

[4] Leland, W. E., et al., "Statistical Analysis and Stochastic Modeling of Self-Similar Data Traffic," *Proc. ITC 14*, Elsevier Science Publishers, Amsterdam, 1994.

[5] Andrade, J., and M. J. Martinez-Pascua, "Use of the IDC to Characterize LAN Traffic," *Proc. 2nd IFIP Workshop on Performance Modelling and Evaluation of ATM Networks*, Bradford, July 1994.

[6] Gusela, R., "Characterizing the Variability of Arrival Processes With Indexes of Dispersion," *IEEE J. Sel. Areas Commun.*, Vol. 9, No. 2, Feb. 1991.

[7] Decreusefond, L., and N. Lavaud, "Simulation of Fractional Brownian Motion and Application to a Fluid Queue," *Proc. ATNAC '96*, Dec. 1996, pp. 285–289.

[8] Hui, J. Y., "Resource Allocation for Broadband Networks," *IEEE J. Sel. Areas Commun.*, Vol. 6, No. 9, 1988, pp. 1598–1608.

[9] Roberts, J., Mocci, U., and Virtamo, J. (Eds), *Broadband Network Teletraffic*, Final Report of Action Cost 242, Springer, 1996.

[10] Addie, R. G., and M. Zukerman, "Performance Evaluation of a Single Server Autoregressive Queue," *Austr. Telecommun. Res.*, Vol. 28, No. 1, 1994, pp. 25–32.

[11] Addie, R. G., and M. Zukerman, "An Approximation for Performance Evaluation of Stationary Single Server Queues," *IEEE Trans. Commun.* Vol. 42, No. 12, Dec. 1994, pp. 3150–3160.

[12] Maglaris, B., et al., "Performance Models of Statistical Multiplexing in Packet Video Communications," *IEEE Trans. Commun.*, Vol. 36, No. 7, July 1988, pp. 834–844.

[13] Mandelbrot, B. B., and J. W. Van Ness. "Fractional Brownian Motions, Fractional Noises and Applications," *SIAM Rev.*, Vol. 10, 1968, pp. 422–437.

[14] Lau, W. C., et al., "Self-Similar Traffic Generation: The Random Midpoint Displacement Algorithm and Its Properties," *Proc. Int. Conf. Communications'95*, 1995.

[15] Neame, T., et al., "Investigation of Traffic Models for High Speed Data Networks," *Proc. ATNAC '95*, Sydney, Dec. 1995.

[16] Addie, G., M. Zukerman, and T. Neame. "Fractal Traffic: Measurements, Modelling and Performance Evaluation," *Proc. IEEE INFOCOM '95*, Boston, Apr. 1995.

[17] Andersen, A. T., "Modelling of Packet Traffic with Matrix Analytic Method," Ph.D. Dissertation, Institute of Mathematical Modelling, Technical University of Denmark, 1995.

[18] Braun, H.-W., and K. C. Claffy, "Web Traffic Characterization: An Assessment of the Impact of Caching Documents From NCSA 022s Web Server," *Computer Networks ISDN Syst.*, Vol. 28, 1999, p. 37–52.

[19] Crovella, M. E., and A. Bestavros, "Self-Similarity in World Wide Web Traffic: Evidence and Possible Causes," *Proc. 1996 ACM SIGMETRICS*, May 1996.

[20] Beran, J., et al., "Variable-Bit-Rate Video Traffic and Long Range Dependence," *IEEE Trans. Commun.*, Vol. 43, No. 2/3/4, 1995, pp. 1566–1579.

[21] Blondia, C., and O. Casals, "Statistical Multiplexing of VBR Sources: A Matrix-Analytic Approach. *Performance Evaluation*, Vol. 16, No. 5, 1992, pp. 5–20.

[22] Frost, V. S., and B. Melamed, "Traffic Modeling for Telecommunication Networks," *IEEE Commun. Magazine*, Mar. 1994, pp. 70–81.

[23] ITU-T Study Group 2, "Question 17/2, User Demand Modeling in Broadband-ISDN," Draft Text for Recommendation E.716, Geneva, May 1996.

[24] Zukerman, M., and P. W. Tse, "An Adaptive Connection Admission Control Scheme for ATM Networks," *Proc. IEEE ICC '97*, Montreal, Canada, June 1997.

[25] Cormen, H., E. Leiserson, and L. Rivest, *Introduction to Algorithms*. Cambridge, MA: The M.I.T. Press, 1990.

[26] Alles, A., "ATM Internetworking," Cisco Systems Inc., 1995.

[27] PNNI SWG 94-0471R13, "ATM Forum PNNI Draft Specifications," ATM Forum, Apr. 1994.

[28] "Traffic Management Specification," Technical report, ATM Forum, 1996.

[29] Rathgeb, E., "Modeling and Performance Comparison of Policing Mechanisms for ATM Networks," *IEEE J. Sel. Area Commun.*, Apr. 1991.

[30] Kuang, L., "On the Variance Reduction Property of the Buffered Leaky Bucket," *IEEE Trans. Commun.* Vol. 42, 1994, pp. 2670–2671.

[31] Lee, D. C., "Effects of Leaky Bucket Parameters on the Average Queueing Delay: Worst Case Analysis," *Proc. IEEE INFOCOM'94*, 1994, pp. 482–489.

[32] Butto, M., E. Cavallero, and A. Tonietti, "Effectiveness of the Leaky Bucket Policing Mechanism in ATM Networks," *IEEE J. Sel. Areas Commun.*, Vol. 9, Apr. 1991, pp. 335–342.

[33] Yamanaka, N., Y. Sato, and K. I. Sato, "Performance Limitation of the Leaky Bucket Algorithm for ATM Networks," *IEEE Trans. Commun.* Vol. 43, 1995, pp. 2298–2230.

[34] Hluchyj, M. G., and N. Yin., "A Second-Order Leaky-Bucket Algorithm to Guarantee QOS in ATM Networks," *Proc. IEEE GLOBECOM'96*, 1996, pp. 1090–1094.

[35] Silvester, J. A., et al., "The Effectiveness of Multi-level Policing Mechanism in ATM Traffic Control," *Proc. IEEE Int. Telecommunications Symp.* 1996, pp. 98–102.

[36] Mayor, G. S., and J. A. Silvester, "The Multi-Level Leaky Bucket Mechanism," *Proc. 4th IEEE Int. Conf. Computer Communications and Networks*, 1995.

[37] Landsberg, P., and C. Zukowski, "A Novel Buffer Sharing Method: Complete Sharing Subject to Guaranteed Queue Minimums," *Proc. First Int. Conf. Computer Communications and Networks*, 1992, pp. 43–48.

[38] Fonseca, N. L. S., and J. A. Silvester, "A Comparison of Push-Out Policies in an ATM Multiplexer," *Proc. 1993 IEEE Pacific Rim Conf. Communications, Computers and Signal Processing*, May 1993, pp. 338–341.

[39] Lin, A. Y-M., and J. A. Silvester, "Priority Queueing Strategies and Buffer Allocation Protocols for Traffic Control at an ATM Integrated Broadband Switching System," *IEEE J. Sel. Areas Commun.*, 1991, pp. 1524–1536.

[40] Sohraby, K., "On the Asymptotic Behavior of Heterogeneous Statistical Multiplexer With Applications," *Proc. IEEE INFOCOM'92*, 1992, pp. 839–847.

[41] Yin, N., S. Q. Li, and T. E. Stern, "Congestion Control for Packet Voice by Selective Packet Discarding," *IEEE Trans. Commun.*, Vol. 38, No. 5, May 1990, pp. 674–683.

[42] Petr, D. W., and V. S. Frost, "Nested Threshold Cell Discarding for ATM Overload Control Minimization Under Cell Loss Constraints," *Proc. IEEE INFOCOM91*, 1991, pp. 1403–1412.

[43] Petr, D. W., and V. S. Frost, "Priority Cell Discarding for Overload Control Minimization Under Cell Loss Constraints," *Proc. IEEE INFOCOM91*, 1991, pp. 1403–1412.

[44] Fonseca, N. L. S., and J. A. Silvester," Estimating the Loss Probability in a Multiplexer Loaded with Multi-Priority MMPP Streams," *IEEE Int. Conf. Communications'93*, May 1993, pp. 1037–1041.

[45] Garcia, J., and O. Casals, "Stochastic Models of Space Priority Mechanisms With Markovian Arrival Processes," *Ann. Operations Res.*, 1992, pp. 271–295.

[46] Fonseca, N. L. S., and J. A. Silvester, "A Multiple Class Buffer Priority Algorithm for the design of B-ISDN Networks," *Proc. First Int. Conf. Computer Communications and Networks*, San Diego, June 1992, pp. 38–42.

[47] Elwalid, A. I., and D. Mitra, "Fluid Models for the Analysis and Design of Statistical Multiplexing with Loss Priorities on Multiple Classes of Bursty Traffic," *Proc. IEEE INFOCOM'92*, 1992, pp. 415–425.

[48] Bae, J. J., T. Suda, and R. Simha, "Analysis of Individual Packet Loss in a Finite Buffer Queue with Heterogeneous Markov Modulated Arrival Processes: A Study of Traffic Burstiness and Priority Packet Discarding," *Proc. IEEE INFOCOM'92*, 1992, pp. 219–230.

[49] Fourneau, J. M., N. Pekergin, and H. Taleb, "An Application of Stochastic Ordering to the Analysis of the Push-Out Mechanism," *Proc. of Performance Modeling and Evaluation of ATM Networks*, Chapman & Hall, London, July 7–9, 1994, pp. 227–244.

[50] Wu, G. L., and J. W. Mark, "A Buffer Allocation Scheme for ATM Networks: Complete Sharing Based on Virtual Partition," *IEEE/ACM Trans. Networking*, Vol. 3, No. 6, Dec. 1995, pp. 660–670.

[51] Gravey, A., and G. Hebuterne, "Mixing Time and Loss Priorities in a Single Server Queue," *Proc. 13th ITC*, Copenhagen, June 1991, pp. 147–152.

[52] Aras, C. M., J. F. Kurose, D. S. Reeves, and H. Schulzrinne, "Real-Time Communication in Packet-Switched Networks," *Proc. IEEE*, Vol. 82, Jan. 1994, pp. 122–135.

[53] Zhang, L., "Virtual Clock: A New Traffic Control Algorithm for Packet Switching Networks," *Proc. ACM SIGCOMM'90*, 1990, pp. 19–28.

[54] Parekh, A., and R. Gallager, "A Generalized Processor Sharing Approach to Flow Control in Integrated Services Networks: The Single-Node Case," *IEEE/ACM Trans. Networking*, Vol. 1, No. 3, June 1993.

[55] Parekh, A., and R. Gallager, "A Generalized Processor Sharing Approach to Flow Control in Integrated Services Networks: The Multiple Node Case," *IEEE/ACM Trans. Networking*, Vol. 2, No. 2, Apr. 1994.

[56] Yaron, O., and M. Sidi, "Generalized Processor Sharing Networks with Exponentially Bounded Burstiness Arrivals," *Proc. INFOCOM'94*, 1994, pp. 628–633.

[57] Bennet, J. C. R., and H. Zhang, "WF2Q: Worst-Case Fair Weighted Fair Queueing." *Proc. IEEE INFOCOM'96*, 1996, pp. 120–128.

[58] Jaffe, J., "A Decentralized Optimal Multiple-User Flow Control Algorithm," *IEEE Trans. Commun.*, Vol. 29, 1981, pp. 954–962.

[59] Bertsekas, D., and R. Gallager, *Data Networks*, 2nd ed., Upper Saddle River, NJ: Prentice Hall, 1992.

[60] Charny, A., "An Algorithm for Rate Allocation in a Packet-Switched Network With Feedback," Master's Thesis, Massachusetts Institute of Technology, 1994.

[61] Jain, R., "Congestion Control and Traffic Management in ATM Networks: Recent Advances and a Survey," *Computer Networks & ISDN Sys.*, Vol. 28, No. 13, 1996, pp. 1723–1738.

[62] Kung, H. T., and R. Morris, "Credit-Based Flow Control for ATM Networks," *IEEE Network Magazine*, Vol. 9, No. 2, Mar. 1995.

[63] Bonomi, F., and K. Fendick, "The Rate-Based Flow Control Framework for the Available Bit Rate ATM Service," *IEEE Network Magazine*, Vol. 9, No. 2, Mar. 1995.

[64] Chen, T., S. Liu, and V. Samalam, "The Available Bit Rate Service for Data in ATM Networks," *IEEE Commun. Magazine*, May 1996, pp. 56–71.

[65] Clark, D., A. Charny, and R. Jain, "Congestion Control with Explicit Rate Indication," *Proc. ICC'95*, Seattle, WA, June 1995, pp. 1954–1963.

Further Reading

Anick, D., D. Mitra, and M. M. Sondhi, "Stochastic Theory of a Data-Handling System with Multiple Sources," *Bell Syst. Tech J.*, 1982, pp. 1871–1894.

Berger, A. W., and W. Whitt, "Traffic Shaping by a Job Buffer in a Token-Bank Rate-Control Throttle," *Commun. Statistics: Stochastic Models*, Vol. 8, 1992, pp. 685–717.

Lee, C., and M. Andersland, "Consecutive Cell Loss Controls for Leaky-Bucket Admission Systems," *Proc. IEEE GLOBECOM'96*, 1996, pp. 1732–1737.

Towsley, D., R. Nagarajan, and J. F. Kurose, "Approximation Techniques for Computing Packet Loss in Voice Multiplexers," *IEEE J. Sel. Areas Commun.* Vol. 9, No. 3, Apr. 1991, pp. 368–377.

5

ATM Switching Systems for Multimedia Services

5.1 Introduction

The asynchronous transfer mode (ATM) was standardized by the International Consultative Committee for Telecommunications and Telegraphy (CCITT), currently called International Telecommunications Union–Telecommunication (ITU-T), as the multiplexing and switching principle for the broadband integrated services digital network (B-ISDN). ATM is a connection-oriented transfer mode based on statistical multiplexing techniques, and it was chosen to provide flexibility in bandwidth allocation and allows a network to carry heterogeneous services ranging from narrowband to wideband services (e.g., video-on-demand, video conferencing, Video-Phone, and video library). However, the challenge is to build fast packet switches capable of supporting the high-bandwidth and high-performance requirements imposed by different services.

In this chapter, we first describe the performance requirements of designing an ATM switch in Section 5.2. A generic ATM switch structure and necessary operation functions are also presented. Section 5.3 discusses several techniques that are used to design an ATM switch. Section 5.4 describes a number of proposed ATM switches in sequence based on their buffering strategies. Section 5.5 presents two multicast switch design approaches: copy network and broadcast-and-select.

The author of this chapter would like to thank his Ph.D. students, J. S. Park and T. Kijkanjanaret, for contributing part of the chapter. He would also like to thank Dr. C. Lam for reviewing the manuscript.

5.2 Performance Requirements and Basic Functions

5.2.1 Design Criteria and Performance Requirements

Several design criteria need to be considered when designing an ATM switch architecture. First, the switch should provide bounded delay and small cell loss probability while achieving a maximum throughput that is close to 100%. Capability in supporting high-speed input lines is also an important criteria for multimedia services, such as video conferencing and Video-Phone. Self-routing and distributed control are essential for implementing large-scale switches. Serving cells based on a first-come, first-served basis provides correct cell sequence at the output ports. Switches with modular structure can easily be scaled up without modifying their structures.

Bellcore has recommended performance requirements and objectives for its Broadband Switching System (BSS) [1]. As shown in Table 5.1, three quality-of-service (QoS) levels and their associated performance objectives are defined: *QoS Class 1, QoS Class 3,* and *QoS Class 4.* QoS Class 1 is intended for stringent cell loss applications including the circuit emulation of high-capacity facilities such as DS3. It corresponds to Service Class A, defined by ITU-T Study Group XIII and the ATM Forum. QoS Class 3 is intended for low-latency, connection-oriented data transfer applications, corresponding to Service Class C in ITU-T Study Group XIII and the ATM Forum. QoS Class 4 is intended for low-latency, connectionless data transfer applications, corresponding to Service Class D in ITU-T Study Group XIII and the ATM Forum.

Table 5.1
Performance Objectives Across the BSS for ATM Connections Delivering Cells to an STS-3c or STS-12c Interface

Performance Parameters	CLP	QoS 1	QoS 3	QoS 4
Cell loss ratio	0	$< 10^{-10}$	$< 10^{-7}$	$< 10^{-7}$
Cell loss ratio	1	N/S†	N/S	N/S
Cell transfer delay (99th percentile)*†	1/0	150 μs	150 μs	150 μs
Cell delay variation (10^{-10} quantile)	1/0	250 μs	N/S	N/S
Cell delay variation (10^{-7} quantile)	1/0	N/S	250 μs	250 μs

*Includes nonqueueing related delays, excluding propagation. Does not include delays due to processing above ATM layer.
†Not specified.
CLP = cell loss priority

The performance parameters used to define QoS Classes 1, 3, and 4 are cell loss ratio, cell transfer delay, and two-point cell delay variation (CDV). The values of the performance objectives corresponding to a QoS class depend on the status of the cell loss priority (CLP) bit (CLP = 0 for high priority and CLP = 1 for low priority), which is initially set by the user and could be changed by a BSS within the connection's path. For QoS Classes 1, 3, and 4, the probability of cell transfer delay greater than 150 μs is guaranteed to be less than (1 – 0.99), that is,

$$\text{Prob[cell transfer delay} > 150\ \mu\text{s]} < (1 - 99\%)$$

or the probability of CDV greater than 250 μs is required to be less than 10^{-10} for QoS Class 1, that is,

$$\text{Prob[CDV} > 250\ \mu\text{s]} < 10^{-10}$$

5.2.2 ATM Switch Structure and Basic Functions

The general structure of an ATM switch is illustrated in Figure 5.1. Input port controllers (IPCs) are synchronized at the cell level so that all cells arrive at the switch fabric with their headers aligned. This simplifies the design of the switch fabric. The cell stream is slotted; we define the time required to transmit a cell across the network to be a *time slot.* IPCs translate the old VPI/VCI to a new VPI/VCI. This operation is performed by table lookup on the incoming VPI/VCI. The routing table also contains routing information specifying the output port of the switch to which the cell is routed. Other information may also be included in the table on a per-connection basis such as the priority, class of service, and traffic type of the connection.

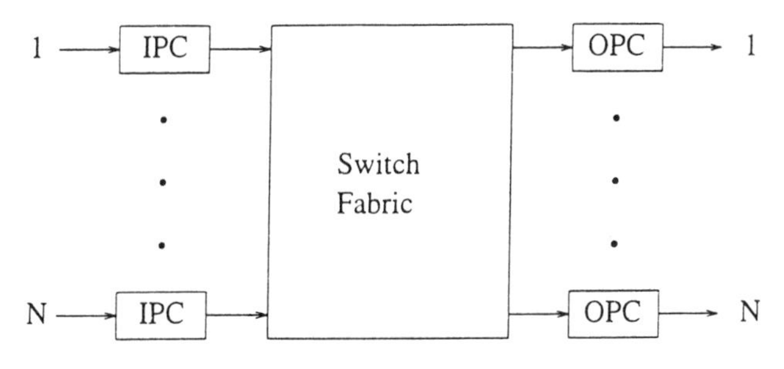

Figure 5.1 General architecture of an ATM switch.

The center switch fabric provides interconnections between any input and output ports. It affects the cost, performance, capacity, scalability, and complexity of the switch system design.

The output port controller (OPC) can only transmit one cell at a time slot. Because cells arrive randomly in the ATM network, it is likely that more than one cell is destined for the same output port. This event is called *output port contention* (or *conflict*). One cell will be accepted for transmission, and the others destined for that output port must either be discarded or buffered. The location of the buffers not only affects the switch performance significantly but also affects the switch implementation complexity. The choice of the contention resolution techniques is also influenced by the location of the buffers.

5.2.2.1 ATM Cell Routing

As mentioned earlier, ATM is connection oriented; that is, an end-to-end connection (or a *virtual channel*) needs to be set up before routing calls. Cells are routed based on two important values contained in the 5-byte cell header: the *virtual path identifier* (VPI) and *virtual channel identifier* (VCI). Each virtual path consists of a number of virtual channels. The number of bits allocated for the VPI depends on the types of interfaces. If the interface is the *user network interface* (UNI), the interface between the user and the first ATM switch, 8 bits are provided for the VPI in each ATM cell along this interface. This means that up to $2^8 = 256$ virtual paths are available at the user access point. On the other hand, if the interface is the *network node interface* (NNI), the interface between the ATM switches, 12 bits are provided for the VPI. This indicates that there are $2^{12} = 4{,}096$ possible virtual paths between ATM switches. In both UNI and NNI, there are 16 bits for the VCI. Thus, there are $2^{16} = 65{,}536$ virtual channels for each virtual path.

The combination of the VPI and the VCI determines a specific virtual connection between two ends. Instead of having the same VPI/VCI for the whole routing path, the VPI/VCI is unique on a per-link basis and is translated at each ATM switch. Specifically, at each incoming link to a switch node, a VPI/VCI may be swapped to another VPI/VCI at the output link using a table called a *header information table* in the ATM switch. This substantially increases the possible number of routing paths.

The operation of routing cells is as follows: Each ATM switch has its own header information table containing four important fields: the old VPI/VCI, the new VPI/VCI, the *output port address*, and the *priority field*. When an ATM cell arrives at an input line of the switch, it is split into the 5-byte header and the 48-byte payload. By using the VPI/VCI contained in the header as the old VPI/VCI value, entries in the header information table are searched for matching with the arriving cell's VPI/VCI. Once a match is found, the old VPI/VCI value is replaced with a new VPI/VCI value. Moreover, the corresponding output port address and priority field are attached to the original 48-byte payload of the cells, which are then sent to the switch fabric. The output port address indicates to which of the output ports the cell should be routed. The three modes of routing operations within the switch fabric are: the *unicast mode,* which refers to the mode in which a cell is routed to a specific output port; the *multicast mode,* which refers to the mode in which a cell is routed to a number of output ports; and the *broadcast mode,* which refers to the mode in which a cell is routed to all output ports. In the unicast mode, $\log_2 N$ bits, where N is the number of input/output ports, are sufficient to indicate any possible output port. However, in the multicast/broadcast modes, N bits, each associated with each output port, are needed for a single-stage switch. The priority field specified according to service requirements enables the switch to transmit cells selectively to the output ports (or discard them when the buffer is full).

The information in the routing table is established by a *call processor.* During the call setup phase, a call processor finds an appropriate routing path between the source and the destination. The VPI/VCI of every link along the path, the output port addresses of the switches, and the priority field are determined and filled into the table by the call processor. The call processor has to ensure that at each switch the VPI/VCI of the cells coming from different connections but going to the same output port are different.

As shown in Figure 5.2, once a call setup is completed by the call processor, the source starts to send a cell whose VPI/VCI is represented by W. As soon as this cell arrives at the first ATM switch, the entries of the table are searched. The matched entry is found with a new VPI/VCI X, which replaces the old VPI/VCI W. The corresponding output port address (whose value is 100) and the priority field are attached to the cell so that the cell can be routed to output port 100 of the first switch. At the second ATM switch, the VPI/VCI of the cell whose value is X, is updated with a new value Y. Based on the output port address obtained from the table, the incoming cell is routed to output port 10. This operation repeats in other switches along the path to the destination. Once the connection is terminated, the call processor deletes the associated entries of the routing tables along the path.

There are two methods of routing cells through an ATM switch fabric: *self-routing* and *label routing.* In self-routing, an output port address field (A) is prepended to each cell at the input port before the cell enters the switch fabric. This field, which has $\log_2 N$ bits for unicast cells or N bits for multicast/broadcast cells, is used to navigate the cells to their destination output ports. Each bit of the output port address field is examined by each stage of the

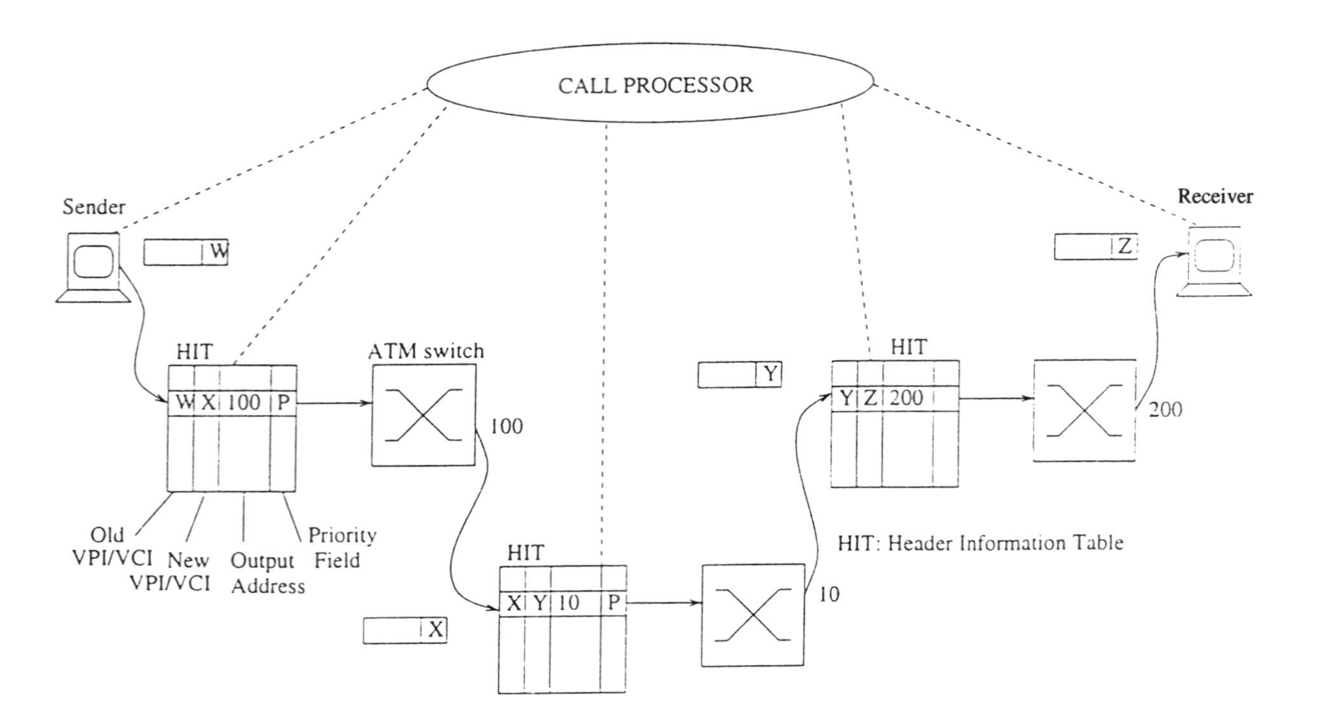

Figure 5.2 VPI/VCI translation along the path.

switch element. If the bit is 0, the cell is routed to the upper output of the switch element. If the bit is 1, it is routed to its lower output. As shown in Figure 5.3, a cell whose output port address is 5 (101) is routed at the input port 2. The first bit of the output port address (1) is examined by the first stage of the switch element. The cell is routed to the lower output and goes to the second stage. The next bit (0) is examined by the second stage and the cell is routed to the upper output of the switch element. At the last stage of the switch element, the last bit (1) is examined and the cell is routed to its lower output corresponding to output port 5. Once the cell arrives at the output port, the output port address is removed.

In contrast, in label routing the VPI/VCI field within the header is used by each switch module to make the output link decision. That is, each switch module has a VPI/VCI lookup table and switches cells to an output link according to the mapping between VPI/VCI and input/output links in the table. Label routing does not depend on the regular interconnection of switching elements as self-routing does and can be used where switch modules are interconnected arbitrarily.

5.2.2.2 Internal Link Blocking

A cell being routed can face a contention problem resulting from two or more cells competing for a single resource. Internal link blocking occurs when multiple cells contend for a link at the same time inside the switch fabric, as shown in Figure 5.4. This usually happens in a switch based on space-division multiplexing where an internal physical link is shared by multiple connections among input/output ports. A blocking switch is a switch suffering from both

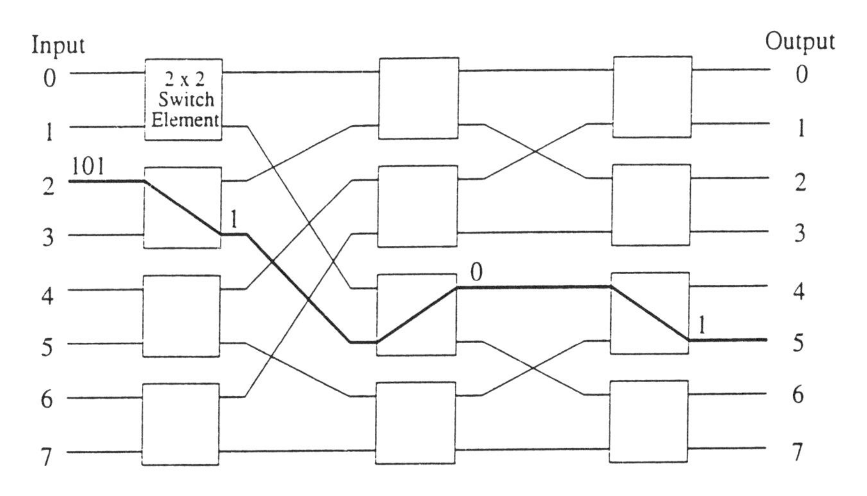

Figure 5.3 An example of self-routing in a delta network.

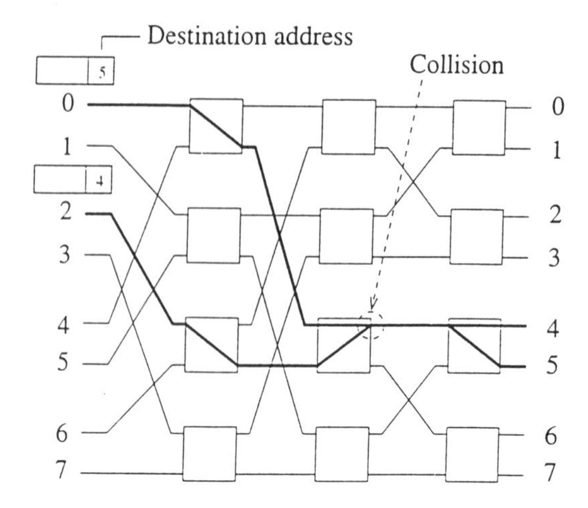

Figure 5.4 Internal blocking in a delta network, two cells destined for output ports 4 and 5 collide.

internal blocking and output port contention. A switch that does not suffer from internal blocking is referred to as *nonblocking.* In an internally buffered switch, contention is handled by placing buffers at the point of conflict. Placing internal buffers in the switch will increase cell transfer delay and degrades the switch throughput.

5.2.2.3 Output Port Contention

Output port contention occurs when two or more cells arrive from different input ports and are destined for the same output port, as shown in Figure 5.5.

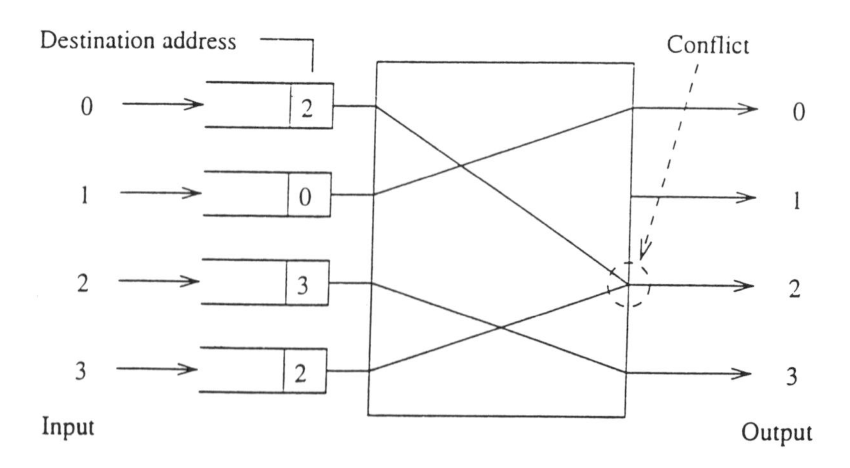

Figure 5.5 Output port contention.

A single output port can transmit only one cell as a time slot, thus the other cells must either be discarded or buffered. In the output-buffered switches, a buffer is placed in each output to store the multiple cells destined for that output port.

5.2.2.4 Head-of-Line Blocking

Another way to resolve output port contention is to place a buffer in each input port, and to select only one cell for each output port among the cells destined for that output port before transmitting the cell. This type of switch is called an *input buffered switch.* An arbiter is necessary to decide which cells should be chosen and which cells should be rejected. This decision can be based on cell priority, time-stamp, or randomness. A number of arbitration mechanisms have been proposed, such as ring reservation, sort and arbitrate, and route and arbitrate. In ring reservation, the input ports are interconnected via a ring that is used to request access to the output ports. For switches that are based on a sorting mechanism in the switch fabric, all cells requesting the same output port will appear adjacent to each other after sorting. In the route and arbitrate approach, cells are routed through the switch fabric and arbiters detect contention at the point of conflict.

A well-known problem arises in an input-buffered switch, called *head-of-line* (HOL) *blocking.* This happens when cells are prevented from reaching a free output because of other cells which are ahead of it in the buffer and cannot be transmitted over the switch fabric. As shown in Figure 5.6, the cell behind the HOL cell at input port 0 is destined for an idle output port 1. But it is blocked by the HOL cell, which failed a transmission due to an output contention. Due to the HOL blocking, the throughput of the input buffered switch is at most 58.6% for random uniform traffic [2,3].

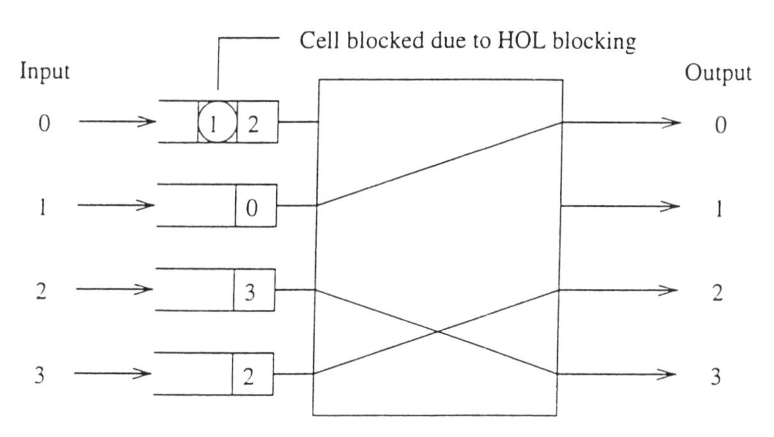

Figure 5.6 Head-of-line (HOL) blocking.

5.2.2.5 Concentration

Concentration arises from the cells in ATM switches destined for the same output, only permitting a subset of cells passing through the switch. Considering an $N \times N$ output buffered switch, the memory must operate at N times (at least) the input rate if zero cell loss in the switch fabric is to be guaranteed. However, this is impractical for a large switch. A concentrator is usually used to reduce the memory speed requirement. An $N \times L$ concentrator may choose up to L cells from N inputs to each output port. Excessive cells are then discarded. By properly engineering the L value, an acceptable cell loss can be achieved under some input traffic distribution.

5.2.2.6 Multicasting

Because of the need of video/audio conference and data broadcasting, ATM switches should have multicast and broadcast capability. Some switching fabrics achieve multicast by first replicating multiple copies of the same ATM cells and then routing each copy to their intended ports. Other switches achieve multicast by utilizing the inherent broadcasting nature of shared medium without generating any copies of ATM cells [4].

5.3 Techniques for Designing ATM Switches

ATM switches can be classified based on their switching techniques into two different groups: *time-division switching* (TDS) and *space-division switching* (SDS). Time-division switching is further divided into *shared memory* type and *shared medium* type. Space-division switching is further divided into *single-path* switches and *multiple-path* switches, which are in turn further divided into several other types, as illustrated in Figure 5.7. In this section, we briefly describe the operation, advantages, disadvantages, and limitations inherent in each type of switch.

5.3.1 Time-Division Switching

In TDS, a single internal communication structure is shared by all cells traveling from input ports to output ports through the switch. The internal communication structure can be a bus, a ring, or a memory. The main disadvantage of this technique is its strict capacity limitation of the internal communication structure. However, this class of switch provides an advantage in that since every cell flows across the single communication structure, it can easily be extended to support multicast/broadcast operations.

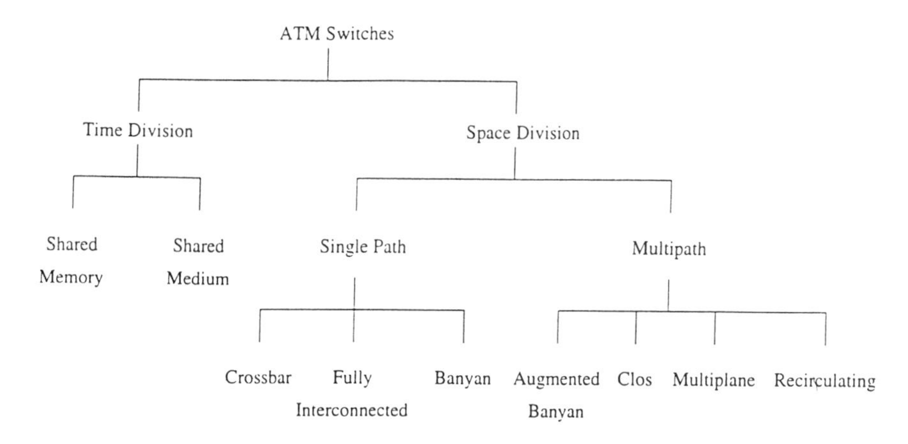

Figure 5.7 Classification of ATM switching architectures.

5.3.1.1 Shared Memory Switch

In a shared memory switch type, as shown in Figure 5.8(a), incoming cells are multiplexed into a single data stream and sequentially written to the appropriate locations of the shared memory depending on their destination addresses. The routing of cells is accomplished by extracting stored cells to form a single output data stream, which is in turn demultiplexed into several outgoing lines. The memory addresses for both writing incoming cells and reading out stored cells are provided by a control module according to routing information extracted from the cell headers.

The advantage of the shared memory switch type is that it provides the best memory utilization since all input/output ports share the same memory. Two different approaches are taken in sharing memory among the ports: *complete partitioning* and *full sharing*. In complete partitioning, the entire

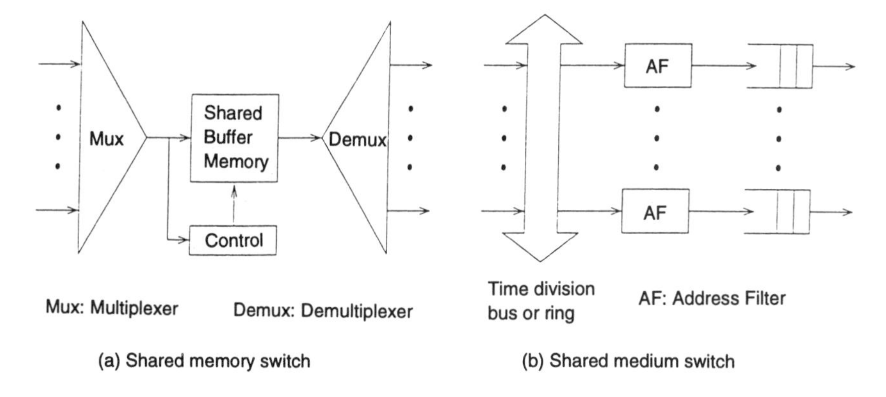

Figure 5.8 Time-division switches.

memory is divided into N equal parts, where N is the number of input/output ports, and each part is assigned to a particular output port. In full sharing, the entire memory is shared by all output ports without any reservation. Some mechanisms, such as limiting an upper and a lower bound on the memory space, are needed to prevent monopolization of the memory by some output ports. The disadvantage of the shared memory switch type is that the memory access speed becomes a limitation when the switch needs to be enlarged. The memory size should be adjusted accordingly to keep the cell loss rate below a specific value.

An example of currently available shared memory switches is Hitachi's 32×32 module [5].

5.3.1.2 Shared Medium Switch

In a shared medium switch type, cells arriving at input ports are multiplexed into a common high-speed medium, such as a bus or a ring, of bandwidth equal to N times the input line rate. The throughput of this shared medium bus determines the capacity of the entire switch. As shown in Figure 5.8(b), each output line is connected to the shared high-speed medium via an interface consisting of an address filter (AF) and an output first-in, first-out (FIFO) buffer. The address filter examines the header part of the incoming cells, and then accepts only the cells destined for itself. This decentralized approach has an advantage in that each output port can operate independently and can be built separately. However, as compared with the shared-memory switch, more hardware logic and more buffers are required to provide the separate interface for each output port.

An example of shared medium switches are NEC's ATOM (ATM Output buffer Modular) switch [6], IBM's PARIS (Packetized Automated Routing Integrated System) switch [7], and Fore System's ForeRunner ASX-100 switch [4].

5.3.2 Space-Division Switching

While in time-division switching a single internal communication structure is shared by all input and output ports, in space-division switching a multiplicity of paths is provided between the input and output ports. These paths operate concurrently so that multiple cells can be transmitted across the switch simultaneously. The total capacity of the switch is thus the product of the bandwidth of each path and the number of paths that can transmit cells concurrently. The upper limit on the total capacity of the switch is therefore theoretically unlimited. However, in practice, it is restricted by physical implementation constraints such as device pinout, connection restrictions, and synchronization considerations.

SDS switches are classified based on the number of available paths between any input/output pair. In *single-path* switches, only one path exists for any input/output pair, while in *multiple-path* switches there is more than one path for an input/output pair.

5.3.2.1 Single-Path Switches

Single-path switches are classified as *crossbar-based* switches, *fully interconnected* switches, and *banyan-based* switches [8], as shown in Figure 5.9.

Crossbar-Based Switches

A crossbar-based switch consists of N^2 crosspoint modules, one for each input and output pair, as shown in Figure 5.9(a). Each crosspoint module consists of an address filter followed by a buffer. Address filters check the destination addresses of the cells and accept the ones whose addresses match its output port address. Among the N buffers connected to an output bus, one buffer will be chosen for transmission in each time slot.

The crossbar-based switch provides a simple and nonblocking structure. However, it is complex in terms of the number of both the address filters and the buffers that grow as a function of N^2, where N is the number of input/output ports. The arbitration mechanism is also more complicated as the switch is enlarged since it has to select one of the stored cells in the N buffers for each output port.

Fully Interconnected Switch

In a fully interconnected switch, the complete connectivity between inputs and outputs is usually accomplished by means of N separate broadcast buses from every input port to all output ports, as shown in Figure 5.9(b). While N^2 dedicated buffers are required in the crossbar-based switch, one at each crosspoint module, only N separate buffers are required in the fully interconnected switch, one at each output port. However, if each of these N output

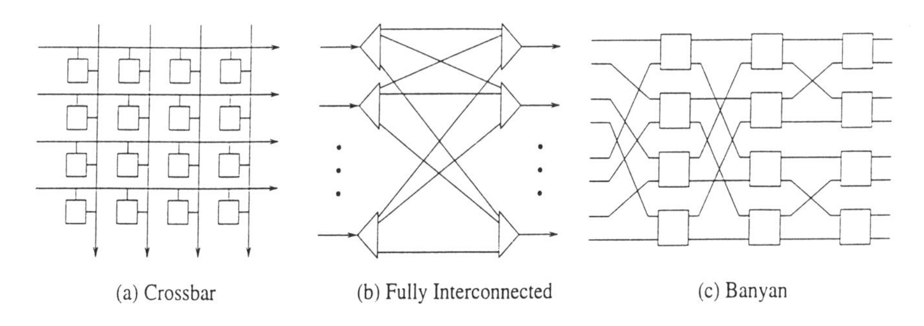

Figure 5.9 Single-path space-division switches.

buffers in the fully interconnected switch is partitioned and dedicated to each input line, the crossbar-based and fully interconnected switch become topologically identical, and thus provide exactly the same performance and implementation complexity.

The fully interconnected switch operates in a manner similar to that of the shared medium switch. A cell from any input port is broadcast to every output port. Thus, cells from several input ports can be simultaneously transmitted to the same output port. Therefore, separate cell filters and dedicated buffers, one for each output port, are required to filter out the misdelivered cells and to temporarily store the properly destined cells.

The fully interconnected switch is different from the shared medium switch in that the speed-up overhead requirement caused by a sequential transmission over the shared medium is replaced by the space overhead requirement of the total N^2 separate broadcast buses. This is considered a disadvantage of the switch type. The advantages of the fully interconnected switch lie in its simple and nonblocking structure, similar to the crossbar-based switch. The knockout switch is the most well-known example of this type of switch [9].

Banyan-Based Switch

Banyan-based switches are a family of self-routing switches constructed from 2×2 switching elements with a single path between any input/output pair. As shown in Figure 5.10, three isomorphic topologies—*delta, omega,* and *banyan* networks—belong to the banyan-based family. All of them offer an equivalent performance.

The banyan-based switch provides several advantages: first, it has a complexity of paths and switching elements of order $N \log N$, which makes it much more suitable than the crossbar-based and the fully interconnected switch, whose complexity is of order N^2, for the construction of large switches. Self-

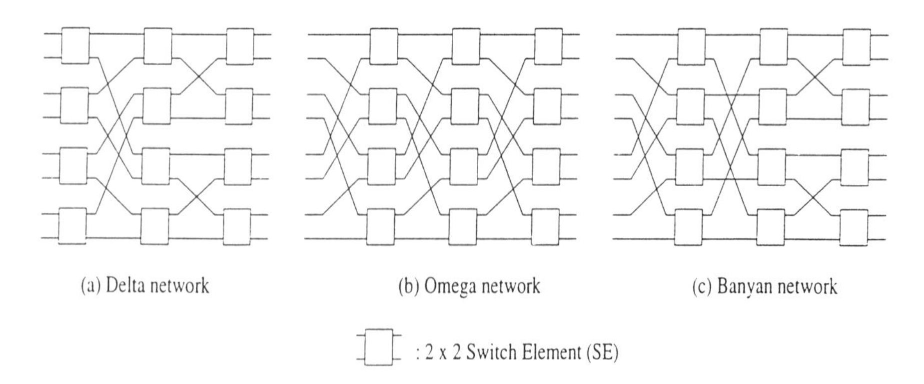

Figure 5.10 Three different topologies of banyan-based switches.

routing property is also an attractive feature in that no control mechanism is needed for routing cells. Routing information is contained within each cell and it is used while the cell is routed along the path. The parallel structure of the switch provides a benefit in that several cells on different paths can be processed simultaneously. Due to their modular and recursive structure, large-scale switches can be built by using elementary switching elements without modifying their structures. This can be appropriately realized by VLSI implementation.

However, the main drawback of the banyan-based switch is that it is an internally blocking switch. Its performance degrades rapidly as the size of the switch increases. The performance may be improved if $M \times M$ $(M > 2)$ switching elements are employed instead of 2×2 switching elements. This leads to the class of *delta-based* switches.

The delta-based switch is a family of self-routing switches constructed from $M \times M$ switching elements with a single path between any input and output port. While the performance of the delta-based switch can be significantly better than that of the banyan-based switch, it is still a blocking switch. The performance of the switch is reduced due to the contention. This can be improved by increasing the speed of internal links within the switch with respect to that of input and output ports or by introducing buffers into the switching elements.

5.3.2.2 Multiple-Path Switches

As shown in Figure 5.11, multiple-path switches are classified as (a) *augmented banyan* switches, (b) *Clos* switches, (c) *multiplane* switches, and (d) *recirculation* switches.

Augmented Banyan Switches

In a regular $N \times N$ banyan switch, cells pass through $\log N$ stages of switching elements before reaching their destinations. The augmented banyan switch, as illustrated in Figure 5.11(a), refers to the banyan switch that has more stages than the regular banyan switch. In the regular banyan-type switch, once a cell is deflected to an incorrect link and thus deviates from a predetermined unique path, the cell is not guaranteed to reach its requested output. Here, in the augmented banyan switch, deflected cells are provided more chances to be routed to their destinations again by using later augmented stages. When the deflected cells do not reach their destinations after the last stage, they are discarded.

The advantage of the augmented banyan switch is that by adding augmented stages, the cell loss rate is reduced. The performance of the switch is improved. The disadvantage of this switch type is its complicated routing

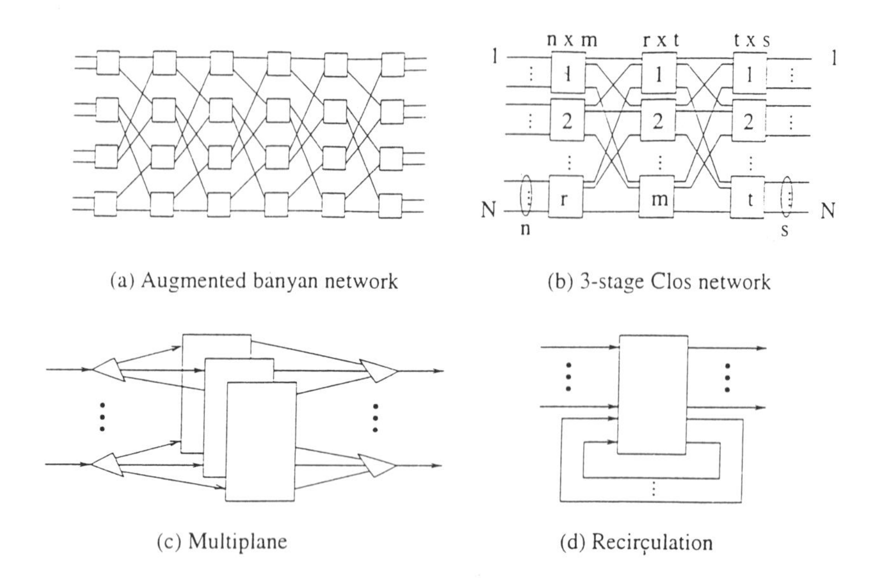

Figure 5.11 Multiple-path space-division switches.

scheme. Cells are examined at every augmented stage to determine whether they have arrived at their requested output ports. If so, they will be sent to the output interface module. Otherwise, they are routed to the next stage and will be examined again. Another disadvantage is that the number of augmented stages needs to be sufficiently large to satisfy desired performance. Adding each augmented stage to the switch causes increased hardware complexity. The tandem banyan switch [10] and dual shuffle exchange switch [11] are examples of the augmented banyan switches.

Three-Stage Clos Switches

The structure of three-stage Clos switches, as shown in Figure 5.11(b), consists of three stages of switch modules. At the first stage, N input lines are broken up into r groups of n lines. Each group of lines goes into each first-stage switch module. There are m outputs in the first-stage switch module; each connects to each middle-stage switch module. Similarly, each middle-stage switch module has t outputs so that it connects to all t third-stage switch modules. At the third stage, N output lines are provided in terms of t groups of s lines.

A consideration with the three-stage Clos switch is that it may be blocking. It should be clear that a crossbar-based switch is *nonblocking*; that is, a path is always available to connect an input port to an output port. This is not always true for the case with the three-stage Clos switch. Figure 5.12 shows a three-stage Clos switch with $N = 9$, $n = 3$, and $m = 3$. The bold lines indicate paths that are already in use. We can see that input port 9 cannot be connected

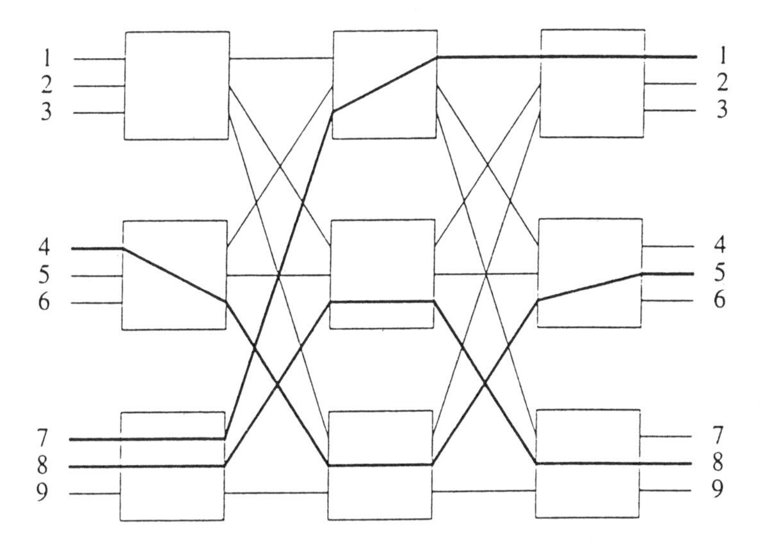

Figure 5.12 Example of internal blocking in a three-stage Clos switch.

to either output port 4 or 6, even though both of these output lines are available.

By increasing the value of m (the number of outputs from each first-stage switch module or the number of middle-stage switch modules), the probability of blocking is reduced. To find the value of m needed for a nonblocking three-stage switch, let us refer to Figure 5.13. We wish to establish a path from input port a to output port b. The worst-case situation for blocking occurs if all of the remaining $n - 1$ input lines and $n - 1$ output lines are busy and are connected to different middle-stage switch modules. Thus a total of $(n - 1) + (n - 1) = 2n - 2$ middle-stage switch modules are unavailable for creating a path from a to b. However, if one more middle-stage switch module exists, an appropriate link must be available for the connection. Thus, a three-stage Clos switch will be nonblocking if

$$m \geq (2n - 2) + 1 = 2n - 1$$

The total number of crosspoints N_x in a three-stage Clos switch when it is symmetric (i.e., when $t = r$ and $s = n$) is

$$N_x = 2Nm + m\left(\frac{N}{n}\right)^2$$

Substituting $m = 2n - 1$ into N_x, we obtain

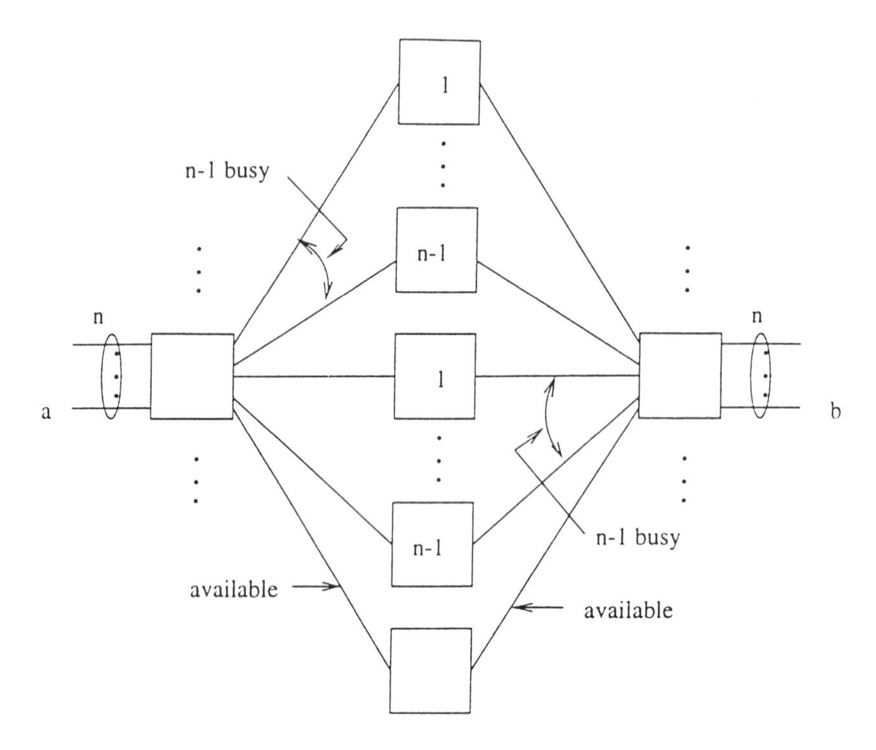

Figure 5.13 Nonblocking condition for a three-stage Clos switch.

$$N_x = 2N(2n - 1) + (2n - 1)\left(\frac{N}{n}\right)^2$$

for a nonblocking switch. For a large switch size, n is large. We can approximate

$$N_x \simeq 2N(2n) + 2n\left(\frac{N}{n}\right)^2 = 4Nn + 2\left(\frac{N^2}{n}\right)$$

To optimize the number of crosspoints, differentiate N_x with respect to n and set the result to 0. The result will be $n \simeq \left(\frac{N}{2}\right)^{1/2}$. Substituting into N_x,

$$N_x = 4\sqrt{2}N^{3/2} = O(N^{3/2})$$

The three-stage Clos switch provides an advantage in that it reduces the hardware complexity from $O(N^2)$ in the case of the crossbar-based switch to

$O(N^{3/2})$ while the switch could be designed to be nonblocking. Furthermore, it also provides more reliability since there is more than one possible path through the switch to connect from any input port to any output port. The main disadvantage of this switch type is that some fast and intelligent mechanism is needed to rearrange the connections in every cell time slot according to arrival cells so that internal blocking can be avoided. This would be the bottleneck when the switch size becomes large. In practice, it is difficult to avoid the internal blocking although the switch itself is nonblocking. Once the contention on the internal links occurs, the throughput is reduced. This can be improved by increasing the number of internal links between switch modules so that there are more paths for routing cells. Increasing the bandwidth of internal links is also helpful in that instead of having one cell for each internal link in each time slot, now more than one cell from the input module that is destined to the same third-stage module can be routed. Another way to reduce the internal blocking is to route cells in a random manner. However, some steps have to be taken carefully at the output ports in order to preserve cell-sequencing.

Multiplane Switches

As shown in Figure 5.11(c), multiplane switches refer to the switches that have multiple, usually identical, switch planes. Multiplane switches are mainly proposed as a way to improve system throughput. By using some mechanisms to distribute the incoming traffic loading, cell collisions within the switches can be reduced. Additionally, more than one cell can be transmitted to the same output port by using each switch plane so the output lines are not necessary to operate at a speed faster than that of the input lines. Another advantage of the multiplane switches is that they are used as a means of achieving reliability since the loss of a complete switch plane will reduce the capacity but not the connectivity of the switches. However, cell-sequencing may be disturbed unless cells belonging to the same connection are forced to use the same plane. The parallel banyan switch and the Sunshine switch [12] are examples of the multiplane switches.

Recirculation Switches

Recirculation switches, as shown in Figure 5.11(d), are designed to handle the output port contention problem. By recirculating the cells that did not make it to their output ports during the current time slot back to the input ports via a set of recirculating paths, the cell loss rate can be reduced. This results in a system throughput improvement. The disadvantage of the recirculation switches is that they require a larger size switch to accommodate the recirculation ports. Also, recirculation may cause out-of-sequence errors. Some mechanisms are needed to preserve the cell-sequencing among the cells in the same connection. The most well-known recirculation switches are the Starlite switch [13] and the Sunshine switch [12].

5.3.3 Buffering Strategies

In this section, we classify ATM switches according to their buffering strategies. Each type of switch is described and its advantages and disadvantages are discussed.

5.3.3.1 Input Buffered Switches

The input buffered switch, as shown in Figure 5.14(a), suffers from the HOL blocking problem and limits the throughput to 58.6% for uniform traffic. To increase the switch's throughput, a technique called *windowing* can be employed, where multiple cells from each input buffer are examined and considered for transmission to the output port. But at most, one cell will be chosen in each time slot. The number of examined cells determines the window size. It has been shown that by increasing the window size to two, the maximum throughput is increased to 70%. Increasing window size does not improve the maximum throughput significantly but increases the implementation complexity of input buffers and the arbitration mechanism. This is because the input buffers cannot use simple FIFO memory any longer, and more cells need to be arbitrated in each time slot.

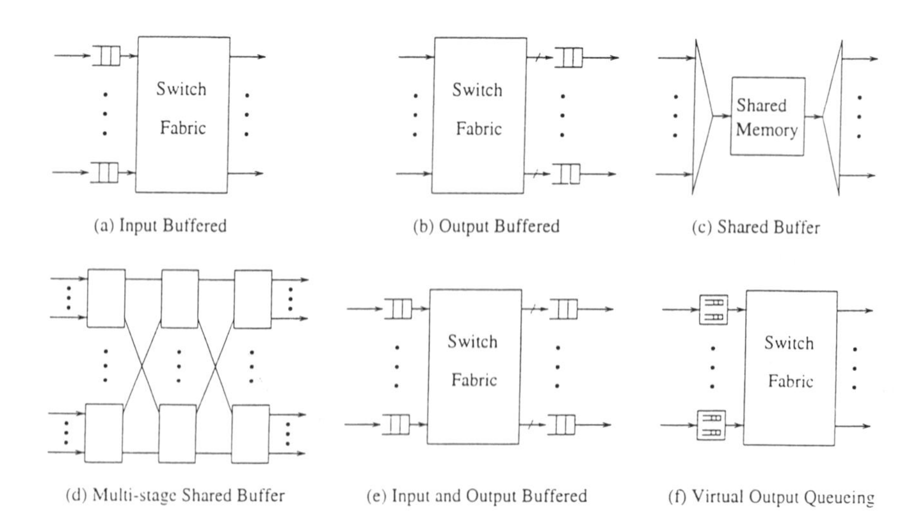

Figure 5.14 Buffering strategies for ATM switches.

5.3.3.2 Output Buffered Switches

The output buffered switch, shown in Figure 5.14(b), allows all incoming cells to arrive at the output port. Because of no HOL blocking, the switch can achieve 100% throughput. However, since the output buffer needs to store N cells in each time slot, its memory speed will limit the switch size. As mentioned before, a concentrator can be used to alleviate the memory speed limitation problem and to obtain a larger switch size. The disadvantage of using the concentrator is the inevitable cell loss in the concentrator. Examples of an output buffered switch are found in Section 5.4.2.

5.3.3.3 Shared Buffer Switches

Figure 5.14(c) shows the shared buffer switch, which is actually the shared memory switch. Its detailed architecture and operations are described in Section 5.4.3.

5.3.3.4 Multistage Shared Buffer Switches

The shared buffer architecture has been widely used to implement small-scale switches because of its high throughput, low delay, and high memory utilization. Although a large-scale switch can be realized by interconnecting multiple shared buffer switch modules, as shown in Figure 5.14(d), the system performance is degraded due to the internal blocking. Examples of this type of switch are presented in Section 5.4.4.

5.3.3.5 Input and Output Buffered Switches

Input and output buffered switches, as shown in Figure 5.14(e), intend to combine the advantages of input buffering and output buffering. In input buffering, the input buffer speed is comparable to the input line rate. In output buffering, there are up to $L (1 < L < N)$ cells that each output port can accept at each time slot. If more than L cells are destined for the same output port, excessive cells are stored in the input buffers instead of discarding them as in the concentrator. To achieve a desired throughput, the speed-up factor L can be engineered based on the input traffic distribution. Since the output buffer memory only needs to operate L times of the line rate, a large-scale switch can be achieved by using input and output buffering. However, this type of switch requires a complicated arbitration mechanism to determine which of L cells among the N HOL cells may go to the output port.

5.3.3.6 Virtual Output Queueing Switches

Virtual output queueing switches are proposed as a way to solve the HOL blocking problem encountered in the input buffered switches. As shown in Figure 5.14(f), each input buffer of the switch is logically divided into N

logical queues. All of these N logical queues of the input buffer share the same physical memory and each contains the cells destined to each output port. HOL blocking is thus eliminated and throughput is increased. However, this type of switch requires a high-speed and intelligent arbitration mechanism. The HOL cells of all logical queues in the input buffers, whose total number is N^2, need to be arbitrated in each time slot, causing the bottleneck of the switch.

5.4 ATM Switch Architectures

In this section, we present the ATM switch architectures proposed in the literature [8,14–17] and classify them according to their buffering strategies: input buffered, output buffered, shared buffer, multistage shared buffer, input and output buffered, and virtual output queueing. Each subsection describes one or two examples belonging to that category and gives some advantages and disadvantages.

5.4.1 Input Buffered Switches

In the input buffered switch, cell accessing the switching fabric is usually arbitrated prior to the transmission according to a certain contention-resolution scheme such as the three-phase algorithm [2]. If more than one cell is destined for the same output port, only one cell is routed. Other cells that fail to get permission are temporarily stored at the input ports. The Batcher-banyan network with the three-phase algorithm is an example of the input buffered switch.

The input buffered switch has the advantage that internal operation speed is only slightly higher than the input line, and thus implementation is easier. However, an input buffered switch has one critical drawback: The throughput is limited to 58.6% because of the HOL blocking phenomena, a well-known effect in which cells destined for idle outputs cannot be served because they are queued behind a blocked cell. The throughput of the input buffered switch is further degraded in the case of the internally blocking switch. Therefore, an input buffered switch usually employs several techniques in order to improve throughput performance.

Three methods have been devised to alleviate HOL blocking effects and thus increase throughput. First, *output expansion* creates extra switch paths and allows more than one cell to be forwarded to each output port during the same time slot. This method, requiring the addition of output buffers to the switch, will be described in detail in Section 5.4.5. Oie, et al. [18] theoretically showed that the maximum throughputs, obtained on the assumption of an infinite number of buffer and random uniform traffic, are 88% and 98% when output expansion ratio $L = 2$ and $L = 3$, respectively. Second, *output channel grouping* divides the switch output ports into a number of groups and allows cells to contend for a group rather than a specific port. Pattavina [19] showed that the maximum throughputs are 88% and 92% when the group sizes are 16 and 32, respectively. Third, *windowing* allows blocked cells behind the HOL cells to contend for output port access. Although output expansion and channel grouping require fundamental alterations of the switching fabric, they are usually preferred over windowing because of the extra contention cycles required to query non-HOL cells, and the throughput improvement is not significant. For instance, when $W = 8$, the maximum throughput is only 88% [20].

5.4.1.1 Three-Phase Algorithm [2]

In the three-phase contention-resolution scheme [2], cells are transferred to the routing network in three phases, as shown in Figure 5.15(a). In phase I, all input ports issue requests in parallel for output ports. In the example of Figure 5.15(a), the HOL cell at input port 1 is destined for output port 3, input port 2 is destined for output port 1, etc. The Batcher sorting network sorts the requests in a nondecreasing order according to the destination addresses. The request will be granted only if the destination address is different from the destination address above in the sorted list. Cells from input ports 2, 4, and 3 are successful. But a cell from input port 1 is unsuccessful because a cell from input port 4 is destined for the same output port 3 and routed to an upper link in the Batcher sorting network. In phase II, the result of the arbitration must be informed to input ports using a fixed connection between the output ports and input ports. In this example, output ports 1, 2, and 4 of the Batcher network inform input ports 1, 2, and 4 of their success and allow these input ports to send an ack cell with destination (2) at input port 1, (4) at input port 2, and (3) at input port 4. These acknowledgment cells will be routed to output ports 2, 3, and 4 without any conflict through the Batcher-banyan network. During phase III, the acknowledged input ports 2, 3, and 4 will forward their HOL cells through the Batcher-banyan network without conflicts. Unsuccessful input port 1 will hold the HOL cell and will try again in the next time slot.

Since the switch fabric itself is used for contention resolution, cells are not being switched at all times and throughput thus decreases. Throughput is also limited due to HOL blocking. The first two phases may be repeated before executing phase III, resulting in improved throughput at the expense of increased arbitration time.

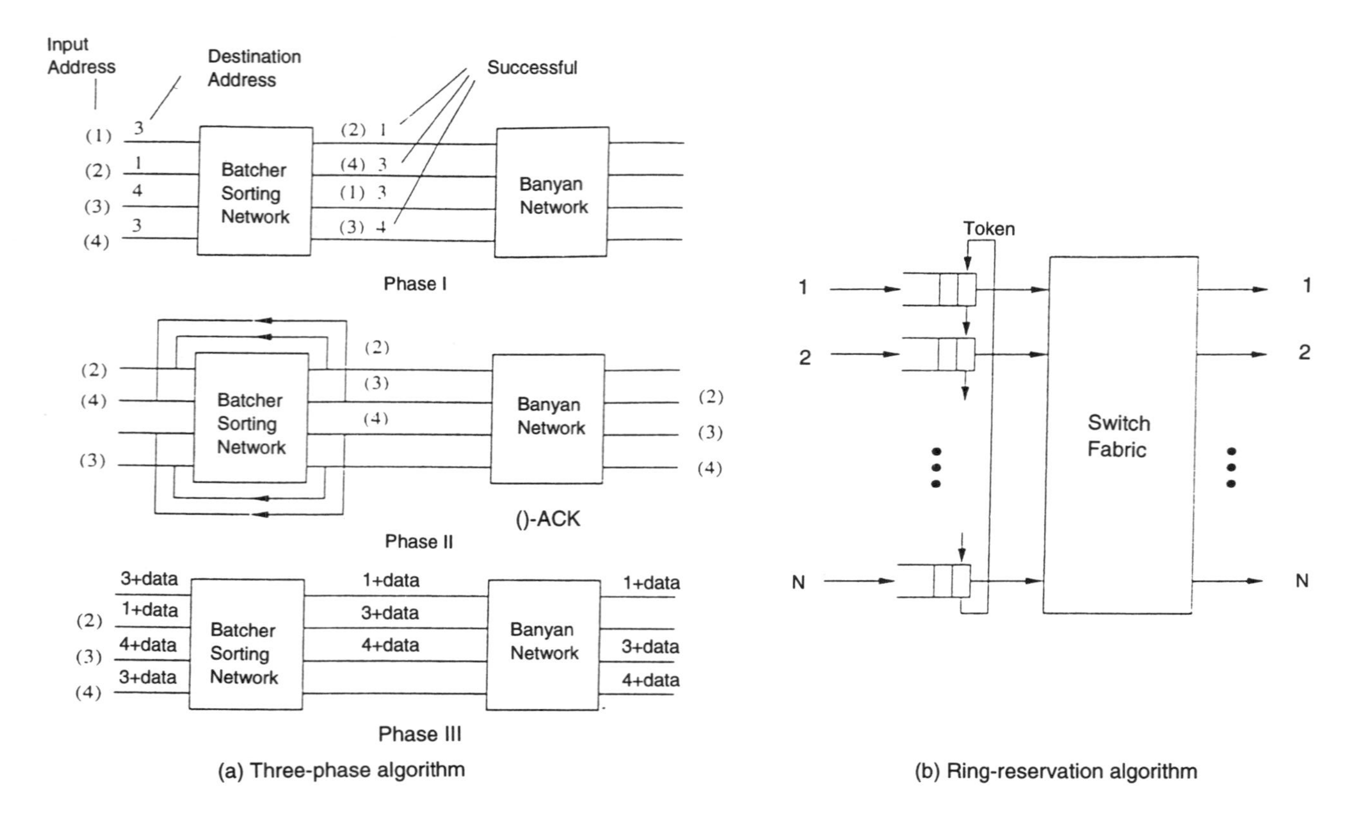

Figure 5.15 The output contention-resolution algorithms in an input buffered switch.

5.4.1.2 Ring Reservation Algorithm [21]

A reservation-based contention-resolution algorithm for the Batcher-banyan network was proposed by Bingham and Bussey [21] in 1988. As shown in Figure 5.15(b), a string of 1-bit tokens is passed around a ring through all input buffers in a pipelined fashion (1 bit per clock), with each bit representing the status of an output port (0 = FREE, 1 = TAKEN). At each input buffer, a counter keeps track of the output port associated with the token bit. If the HOL cell is destined for the port corresponding to the present token, and the token is TAKEN, the cell will be blocked. If the ports match and the token is FREE, the cell is granted admission, and the token is flipped to TAKEN. If the cell is destined for a different port, the token is passed unchanged.

Figure 5.16 shows an implementation of the scheme. There are six HOL cells, each associated with a number indicating the output port address. For each input port *i*, we have a 1-bit token x_i and a counter z_i, which is initialized to *i* at the start of an arbitration cycle. All the token bits x_i's are linked in a circular shift register, which is set to zero (FREE) before starting the arbitration.

For each time slot, each cell's output port address is compared with the corresponding counter value. If they both are matched, a check is used to indicate that the cell can be sent to that output port and the token is set to be one (TAKEN) indicating that the output port is reserved for this cell. After finishing the operations in each time slot, the tokens are shifted up 1 bit and all counter values associated with every port are modulo-increased by one.

As shown in Figure 5.16, at the first time slot, the output port addresses of cells from input ports 1 and 5 are matched, some checks are used to indicate that the cells can be sent to these output ports. Token bits x_1 and x_5 are set to one to indicate that output ports 1 and 5 are already reserved. All the token bits are shifted up 1 bit and the counter values are also modulo-increased by one for the second time slot. No matches are found at the second and the third time slots. At the fourth time slot, the output port addresses of cells from input ports 0 and 2 are matched. Since output port 5 is already reserved for the cell in the previous time slot, which is indicated by the value of the token bit x_2, the cell at input port 2 cannot be sent. Similarly, for the fifth and the sixth time slots, cells at input ports 3 and 4 cannot be sent to output ports 1 and 3, respectively, since those output ports are already reserved in the previous time slots. At the end, cells from the input ports that are checked are the ones winning the contention.

In this example, since there are six input ports, the arbitration cycle can be completed within six time slots. This scheme uses the serial mechanism and, in general, the arbitration cycle can be done within *N* time slots, where *N* is the number of input/output ports of the switch. This will become the

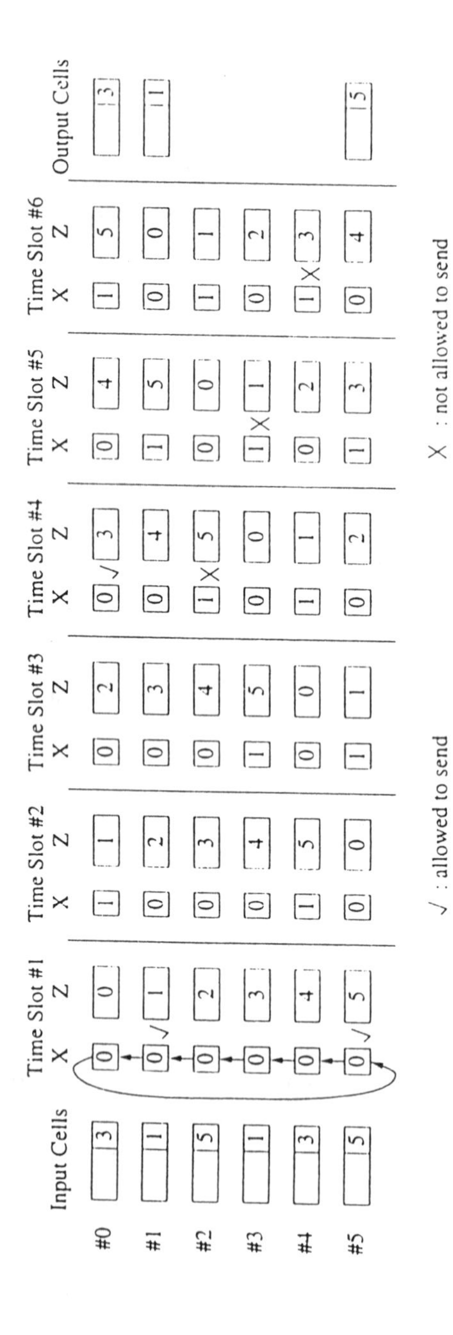

Figure 5.16 The implementation of the ring reservation scheme.

bottleneck when the number of ports of the switch is large. However, by arbitrarily setting the appropriate values for the counters prior to the arbitration, this scheme provides fairness among the input ports. Another advantage of this scheme is that it can be employed at the input of any type of switch fabric.

5.4.2 Output Buffered Switches

Output queueing has the optimal or best possible delay-throughput performance for all traffic distributions, which is particularly important for future ATM applications in which traffic behavior will be difficult to predict. To implement output queueing, an N-times speed-up fabric is required because all N arriving cells may be destined for the same output. Note that speedup can be implemented with increased clock rate or parallel buses.

Among output buffered switches are the knockout switch [9], the recursive modular switch [22], the GAUSS switch [16], the ATOM switch [6], the MOBAS switch [23], the shared concentration and output queueing (SCOQ) switch [24], and the tandem banyan switch [10].

5.4.2.1 Knockout Switch

The knockout switch [9], as shown in Figure 5.17(a), exploits a key observation: In any practical switching system, cell loss within the network is unavoidable. This architecture uses a fully interconnected topology to broadcast passively all incoming cells to all outputs. Each output port has a bus interface that performs several functions: filtering, concentration, and finally buffering. A concentrator achieves an N-to-L concentration. If there are k cells arriving in a time slot for a given output, these k cells, after passing through the concentrator, will emerge from the concentrator on outputs 1 to k, when $k < L$. If $k > L$, then all L outputs of the concentrator will have cells and $k - L$ cells will be dropped within the concentrator. A concentrator with cell loss probability smaller than 10^{-6} at a load of 0.9 can be achieved with $L = 8$ for an arbitrary large value of N. Larger values of L are required under imbalanced traffic models [25]. Despite its excellent performance, the knockout switch is complex for large switch sizes. It requires N^2 disjoint physical paths, N^2 address filters, and N N-to-L concentrators.

5.4.2.2 Recursive Modular Switch

A recursive modular switch [22] was proposed to reduce the complexity of the knockout switch. The switch uses the distributed knockout principle, as shown in Figure 5.17(b). While cell filtering and concentration functions are performed centrally at each output port in the knockout switch, they are performed in a distributed way in the recursive modular switch by small switching elements

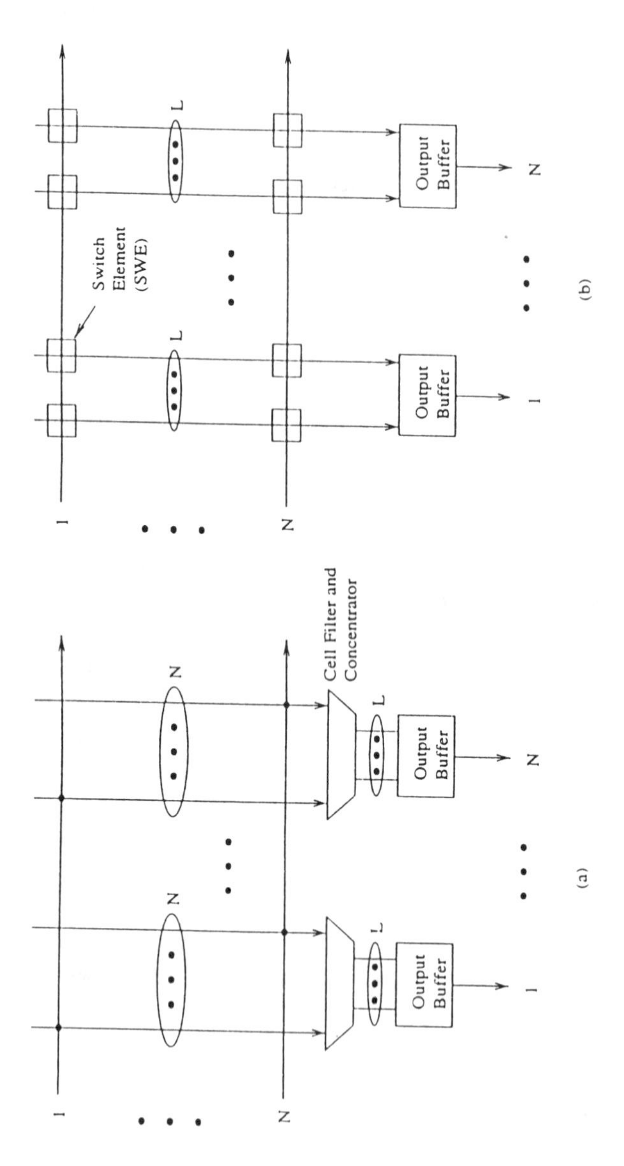

Figure 5.17 (a) The knockout switch and (b) the recursive modular switch using the distributed knockout principle.

(SWEs) located at the intersections of the crossbar lines. The SWEs examine incoming cells from horizontal lines and route them to one of the L vertical lines of each output port. The number of vertical lines is reduced from N^2 to LN, and concentration within each output bus interface is no longer needed. This is, however, at the expense of LN^2 SWEs. It was pointed out that because of the crossbar structure of SWEs this switch design has a regular and uniform structure and has relaxed synchronization requirements. Several methods for reducing L were also discussed. The recursive modular switch is modified to cope with the multicast capability in [23], in which it is shown that a switch designed to meet the performance requirement for unicast calls will also satisfy multicast call performance.

5.4.3 Shared Buffer Switches

Several different architectures of shared buffer switches have been proposed in recent years [26]. In this section, we consider four shared memory schemes: linked-list, hybrid shared and dedicated output buffered, content addressable memory based, and space-time-space based.

5.4.3.1 Linked-List-Based Shared Buffer Switch [5]

The first group of shared buffer switches employs a linked list to organize logically the buffering system. In this switch, the buffer memories for output queueing are completely shared by all the switch output ports and can be assigned to any output port as required by traffic conditions. The switch architecture is illustrated in Figure 5.18, and its basic operation is as follows. First, cells are converted from serial to parallel format, and the cell headers are analyzed and appropriately translated by header converters. Subsequently, cells from all input ports are multiplexed and written into the shared buffer memory. The buffer memory is logically organized into N linked lists, one for each output line. A linked list is a set of chained memory locations that indicates the place where successive cells for a particular output are stored, maintaining the cell sequence. A pair of address registers (one for write address register (WAR) and one for read address register (RAR)) controls the proper operations in the shared memory. WAR_i indicates the end of the chain and so designates an empty memory location where the next cell that arrives for output port i can be stored. RAR_i indicates the beginning of the list, the location of the first cell that can be delivered to output port i. Both the cells and the linked list structures are stored in the same buffer memory. For each input cell, an appropriate WAR is accessed to get the memory location for that cell. Simultaneously, a new address for an empty buffer location is read from the idle address FIFO buffer (IAFB), which keeps a pool of empty buffer locations, to update

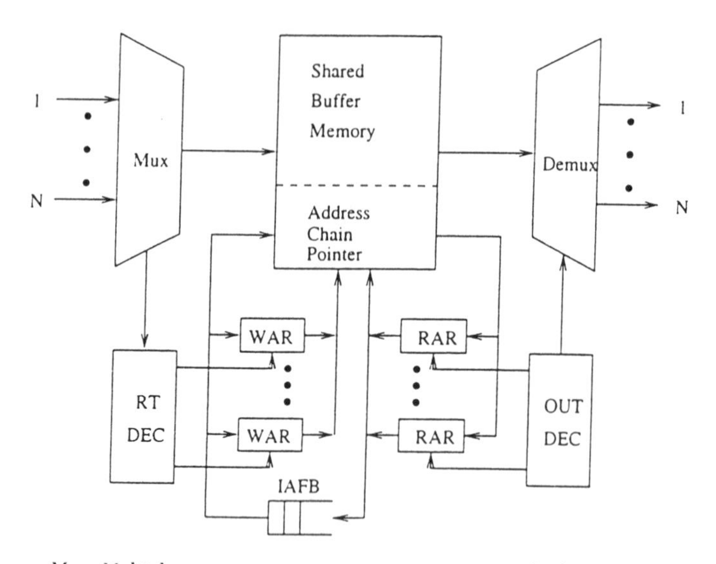

Figure 5.18 A linked-list-based shared buffer switch.

the content of the WAR. Analogously, at each switching cycle a cell from each linked list is identified through the content of RAR, retrieved, demultiplexed, and transmitted. At the same time, the contents of the RARs rejoin the idle address buffer and are replaced by the next chain addresses obtained from the same addresses as the output cells, thus updating the pointers. A complete study on this switch's approaches, including prioritized scheme, multicast function, and performance issues and implementation details, can be found in [5].

5.4.3.2 Hybrid Shared Buffer Switch [27]

In this approach, the control mechanism to route the cell in the shared memory to its output port uses N dedicated FIFO buffers, one for each output port (Figure 5.19). Thus, instead of arranging cells that have the same output port destination in a linked list by the address chain pointer, their addresses are stored in the FIFO buffer. The performance of the switch is the same as that of the shared buffer switch using linked lists, but the necessary amount of buffering is slightly higher due to the buffering required for the FIFO queues. Nevertheless, the implementation is simple and the multicasting and priority control are easy to implement. The hybrid approach provides better reliability than the linked-list approach in that if the cell address stored in FIFO is

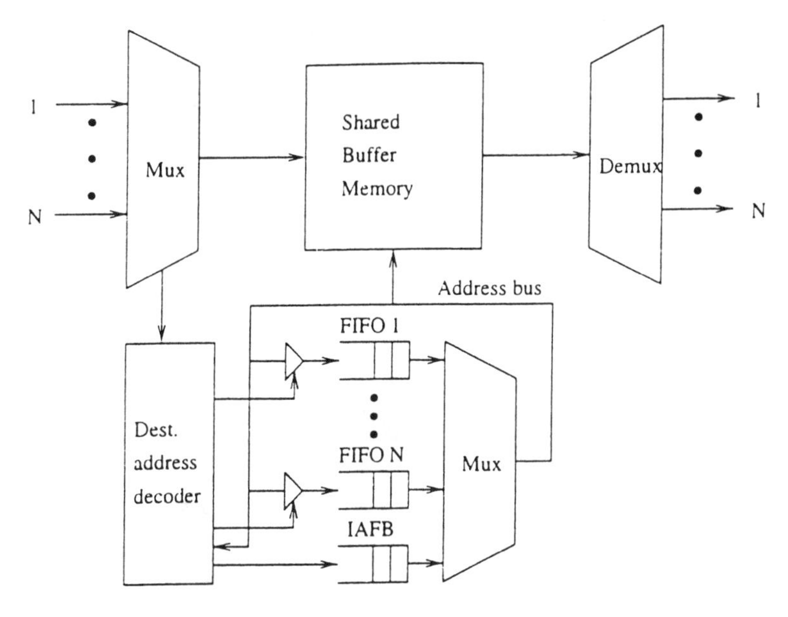

Figure 5.19 A hybrid shared buffer switch.

corrupted due to hardware failure, only one cell is affected. However, in the linked-list approach, if the pointer in the linked list is corrupted, cells in the remaining list will be either routed to an incorrect output port or never be accessed.

5.4.3.3 CAM-Based Switch [28]

Another interesting approach, proposed by Gulak et al. [28], replaces the linked lists by a content addressable memory (CAM), thus simplifying maintenance of cell queues. The basic concept of this architecture is shown in Figure 5.20. Both write and read operations are processed on a round-robin basis. The header of the incoming cell includes the information of the destination port, say, i. The write sequence RAM (WS-RAM) is addressed by i, and returns a value $WS[i] = s$, corresponding to the position of the cell within a virtual queue in the CAM-RAM. This pair (i, s) identifies the cell uniquely while reading. The tag consists of i and s variables, and additionally of a free-busy bit, and is written at the first free location of the CAM. The associated cell is queued at the associated RAM buffer. The updating of the write sequence value $WS[i] = s + 1$ ends the write phase. Reading starts from establishing the

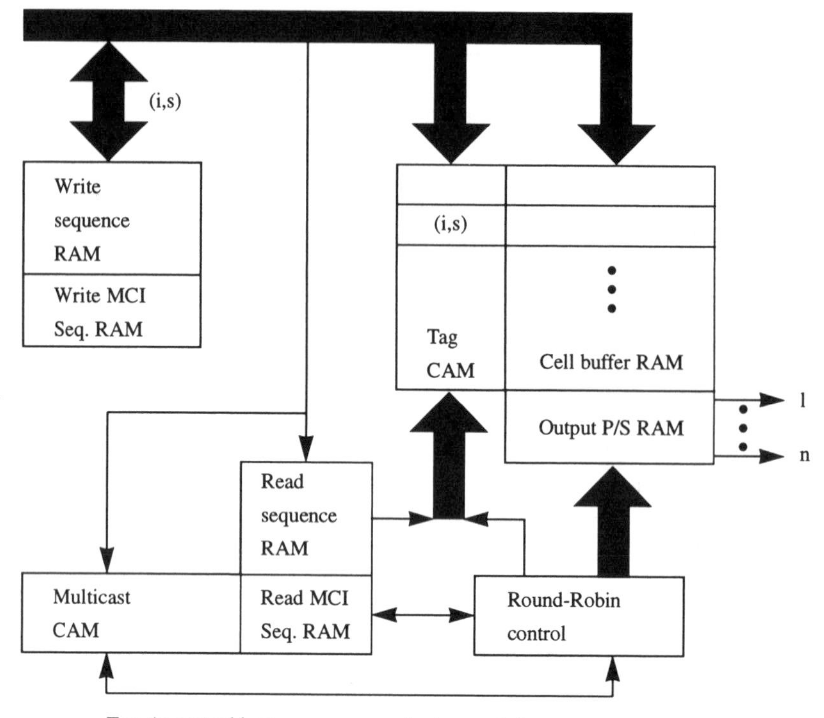

Figure 5.20 A CAM-based shared buffer switch.

position of the cell in the corresponding queue, read sequence RS[i] = s. Then the CAM is searched against the tag (i, s), and the cell associated with it is read from the RAM. Finally, the (i, s) buffer location is free and the RS[i] is incremented. For better performance, the switch operations may be pipelined.

5.4.3.4 STS-Based Shared Memory Switch [29]

Oshima et al. proposed an ATM switch design based on space-time-space (STS) shared buffering scheme (Figure 5.21). In this architecture, the usual multiplexing and demultiplexing stages are replaced by crosspoint space switches, so that the memory access speed can be relaxed. Multiple buffer memories are shared by all input/output ports via the crosspoint space switches. The input side crosspoint switch connects each input port to one shared buffer memory (SBM) module. The output port destination of incoming cells is inspected by header detectors at the input ports. This information is used by a write address control block to set the state of the input side space switch, thus connecting the input ports to the appropriate buffer memories. While writing incoming cells to the memory modules, the least-occupied memory

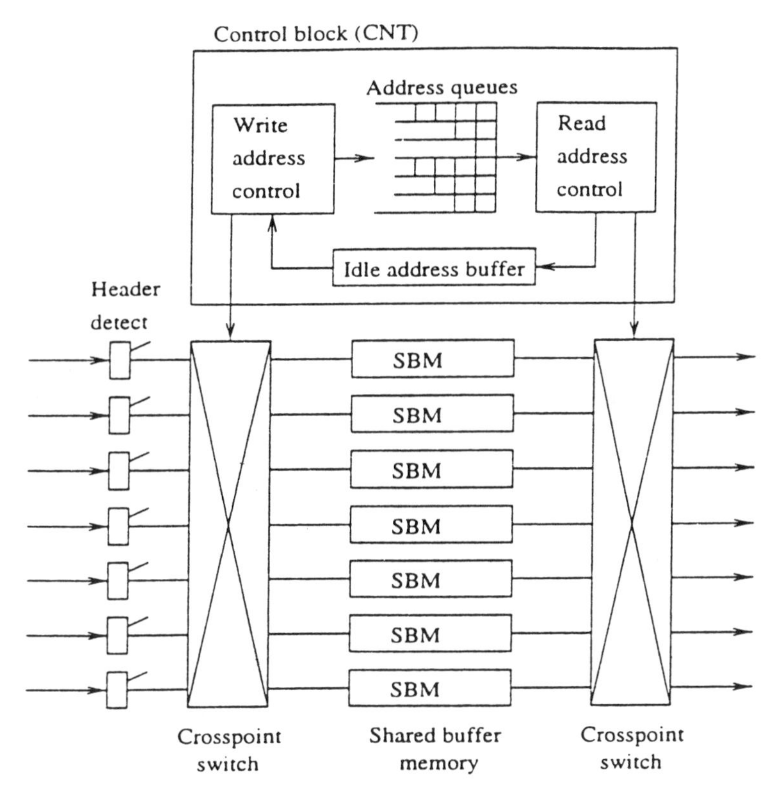

Figure 5.21 A space-time-space–based shared buffer switch.

module is chosen in order to share the buffer effectively, which increases implementation complexity. A read address control block selects the first address from each address FIFO queue and commands the second crosspoint switch to deliver the cells to the requested output line. The idle address buffer maintains empty location addresses for reuse.

In this approach, it may occur that two or more cells with different output port destinations should be read from the same buffer module at the same time. This fact leads to contention and consequently to the degradation of the system throughput. Two methods can be used to alleviate the effects produced by this type of contention. One is to increase the number of SBMs, the other is to accelerate the memory read-out process. Both methods are evaluated in [29] and it is concluded that the speed-up method gives better results. A speed-up of 3 is chosen; therefore, each SBM has to be able to perform one write and three read operations in one time slot.

5.4.4 Multistage Shared Buffer Switches

The shared buffer structure has been widely used for the realization of a small size switch. The next step is to determine how to interconnect these switch modules to have a large-scale switch. Among multistage shared buffer switches are the Washington University gigabit switch [30], concentrator-based growable switch architecture [31], Multinet switch [32], Siemens switch [33], and Alcatel switch [34].

5.4.4.1 Washington University Gigabit Switch [30]

Turner proposed an ATM switch architecture called the Washington University gigabit switch (WUGS). The overall architecture is shown in Figure 5.22. It consists of three main components; the input port processors (IPPs), the output port processors (OPPs), and the central switching network. The IPP receives cells from the incoming links, buffers them while awaiting transmission through the central switching network, and performs the virtual path/circuit translation required to route cells to their proper outputs. The OPP resequences cells received from the switching network and queues them while they await transmission on the outgoing link. Each OPP is connected to its corresponding IPP, providing the ability to recycle cells belonging to multicast connections. The central switching network is made up of switching elements (SEs) with eight inputs and eight outputs and a common buffer to resolve local contention. The SEs switch cells to the proper output using information contained in the cell header or distribute cells dynamically to provide load balancing. Adjacent switch elements employ a simple hardware flow control mechanism to regulate the flow of cells between successive stages, eliminating the possibility of cell loss within the switching network. The switching network uses a Benes network topology. The Benes network extends to arbitrarily large configurations by way

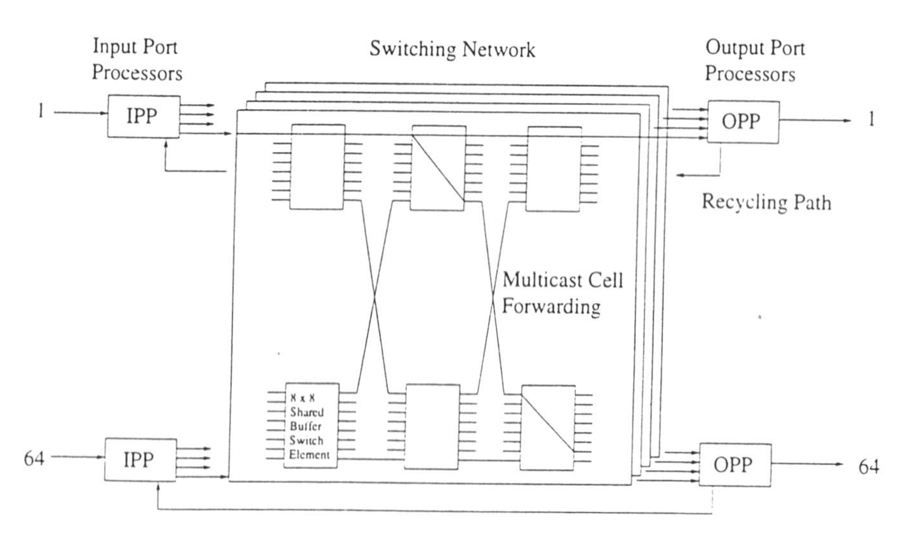

Figure 5.22 Washington University gigabit switch.

of a recursive expansion. Figure 5.22 shows a 64-port Benes network. A 512-port network can be constructed by taking eight copies of the 64-port network and adding the first and fifth stage on either side, with a 64 switch elements copy. Output *j* of the *i*th switch element in the first stage is then connected to input *i* of the *j*th 64-port network. Similarly, output *j* of the *i*th 64-port network is connected to input *i* of the *j*th switch element in the fifth stage. Repeating in this way, any large size network can be obtained. For $N = 8^k$, the Benes network constructed using 8-port switch elements has $(2k - 1)$ stages. Since $k = \log_8 N$, the number of switch elements scales in proportion to $N \log_8 N$, which is the best possible scaling characteristic. In [30], it is shown that when dynamic load distribution is performed in the first $(k - 1)$ stages of the Benes network, the load on the internal data paths of the switching network cannot exceed the load on the external ports; that is, the network achieves ideal load balancing. This is true for both point-to-point and multipoint traffic.

5.4.4.2 Concentrator-Based Growable Switch Architecture [31]

A concentrator-based growable switch architecture is shown in Figure 5.23 [35]. The 8×8 output ATM switches are each preceded by an $N \times 8$ concentrator. The $N \times 8$ concentrator is preceded by a front-end broadcast network (i.e., a memoryless cell distribution network). The concentrators are preceded by address filters (AFs) that only accept cells that are destined to its group of dedicated outputs. The valid cells in each concentrator are then buffered for

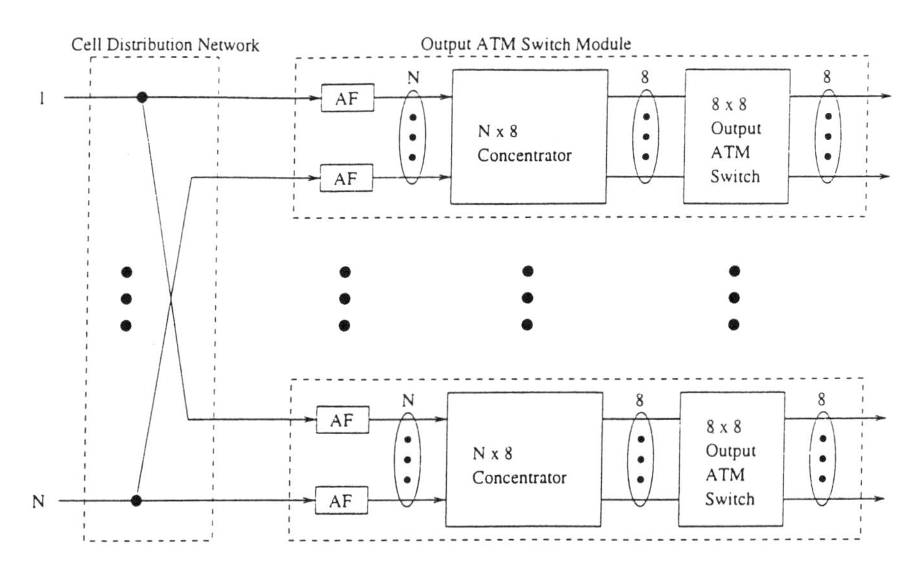

Figure 5.23 A concentrator-based growable switch architecture.

a FIFO operation. In other words, each concentrator is an N-input 8-output FIFO buffer. The actual ATM cell switching is performed in the ATM output switch. Obviously, this is increasingly challenging as N becomes large. For instance, a 512×8 concentrator can be built using a column of eight 64×8 concentrators followed by another 64×8 concentrator. At 2.5 Gbps per port, a 64×64 switch has a capacity of 160 Gbps, and a 512×512 switch has a capacity of 1.28 Tbps.

5.4.5 Input and Output Buffered Switches

Output buffering (including shared memory output buffering) has been proven to provide the best delay and throughput performance. As the switch grows up to a certain size (e.g., 256 input and output ports), memory speed may become a bottleneck or the technology used to implement such memory may become too costly. One way to eliminate memory's speed constraint is to temporarily store some cells destined for the same output port at the input buffers. Input buffering's well-known HOL blocking drawback can be improved by speeding up the internal links' bandwidth (e.g., two to four times that of the input line's) and buffering excessive cells at the output ports. The input-and-output buffering approach thus provides satisfactory performance and eliminates memory speed limitation. Examples of input-and-output buffered ATM switches are the Abacus switch [36], NTT's internal speed-up crossbar switch [37], memory-space-memory (MSM) switch [38], and Lee's modular switch [39].

The challenge of implementing input and output buffered switches is the output port contention resolution of the input cells destined for the same output port (or the same module). Such function is usually handled by an arbiter. The bottleneck caused by the memory speed is now shifted to the arbiter. If parallel processing and pipeline techniques can be intelligently applied to implement the arbiter, a large-scale switch will be feasible.

5.4.5.1 Abacus Switch [36]

Chao et al. proposed the *Abacus* switch, which has a scalable multicast architecture using input and output buffers. It eliminates the possibility of cells being discarded due to the loss of contention in the switch fabric.

As shown in Figure 5.24, the proposed Abacus switch consists of input port controllers (IPCs), a multicast grouping network (MGN), multicast translation tables (MTTs), small switch modules (SSMs), and output port controllers (OPCs). The switch performs cell replication and cell routing simultaneously. Cell replication is achieved by broadcasting incoming cells to all routing modules (RMs), which then selectively route cells to their output links. Cell routing is

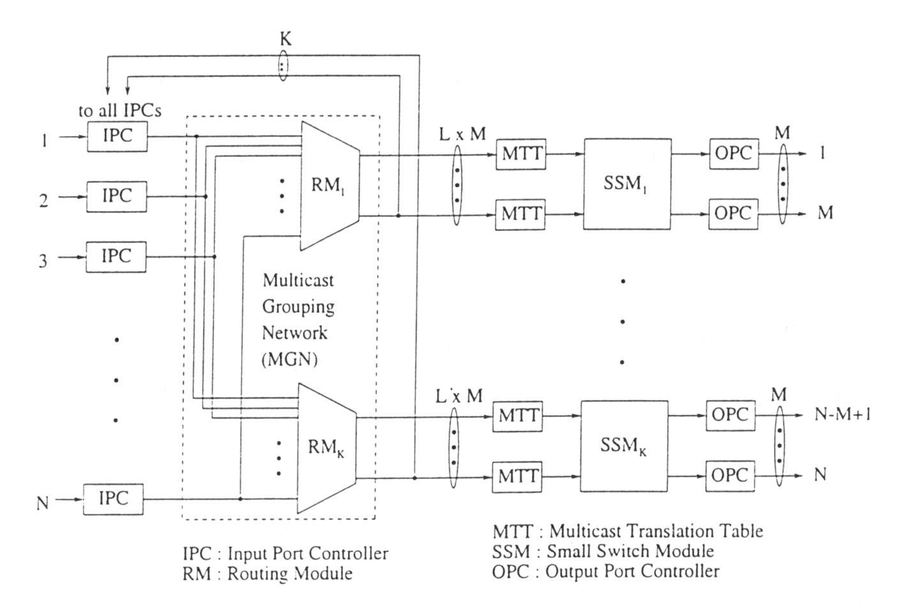

Figure 5.24 The architecture of the Abacus switch.

performed distributedly by an array of switch elements (SWEs). The concept of sharing routing links (also called channel grouping [19]) is also applied to construct the MGN in order to reduce hardware complexity, where every M output ports are bundled in a group. For a switch size of N input ports and N output ports, there are K output groups ($K = N/M$). The MGN consists of K routing modules; each of them provides $L \times M$ routing links to each output group, and L is defined as a group expansion ratio, the ratio of required routing links to the group size. Cells from the same virtual connection can be arbitrarily routed to any one of the $L \times M$ routing links and their sequence integrity will be maintained. Based on a novel arbitration mechanism, up to $L \times M$ cells from N IPCs can be chosen in each RM. Cells that lose contention are temporarily stored in an input buffer and will retry in the next time slot. On the other hand, cells that are successfully routed through RMs will further be routed to proper output port(s) through the SSMs.

We can engineer the group expansion ratio L in such a way that the required maximum throughput in a switch fabric can be achieved. Performance study shows that the larger M is, the smaller L is required to achieve the same maximum throughput. For instance, for a group size M of 16 and input traffic with an average burst length of 15 cells, L has to be at least 1.25 to achieve a maximum throughput of 0.96. But, for a group size M of 32 and the same input traffic characteristic, L can be as low as 1.125 to achieve the same throughput.

The IPCs terminate input signals from the network, look up necessary information in a translation table, resolve contention among cells that are destined to the same output group, buffer those cells losing contention, and attach routing information in front of cells so that they can be routed properly in the MGN.

Each RM in the MGN contains a two-dimensional array of switch elements and an address broadcaster (AB), as shown in Figure 5.25. The SWE routes cells from west and north to east and south, respectively, when it is at cross state, or to south and east, respectively, when it is at toggle state. The SWE's state is determined from the comparison of adddress bits and priority bits of cells from west and north. The AB generates dummy cells that carry proper output group addresses. This permits the SWE not to store the output group address information, which simplifies the circuit complexity of the SWE significantly and results in higher VLSI integration density. The detailed operations of the SWE and the AB can be found in [22,23]. In addition to routing cells, the RMs also sort cell priorities at the output links, which facilitates the new multicast contention resolution algorithm. Each routing module in the MGN has N horizontal input lines and $L \times M$ vertical routing links. These routing links are shared by the cells that are destined for the same output group (i.e., the same small switch module). Each input line is connected to all routing modules so that cells from any input line can be broadcast to all K output groups.

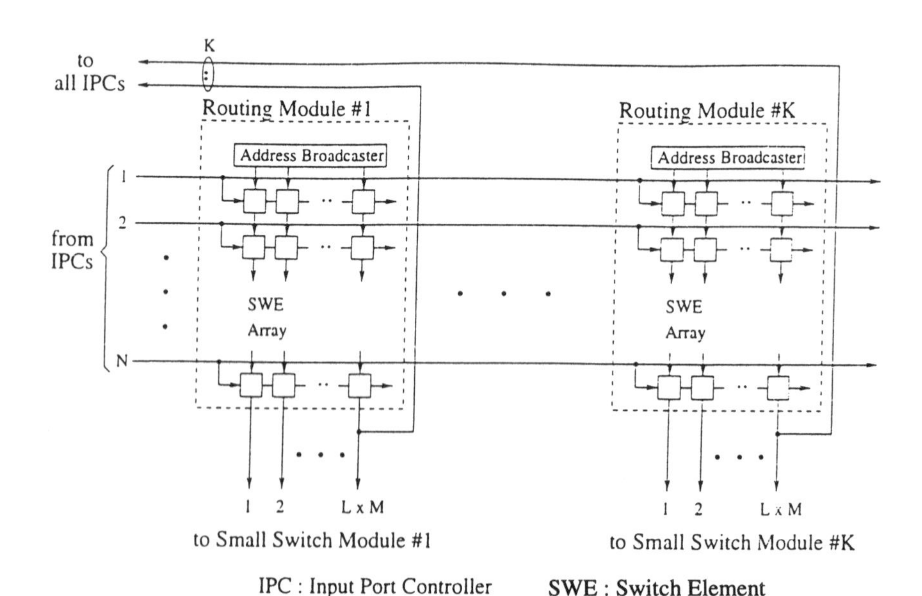

Figure 5.25 The multicast grouping network.

Cells from multicast calls are first replicated and routed by the MGN to multiple SSMs. Before the copied cells are further replicated and routed by the SSMs, their routing field will be updated by the multicast translation tables (MTTs) with proper routing information that is used by the SSMs. Each SSM has $L \times M$ inputs and M outputs. The SSMs must have multicast capability and output buffering structure. The latter is required to maintain the cell sequence for cells distributed among the $L \times M$ links. One example for such SSMs is Hitachi's 32 × 32 shared buffer ATM switch [5]. The output port controller (OPC) updates each multicast cell with a new VPI/VCI and sends the cell to the network.

The RM can not only route cells to proper output groups, but also, based on the cells' priority levels, choose up to $L \times M$ cells that are destined for the same output group. The HOL cell of each input port is assigned a *unique* priority level that is different from the others. After cells are routed through an RM, they are sorted at the output links of the RM according to their priority levels from left to right in a descending order. The cell that appears at the rightmost output link has the lowest priority level among the cells that have been routed through this RM. This lowest priority information is broadcasted to all IPCs. Each IPC will then compare the local priority level (LP) of the HOL cell with a feedback priority, say, FP_j, to determine if the HOL cell has been routed through the RM_j. Note that there are K feedback priorities, $FP_1, \ldots, FP_K$. If the feedback priority level (FP_j) is lower than or equal to the local priority level (LP), the IPC determines that its HOL cell has reached one of the output links of the RM_j. Otherwise, the HOL cell must have been discarded in the RM_j due to loss of contention and will be retransmitted in the next time slot. Since there are K RMs in total, there will be K lines broadcast from K RMs to all IPCs, each carrying the lowest priority information in its output group.

At the beginning of the time slot, each IPC sends its HOL cell to the MGN. Meanwhile, the HOL cell is temporarily stored in a one-cell size buffer during its transmission. After cells have traversed through the RMs, priority information, FP_1 to FP_K (the priority of the rightmost link of each RM), is fed back to every IPC. Each IPC will then compare the feedback priority level FP_j, $j = 1, 2, \ldots, K$, with its local priority level, LP. Three situations can happen. First, $MP_j = 1$ and $LP < FP_j$ (recall that the smaller the priority value, the higher the priority level), which means the HOL cell is destined for the jth output group and has been successfully routed through the jth RM. The MP_j bit is then set to 0. Second, $MP_j = 1$ and $LP > FP_j$, which means the HOL cell is destined for the jth output group but discarded in the jth RM. The MP_j bit remains 1. Third, $MP_j = 0$, the jth bit of the HOL cell's multicast pattern can be equal to 0, which means the HOL cell is not destined for jth output group. Then, the MP_j bit remains 0.

After all MP_j bits (j = 1, 2, . . . , K) have been updated according to one of the preceding three scenarios, a signal indicating whether the HOL cell should be retransmitted, *resend,* will be set to 1 if one or more bits in the multicast pattern remains 1. The *resend* signal is initially set to 0. If multicast pattern bits are all 0, meaning the HOL cell has been successfully transmitted to all necessary output groups, the *resend* signal will be disasserted. The IPC will then clear the HOL cell in the one-cell buffer and transmit the next cell in the input buffer in the next time slot (if any).

5.4.5.2 Internal Speed-Up Crossbar Switch [37]

The internal speed-up crossbar switch, proposed by NTT, uses input and output buffering, and has a center bufferless routing network, as shown in Figure 5.26. The switch architecture properly apportions two functions; cell buffers require only low-speed devices such as CMOS or Bi-CMOS VLSI, while high-speed cell routing is carried out by a bufferless routing network using high-speed devices such as GaAs or bipolar VLSI. Therefore, the switch architecture can offset the speed limitation of cell buffers and allows cell buffer

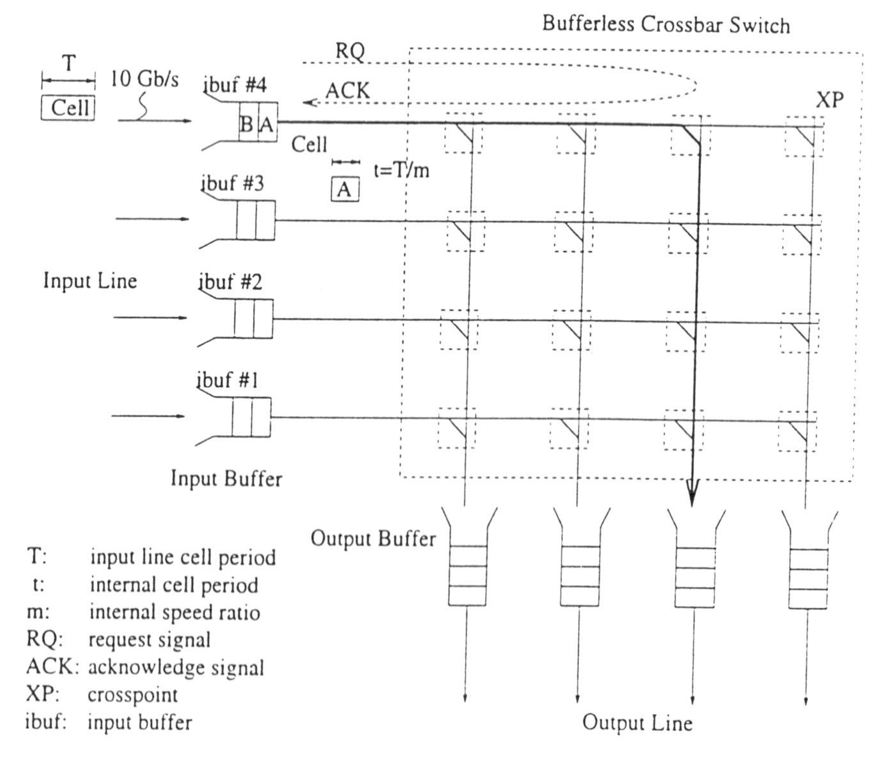

Figure 5.26 The NTT crossbar switch with internal speed-up.

size to increase. High switch throughput is achieved by internal speed-up. State-of-the-art high-speed technologies must be employed to realize cell routing speeds over 10 Gbps in a bufferless routing network.

A crossbar network with short and simple interconnections between adjacent crosspoints is adopted as the bufferless routing network. Each input and output buffer is the FIFO memory. To avoid HOL blocking and to increase the switch throughput, cells are transmitted in the routing network at m times the input line rate. The switch adopts a simple arbitration algorithm based on the three-phase algorithm, which consists of the request phase, acknowledge phase, and cell transmission phase. These three phases operate in a pipelined manner. The switch throughput, calculated as the product of switch size and input loading, increases with the internal speed-up ratio (m) and saturates when m reaches 1.6 for random traffic.

5.4.6 Virtual Output Queueing Switches

Virtual output queueing switches store incoming cells in the buffers at the input ports. The input buffer has N logical queues, each associated with an output port. An arbiter will choose at most N cells from the total N^2 logical queues to transmit to N output ports. For instance, LaMaire et al. [40] proposed a two-dimensional round-robin (2DRR) scheduler to arbitrate the contention among N^2 logical queues, which provides high throughput and fair access in an ATM switch. Two other examples of such a switch are the Tiny Tera switch [41] and Nortel switch [42], which are described next.

5.4.6.1 Tiny Tera Switch [41]

The Tiny Tera switch consists of three main parts: a parallel sliced self-routing crossbar switch, a centralized scheduler for configuring the crossbar, and 32 port interfaces, each operating at 10 Gbps. When a cell arrives at a port interface, it is buffered in an input queue according to its destination, priority class, and unicast or multicast. The cell awaits a decision by the scheduler (i.e., the arbiter) allowing it to traverse the crossbar switch fabric. At the beginning of each time slot, the scheduler examines the contents of all input queues, decides on the configuration of the crossbar, and chooses a set of conflict-free connections between inputs and outputs. The scheduling decision is passed back to the port interfaces, which communicate the configuration information with the crossbar slices and then transmit cells into the crossbar. Cells leaving the crossbar are buffered in the output queues where they await transmission to the external line.

The port interface uses separate structures for buffering unicast and multicast cells, as shown in Figure 5.27. For unicast cells, the port interface

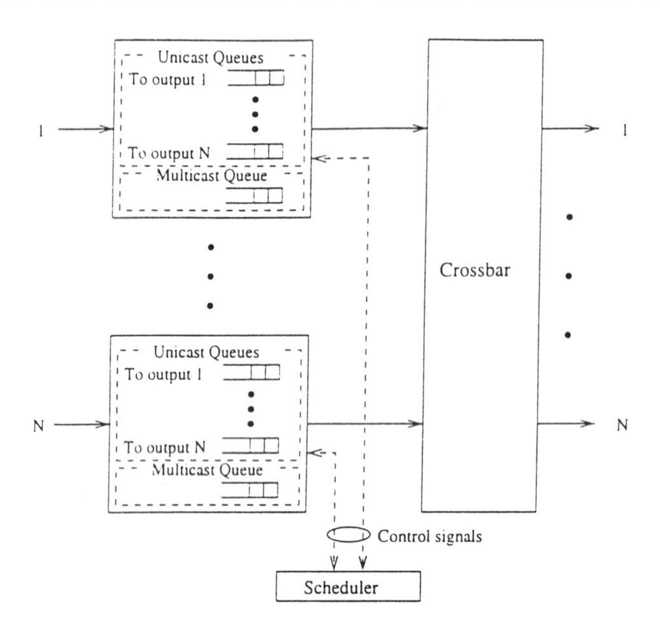

Figure 5.27 The Tiny Tera switch.

maintains a separate logical queue for each output and eliminates HOL blocking. However, it is impractical to eliminate HOL blocking for multicast cells; to do so would require each input to maintain a separate queue for each possible set of destinations. For a 32-port switch, this would mean maintaining $(2^{32} - 33) > 4$ billion different queues. In fact, it has been found that there is little benefit obtained from maintaining multiple queues unless the number of queues approaches 2^{32}. Thus, the Tiny Tera switch only uses one logical queue for all multicast cells in the input buffer.

5.4.6.2 Nortel Switch [42]

A high-capacity ATM switch based on advanced electronic and optical technologies is introduced in [42]. The main components of the switch are input buffer modules, output buffer modules, a high-speed switching core, and a central control unit, as shown in Figure 5.28. The core switch contains a 16×16 cross-connect network using optical links running at 10 Gbps.

The central control unit receives requests from input buffer modules and returns grant messages. Each request message indicates the number of queued cells in the input buffer module, which is later used to determine the size of the burst allowed to transmit to the switch fabric. A connection can only be made when both input and output ports are free. A control bus is used by the free input ports to broadcast their requests, and by the free output ports to return grant messages.

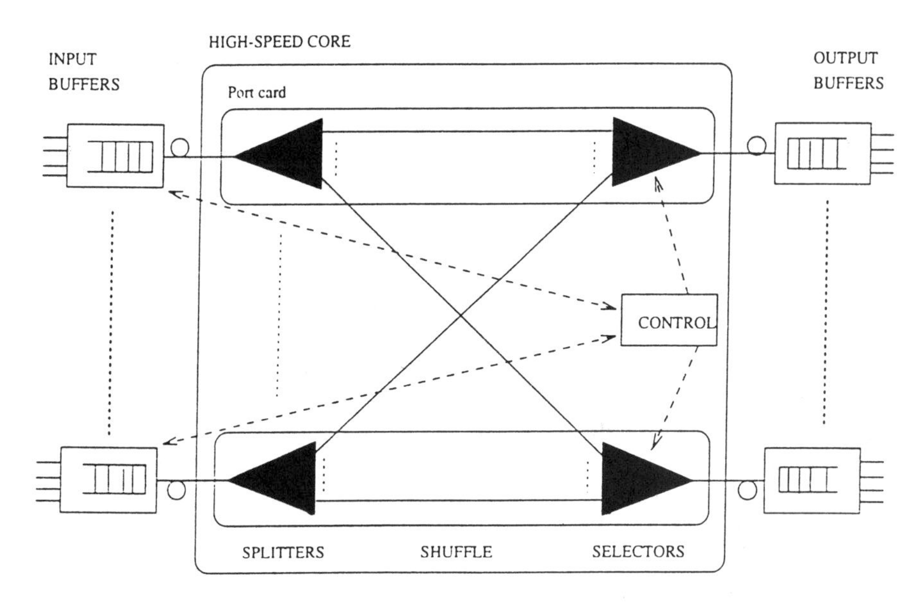

Figure 5.28 A high-capacity ATM switch proposed by Nortel.

An arbitration frame consists of 16 cell time slots for the 16 × 16 core switch. In each slot, the corresponding output port polls all 16 inputs. For example, in time slot 1, output port 1 (if it is idle) will choose the input that has the longest queue destined for output port 1. If the input is busy, another input port that has the second longest queue will be examined. This operation repeats until a free input port is found. If a match is found (free input, free output, and outstanding request), a connection is made for the duration corresponding to the number of cells queued for this connection. So, this switch is a burst switch, not a cell switch. In time slot 2, output port 2 repeats the operation. The switch capacity is limited by the speed of the central control unit. If it is slow, cell streams can have a long waiting time in the input buffer modules under a high traffic load.

5.5 Multicast ATM Switches

B-ISDNs are expected to provide multipoint connections in addition to point-to-point connections. ATM switches should have multicast capability to cope with a wide range of applications such as voice/video teleconferencing, entertainment video, LAN bridging, and distributed data processing. There are basically two different approaches to performing the multicast function: broadcast-and-select and copy network.

Broadcast-and-select type multicast switches broadcast the multicast cell to every output port over the time-division shared medium (bus or ring) or a space-division network. Many switches such as the ATOM switch [6], the multicast knockout switch [43], the GAUSS switch [44], the MOBAS switch [23], and the Abacus switch [36] adopt this approach. Switches that use the copy network to replicate cells before routing them include the ones described in [45–47].

When designing a multicast switch, addressing is one of the more difficult and challenging problems. Because there are 2^N patterns for each multicast cell, there may be up to N bits of information in the routing field of each cell. For instance, for broadcast-and-select type multicast switches, the bit map addressing scheme is usually employed. To reduce the cell header overhead, a multiple-stage structure is applied to implement the switches. The number of addressing bits is reduced from $O(N)$ to $O(\sqrt[n]{N})$, where n is the number of stages. The cascaded-type multicast switches, which consist of a copy network and a routing network, solve the multicast addressing problem by separating the cell replication function from the cell-routing function. The copy network generates the copies requested by incoming cells, and the point-to-point switch routes the replicated copies to their final destinations. The copy network uses $O(\log_2 N)$ bits information for cell replication, and the routing network uses $O(\log_2 N)$ bits information for routing each copy of replicated cells.

5.5.1 Call Splitting

Several call scheduling disciplines have been proposed when considering the multicast function, such as one-shot scheduling, strict-sense call splitting, and wide-sense call splitting [48]. One-shot scheduling requires all the copies of the same cell to be transmitted in the same time slot. Strict-sense (SS) call splitting and wide-sense (WS) call splitting permit the transmission of the cell to be split over several time slots. A matrix can be used to describe the operation of the scheduling algorithm, as shown in Figure 5.29, where each row and column correspond to the input line and output line of the switch, respectively. With the multicast feature each row may contain two or more 1's rather than only one as in the unicast case. At each time slot only one cell can be selected from each column to be transmitted to the appropriate output port.

5.5.1.1 One-Shot Scheduling

Under this scheduling policy, all copies of the same cell must be switched successfully in one time slot. If at least one copy loses the contention for an output port, the original cell waiting in the input queue must try again in the next time slot. It is obvious that this strategy favors cells with fewer copies

Input \ Output	1	2	3	4	5
1	1	1	0	0	1
2	0	1	0	0	0
3	0	0	0	1	0
4	0	1	1	0	0
5	0	0	1	1	0

Transmission request matrix

(1: copy request, 0: no request)

(a) One-shot

X	X	0	0	X
0	1	0	0	0
0	0	0	X	0
0	1	1	0	0
0	0	1	1	0

(b) Strict-sense call splitting

X	1	0	0	1
0	X	0	0	0
0	0	0	X	0
0	1	X	0	0
0	0	1	1	0

(c) Wide-sense call splitting

X	1	0	0	X
0	X	0	0	0
0	0	0	X	0
0	1	X	0	0
0	0	1	1	0

(X: accepted request, 1: rejected request)

Figure 5.29 Call scheduling disciplines.

(i.e., favors calls with fewer recipients), and more often blocks multicast cells with more copies.

5.5.1.2 Strict-Sense Call Splitting

As we know in the case of one-shot discipline, if only one copy loses contention, the original cell (all its copies) must be retransmitted again in the next time slot. Thus, the one-shot discipline degrades the throughput performance of the switch. This drawback suggests transmitting copies of the multicast cell independently, but only one copy in each time slot, which is strict-sense (SS) call splitting. In SS call splitting, a multicast cell waits in the input queue contending for an output port in the following time slots until all of its copies have been transmitted. If a multicast cell has K copies to transmit, it needs at least K time slots to be transmitted. Statistically this algorithm results in a low throughput when the switch gets underloaded during light traffic, since this algorithm does not change dynamically with the traffic pattern.

5.5.1.3 Wide-Sense Call Splitting

The case of light traffic should be considered when only one active input line has a multicast cell to be transmitted and all output ports are free. In this case,

SS call splitting only allows one cell to be transmitted per time slot, and a low utilization results. Wide-sense (WS) call splitting was proposed to allow more than one copy from the same multicast cell to gain access to output ports simultaneously as long as these output ports are free.

It is clear that one-shot discipline has the lowest throughput performance among all three algorithms; however, it is easy to implement. SS call splitting performs better than one-shot in the heavy traffic case, but it seems rigid in the light traffic case and results in a low utilization. WS call splitting has the advantage of the full use of the output trunks, allowing the multicast cell to use all free output ports, which results in a higher throughput.

5.5.2 Copy Network

Lee proposed a nonblocking copy network [46] in which cell replications are accomplished by an encoding process and a decoding process. As shown in Figure 5.30, it consists of a runing adder network (RAN), a set of dummy address encoders (DAEs), a concentrator, a broadcast banyan network (BBN), and a group of trunk number translators (TNTs). The encoding process is carried out by the RAN and a set of DAEs at the output of the RAN. The sequence of running sums of the copy numbers requested by incoming multicast cells is first calculated in the RAN. Then pairs of adjacent running sums are used by the DAEs to form a pseudo-header for each multicast cell. The header contains a dummy address interval specified by the minimum (MIN) and the maximum (MAX) of the output port address interval and an index reference (IR). In the decoding process, the BBN will perform the cell replications by the Boolean interval splitting algorithm based on the dummy address interval (MIN, MAX) and the TNTs will determine the destination address of each replicated cell based on its copy index and broadcast channel number.

These encoding and decoding processes are illustrated by an example in Figure 5.30. Input ports 1, 4, and 5 are idle and others have active cells with given copy requests. Based on a running sum calculated at the running adder network, only cells at input ports 0 and 2 are forwarded to the concentrator. Cells whose running sum exceeds the capacity of the copy network, for example, cells from ports 3, 6, and 7, are not forwarded to the broadcast banyan network. Figure 5.30 shows that the MIN of each interval is set equal to the running sum of its upper adjacent port, and the MAX is equal to its own running sum decreased by one. The index reference is set equal to the corresponding MIN value. After being concentrated, a cell from input port 0 is then copied to output ports 0 and 1, while the other from input port 2 is copied to output ports 2, 3, and 4. Upon receiving a replicated cell, each output port of the BBN then calculates the distinct copy index (CI) by subtracting the index

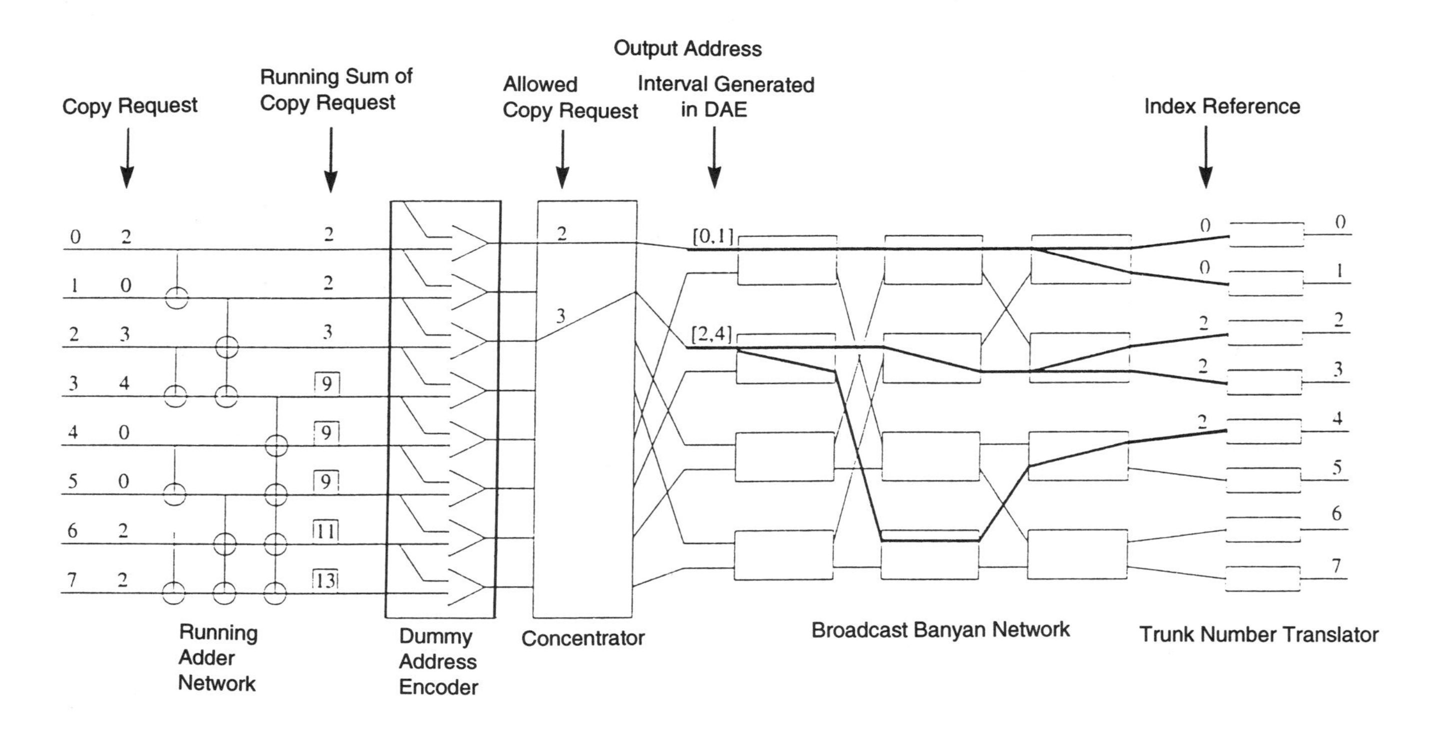

Figure 5.30 A nonblocking copy network.

reference (IR) from its own output port number. In this way, three copies from input port 2 are uniquely identified by the group name, called broadcast channel number (BCN), together with the copy index, 0, 1, and 2, respectively.

One of the problems inherent in this copy network is the overflow of the copy requests in the RAN, which occurs when the total number of copy requests exceeds the capacity of the copy network. The overflow problem may degrade the throughput of the copy network and introduce unfairness among incoming multicast cells. The utilization of the copy network would be improved if partial service of copy request, called call splitting, could be implemented when overflow occurs. As illustrated in Figure 5.30, if input port 3 receives a partial service with call splitting, a total of eight copies rather than five copies (from ports 0 and 2) could be generated in one time slot.

Another undesirable consequence of overflow is the input fairness problem, which arises due to the fixed structure of the RAN. Since the calculation of a running sum always starts from input port 0 in every time slot, lower numbered input ports have higher service priorities than the higher numbered ports, especially when the overflow occurs frequently. A dual running adder network [46] can alleviate this problem but does not provide a complete solution. The method described in [49] is not practical in implementation, because it requires rather complicated hardware. A simple scheme is proposed in [50] to handle the overflow problem, but the input fairness problem remains unsolved.

Another concern is the output fairness problem. This arises at the output of the copy network due to underflow, which occurs when the sum of the copy requests over all HOL input cells is frequently less than the capacity of the copy network. The dummy address interval assigned by DAEs to each multicast cell specifies the output from which the replicated cells will emerge. Since address assignment constantly starts from output address 0, replicated copies will appear at the lower numbered outputs much more frequently than the higher numbered outputs. As a result, the distribution of replicated cells over the inputs of the following point-to-point switch becomes imbalanced and may adversely affect the overall performance of the switch. The modular architecture proposed in [51,52] can alleviate the overflow and unfairness problems at inputs and reduce the memory requirement, but the output unfairness issue is still not settled. The ring reservation scheme employed in [53] is a sequential processor, and its capacity may be limited by the fixed cycle time of the ATM.

Byun and Lee proposed some mechanisms to solve these inherent system problems in [47]. As shown in Figure 5.31, the structure of the modified copy network is basically the same as the copy network. The main difference lies in the cyclic running adder network (CRAN) and a shifter. The cylindrical

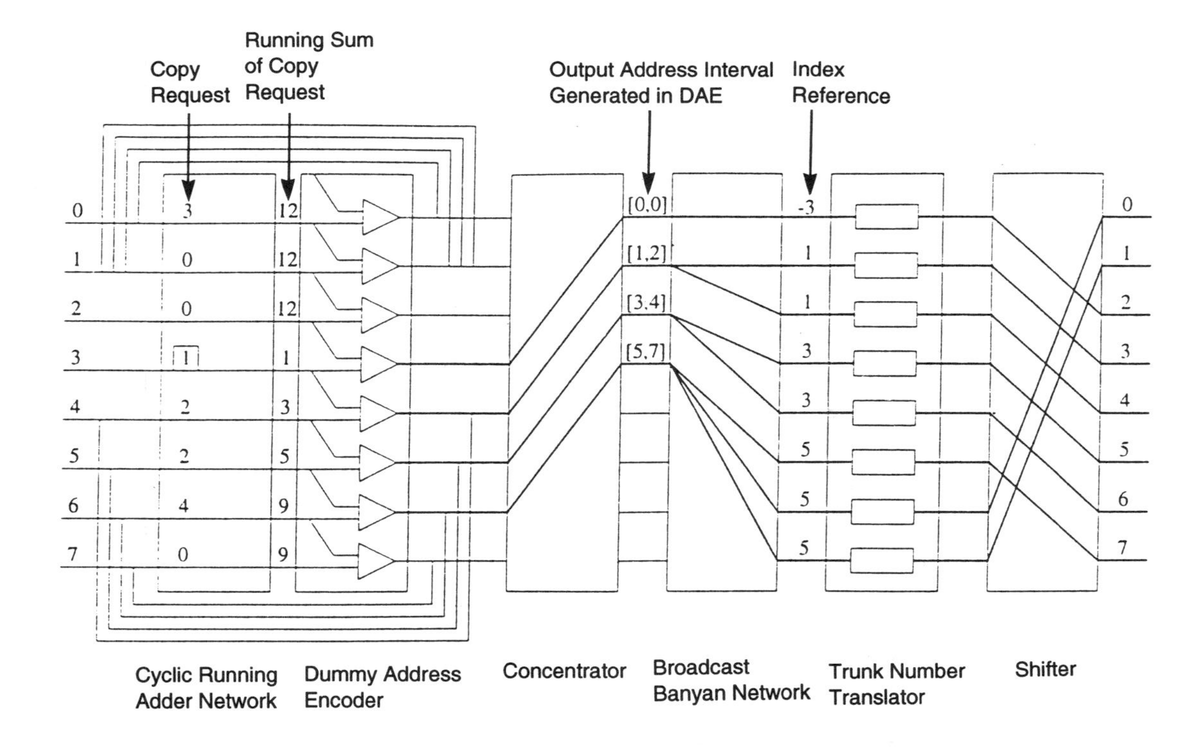

Figure 5.31 The adaptive nonblocking copy network.

structured CRAN can calculate the running sums starting from any input port in a cyclic manner. The starting point of computing the running sums in every time slot is determined adaptively by the overflow condition of the previous time slot. This decision will be made at outputs of CRAN and returned to the input ports through a feedback loop. A shifter will resolve the output fairness problem simply by shifting the replicated cells cyclically, such that copies of multicast cells will be uniformly distributed over the inputs of the routing network. This modified copy network has a complete parallel architecture; the complexity of every component interconnection network involved is $O(N \log N)$.

5.5.3 Multicast Shared Memory Switches

Of the various ATM switch architectures, the shared memory switch provides superior benefits compared to others. Since its memory space is shared among its switch ports, it achieves high buffer utilization efficiency. Under conditions of identical memory size, the cell-loss probability of the shared memory switches is smaller than that of output buffered switches. Additionally, multicast operations can be easily supported by this switch architecture. In this section, some classes of multicast shared memory switches are described along with their advantages and disadvantages.

5.5.3.1 Shared Memory Switch With Cell-Copy Circuit

In this switch scheme, a multicast cell is replicated into multiple copies for all its multicast connections by a cell-copy circuit located before the routing switch. Bianchini and Kim [53] proposed a nonblocking cell-copy circuit for multicast purposes in an ATM switch. Each copy is stored in the shared buffer memory (SBM). The address for each copy is queued in output address queues (AQs).

Figure 5.32 shows an example of a shared memory switch with a cell-copy circuit. A multicast cell arrives and is replicated into three copies for output ports 0, 1, and 3. In this scheme, after replicating the multicast cell, each copy is treated the same way as a unicast cell. It provides fairness among ATM cells. The biggest disadvantage of this approach is that, in the worst case, N cells might get replicated to at most N^2 cells. Thus, the number of replicated cells to the SBM in each switch cycle could be $O(N^2)$. Since only at most N cells could be transmitted, this would result in storing $O(N^2)$ cells in each cycle. For a finite memory space, it would result in considerable cell loss. The replication of multicast cells would also require $O(N^2)$ cells to be written to the SBM, which will limit the switch size from the memory speed. Moreover, adding the cell-copy circuit in the switch increases hardware complexity.

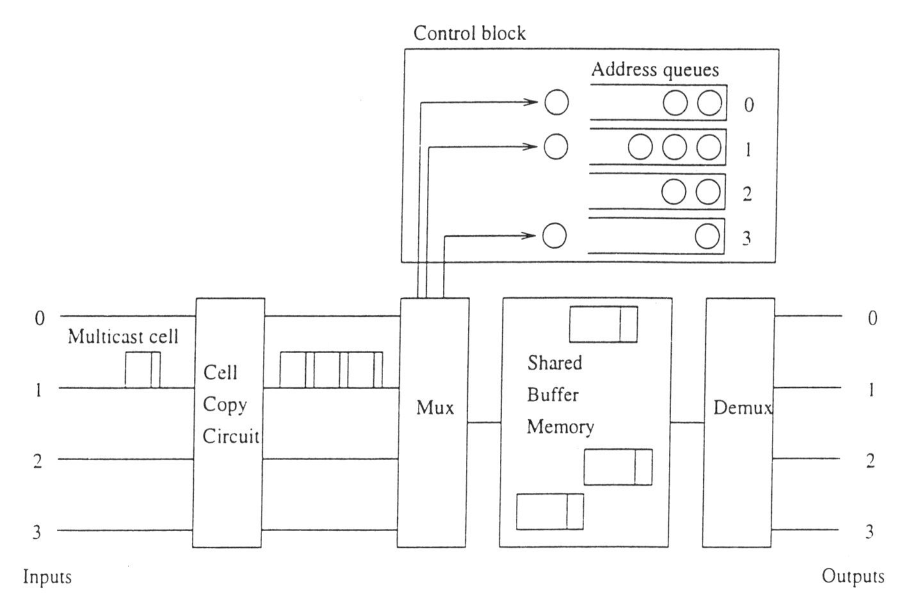

Figure 5.32 Shared memory ATM switch with cell-copy circuit.

5.5.3.2 Shared Memory Switch With Address-Copy Circuit

Saito et al. [54] proposed a new scheme that requires less memory for the same cell loss probability. In this scheme, an address-copy circuit is used in the controller. When a multicast cell arrives, only a single copy is stored in the shared buffer memory (SBM). Its address is copied by the address-copy circuit and queued into multiple address queues (AQs), as shown in Figure 5.33. The multicast cell will be read multiple times to different output ports. Once completed, the cell's address becomes available for the next arriving cells.

In this architecture, multicast cell counters (MCCs) are required to hold the number of copies of each multicast cell. The MCC values decrease by one each time cells are read out from the SBM.

Figure 5.34 illustrates an example of a 3×3 ATM switch. In this example, a multicast cell arriving at input port 1 is destined for output ports 1 and 2. The multicast cell is written into A0 in the SBM, and its address is provided by idle address FIFO. A0 is copied and queued in address queues 1 and 2. At the same time, the MCC of A0 is set to two. Another multicast cell stored at B3 is destined for output ports 0 and 1. As it is transmitted to both output ports 0 and 1, the associated MCC becomes zero and the address B3 is released to the idle address FIFO. The address C1 is not released until the cell is transmitted to output port 0.

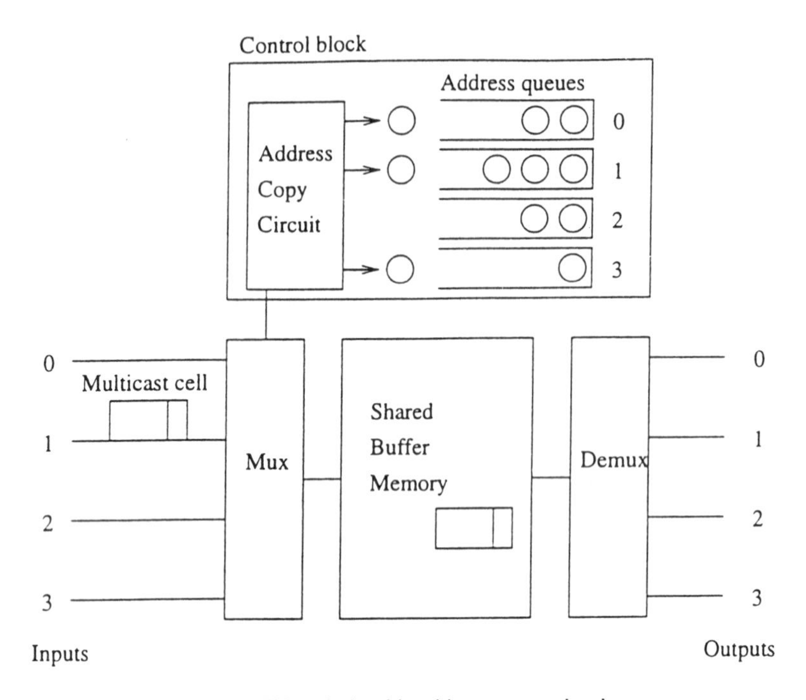

Figure 5.33 Shared memory ATM switch with address-copy circuit.

Although in each time slot the maximum number of cells to be written into the buffer is N, the number of replications of cell addresses for incoming cells could be $O(N^2)$ in the broadcast case, which still limits the switch size.

5.6 Summary

In this chapter, we first described the performance requirements of an ATM switch and several design criteria that need to be considered when designing an ATM switch system. The major functions that an ATM switch has to perform are routing incoming cells to proper output port(s) and resolving output port contention, where multiple cells are destined for the same output port. Due to the high-speed input lines, cell routing is normally done in a distributed manner, called *self-routing*. The contention among incoming cells is resolved by placing a buffer to temporarily store cells that lost contention to others.

The ATM switch architectures can be classified according to their buffering strategies: internally buffered, recirculation buffered, crosspoint buffered, input buffered, output buffered, shared buffered, multistage shared buffered, input and output buffered, and virtual output queueing. Due to HOL blocking,

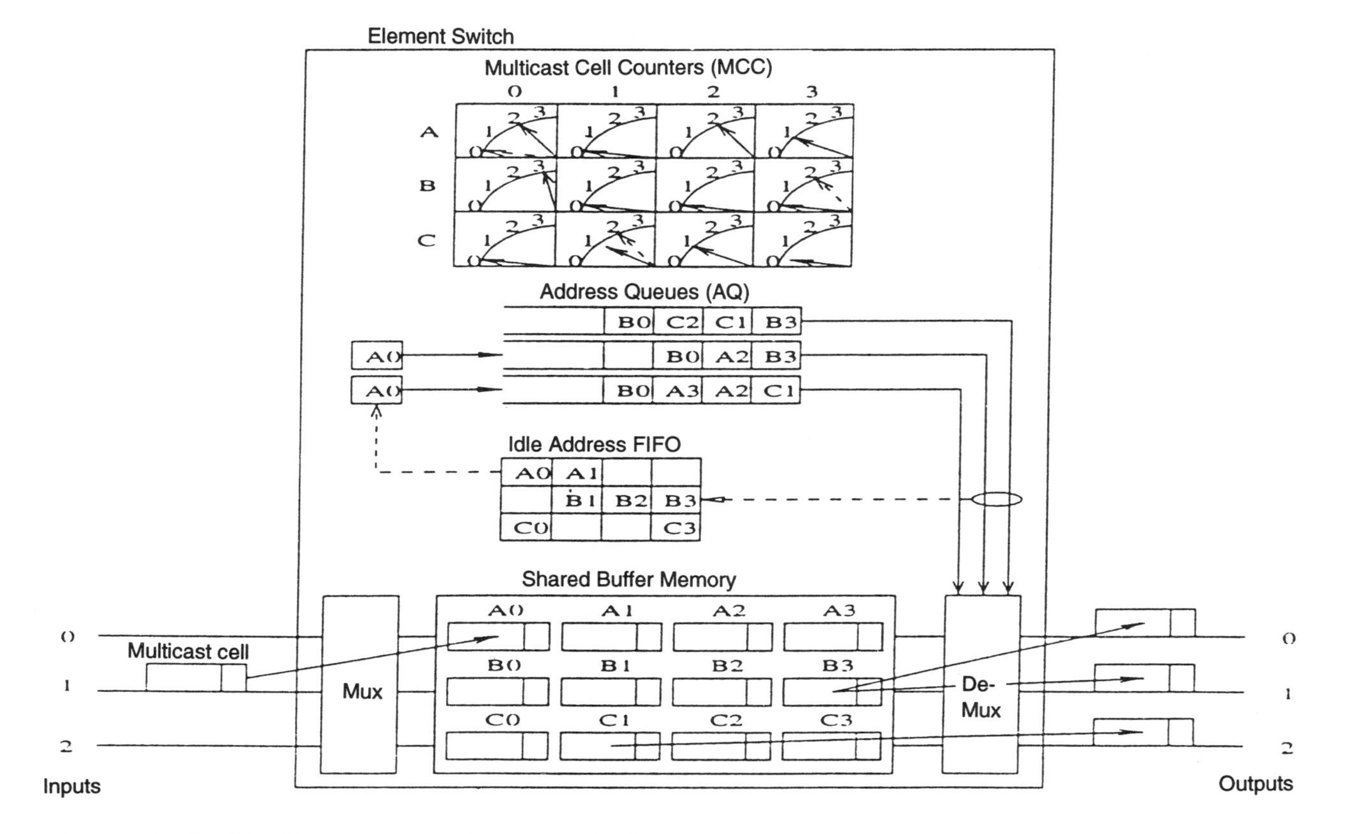

Figure 5.34 Example of multicast function in an address-copy circuit.

the input buffered switches can only have maximum throughput up to 58.6% under random uniform traffic distribution. The output buffered switches have been proven to provide the best delay/throughput performance for arbitrary traffic distribution. However, due to memory speed limitations, the switch does not scale to a large size. The shared memory switches have the same performance and switch size limitation as output buffered switches. But they give the best memory utilization among all switch architectures. The input and output buffered switches can be implemented to a large size without having the memory speed bottleneck. However, their output port contention resolution can be very complex and challenging. The virtual output queueing switches store cells at input buffers but in different logical queues that are associated with output ports. It has been proven that high throughput can be achieved with the expense of sophisticated contention resolution schemes (especially for multicasting).

Basically two different approaches are used to perform the multicast function in an ATM switch: broadcast-and-select and copying cells. Broadcast-and-select type switches broadcast the multicast cell to all output ports, and each output port filters the incoming cells using an address filter. A copy network replicates multicast cells before routing them in the following point-to-point switching network. The former architecture is simple but limited to a certain size. The latter scales better but is more complicated and difficult to implement.

Introduction of optical fibers to communication networks has resulted in a tremendous increase in the speed of data transmitted. The virtually unlimited bandwidth of optical fibers comes from the carrier frequency of nearly 200 THz. However, the capacity of optical fibers is not used properly since the capacities of other network elements, such as electronic switches, cannot compete with the capacity of optical fibers. Considering the intense interest in B-ISDN applications and increased line speed, researchers have devoted their efforts to implementing more powerful ATM switches, such as photonic ATM switches that will play an important role in the future ultraband integrated services network.

References

[1] Bellcore, "Broadband Switching System (BSS) Generic Requirements, BSS Performance," GR-110-CORE, Issue 1, Sept. 1994.

[2] Hui, J., and E. Arthurs, "A Broadband Packet Switch for Integrated Transport," *IEEE J. Sel. Areas Commun.*, Vol. 5, No. 8, Oct. 1987, pp. 1264–1273.

[3] Karol, M. J., M. G. Hluchyj, and S. P. Morgan, "Input Versus Output Queueing on a Space Division Packet Switch," *IEEE Trans. Commun.,* Vol. COM-35, No. 12, Dec. 1987.

[4] FORE Systems, Inc., "White Paper: ATM Switching Architecture," Nov. 1993.

[5] Kozaki, T., et al., "32 × 32 Shared Buffer Type ATM Switch VLSI's for B-ISDN's," *IEEE J. Sel. Areas Commun.,* Vol. 9, No. 8, Oct. 1991, pp. 1239–1247.

[6] Suzuki, H., et al., "Output-Buffer Switch Architecture for Asynchronous Transfer Mode," *Proc. IEEE ICC '89,* June 1989, pp. 99–103.

[7] Cidon, I., et al., "Real-Time Packet Switching: A Performance Analysis," *IEEE J. Sel. Areas Commun.,* Vol. 6, No. 9, Dec. 1988, pp. 1576–1586.

[8] Newman, P., "ATM Switch Design for Private Networks," *Issues in Broadband Networking,* 2.1, Jan. 1992.

[9] Yeh, Y. S., M. G. Hluchyj, and A. S. Acampora, "The Knockout Switch: A Simple, Modular Architecture for High-Performance Switching," *IEEE J. Sel. Areas Commun.,* Vol. 5, No. 8, Oct. 1987, pp. 1274–1283.

[10] Tobagi, F. A., T. K. Kwok, and F. M. Chiussi, "Architecture, Performance and Implementation of the Tandem Banyan Fast Packet Switch," *IEEE J. Sel. Areas Commun.,* Vol. 9, No. 8, Oct. 1991, pp. 1173–1193.

[11] Liew, S. C., and T. T. Lee, "*N* log *N* Dual Shuffle-Exchange Network With Error-Correcting Routing," *Proc. IEEE ICC '92,* Chicago, IL, June 1992, Vol. 1, pp. 1173–1193.

[12] Giacopelli, J. N., et al., "Sunshine: A High-Performance Self-Routing Broadband Packet Switch Architecture," *IEEE J. Sel. Areas Commun.,* Vol. 9, No. 8, Oct. 1991, pp. 1289–1298.

[13] Huang, A., and S. Knauer, "STARLITE: A Wideband Digital Switch," *Proc. IEEE GLOBECOM '84,* Dec. 1984, pp. 121–125.

[14] Awedeh, R. Y., and H. T. Mouftah, "Survey of ATM Switch Architectures," *Computer Networks ISDN Syst.,* Vol. 27, 1995, pp. 1567–1613.

[15] Chen, X., "A Survey of Multistage Interconnection Networks in Fast Packet Switches," *Int. J. Digital Analog Commun. Syst.,*Vol. 4, 1991, pp. 33–59.

[16] Ahmadi, H., and W. E. Denzel, "A Survey of Modern High-Performance Switching Techniques," *IEEE J. Sel. Areas Commun.,* Vol. 7, No. 7, Sept. 1989, pp. 1091–1103.

[17] Tobagi, F., "Fast Packet Switch Architectures for Broadband Integrated Services Digital Networks," *Proc. IEEE,* Vol. 78, No. 1, Jan. 1990, pp. 133–167.

[18] Oie, Y., et al., "Effect of Speed Up in Nonblocking Packet Switch," *Proc. ICC '89,* June 1989, pp. 410–414.

[19] Pattavina, A., "Multichannel Bandwidth Allocation in a Broadband Packet Switch," *IEEE J. Sel. Areas Commun.,* Vol. 6, No. 9, Dec. 1988, pp. 1489–1499.

[20] Hluchyj, M. G., and M. R. Karol, "Queueing in High-Performance Packet Switching," *IEEE J. Sel. Areas Commun.,* Vol. 6, No. 9, Dec. 1988, pp. 1587–1597.

[21] Bingham, B., and H. Bussey, "Reservation-Based Contention Resolution Mechanism for Batcher-Banyan Packet Switches," *Electron. Lett.,* Vol. 24, No. 13, June 1988, pp. 772–773.

[22] Chao, H. J., "A Recursive Modular Terabit/sec ATM Switch," *IEEE J. Sel. Areas Commun.,* Vol. 9, No. 8, Oct. 1991, pp. 1161–1172.

[23] Chao, H. J., and B. S. Choe, "Design and Analysis of a Large-Scale Multicast Output Buffered ATM Switch," *IEEE/ACM Trans. Networking,* Vol. 3, No. 2, Apr. 1995, pp. 126–138.

[24] Chen, D. X., and J. W. Mark, "A Buffer Management Scheme for the SCOQ Switch Under Nonuniform Traffic Loading," *IEEE/ACM Trans. Commun.,* Vol. 42, No. 10, Oct. 1994.

[25] Yoon, H., M. T. Liu, and K. Y. Lee, "The Knockout Switch Under Nonuniform Traffic," *Proc. IEEE GLOBECOM '88,* Hollywood, FL, Nov. 1988, pp. 1628–1634.

[26] Jajszczyk, A., M. Roszkiewicz, and J. Garcia-Haro, "Comparison of ATM Shared-Memory Switches," *Proc. ISS '95,* Berlin, Apr. 1995, pp. 409–413.

[27] Lee, H., et al., "A Limited Shared Output Buffer Switch for ATM," *Proc. Fourth Int. Conf. on Data Commun. Sys. and Their Performance,* Barcelona, June 1990, pp. 163–179.

[28] Schultz, K. J., and P. G. Gulak, "CAM-Based Single-Chip Shared Buffer ATM Switch," *Proc. IEEE ICC '94,* New Orleans, LA, May 1994, pp. 1190–1195.

[29] Oshima, K., et al., "A New ATM Switch Architecture Based on STS-Type Shared Buffering and Its LSI Implementation," *Proc. ISS '92,* 1992, pp. 359–363.

[30] Turner, J. S., "An Optimal Nonblocking Multicast Virtual Circuit Switch," *Proc. IEEE INFOCOM '94,* 1994, pp. 298–305.

[31] Eng, K. Y., and M. J. Karol, "State of the Art in Gigabit ATM Switching," *Proc. IEEE BSS '95,* Poland, Apr. 1995, pp. 3–20.

[32] Kim, H. S., "Design and Performance of Multinet Switch: A Multistage ATM Switch Architecture With Partially Shared Buffers," *IEEE/ACM Trans. Networking,* Vol. 2, No. 6, Dec. 1994, pp. 571–580.

[33] Fischer, W., et al., "A Scalable ATM Switching System Architecture," *IEEE J. Sel. Areas Commun.,* Vol. 9, No. 8, Oct. 1991, pp. 1299–1307.

[34] Banniza, T. R., et al., "Design and Technology Aspects of VLSI's for ATM Switches," *IEEE J. Sel. Areas Commun.,* Vol. 9, No. 8, Oct. 1991, pp. 1255–1264.

[35] Eng, K. Y., M. J. Karol, and Y. S. Yeh, "A Growable Packet (ATM) Switch Architecture: Design Principles and Applications," *IEEE Trans. Commun.,* Vol. 40, No. 2, Feb. 1992, pp. 423–430.

[36] Chao, H. J., B. S. Choe, J. S. Park, and N. Uzun, "Design and Implementation of Abacus Switch: A Scalable Multicast ATM Switch," *IEEE J. Sel. Areas Commun.,* Vol. 15, No. 5, June 1997, pp. 830–843.

[37] Genda, K., et al., "A 160-Gb/s ATM Switching System Using an Internal Speed-Up Crossbar Switch," *Proc. IEEE GLOBECOM '94,* Nov. 1994, pp. 123–133.

[38] Chiussi, F. M., and F. A. Tobagi, "A Hybrid Shared-Memory/Space-Division Architecture for Large Fast Packet Switches," *Proc. IEEE ICC '92,* 1992, pp. 904–911.

[39] Lee, T. T., "A Modular Architecture for Very Large Packet Switches," *IEEE Trans. Commun.,* Vol. 38, No. 7, July 1990, pp. 1097–1106.

[40] LaMaire, R. O., and D. N. Serpanos, "Two-Dimensional Round-Robin Schedulers for Packet Switches With Multiple Input Queues," *IEEE/ACM Trans. Networking,* Vol. 2, No. 5, Oct. 1994, pp. 471–482.

[41] McKeown, N., P. Varaiya, and J. Walrand, "Scheduling Cells in an Input-Queued Switch," *Electon. Lett.,* Vol. 29, No. 25, Dec. 1993, pp. 2174–2175.

[42] Munter, E., J. Parker, and P. Kirkby, "A High Capacity ATM Switch Based on Advanced Electronic and Optical Technologies," *IEEE Commun. Magazine,* Vol. 33, No. 11, Nov. 1995, pp. 64–71.

[43] Eng, K. Y., M. G. Hluchyj, and Y. S. Yeh, "Multicast and Broadcast Services in a Knockout Packet Switch," *Proc. IEEE INFOCOM '88,* Mar. 1988, pp. 29–34.

[44] De Vries, R. J. F., "ATM Multicast Connections Using the Gauss Switch," *Proc. IEEE GLOBECOM '90,* 1990, pp. 211–217.

[45] Turner, J. S., "Design of a Broadcast Packet Switching Network," *IEEE Trans. Commun.,* Vol. 36, No. 6, June 1988, pp. 734–743.

[46] Lee, T. T., "Nonblocking Copy Networks for Multicast Packet Switching," *IEEE J. Sel. Areas Commun.,* Vol. 6, No. 9, Dec. 1988, pp. 1455–1467.

[47] Byun, J. W., and T. T. Lee, "The Design and Analysis of an ATM Multicast Switch With Adaptive Traffic Controller," *IEEE/ACM Trans. Networking,* Vol. 2, No. 3, June 1994, pp. 288–298.

[48] Chen, X., and J. F. Hayes, "Call Scheduling in Multicast Packet Switching," *Proc. IEEE ICC '92,* 1992, pp. 895–899.

[49] Lee, T.-H., and S. J. Liu, "A Fair High-Speed Copy Network for Multicast Packet Switch," *Proc. IEEE INFOCOM '92,* Florence, Italy, May 1992, pp. 886–894.

[50] Turner, J. S., "A Practical Version of Lee's Multicast Switch Architecture," *IEEE Trans. Commun.,* Vol. 41, Aug. 1993, pp. 1166–1169.

[51] De Zhong, W., Y. Onozato, and J. Kaniyil, "A Copy Network With Shared Buffers for Large-Scale Multicast ATM Switching," *IEEE/ACM Trans. Networking,* Vol. 1, Apr. 1993, pp. 157–165.

[52] De Zhong, W., Y. Onozato, and J. Kaniyil, "Performance Enhancement in Recursive Copy Networks for Multicast ATM Switching: A Simple Flow Control Scheme," *IEICE Trans. Commun.,* Vol. E77-B, No. 1, Jan. 1994, pp. 28–34.

[53] Bianchini, Jr., R. P., and H. S. Kim, "Design of a Nonblocking Shared-Memory Copy Network for ATM," *Proc. IEEE INFOCOM '92,* 1992, pp. 876–885.

[54] Saito, H., et al., "Multicast Function and Its LSI Implementation in a Shared Multibuffer ATM Switch," *Proc. IEEE INFOCOM '94,* 1994, pp. 315–322.

6

Transport Network Architectures and Technologies for Multimedia Services

6.1 Introduction

To deliver multimedia services efficiently, cost effectively, and within acceptable performance levels, a robust transport system is required. This transport system ensures the connection between a customer and a network provider, among customers, and among network providers. It includes both the access system and the trunk transport system. This chapter deals with transport network architectures and technologies for multimedia services and discusses various access alternatives and various trunk issues including the role of SDH/SONET and ATM. It is divided into three major parts. The first part starts with a discussion of customer-premises equipment and architectures followed by a discussion on access transport network architectures and technologies. The third part deals with trunk transport network architectures and technologies.

6.2 Customer-Premises Equipment and Architectures

The term *customer-premises equipment* (CPE) refers to equipment owned and operated by a customer although the term is sometimes used to include any equipment within the customer's premises, whether or not it is owned and operated by the customer. The term *CPE architectures* refers to the how CPEs are connected such as point-to-point or point-to-multipoint topologies. CPE is required to deliver various services to customers where the architecture used and its degree of complexity vary depending on the service provided and its associated features.

The telephone was invented by Alexander Graham Bell in the late 1800s. The first residential customers had a telephone set connected in a point-to-point manner to other customers using a single bare copper wire. In that case, CPE simply consisted of a telephone set, and its associated architecture consisted of a pair of copper wire. In the early 1900s, central offices were put in place and customers were connected via twisted-pair copper wire. Over time, customer needs changed, and that led to a change in the telephone set design. Telephones became more and more sophisticated with many features and many designs. For example, speaker phones and cordless phones became popular in the 1980s. Today, telephones have evolved to include very advanced built-in features such caller identification displays, automatic dialers, voice recognition, and spread-spectrum transmission and are widely available and affordable.

The first instance of multimedia CPE in the residential market was the television set. In the telephone industry, the first multimedia CPE came in the 1960s when the picture phone was introduced. It allowed calling customers to view the called party and vice versa. Although the introduction of the picture phone was not very successful, this lack of success cannot be attributed to the lack of interest or demand for multimedia. The demand was always present and has been evolving over time. This can be demonstrated by the convergence among various technologies. For instance, the telecommunications industry is converging with the personal computer, creating a larger demand for multimedia CPE. The penetration of personal computers into the residential market is constantly increasing and with the explosion of Internet services such as electronic mail and World Wide Web access, the personal computer opened new avenues for the telecommunications services. This is also true for the television set. Using a set-top box, residential customers can order video programs and even gain access to the Internet. In keeping up with this demand, multimedia CPE is evolving and architectures to support the new service demands are also evolving from a single copper pair to customer-premises networks, which consist of copper wire, coaxial cables, and optical fiber connected in various point-to-point and point-to-multipoint configurations and topologies.

So far, our discussion has focused on residential customers. From an access point of view, the emphasis is on the residential customers because of the importance of relative cost targets and the relative number of customers. For business customers, it is much easier to justify higher investment because of the expected higher revenue. In fact, in most instances business customers are served directly by microwave links or direct optical fiber connections. Therefore, the access portion of this chapter will focus on residential customers. However, it is important to note that for business customers, multimedia equipment for video conferencing has been more widespread. Also, the business customer-premises networks are much more sophisticated than those for residential customers and range from copper wiring to fiber local-area networks (LANs). The next section discusses access transport architectures and technologies for multimedia services (e.g., telephony, ISDN, Internet, and video).

6.3 Access Transport Network Architectures and Technologies

This section discusses how the access network has evolved over the years and how multimedia services are expected to be delivered to customers by means of various technologies. It covers various architectures and topologies such as fiber-to-the-home (FTTH), fiber-to-the-curb (FTTC), hybrid fiber coax (HFC), asymmetric digital subscriber line (ADSL), and wireless technologies such as multichannel multipoint distribution service (MMDS) and local multipoint distribution service (LMDS). The use of SONET/SDH and ATM technologies in the access portion of the network are also addressed in this section.

6.3.1 Evolution of the Access Network

The access network is defined as the facilities that connect customers with the central office as shown in Figure 6.1. It started as copper wire pairs and evolved

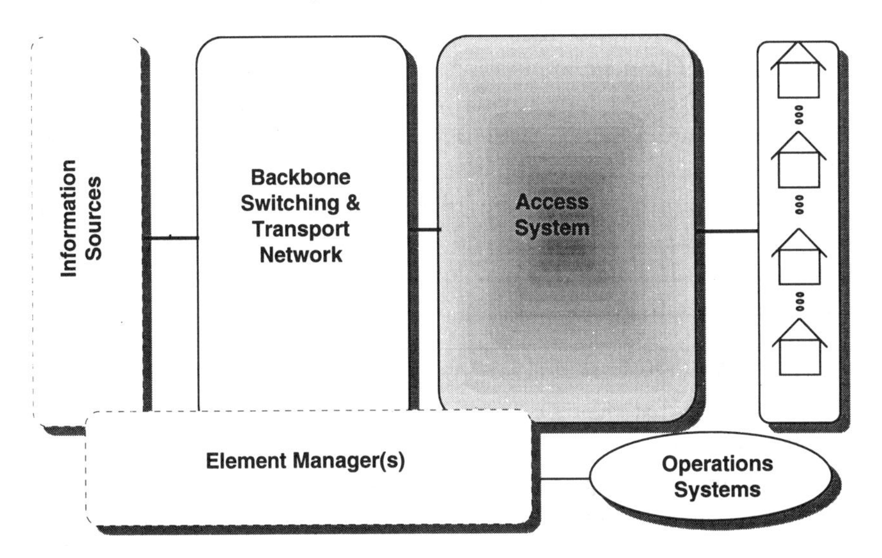

Figure 6.1 High-level view of an end-to-end network.

into analog then digital electronics using copper wire, optical fiber, and coaxial cable. The figure also shows a backbone switching and transport network, which consists of switching systems, cross-connects, multiplexers, and related SONET/SDH and ATM equipment. This network is discussed in more detail in Section 6.4. The information sources provide programs such as movies or other multimedia content that is delivered to the customers through the backbone and access networks. Also shown in the figure are element managers and operations systems that are needed to manage the network.

6.3.1.1 Evolution of the Loop

As mentioned earlier, the loop consisted of bare copper wire at the invention of the telephone in the late 1800s. As more and more customers subscribed to telephone service, a central location was used to connect customers. This location later contained the switch and is known today as the *central office, wire center,* or *local exchange.* Major evolution in the loop did not occur until the 1950s. Prior to that, the improvements that were made for the loop consisted of better wire and cable and more cost-effective wire. In the 1950s, electronics were introduced in the loop to improve transmission. For instance, amplifiers were used to compensate for losses on long loops. Also, analog carriers were introduced in the late 1940s and are discussed next.

6.3.1.2 Introduction of Analog Carriers

First-generation amplitude-modulated analog carriers were used in the loop in the late 1940s. In the 1950s, these were improved and were designed and built using transistors. However, their cost was still high and their reliability was not satisfactory. The main application for such carriers was for rural environments where loops were long (typically much greater than 12,000 ft). In the 1960s, analog carrier system designs were improved, and many companies manufactured analog carrier equipment. Over time, the cost of these carriers decreased significantly, and they are still deployed in some areas. However, the introduction of digital technology and its rapid penetration into the telephone network created demand for a new loop carrier technology, the digital loop carrier systems as discussed next.

6.3.1.3 Introduction of Digital Carriers

At the same time that analog carrier systems were being manufactured, digital technology started to have a major impact on the network. Also, while analog carriers were used for long loops, studies showed that digital systems might be cost effective for shorter loops. The first digital system was introduced in 1962. In the 1970s, digital loop carrier systems became more widespread. These systems were called *universal* because of their capability to work with any switching system. Universal systems require a central office terminal that is used to adapt the carrier system to any switching system.

In the 1980s, integrated digital loop carriers were introduced in which the requirements and specifications of these systems allowed them to be separate from the switch so that exchange carriers could be mixed and matched between switches and digital loop carrier systems. Although these systems were designed to serve narrowband services initially, systems are available that can serve ISDN and higher bandwidth services and are expected to have ATM capabilities. It is envisioned that digital loop carriers will converge to multimedia to support many services and technologies ranging from ADSL to fiber-based services and technologies. Digital loop carriers made a significant impact on today's and future access architectures in that these architectures are based on the digital loop carrier platform (i.e., pair gain and concentration of services).

6.3.1.4 Fiber Optics Penetration Into the Loop

Telephone companies across the world have been aggressively deploying optical fiber in the feeder portion of their network and have started to invest large amounts of money in the deployment of optical fiber in the distribution portion. Figure 6.2 illustrates the feeder and distribution portions of a typical telephone company's network [1,2].

Although some telephone companies are more aggressive in their plans than others, they all see fiber as the future transmission medium to serve all of their customers. This is because it gives them a very large bandwidth medium that allows them to offer existing telecommunications services (i.e., POTS,[1] coin telephone, voice frequency special services, ISDN basic and primary rate access, etc.), as well as build an infrastructure that enables them to offer future broadband and multimedia services to customers (e.g., a menu of video services including video-on-demand, impulse pay per view, pay per view, tele-education, video telephony, etc.). Access systems that will allow delivery of the services mentioned above are discussed later in this section.

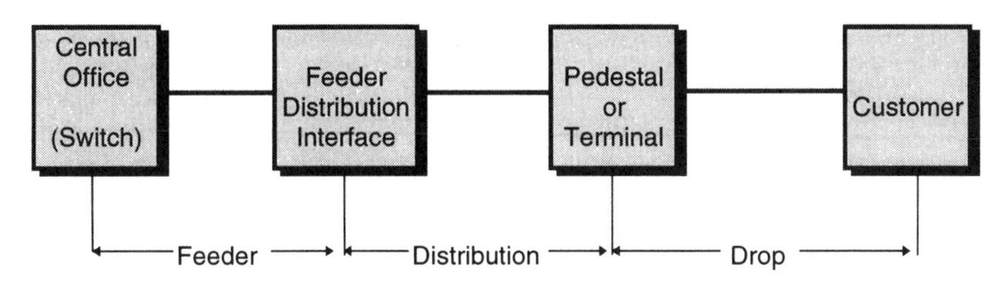

Figure 6.2 Telephone network terminology.

1. POTS or plain old telephone service refers to single-party message telephone service.

6.3.1.5 Role of ATM

In parallel to the work that has been done on access systems, work was also done on defining flexible bandwidth transport in the network using asynchronous transfer mode (ATM) technology. ATM is viewed as the transport vehicle for many multimedia services such as broadband ISDN (B-ISDN) services, where ATM cells may be transported in a SONET/SDH frame. The architecture used to deliver these ATM-based services represents a target architecture that is somewhat different than the current access architectures. The reason for this difference is that the B-ISDN architecture was initially intended to provide high-bandwidth services to business customers, whereas access systems were intended to deliver existing telecommunications services to residential and small business customers initially. In the long run, the two architectures are expected to converge. It is expected that ATM-based services will become more and more prevalent, and every network architecture will be designed or modified to carry and offer ATM services.

A lot of work has been done on access systems and on ATM separately, and some work has been done on combining the two architectures/technologies. ATM-based multimedia services are expected to be delivered to customers all the way to the set-top box over various access architectures, including FTTH, FTTC, and HFC. Because of the flexibility of ATM and its capabilities, it can provide users with enormous information transfer capacity that can be flexibly drawn on to meet existing and future service needs. Switching and transmission requirements for emerging multimedia services cannot be known precisely, but the flexibility provided by ATM will provide switching capabilities that are independent of the type and rate of the services being supported. Because it provides virtual channels rather than dedicated channels, it allows better use of the physical resources with high performance and provides variable information transfer rates. Therefore, ATM combined with various access systems will provide the basis for providing multimedia services.

6.3.1.6 Service Evolution

Services have evolved a great deal over the years. As mentioned earlier, POTS or single-party messaging telephone service started in the late 1800s. The service evolved over time and gained many enhanced features such as call waiting, call forwarding, caller number delivery, on-hook transmission, and distinctive ringing. Also, services evolved from POTS to ISDN to multimedia services, such as video programs. The following list helps define the video services that might have the potential to be offered as multimedia services. They are offered here only as examples and are in random order with no priority intended.

- *Movie-on-demand.* This service allows users to select any movie contained in a large library of movies with a relatively short delay between the time the user requests the movie and the time the movie begins.
- *Interactive video games.* This service allows the user to play electronic video games through a LEC's network. This would give customers convenient access to a larger selection of video games than they would typically have available to them.
- *Interactive video services.* This service provides the user with control over the program being viewed. For example, the user could control the camera angle used to view a sporting event, take part in a game show, remotely participate in a story line, or request additional information on news stories or sporting events.
- *Tele-education.* This service brings educational courses (recorded or live) to the user.
- *Multimedia library.* This service, which could be viewed as a subset of tele-education, allows the user to interactively search for and view/read/listen to educational multimedia information.
- *Broadcast video.* This service allows the user to access and rapidly switch among a range of channels provided (to the user) by one or more sources.
- *Transaction services type 1.* These services allow the user to take part in interactive transactions, such as banking, browsing for and purchasing services/products, or making reservations.
- *Transaction services type 2.* A more limited set of transaction services than type 1, these services focus on catalog and grocery shopping.
- *Targeted advertising.* This service allows advertisers to choose which individuals/groups see their advertisements as well as tailor the version of the advertisement to the specific individuals/groups.
- *Video mail box.* This service allows users to register their interest profiles and receive solicited or unsolicited video mail messages from advertisers.
- *Telecommuting.* This service allows users to maintain visual/audio communication as an adjunct to working from home.
- *High-quality video telephone.* This service allows appropriately equipped users to make switched video/audio calls to each other.
- *Subscription television.* This service is a "premium" broadcast type service to which access is restricted to subscribers.

- *Past television.* This service would allow users to "recall" or "replay" television programs that they might have missed (e.g., the previous day's episode of a favorite soap opera).

Services are expected to continue to evolve, particularly in light of the evolution of technology and the explosion of the Internet and the World Wide Web. Next, we discuss access technology alternatives.

6.3.2 Access Technology Alternatives

Many access technologies can be used to offer multimedia services. However, technical and economic (among other) considerations will narrow down the choices for a particular application. Access technologies that may be used to deliver multimedia services include but are not limited to FTTH, FTTC, HFC, ADSL, and radio technologies such as MMDS and LMDS [1–3]. Other technologies that are not discussed in this chapter also exist and in many cases, these technologies may be combined to offer the required services. Depending on the application, the combination ranges from simple overlay to integration of many major subsystems and components. Access technologies are discussed next.

6.3.2.1 Fiber-to-the-Curb Systems

A FTTC system is defined to consist of a host digital terminal (HDT) and subtending optical network units (ONUs) that are managed by the HDT. An optical distribution network (ODN) connects ONUs to the HDT and provides the optical pathways by which they communicate. This is shown in Figure 6.3. The ONUs are shared among a small number of customers and may be located on a pole, pedestal, or inside a building.

The ONU terminates optical fibers from the ODN and processes the optical signals. It is the network element that provides the tariffed telecommunications as well as video service interfaces for multiple residential and small business customers. Services are communicated over metallic twisted pairs and coaxial cable (possibly fiber cable or wireless) drop to a network interface, where they are handed off to the customer's network. The service interfaces cited are electrical; none are optical yet. Thus, an important function of the ONU is an electrical-to-optical (E/O) conversion. The ONU performs other functions, such as analog-to-digital (A/D) conversion of voice frequency signals, multiplexing of individual services onto the high-speed optical facility, and maintenance of both individual services and the optical transmission facility.

The ODN physically connects the ONUs to the HDT and contains only passive optical components. The ODN consists of single-mode optical fiber and

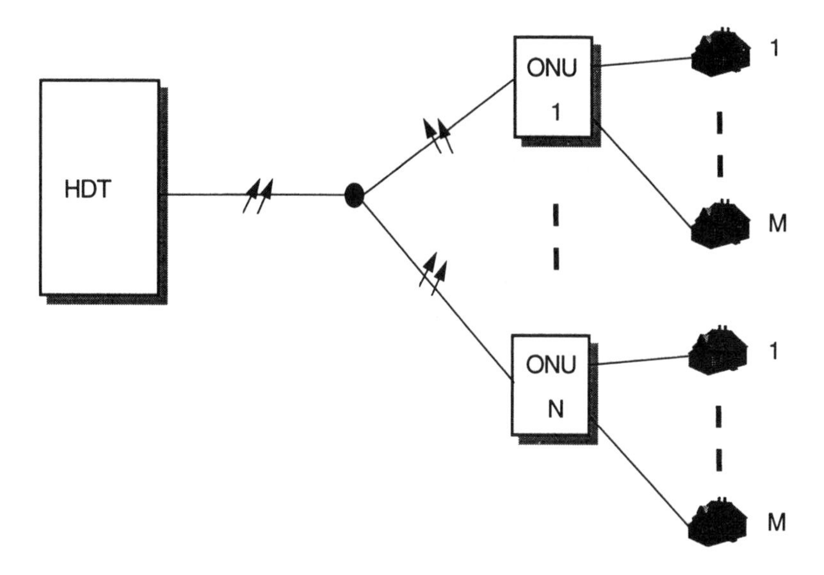

Figure 6.3 FTTC reference model.

various optical components, which could include connectors, power splitters/couplers, and wavelength-division multiplexing (WDM) devices. The ODN does not contain O/E devices. Thus, O/E conversions take place only at either end of the ODN in the ONUs and the HDT. Note that optical amplifiers that boost the optical signal level without O/E conversion could be installed on longer fibers. The ODN may be implemented in either a point-to-point configuration (where a unique fiber facility extends from an optical line unit at the HDT to an ONU) or a point-to-multipoint configuration (where a branching component located at some point within the ODN couples an HDT optical line unit to multiple ONUs).

The HDT manages the subtending ONUs and plays the major role in interfacing the FTTC system to the remainder of the local exchange carrier transmission and operations network. The HDT is located either at a remote site in the outside plant or in the central office. The HDT may either be a separate physical piece of equipment or it may be combined with other functionality in a hybrid network element. An HDT, as described here, does not contain per-line equipment; rather, it contains the equipment necessary to interface with the remainder of the loop access network and the optical line units necessary to interface with ONUs.

The HDT may concentrate traffic from all of its subtending ONUs for efficient feeder transport and present a highly utilized interface to the local switching system. The HDT may separate (groom) locally switched traffic from nonswitched and nonlocally switched traffic so that the latter can be routed away from the local switch fabric. Additionally, the HDT may manage signaling and supervision communications from the ONUs as well as manage provisioning and maintenance operations for both itself and its subtending ONUs. Also, the HDT performs many functions to support multimedia services, including routing, signaling, and management.

6.3.2.2 Fiber-to-the-Home Systems

FTTH systems are similar to FTTC systems. The main difference is that the ONU is not shared among many customers. Each customer has a dedicated ONU. Signals are passively split and placed on fiber drops to an ONU at each customer location where the conversion and processing are completed. FTTH systems provide narrowband services and multimedia services. A FTTH system is shown in Figure 6.4.

6.3.2.3 Hybrid Fiber Coax Systems

HFC systems use analog transmission where digital telecommunications services and multimedia services are subcarrier multiplexed into 6-MHz channels at the host (located in the CO). Services are then combined with analog video channels. At a fiber node, the optical signals are received, converted to RF, amplified, and placed on coax buses for transport to coaxial network units (CNUs). The CNUs may be shared among several customers or may be dedicated (one CNU per customer).

The types of services that can be provided by HFC systems are similar to those for FTTC and FTTH systems. However, the volume of service that can be carried may be quite different, especially for two-way services. The upstream bandwidth available on a coaxial system is limited. The amount of

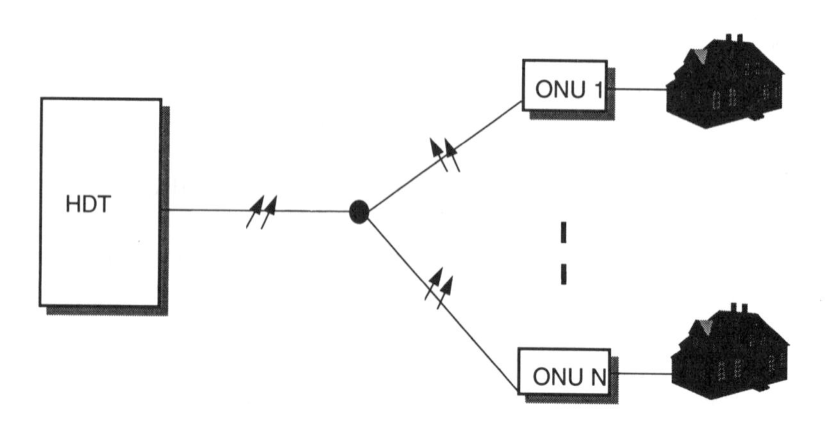

Figure 6.4 FTTH reference model.

bandwidth available is strongly dependent on the fiber node (or optical node) sizes used in these systems. An HFC system is shown in Figure 6.5.

6.3.2.4 Asymmetric Digital Subscriber Line–Based Systems

ADSL technology uses existing copper loops. ADSL-3 provides up to 6-Mbps multimedia paths, which can be used flexibly to provide a variety of service options. The ADSL bit stream resides in the bandwidth above analog POTS and is usable out to a range of about 3.6 km. The return channel for signaling and control is provided by a bidirectional control channel (e.g., 64 Kbps).

The ADSL terminal unit-CO card combines the analog and digital signals for transport to the ADSL terminal unit-remote card at the customer's premises. These cards provide the adaptive filtering and other functions necessary to ensure reliable coexistence of both services over the same copper pair. ADSL-1 (1.5 Mbps) is currently available as a commercial product. ADSL-3 (6 Mbps) is just now becoming a commercial product.

ADSL-based systems can also provide broadcast channels using real-time MPEG encoders, for which prices are dropping and video quality is improving. One concern with ADSL-based systems is that the amount of bandwidth that they provide to each customer is not as high as fiber systems, but for current applications and the foreseeable future, this bandwidth is sufficient to deliver requisite services. The biggest advantage of these systems is that they can provide digital video to customers very quickly and economically because they use the existing telephone lines and that they can be synergistically deployed with other systems, such as FTTC and FTTH systems.

6.3.2.5 Radio-Based Systems

In addition to the fiber and coax broadband access systems, there are radio-based systems that can deliver multimedia services. The discussion in this chapter is limited to two such systems, *multichannel multipoint distribution service* (MMDS) and *local multipoint distribution service* (LMDS).

MMDS is a wireless cable system that covers a 35-mile radius with a single broadcast antenna. In general, there is a single head-end facility adjacent to the broadcast antenna that gathers the video signals for redistribution to subscribers. A subscriber will need an antenna to pick up the video signal and a set-top box to convert the signal to an NTSC signal for the television.

Currently analog MMDS systems carry as many as 33 analog video channels. The FCC channel allocation of MMDS allows 10 channels to be used for commercial use, 20 for educational use, and 3 for private use. MMDS service providers can negotiate use of the noncommercial channels from the current owners; usually, however, the MMDS provider needs to carry educational video for certain periods of the day. With digitization, MMDS may

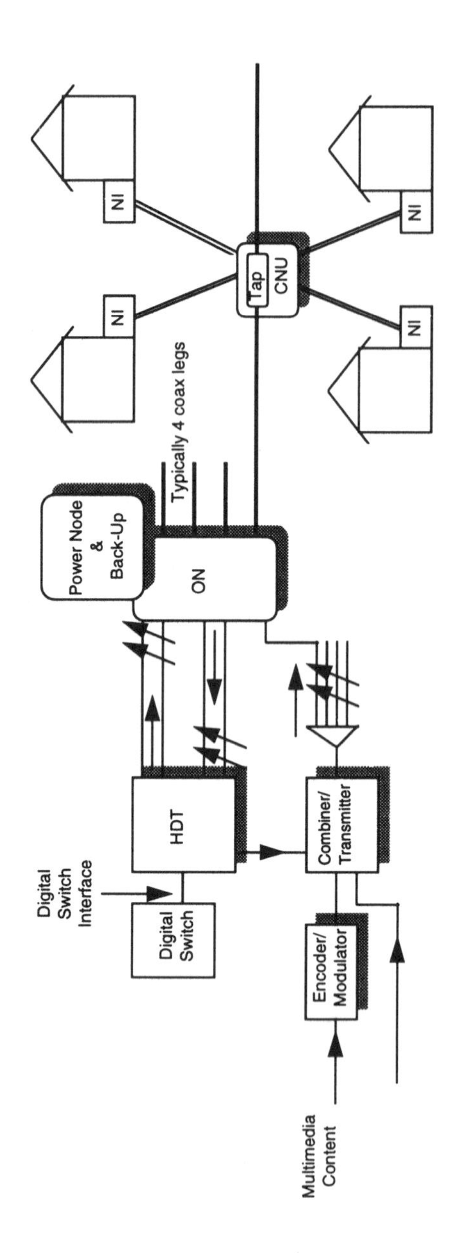

Figure 6.5 HFC system.

support as many as 160 video channels. MMDS is best suited for one-way broadcast of local, national, and near video-on-demand channels as well as one-way data channels.

LMDS is a wireless cable system that employs transmitters in a 3-mile radius of cells to cover their serving area. In general, there is a single head-end facility that gathers the video signals for distribution to the LMDS cell antennas, which broadcast the signal to subscribers. A subscriber will need a small antenna to pick up the video signal and a set-top box to convert the signal to an NTSC signal for the television. The subscriber antenna must be within the line of site of the transmitting antenna in order to receive the signal. Any foliage or buildings between the transmitting and receiving antennas will block the signal. There is currently one analog LMDS system that carries 49 analog video channels to customers. The FCC is currently deciding how to allocate the LMDS spectrum for commercial use. With digitization, LMDS may support several hundred video channels. However, with adjacent cell interference, about 250 channels are expected to be supported by each system. LMDS is best suited for one-way broadcast of local, national, and near video-on-demand channels as well as one-way data channels. Because the LMDS cell size is relatively small, it is also possible to support interactive video services such as video-on-demand. Additional research is required to identify and overcome technical issues with an interactive LMDS system.

6.3.3 Access System Comparison

It is difficult to compare access systems in a fair manner. This is because they all have different capabilities that are suited for some applications better than others. In order to be as fair as possible, common assumptions must be used. However, because various characteristics of the alternatives are fundamentally different, there are places where features or capacity limits either provide advantages or impose restrictions. This section provides a high-level comparison of the various alternatives and attempts to be as fair as possible. The reader is therefore cautioned not to assume that the comparisons given here always apply. These comparisons are generic and need to be reevaluated for a particular application or study since they are assumption driven, and the assumptions should be revisited and customized on a case-by-case basis. Having said that, we cover services, economics, performance, and evolution for various access alternatives.

6.3.3.1 Service Comparison

All access technologies discussed in this chapter have the capability to deliver telecommunications and multimedia services to customers. However, these services are not always integrated in the sense that some services may be delivered on separate facilities. An example of this is ADSL, where the existing narrowband service remains on the embedded copper loop.

The bandwidth provided to each customer varies among the access alternatives. With ADSL, it may be as high as 6 Mbps if 12,000-ft loops are assumed or much higher for shorter loops. (Much work is currently being conducted in the standards groups on VDSL or very-high-speed digital subscriber lines, where rates of 50 and 150 Mbps are being examined for use on drops.) The shared nature of the coaxial cable in HFC networks may ease initial implementation of multimedia services such as video telephony, but the available upstream bandwidth on the coax is limited. This limited bandwidth, coupled with the presence of ingress noise in the upstream band, will make it difficult to serve two-way services to a large number of customers from one serving optical node. The quality of service may be reduced if traffic engineering is applied and concentration is utilized so that a larger number of customers share the available bandwidth.

FTTH and FTTC systems allow delivery of high-bandwidth services. However, the implementation used in the fiber distribution of the FTTC or FTTH system may limit this bandwidth. For example, passive optical networks (PONs) use TDMA to allow multiple ONUs to share the fiber and the HDT optical transmitter. This protocol limits the bandwidth in the upstream direction (i. e. from the customer toward the network). Point-to-point topologies are not bandwidth limited but require much more fiber and do not allow sharing of costly optical transmitters and related equipment.

6.3.3.2 Relative Cost Comparison

Figure 6.6 provides per-subscriber cost comparisons for FTTC, FTTH, HFC, and ADSL capable of providing telephony and interactive multimedia services.

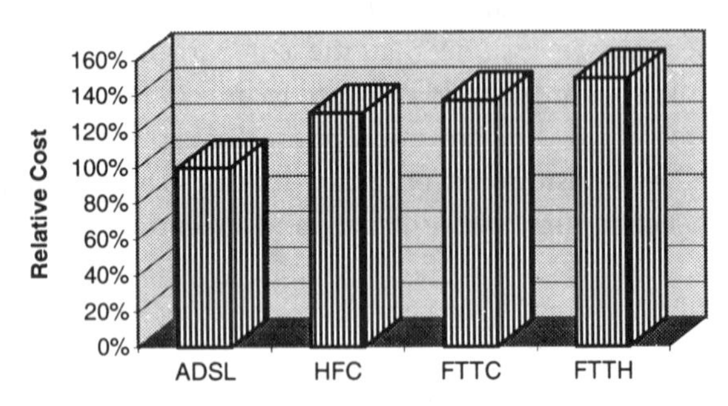

Figure 6.6 Example of relative installed first cost access system alternatives comparison.

These cost comparisons have been normalized against the cost of one particular implementation of ADSL. Note that the cost of ADSL may also vary greatly depending on whether the customers are served out of the central office, a remote location, or a combination of both. These costs use many built-in customer and service requirements assumptions and will vary if the assumptions are changed. They are provided here for illustrative purposes and show that the cost of ADSL is lower than that of other alternatives. The figure also shows that the cost of FTTH is highest here, and this is mainly due to the dedication of certain electronics to customers.

Another important thing to note here is that the costs presented are installed first costs and do not reflect any operational costs. Operational costs take into account the cost of operating and maintaining various architectures. These costs are not covered here but need to be determined and clearly understood before any economics conclusions can be drawn.

6.3.3.3 Performance Comparisons

The access alternatives presented herein can all be built to be robust and reliable. For example, redundant equipment can be built to provide higher reliability and availability objectives. Sufficient backup power can also be provided with them to compensate for loss of commercial AC power. Standby optical links could be provided for digital transport between the HDT and fiber nodes to enhance reliability. The service lifetime and reliability of coaxial cable plant may be a concern. The combined impact of filled connectors, jacketed cable, and good construction techniques aimed at preventing the intrusion of moisture into the cable should help prevent failures. The reliability of ADSL is expected to be essentially the same as current POTS service. The presence of foliage in the signal path may affect the signal and presents concerns for MMDS and LMDS systems. With multimedia services, the network provider or service/application provider is concerned about the security of the network while the end user or customer is concerned about privacy. In connection with multimedia services, various combinations of interdiction, encryption, and scrambling are used by the industry today. The addition of these functions will add costs to the access systems, and the costs will vary depending heavily on how and where these functions are provided.

6.3.3.4 System Evolution and Upgrades

The various access systems might evolve in any of various ways. If we take FTTC and FTTH systems, for example, we see that the initial focus has been on FTTC, where the network element closest to the customer, the ONU, serves multiple customers. These systems will deliver telecommunications and multimedia services over individual metallic interfaces. One goal is to make any plant upgrades as transparent to customers and their existing equipment and wiring as possible. Another evolution option is to envision the HDT and ONU connected using a combination of optical fiber and coaxial cable. The coaxial cable may be used initially and a system upgrade will reduce the coax and increase the fiber. It is very possible that many access systems will coexist initially and will converge toward a more integrated system as service demands are more stable and as system costs come down. Also, over time, as multimedia services become more and more widely available, more optimal service interfaces will be defined to take advantage of an all-fiber distribution network. Ultimately, it is possible that service to residential and small business customers could be delivered over a single, high-bandwidth optical fiber interface.

As mentioned earlier, several access methods to provide existing services to customers are possible today. These alternatives work with both digital and analog switches. However, parallel development of various access systems and the introduction of SONET/SDH technology in the feeder have led many suppliers to develop new multifunctional access terminals. Such terminals, referred to as *multiple technologies access terminals* or *integrated access terminals,* may have a SONET/SDH interface to the feeder portion of the network (e.g., performing an add/drop multiplexer functionality) and both fiber and copper terminations on the distribution side. In the future, these terminals will also support other technologies (such as ATM).

As with the other issues considered in this chapter, system evolution issues differ from architecture to architecture. The factor that determines the business need for system evolution is services evolution. Perhaps the most prevalent view of services evolution involves increasing interactivity and an open network architecture providing customers with access to multiple services and content providers. This view assumes that services and content are carried over the backbone network in digital format. Each access architecture should be evaluated based on its ability to interwork within this view of the evolving network. Two specific issues that must be considered are access to services and content provided by many content or information providers and two-way services and content requiring increased interactivity.

Another concern with the evolution of access systems is the limited bandwidth per customer for both upstream and downstream directions (this varies from architecture to architecture, where it might not pose a problem for some as it does for others). This will limit the services that the systems will be able to provide as the services evolve. However, ATM-based services are expected to prevail and the access architectures that cannot support these services will be replaced by ones that can.

6.3.4 Application of SONET/SDH and ATM to the Access Network

So far, the use of SONET/SDH and ATM have been proposed in the access network for residential customers, but have not been implemented mainly due to the cost associated with such equipment. Equipment cost, including overhead processing, cannot be economically justified for the access yet because the functionality (e.g., automatic protection switching), driven by service demands and service revenues, is not required in the access for the mass market consumers. However, there has been an attempt to drive both the SONET/SDH and ATM technologies closer and closer to the residential customers. As an example, some FTTC systems have been designed using a SONET-like frame structure for transmission between the ONU and the HDT. This structure will allow graceful evolution to a full SONET/SDH network whenever it can be justified economically. Also, set-top boxes for use with interactive video services as well as data services are starting to implement ATM-based transmission all the way to the residential customer. Because of the flexibility that ATM provides, the use of ATM for access has proven to be economical in some cases provided that not all ATM header functionality is implemented. SONET/SDH and ATM as described in Sections 6.4.2 and 6.4.3, respectively, of this chapter provide functionality required in the access from the physical layer to the application and service layers.

It is envisioned that as service demands increase and can justify the use of a more feature-rich and "standardized" access network, SONET/SDH and ATM will be implemented in the last mile. For instance, as Internet access services become more widespread among residential customers and as they become used for "reliable and secure" services such as home banking, then it is expected that network providers will be able to justify the use of backbone technologies in the access network. It is expected that this transition will be graceful and will happen slowly during the next 20 years.

The next sections describe the backbone network and will allow the reader to understand the applicability of SONET/SDH and ATM.

6.4 Trunk Transport Network Architectures and Technologies for Multimedia Services

6.4.1 Trunk Transport Networks

6.4.1.1 Transport Network Layered Architecture

The capacity of transmission systems has been increasing rapidly with the advancement of recent optical technology. Transport node throughput has also been enhanced with the development of SDH/SONET and ATM technologies, and it is expected to further advance our ability to capitalize on new emerging optical technologies. New services need to be developed and provided through the network in pace with the increasing demands. A telecommunications network and the services provided by the network should gracefully evolve to incorporate the advances in technologies and to respond to diverse demands.

The increased transmission capacity, on the other hand, causes severe damage on single transmission link/system failure. The development of effective network restoration techniques against such network failures is, therefore, becoming more and more important. Moreover, the increasing volume of data transmission between computer systems urges rapid network restoration to minimize social economic loss. The cost of the network operations needed to satisfy the requirement, however, must be minimized. Flexibility of the network is another key issue, since it permits a variety of cost-effective end-customer controls and provides the network with adaptability in the face of unknown service requirements.

A transport network architecture that supports rather diverse demands including those already mentioned can be developed by layering the network on the basis of network functionality. This layered transport network architecture simplifies the design, development, and operation of the network, and allows smooth network evolution. The layered concept also makes it easy for each network layer to evolve independently of the other layers by capitalizing on the introduction of new technology specific to each layer. The layering concept has been extensively discussed within ITU-T for the SDH transport network [4].

Figure 6.7 illustrates a layered network structure for a public telecommunication network. It comprises a service network layer and a transport network layer. The service network layer consists of different service networks, each of which is dedicated to a specific service. This layer provides circuits or channels. A transport network is realized with paths and physical transmission media, and is less service dependent. Thus, the telecommunication network can be divided into three layers from the viewpoint of functions: a physical media layer, a transmission path layer, and a communication circuit layer, as shown in Figure 6.7. Operation, administration, and maintenance (OA&M) of these layers is performed by dedicated OA&M systems, and these systems work cooperatively as required.

In connection-oriented communication, an end-to-end connection is established/released dynamically or on the basis of short-term provisioning between users via service nodes. In connectionless communication, a block of data (packet, cell, frame, etc.) is transferred to the neighboring service node without establishing a connection. The circuit layers are dedicated to specific

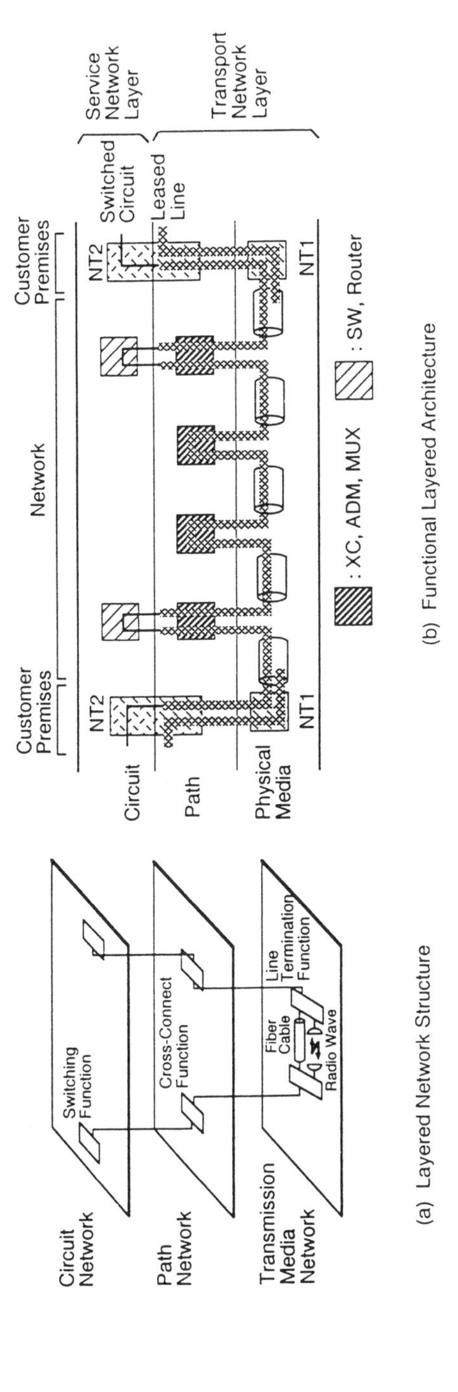

Figure 6.7 Layered architecture of public telecommunication network (*After:* [5]).

services such as the public switched telephone service, packetized data communication service, frame relay service, and ATM cell relay service.

The transmission media, which interconnects nodes and/or subscribers, is constructed based on long-term provisioning; geographical conditions are taken into consideration. The point-to-point transmission capability has increased significantly during this decade through the introduction of optical fiber transmission; 10-Gb transmission systems are now being introduced. However, traffic demands between two service nodes are not always large enough to fill up the bandwidth of a high-speed transmission system that connects the two service nodes directly.

The path layer bridges the circuit and transmission media layers and provides a common signal transport platform supporting different service-dependent circuit networks. A path can be a grouped circuit serving as the unit of network operation, design, and provisioning and can be an object to be manipulated for restoring node and transmission line/system failures. Here, a path used as the unit of a group of connections between service nodes (or node systems) is called a *service access path.*

For service access, the appropriate path bandwidth is determined by the service bandwidth and the traffic demands and is usually not very large; in the existing SDH network for telephony, it is 1.5 Mbps (twenty-four 64-Kbps channels) in North America and Japan and 2.0 Mbps (thirty-two 64-Kbps channels) in Europe. Due to recent transmission technology advances, available transmission capacity has been increased to 2.5 Gbps or more. To maximally exploit that large-capacity transmission capability, it is necessary to realize cost-effective access to the transmission system (called *trans-access,* hereafter) and effective grooming capability at the nodes. Thus, in addition to the service access path, another path stage for trans-access may need to be introduced to create economical networks. For example, in SDH networks higher layer paths of 49 Mbps (VC-3) or 150 Mbps (VC-4) are usually used for this purpose as detailed in a later section. The higher level paths accommodate service access paths. The higher level paths are accommodated within a transmission media network.

Path layer functions can be performed with intelligent ADM (add/drop multiplexer) systems and/or digital cross-connect systems and control systems. Network flexibility can be enhanced with path layer control. The hierarchical path structure is depicted schematically in Figure 6.8.

6.4.1.2 Trunk Transport Network Architecture

Figure 6.9 depicts a typical example of the transport network architecture. The transport network is divided into several layers. This multiple-layer arrangement reduces the number of cross-connect/ADM nodes that are traversed by a path.

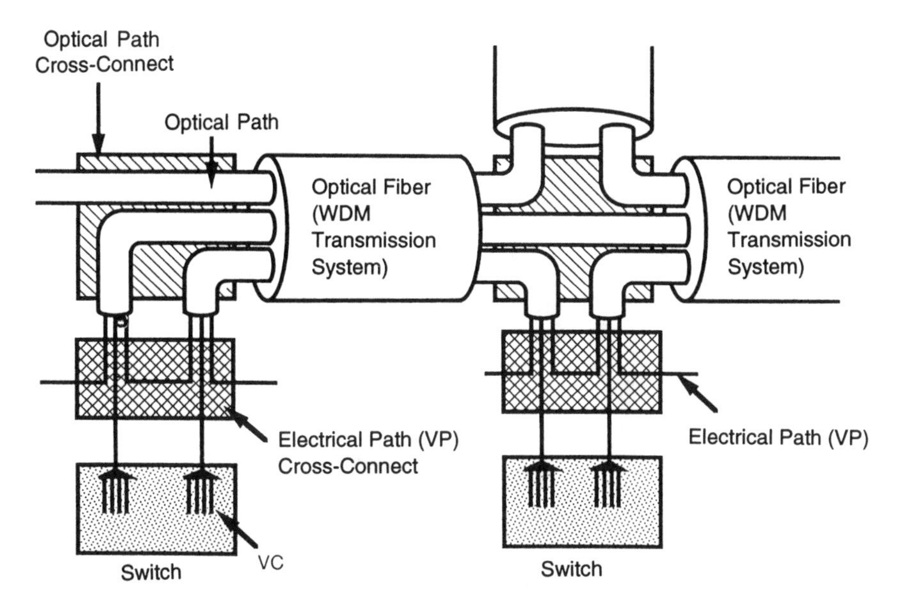

Figure 6.8 Schematic illustration of hierarchical path structure and cross-connect function.

A nationwide backbone network usually consists of digital cross-connect systems (DCSs) connected with high-speed (gigabit class) optical transmission links in a mesh configuration. A backbone network usually consists of many nodes and the distances between adjacent nodes are relatively large. So the transmission cost portion is relatively high and, as a result, the efficient use of the transmission line capacity is very important. Furthermore, the growth rate of point-to-point traffic demands can be very high. This requires high connectivity between nodes, grooming capability at nodes, and minimization of spare capacity needed for network restoration from failures. A mesh network configuration with cross-connect systems, therefore, can be an effective solution.

Regional networks may consist of one or more network layers. Each regional network consists of a relatively small number of nodes and the distance between adjacent nodes is relatively small. Therefore, the reduction in transport node cost and network operation cost that can be attained by simplifying the network architecture is of greater importance than the reduction in transmission cost achieved by enhancing the transmission line utilization. Furthermore, the growth rate of traffic demand is quite stable and relatively slow in a regional network. Therefore, a ring structure with ADMs is often favored because this configuration enables rapid and simple restoration from failures. (Network restoration techniques are discussed in later sections.) Service nodes have been introduced in some of the layers depending on the services provided. For public

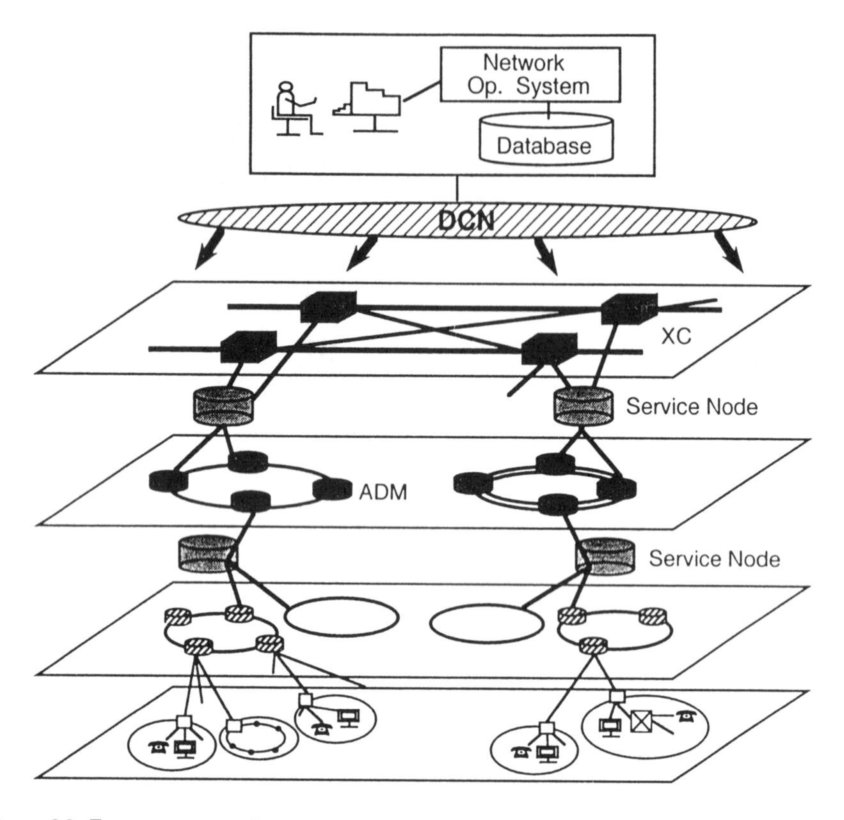

Figure 6.9 Transport network example.

telephone services, at least two stages of hierarchical switching systems, local and transit switches, have been implemented. The chosen access transport network architecture depends on the technologies used as explained in Section 6.3.

Path restoration and section protection (APS or automatic protection switching) are used for network reliability enhancement. The APS switches the signal from a failed working fiber to the protection fiber; 1:*n* or 1 + 1 protection can be used. In the 1 + 1 configuration, two copies of the signals are transmitted through different fibers and the receiver selects one of them as the working channel. The 1 + 1 type of protection can be done nonrevertively, which means that signal is not returned to the original fiber when it is restored. In 1:*n* protection, when any one working fiber fails it is switched to the protection fiber, and when the failure is restored the switched channel is returned to the original working fiber. APS provides service protection against equipment and line failure of a single working fiber, but at the network level it cannot protect against multiple failures. In a common 1:*n* protection structure, the protection fiber is placed in the same route as that of working fibers, and so it is ineffective against route failure (cable cut).

On the other hand, path restoration can cope with route failure and multiple failures, but it requires more sophisticated operation systems. For 100% path restoration from any one link failure, the number of restoration paths must be 30% to 50% of the active path number, the exact value depends on the network topology and other parameters.

6.4.1.3 Transport Network Technologies

Optical Transmission Technology

Since the first demonstration of laser emission in 1960, tremendous efforts have been made to realize fiber optic transmission systems. Remarkable progress has been seen, particularly in this decade. Notable developments include low-loss single-mode fibers and optical sources in the 1.5-μm wavelength region, and ultra-high-speed electronic and photonic circuits.

Optical fiber transmission was first introduced in early the 1980s and since then the transmission capacity has been increased by more than one order of magnitude per decade [6]. This resulted in a significant transmission cost reduction during the last decade. The continued enhancement of electronic devices, Si-based high-speed electronics, and GaAs-based devices now permits operation speeds of several tens of gigabits per second.

Another technology that must be emphasized is the optical amplifier. In the last few years, significant progress has been recorded in the development of erbium-doped-fiber amplifiers (EDFAs). This technology greatly increases the application area of optical technologies by overcoming the optical loss of transmission, devices, and optical components, as explained in Section 6.4.4. Very recently, wavelength-division multiplexing (WDM) technologies have reached a level that allows the commercial introduction of 16-channel WDM point-to-point transmission systems with 200-GHz channel spacing. This WDM technology will enable further expansion of the transmission capacity and network flexibility.

The preceding developments will accelerate the penetration of high-speed multimedia services. Thus, the continually evolving optical fiber transmission technologies will form an important portion of the base on which the future ubiquitous high-speed multimedia network should be created.

Transport Node Technology

As for the path layer technology, cross-connect systems were first put into commercial use by NTT in Japan in 1980 for the PDH. They employ DS-2 (6.3-Mbps) interfaces with line facilities, and the unit of cross-connection with time slot interchange techniques is 384 Kbps, or six telephone channels. In 1981, DACS I was introduced by AT&T in North America. It was designed to terminate up to 128 DS-1 (1.5-Mbps) channels and the unit of cross-connection is DS-0 (64 Kbps). It is used to support channel processing for special services and provides a convenient testing environment for T1 carriers.

SDH paths were introduced in 1989 [7] and are being widely deployed in the 1990s throughout the world. The introduction of SDH has simplified the existing plesiochronous digital hierarchy (PDH) path networks and allowed direct multiplexing/demultiplexing of paths so that the digital cross-connect/ADM function and transmission signal monitoring capabilities can be easily implemented (see Section 6.4.2).

The path is fundamentally a logical concept whose attributes are the bandwidth, route, and QoS that it provides. In SDH networks, a path is tightly linked to the physical interface structure for transmission. This linkage produces inefficiencies that can become a serious handicap in the multimedia environment. For ATM networks, the virtual path (VP; see Section 6.4.3) was standardized in ITU-T in 1990 [8]. The VP strategy enables the fully logical realization of path functions and can enhance network capability through its flexibility. The VP benefits are discussed in Section 6.4.3. VPs began to be introduced in the early 1990s in various countries for the provisioning of ATM services and to provide a common transport mechanism to support frame relay service and SMDS service.

The throughput of cross-connect systems has been increasing in line with the introduction of SDH and ATM, and transport node cost per bit has been accordingly reduced. The ATM technology not only reduces transport node cost but also greatly enhances the flexibility of the path network, which will allow multimedia networks to be constructed.

One important point is that existing path layer approaches utilize electrical technologies. In Section 6.4.4, a new optical technology that can enhance transport node throughput is discussed.

6.4.2 SDH/SONET Network Technologies

6.4.2.1 SDH Technology

Networks installed before 1990 are based on the plesiochronous digital hierarchy (PDH) [9]. In PDH networks, the bit rate of each tributary signal is not completely synchronized with the multiplexing equipment, although it is controlled within a specific limit. Accordingly, the multiplexer reads each tributary at the highest allowed clock speed and, when there are no bits in the input buffer, it adds stuffing bits to adapt the input signal rate to the higher clock speed (this is called *positive stuffing*). The individual tributaries are synchronized with the multiplexer at each multiplexing step by utilizing positive bit stuff justification. The multiplexer also has a mechanism to indicate that it has performed stuffing, so that the demultiplexer can identify which bits to throw out. This type of multiplexing is referred to as *plesiochronous*—that is to say *almost* synchronized.

The asynchronous nature of this multiplexing scheme creates several difficulties. For example, to access a 1.544-Mbps signal for rerouting or testing, the entire line signal structure must be demultiplexed step by step down to the 1.544-Mbps level. This entails significant processing overhead and causes inefficiencies. In PDH, limited network management and maintenance capabilities are possible since relatively few overhead bits are available. Furthermore, different frame structures are defined at different bit rates and no worldwide standard exists, so interworking between equipment supplied from different manufacturers cannot be expected. The digital hierarchies of Europe, North America, and Japan are different as shown in Figure 6.10.

If the network is completely synchronized, direct multiplexing/demultiplexing of transmission signals (individual tributary signals) can be done easily through any network device. Also, any low-speed multiplexing level signals can be accessed so digital cross-connect function and transmission signal monitoring

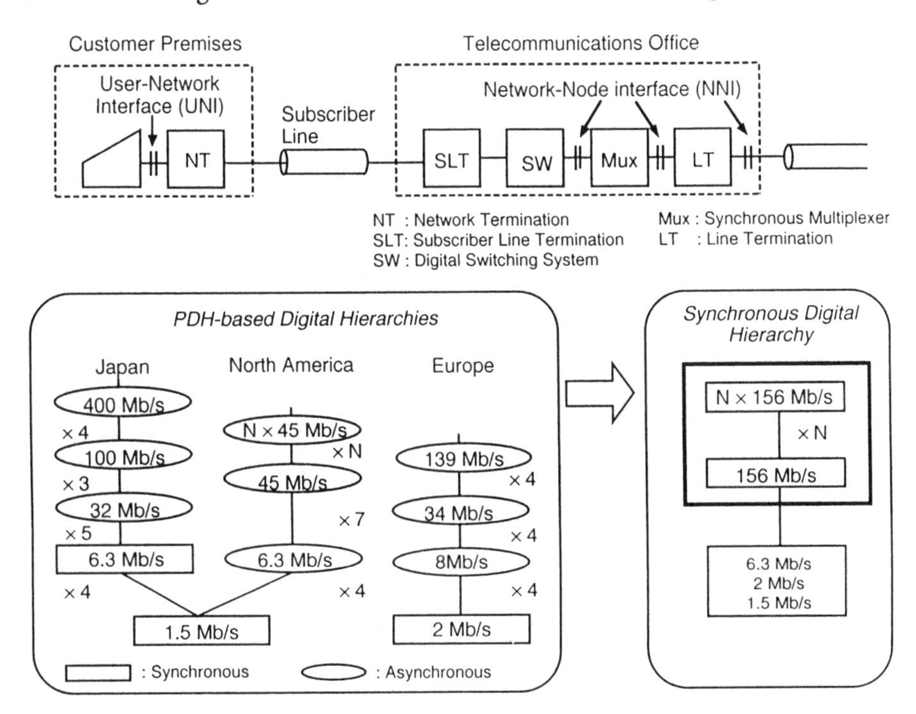

Figure 6.10 PDH and SDH digital hierarchies and network reference point.

capabilities are more easily implemented. A telecommunication network can, therefore, be handled as just a single digital circuit pack, and advanced network functions can be realized simply. This background has led CCITT to develop the SDH NNI (network node interface) standards [4,10,11] for a synchronous network.

This worldwide standard allows us to create a unified telecommunication network infrastructure, and interconnection of network equipment from different vendors becomes possible. SDH is designed to accommodate various existing signal types via virtual containers. The SDH network can, therefore, overlay existing networks and flexibly support new service signals. Ample overhead bytes (more than 4% of the total is available) allows enhanced network OA&M functions that ensure network integrity. The transport network layered concept is effectively realized with SDH as explained in the next sections.

6.4.2.2 SDH Transport Network Layered Architecture

Figure 6.11 depicts the layered model of the transport network and shows the layer relationships for the SDH-based transport network [12]. Different service networks are supported by SDH transport layers. Transmission medium layer networks are dependent on the transmission medium such as optical fiber or radio waves. Transmission media layer networks are divided into section layer

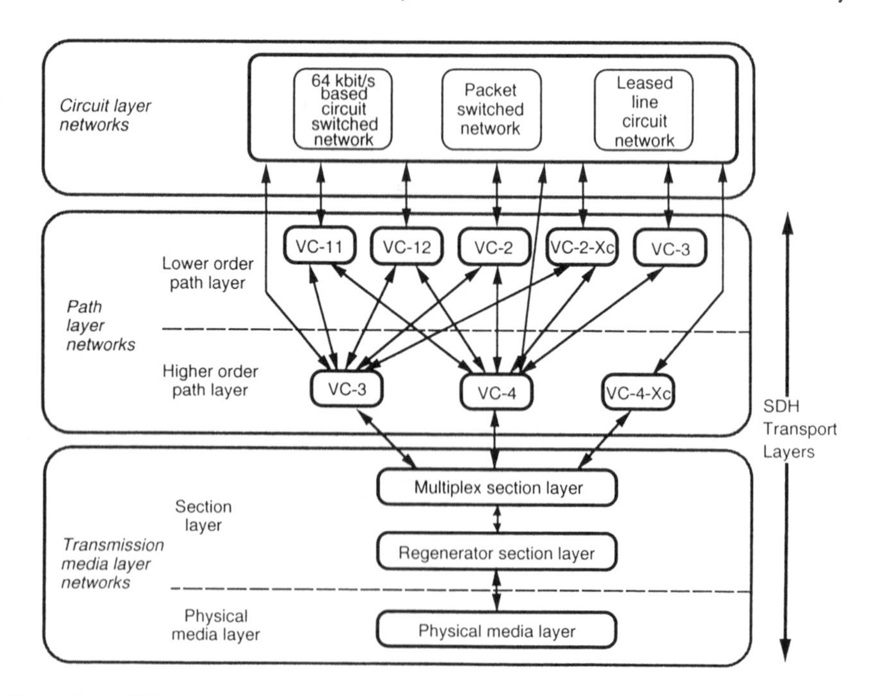

Figure 6.11 SDH-based transport network layered model.

networks and physical media layer networks. Section layer networks are divided into two parts: a multiplex section (MS) layer network and a regenerator section (RS) layer network. The multiplex section layer is concerned with the end-to-end transfer of information between locations that route or terminate paths. The regenerator section layer is concerned with the transfer of information between regenerators and locations that route and terminate paths. Section and path layers have their own overheads for transmitting OA&M information, thus allowing enhanced network OA&M functions.

Path layer networks are divided into higher order path layer networks and lower order path networks (Figure 6.11). The higher order paths consists of VC-3, VC-4, and VC-4-Xc paths, while the lower order paths consists of VC-11, VC-12, VC-2, VC-2-Xc, and VC-3 paths. Trans-access, as discussed earlier, can be performed with the higher order paths, while service access can be performed with the lower order paths. Here, a *virtual container* (VC) is an information structure that consists of information payload and path overhead (POH). VC-11 can contain a hierarchical bit rate of 1.544 Mbps; VC-12, 2.048 Mbps; VC-2, 6.312 Mbps; VC-3, 44.736 and 34.368 Mbps; and VC-4, 139.264 Mbps. The lower order virtual container comprises a single container plus path OH. The higher order virtual container comprises either a single container, n (where n = 3,4), or an assembly of tributary unit (lower order VC plus tributary unit pointer [11]) groups together with an OH appropriate for the level [13].

PDH networks have to handle several path stages. In contrast, SDH networks have only two kinds of paths: higher order and lower order paths. SDH path networks are, thus, simple compared to those of PDH. PDH networks need multistage multiplexers (MUXs) to match the hierarchy. In SDH networks, only lower and higher order multiplexers or digital cross-connect systems are necessary to attain high network efficiency. This simplification offers increased link utilization (link cost reduction) and transport network node cost reduction. For example, in PDH networks when the average multiplexing efficiency attained by single-stage multiplexing is 0.8, four-stage multiplexing results in the link utilization of only 0.4 ($= 0.8^{0.4}$).

6.4.2.3 SDH Network Node Interface

A possible network configuration and locations of the SDH NNI are illustrated in Figure 6.10. NNI specifications need to be standardized to enable interconnection of SDH network elements for the transport of different types of payloads [14].

Synchronous transport module level N (STM-N) is the information structure of level N for SDH NNI. The STM-N represents the substance of synchronous transmission and SDH, since it incorporates the layer concept, synchronization mechanism employing a pointer, and the visibility of DS-0. The pointer isolates the VC from the transmission frame or the higher order VC, and allows wander and jitter to be accommodated. The STM-N consists of section overhead information fields, administrative unit pointer(s) and information payload, organized in a block frame structure that repeats every 125 μs. The first level of the synchronous digital hierarchy is 155.520 Mbps and the higher SDH bit rates are defined as integer (N) multiples of the first-level bit rate. Presently, the standardized values of N are 1, 4, 16, and 64.

The two-dimensional representation for the STM-1 signal frame consists of 9 rows by 270 columns giving a total signal capacity of 2,430 bytes per 125-μs frame, which corresponds to the bit rate of 155.52 Mbps (see Figure 6.12). There are two reasons for the 9-row structure: a 9-row periodic structure within a 125-μs frame can reduce the memory needed for signal processing in SDH transmission equipment, and the 9-row structure is suitable for accommodating the tributaries of 1.544 and 2.048 Mbps, since a 1.544-Mbps signal is accommodated within 27 bytes (64 Kbps × 9 rows × 3 columns = 1.728 Mbps) and a 2.048-Mbps signal is accommodated within 36 bytes (64 Kbps × 9 rows × 4 columns = 2.304 Mbps).

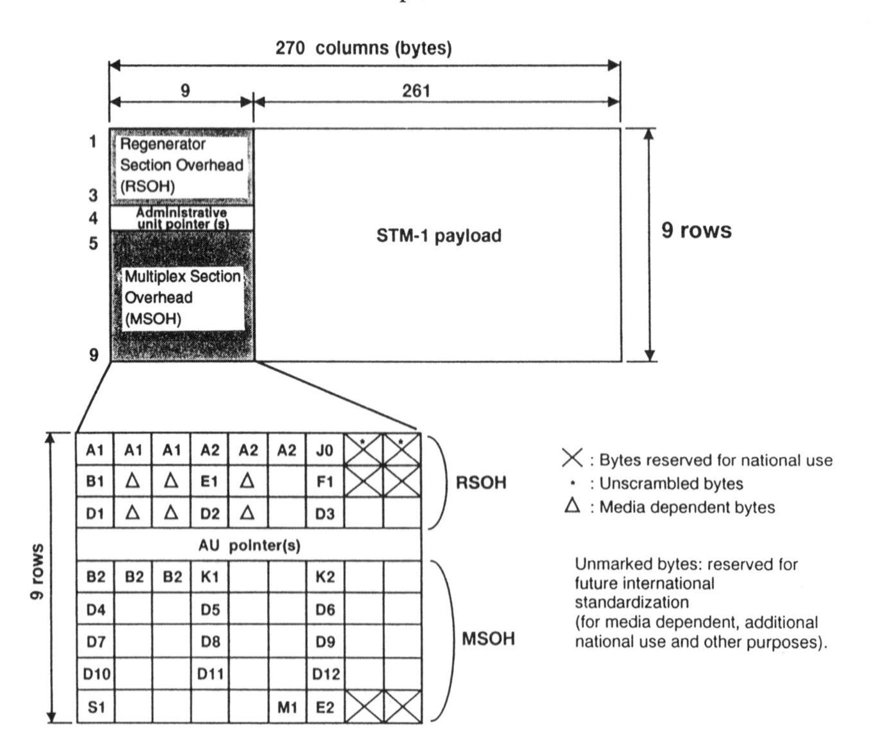

Figure 6.12 STM-1 frame structure.

STM-1 consists of a payload that contains one VC-4 or three VC-3s, a section overhead (SOH), and administrative unit pointer(s). An STM-N (where $N = 4$, 16, and 64) signal is assembled by byte-interleaving N parallel frame-synchronized STM-1 payloads and administrative unit (AU) pointers. The first $9 \times N$ columns are occupied by SOH and AU pointers. The remaining columns are occupied by the $N \times$ (VC-4 or 3 VC-3) signals associated with the N individual STM-1 signals.

Section Overhead

SOH contains frame synchronization signals and OA&M-related signals. It pertains only to an individual transport system and is not transferred with the VC between transport systems. SOH is divided into regenerator section overhead (RSOH; rows 1–3) and multiplex section overhead (MSOH; rows 5–9) as shown in Figure 6.12 for STM-1. The function of each overhead bytes is as follows: for framing purposes (A1 and A2), regenerator section trace (J0), regenerator section error monitoring (B1), regenerator section engineering order wire (E1), user channel (F1), data communication channel (DCC; D1, D2, D3), multiplex section error monitoring (B2), automatic protection switching channels (K1, K2), multiplex section DCC (D4–D12), MS engineering order wire (E2), synchronization status (S1), and MS remote error indication (M1). Some overhead bytes of the STM-4 overhead are identical in number to those of STM-1, but others are quadrupled from those of STM-1.

Administrative Unit-n (AU-n)

An administrative unit consists of a higher order virtual container and an administrative unit pointer, which indicates the offset of the payload frame start from the multiplex section frame start. Two administrative units are defined: AU-4 and AU-3. AU-4(3) consists of a VC-4(3) plus an administrative unit pointer that indicates the phase alignment of the VC-4(3) with respect to the STM-N frame. The VC-n associated with each AU-n does not have a fixed phase with respect to the STM-N frame. The AU-n pointer indicates the location of the first byte of the VC-n.

AU-n Pointer

The AU-n pointer provides a method that allows flexible and dynamic alignment of the VC-n within the AU-n frame. The VC-n is allowed to "float" within the AU-n frame. This means that VC-n may begin anywhere within the AU-n payload. Usually, the VC-n begins in one AU-n frame and ends in the next. This mechanism minimizes the processing delay needed for multiplexing. Generally, the frame phases transferred to a node from different nodes do not coincide with each other due to differing amounts of transmission delay. To align the phase of different signal frames, frame memory is utilized and, therefore, delays of half the frame time are incurred on average (one frame time maximum). This method demands large memories if the transmission speed increases, which increases the power consumption of the system and hinders system downsizing. The pointer technology resolves these problems. It was first proposed in SONET [15] and then adopted as an ITU-T standard. The AU-4 pointer is contained in the three bytes of H1, H2, and H3.

Mapping of ATM Cell

As the physical-layer technology of the ATM transport network, two options have been proposed: cell-based and SDH-based ATM transport networks [16]. Detailed standards have been issued for the SDH-based ATM transport network [14,17]. This is because exploiting already developed SDH technologies realizes the economical introduction of ATM techniques with minimum delay but maximum efficiency through the utilization of SDH network operation capabilities. Thus, ATM networks can be introduced by overlaying SDH networks.

VC-4 and VC-4-Xc are utilized to convey ATM cells (53 bytes), which are mapped into a C-4 (C-4-Xc) with its octet boundaries aligned with the C-4 (C-4-Xc) byte boundaries. The C-4 (C-4-Xc) is then mapped into VC-4 (VC-4-Xc) together with the VC-4 OH (VC-4-Xc OH and (X-1) columns of fixed stuff), as shown in Figure 6.13. ATM cells may cross a C-4 (C-4-Xc) boundary since no integer multiple of cells matches the C-4 (C-4-Xc) capacity (44.1509 . . . cells/C-4).

6.4.2.4 SONET and SDH

Synchronous Optical Network (SONET) is a North American standard as approved by ANSI in 1988. SONET became the basis of SDH as standardized later in CCITT. SONET was first proposed by Bellcore in 1984 as an optical signal transmission interface to resolve multivendor and multiowner associated problems. It offers not only an interoperability, but also efficient multiplexing, sophisticated OA&M capabilities, and enhanced network integrity. The first objective of SONET is to define a synchronous transport signal at the DS-3 (44.736-Mbps) level, STS-1.

In the mid-1980s, CCITT (now ITU-T) began the process of standardizing SDH for the NNI. North America proposed the use of 50 Mbps based on STS-1 of SONET, while the Conference of European Post and Telecommunications Administrations (CEPT) and Japan proposed to adopt 150 Mbps. CCITT wanted a worldwide unique digital hierarchy for constructing the global communication network of B-ISDN. Therefore, SDH was standardized (Recommendations G. 707, 708, and 709) with a basic rate of 155.520 Mbps, which can accommodate North American, European, and Japanese digital

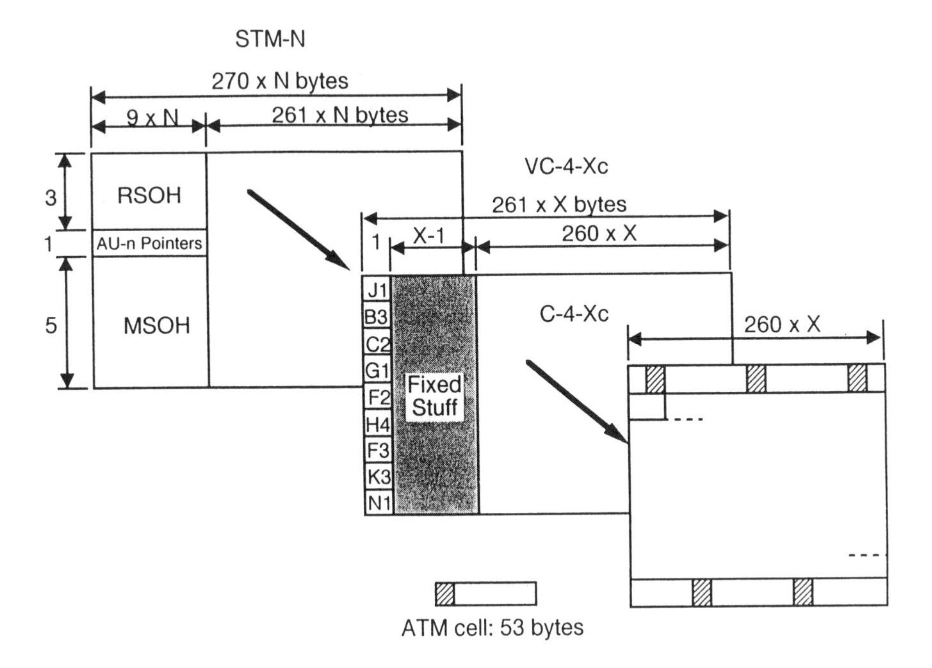

Figure 6.13 VC-4-Xc structure and mapping of ATM cells into the VC-4-Xc.

hierarchies. As explained before, one of the important points of SDH is that it enables the efficient multiplexing of various existing service bit rates through the introduction of the virtual container concept. With this technique, the DS-3 signals of North America can be effectively accommodated within the SDH hierarchy signals. The digital hierarchies of SONET and SDH are different. The base rate of SONET is 51.84 Mbps (STS-1), and the higher rate signals (STS-N) are *N* multiples of the basic rate. Currently used *N* values in standards are 1, 3, 12, 24, 48, and 192. On the other hand, the basic rate of SDH is 155.520 Mbps (STM-1) and the higher rate signals (STM-N) are *N* multiples of the basic rate; currently the *N* values are 1, 4, 16, and 64. STM-N is the equivalent of the concatenated signal STS-3Nc.

As can be recognized from the standardization process of synchronous digital transmission, SDH and SONET are very similar but not identical. They have similar sets of overheads and functions, but are not identical. Further, the multiplexing structures whose signal rate is lower than STS-3 or STM-1 are different. In SONET, five tributaries, DS-1 (1.544 Mbps), DS-1E (2.048 Mbps), DS-1C (3.152 Mbps), DS-2 (6.312 Mbps), and DS-3 (44.736 Mbps), are multiplexed into STS-1 utilizing one type of intermediate signal called the *virtual tributary* (VT), while STM-1 requires the systematic multiplexing of all tributaries from DS-1 to DS-4 utilizing such intermediate signal units as C, VC, TU, TUG, AU, AUG, as explained before. The VC is equivalent to VT.

The terminology in the layered architecture also differs between SDH and SONET. The physical medium, regenerator section, multiplex section, and path layer of SDH are called, respectively, the photonic, section, line, and path layer in SONET.

6.4.2.5 DCS Mesh Network

The DCS provides a centralized point for grooming, testing and cross-connection of paths as explained in Section 6.4.1. Thus, it offers a flexibility point for traffic management and restoration, where necessary operation communication with operation systems is provided through DCC, or standardized Q3 interfaces to the data communication networks. Along with the multiplex section protection as described in another section, a DCS mesh network can provide the path network reconfiguration capability, which enables much more flexible path provisioning than is possible in an ADM ring network.

The DCS mesh configuration enables higher link capacity sharing and more effective path restoration against fiber cable cut and node equipment failure, since the connectivity can be higher than that of the ring structure. As a result it does not require dedicated standby protection facilities unlike a ring network (as discussed later). Figure 6.14 shows an example of DCS restoration. Demands between switches A and B are normally routed over two links, link

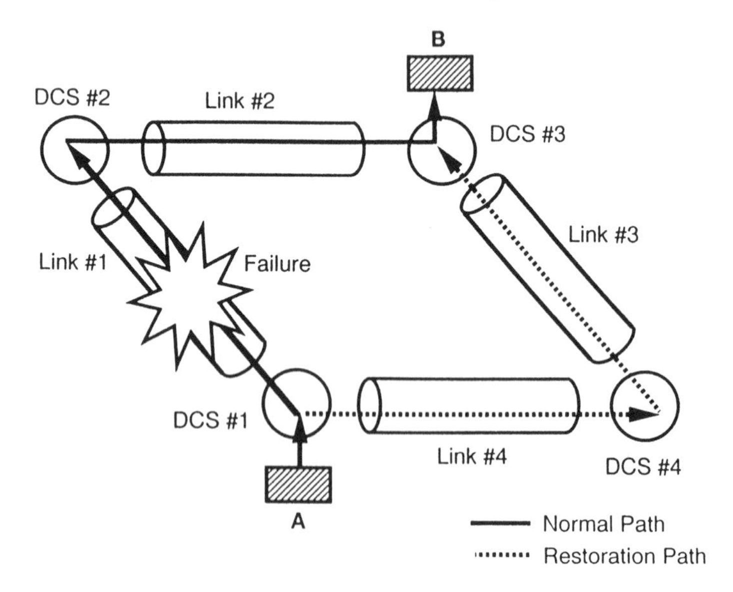

Figure 6.14 DCS mesh network and restoration.

1 and link 2, but are rerouted over link 4 and link 3 through DCS 4 if a cable cut occurs on link 1. The efficient use of spare capacity for restoration is attained through network operation systems that are more sophisticated than that of the ring architecture. The path restoration speed is of the order of minutes for 48 VC-3 paths (accommodated within a 2.5-Gbps transmission system) when the restoration is effected by a centralized control scheme. This restoration time can be significantly reduced by using the distributed control approach. The DCS mesh architecture is widely adopted for backbone networks that span a large geographical area with a large number of transport nodes.

6.4.2.6 ADM Ring Network

Owing to the development of high-speed optical transmission systems and the add/drop multiplexing capability of high-speed signals that were made possible with SDH/SONET technology, the ADM self-healing ring network became practical. ADM ring networks provide a very simple fiber capacity sharing mechanism and protection mechanism against network failure, thus enabling very cost-effective networking with high reliability [18,19]. Because of its simple architecture, the time required for protection switching can be very fast, say, 50 ms. These advantages led to the intensive introduction of the network, particularly in regional networks where the number of nodes in each area is relatively small.

The simplest ring architecture is the *unidirectional path switched ring* (UPSR) as it is called in SONET. In a unidirectional ring under normal working conditions, every node on the ring transmits in the same direction, either clockwise or counterclockwise. Therefore, two-way traffic (up- and downstream traffic) between any two nodes traverses the entire length of ring. Two unidirectional rings are used to realize a path-switched ring, as shown in Figure 6.15. A path in the transmitting side is duplicated and sent to both fibers in opposite directions. The receiving end always receives two identical path signals with different delay and selects the path of the default ring. The performance of each path (STS-1 in SONET) is monitored through its POH, and if deterioration beyond a specific level is detected, the service is switched to the alternate path on the other fiber. No end-to-end communication is required. Each path is routed independently of the other, and a route change on one has no effect on the other path. Therefore, it is a form of 1 + 1 protection switching and so requires 100% more capacity for restoration. This unidirectional ring architecture achieves the maximum efficiency when the logical (traffic) network topology is a star with a hub node and the other nodes connected to the hub.

Another important architecture is the *bidirectional line switched ring* (BLSR). The ring is protected through a line protection switching operating

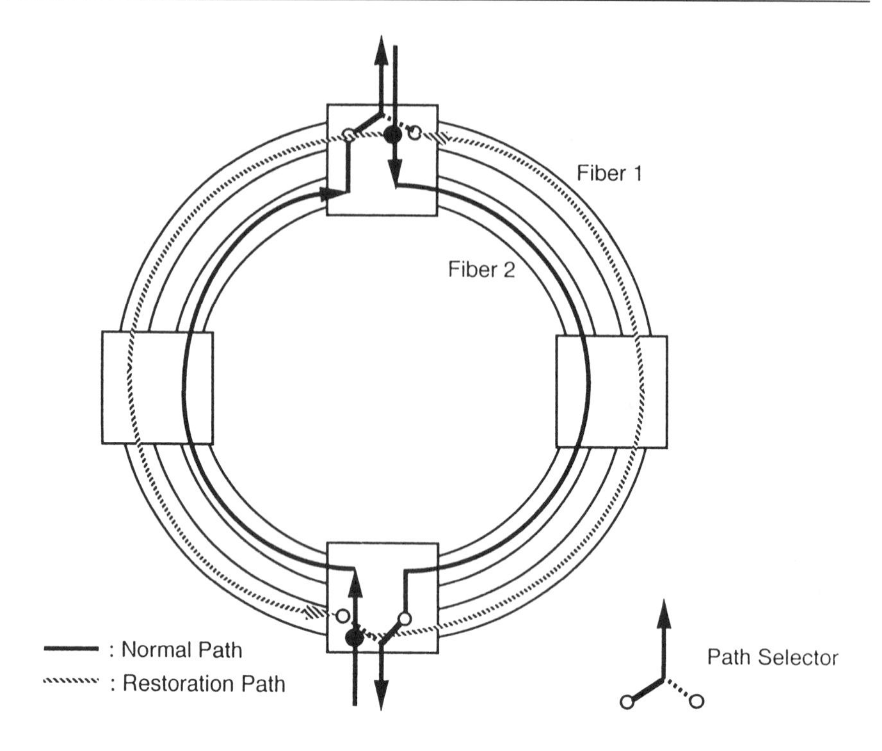

Figure 6.15 Unidirectional path-switched ring.

at the SONET line layer (which is equivalent to the multiplex section layer in SDH). The BLSR allows for the protection bandwidth to be shared among the working paths, which enables bandwidth from node to node to be reused around the ring and so achieves higher line utilization in many traffic demand situations. There are two types of BLSRs: two-fiber (Figure 6.16) and four-fiber (Figure 6.17). In a bidirectional ring under normal working conditions, two-way traffic between any two nodes is carried in opposite directions around the ring, so traffic between any two nodes occupies both directions of the portion of the ring between the two nodes, normally over the shorter length portion. In contrast to the unidirectional ring, two-way traffic between any two nodes does not traverse the entire ring. It should be noted that standby fibers (four-fiber) or capacities (two-fiber) can be used to carry lower priority traffic that will be dropped during the network restoration process.

In a two-fiber BLSR, each of the fibers in the spans carries both working and protection paths; in each fiber, half are assigned as working capacity and the other half as protection capacity (see Figure 6.16(a)). A failure is indicated in the MSOH on all spans and necessary switching actions are carried out (see Figure 6.16(b)). This includes access control to the shared protection paths.

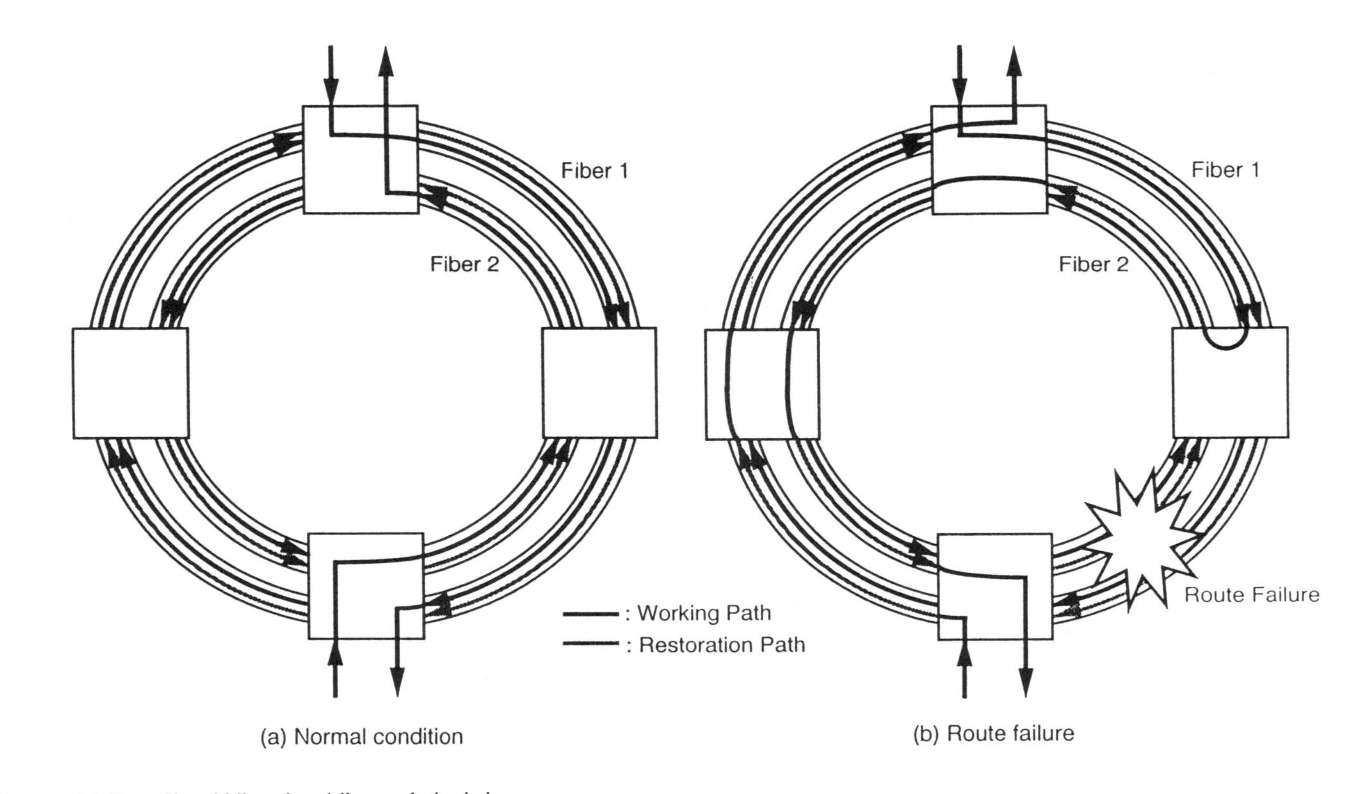

Figure 6.16 Two-fiber bidirectional line switched ring.

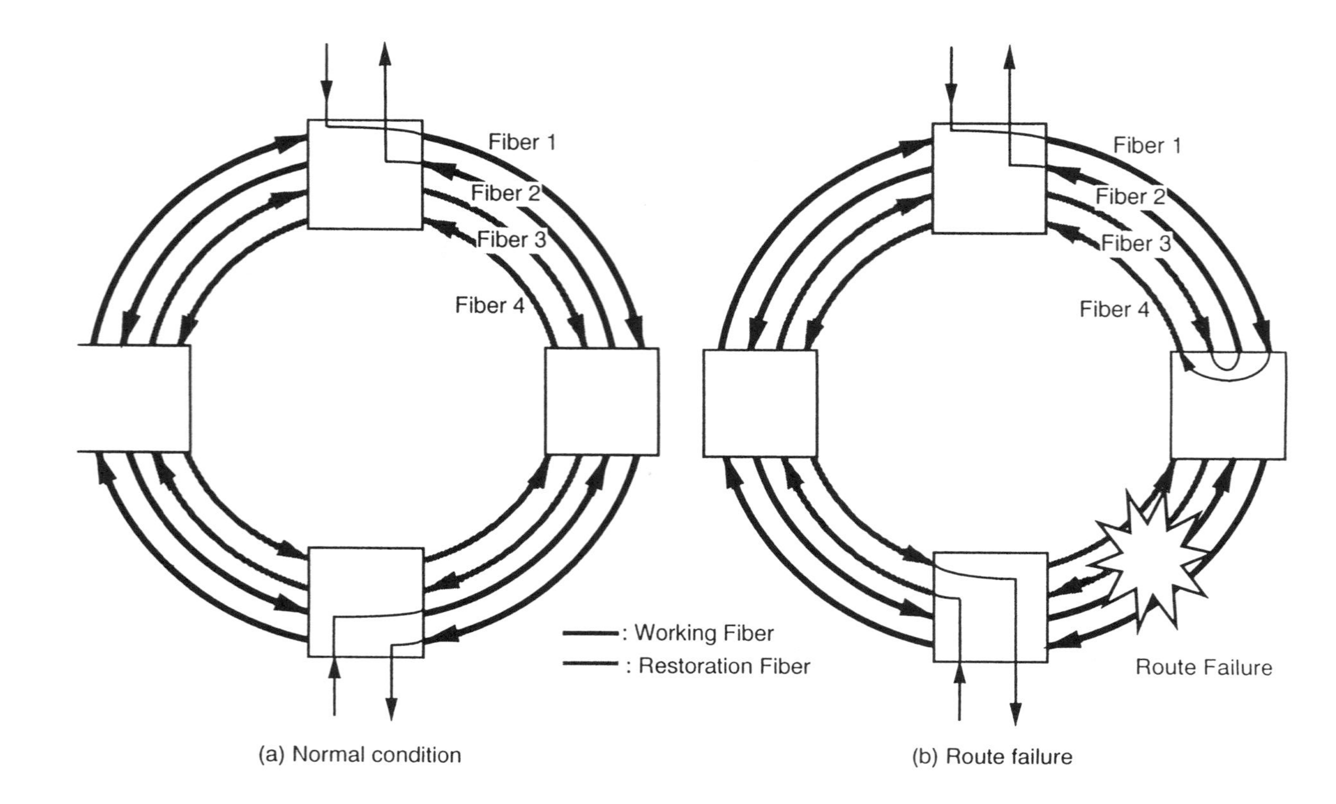

Figure 6.17 Four-fiber bidirectional line switched ring.

Switching takes place at the nodes on each side of the break so that the working channels are looped back to the protection channels on the fiber in the opposite direction of transmission and all the traffic is restored. To perform protection switching against network failure, time slot interchange capability is needed to switch signals from the time slots of the affected fiber to the time slots reserved in the other fiber.

In a four-fiber BLSR, two of the fibers carry the working channels in opposite directions and the other two carry the protection channels (see Figure 6.17(a)). In the event of failure, switching takes place at the nodes on each side of the break so that both working fibers are looped back on the corresponding protection fibers in the opposite direction of transmission to restore all traffic (see Figure 6.17(b)).

The benefits of the two-fiber BLSR are lower fiber requirements and lower initial cost due to reduced network element hardware; however, if traffic grows to require another two-fiber BLSR, it will be more expensive than one four-fiber BLSR due to the duplication of common equipment. This BLSR architecture becomes more efficient as more disjoint signals are accommodated; in other words, it is more effective when the traffic distribution among nodes becomes more uniform. In the restoration process, traffic is switched from the working to the protection channel and communication among the nodes around the ring is necessary so that the protection capacity can be activated properly.

6.4.3 ATM Network Technologies

This section describes asynchronous transfer mode technologies from the viewpoint of transport network aspects. They have been extensively developed since the early 1990s and their development is still in progress. ATM is recognized as a key technology for creating multimedia communications networks. In this section, we first discuss the important points of ATM and clarify the benefits, in comparison to synchronous transfer mode (STM), in terms of multimedia transport network development. The virtual path is elaborated because it makes the best use of inherent ATM capabilities and plays an important role in realizing a powerful transport network [20]. ATM network resource management principles and ATM network hardware systems are also discussed.

6.4.3.1 ATM Network Layered Architecture

Two types of ATM transport technologies were discussed in ITU-T: SDH-based ATM and cell-based ATM [16]. International standardization of SDH-based ATM has been substantially completed and it is now widely introduced in our networks. Therefore, SDH-based ATM is discussed hereafter. The ATM transport network consists of two layers, the ATM layer and the physical layer, as shown in Figure 6.18. The ATM layer is divided into two levels: the VC level and the VP level. The physical layer is divided into three levels: the transmission path level, digital section level, and regenerator section level. In SDH-based ATM, cells are mapped within a VC or a concatenated VC as explained in Section 6.4.1. Figure 6.19 shows an example of the ATM network structure and corresponding ATM transport network layer functions. The VP and VC play similar roles to the digital path and circuit, respectively, in STM networks.

Simplification of the network architecture coupled with simplification of node processing is the key to developing a cost-effective and flexible network. The virtual path concept is based on ATM capabilities and provides the network with a powerful transport mechanism for multimedia. The concept was first proposed in 1987 [21,22] and was approved as an international standard in CCITT (now ITU-T) in 1990 [8]. The VP concept allows management of

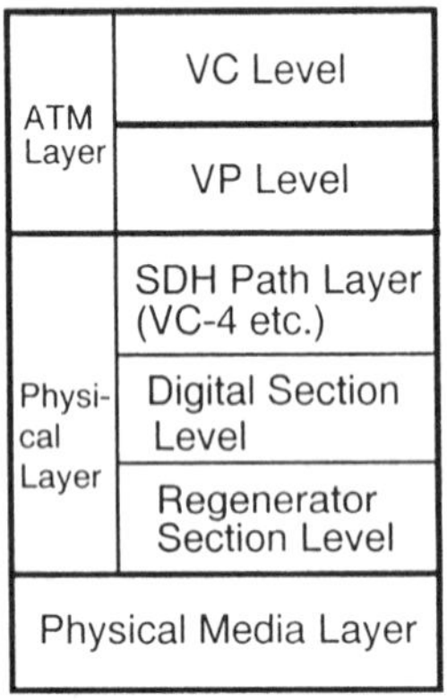

ATM Layer	VC Level
	VP Level
Physical Layer	SDH Path Layer (VC-4 etc.)
	Digital Section Level
	Regenerator Section Level
Physical Media Layer	

Figure 6.18 SDH-based ATM transport network layered model.

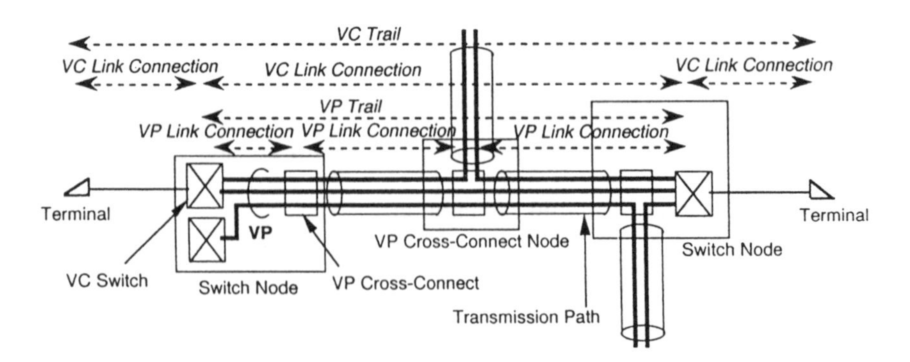

Figure 6.19 An example of ATM network configuration and the architectural functions.

virtual channels by grouping them into bundles. Consequently, virtual channels can be transported, processed, and managed in bundles, which permits many advantages such as reduced node costs and simplification of the transport network architecture. These advantages are discussed in comparison with the STM-based network in the next section.

The path is fundamentally a logical concept in the sense that a path is a bundle of circuits. In STM-based networks, however, only positioned paths carried within a framed interface are possible. In STM networks, specific time slots within the 125-μs TDM frame are assigned to each path. Positioned paths are identified by their time position within a TDM frame, and the capacity of the path is deterministic only. Thus, a path is tightly linked to the physical interface structure for transmission. STM paths, therefore, exhibit a hierarchical structure that reflects the transmission system hierarchy. This linkage produces inefficiencies in the multimedia environment where a wider range of transmission bit rates and various traffic characteristics (continuous/bursty) need to be handled.

In ATM networks, virtual paths are identified by the labels (VPI, virtual path identifier) carried in each cell header. This mechanism enables the separation of logical path aspects from the physical transmission interface structure, so the fully logical realization of path functions becomes possible. The capacities of VPs are either deterministic or statistical. A virtual path is established between VP termination points (VPTs) using the ATM cross-connect/ADM systems. The virtual paths are cross-connected as the ATM cross-connect system performs cell switching according to the VPI. VP terminating functions include path usage monitoring and control. The ADM system is based on the same mechanism as the cross-connect system.

6.4.3.2 Comparison Between SDH and ATM Networks

The introduction of VPs will impact transport network architecture and network management. Important points are explained next in comparison with SDH.

Path Networks and Transport Node Architecture

The direct multi/demultiplexing capability of VPs into and out of transmission links eliminates hierarchical path structures and, therefore, eliminates the hierarchical digital cross-connect systems needed by STM. An example of path layer simplification and the corresponding resultant cross-connect node structure simplification is shown in Figure 6.20. This simplification reduces not only node cost, but also transmission cost due to the improved link utilization (discussed later) offered by VPs, which enables total network cost to be reduced.

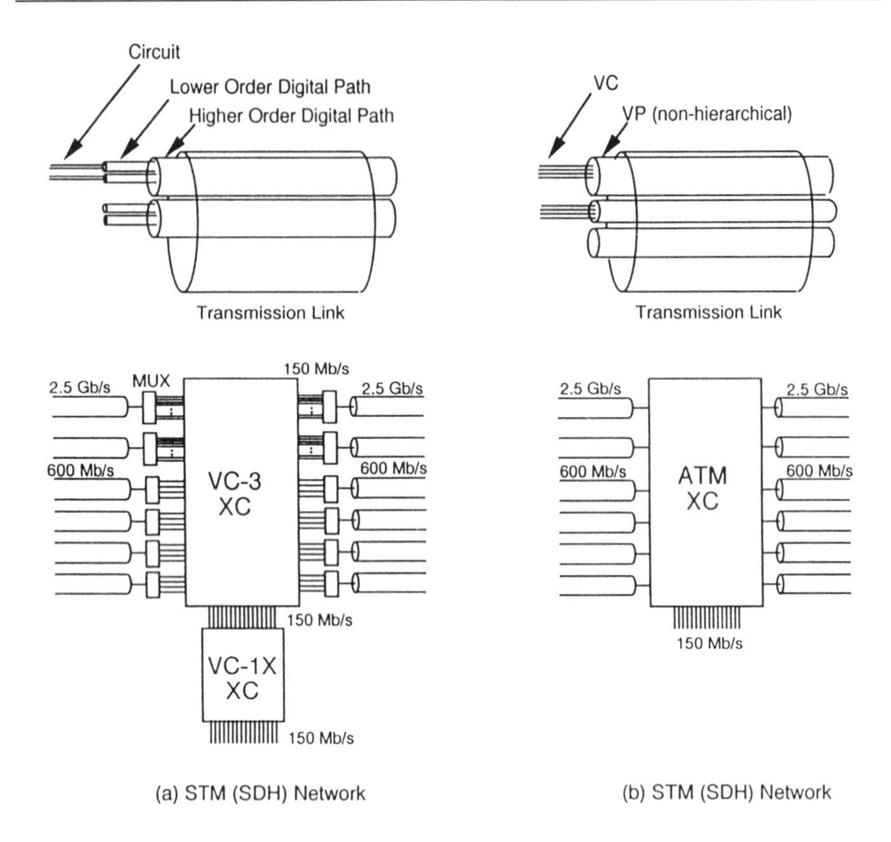

Figure 6.20 Simplification of path network and transport node structure (*After*: [23]).

Path Accommodation

Paths are accommodated within the transmission facility network using cross-connect/ADM systems. Figure 6.21 illustrates a path accommodation example in STM and ATM networks. For the STM network, hierarchical path stages exist and digital path accommodation design is needed to optimize the cross-connection stages of paths at nodes in terms of the total network cost, that is, transmission cost plus cross-connecting cost. This is because not all nodes in the network will be equipped with all cross-connect system stages. The hierarchical path structure and limited number of cross-connect system stages tend to result in poor link utilization. If the utilization efficiency of paths at each stage is 0.7, three multiplexing stages yield the link utilization of 0.7^3, that is, only 34%.

In ATM networks, on the other hand, there is only one stage of VP and the direct multiplexing of VPs becomes possible with ATM cross-connect systems. This facilitates VP accommodation design as seen in Figure 6.21, and

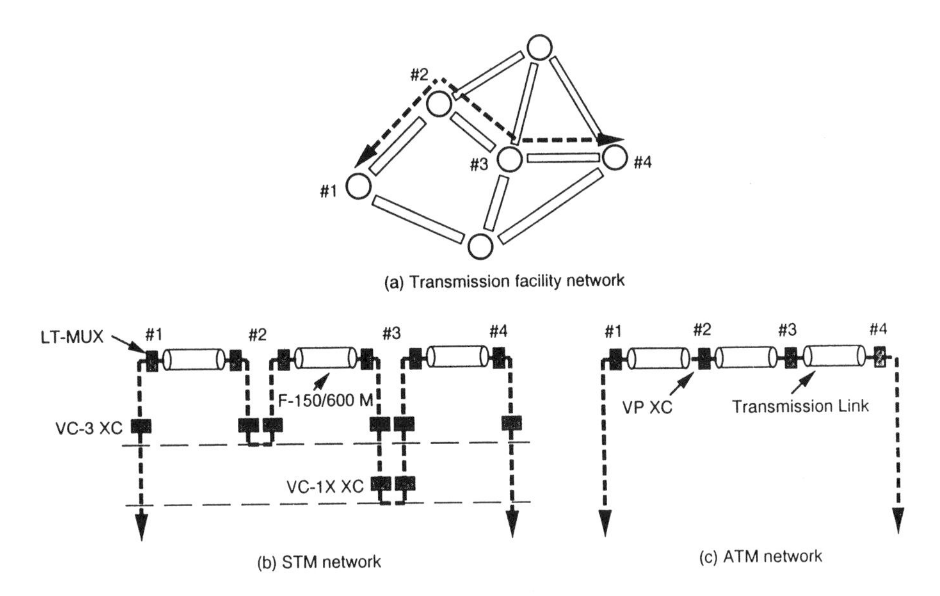

Figure 6.21 Comparison of path accommodation in STM and ATM networks (*After*: [24]).

results in increased link utilization, since the link utilization is determined by the single-stage multiplexing efficiency. ATM, however, requires sophisticated network resource management due to the asynchronous nature of cell multiplexing. ATM network resource management principles are explained in a later section.

Dynamic Path Network Control

Dynamic/adaptive path network reconfiguration will be used to create highly reliable multimedia networks and to enhance the grade of services that the network can provide. With this capability, network operating companies can increase their responsiveness to changing customers demands and can provide customers with control capability over their own closed networks. Path networks can be reconfigured in response to changes in the network state by controlling two parameters: bandwidth and route. Bandwidth control is needed to respond to traffic changes. Route alteration is required for traffic changes and network failures.

The virtual path network offers enhanced reconfiguration capability with reduced cost, adapting to unexpected traffic variations and network failures. This is due to the following unique characteristics of VPs:

- VP bandwidth is determined nonhierarchically, which enables optimized or improved path routing and bandwidth control.

- Direct multiplexing of VPs with different bandwidths becomes possible by employing a single kind of ATM cross-connect/ADM.
- No processing is necessary at nodes along the paths between VP terminators for path bandwidth allocation or alteration. The time slot reallocation within TDM frames necessary in STM networks is eliminated.

The last point is explained as follows. In STM networks, the alteration of path capacity and route requires time slot reallocation in the TDM frames at all STM cross-connects along the path. On the other hand, in ATM networks, a VP route is established by setting the routing tables of the ATM cross-connect systems. The path capacity allocation in the transmission facility network is done logically. The VP capacity is logically reserved in the network through management procedures. The ATM cross-connect nodes between the VPTs do not engage in VP capacity management. Therefore, alteration of VP capacity requires no processing at cross-connect nodes along the VP. This independence of route and capacity control can simplify path capacity control process.

In ATM networks, switches will be connected by VPs. Offered traffic between switches varies with the time of the day, the time difference within the country, and for unexpected reasons. Moreover, offered traffic fluctuates from minute to minute. If the link capacity is dynamically shared among VPs, the total VP bandwidth required to attain a specific call blocking probability (or cell loss rate) will be reduced. This will reduce the minimum link capacity required to accommodate the paths.

Fallback VP Application

The VP benefits just described can be applied to the protection routing of VPs, as shown in Figure 6.22. In this application, protection routes can be preestablished without reserving capacity [26]. That is, the VPI numbers associated with each link along the protection path routes are reserved and preassigned in the path connect tables at cross-connect nodes along the paths. A certain link capacity should be reserved for protection path use, but the reserved capacity can be shared among many possible protection paths accommodated by the link. When a network failure is detected, the restoration path bandwidths are reserved logically through a centralized management process, and cells are switched to the protection path route by providing new VPI values specific to the protection path. Of the transmission facilities, only the originating virtual path terminator (in the case of path route switching at the originating endpoint of path; this is schematically illustrated in Figure 6.22) or the ATM cross-

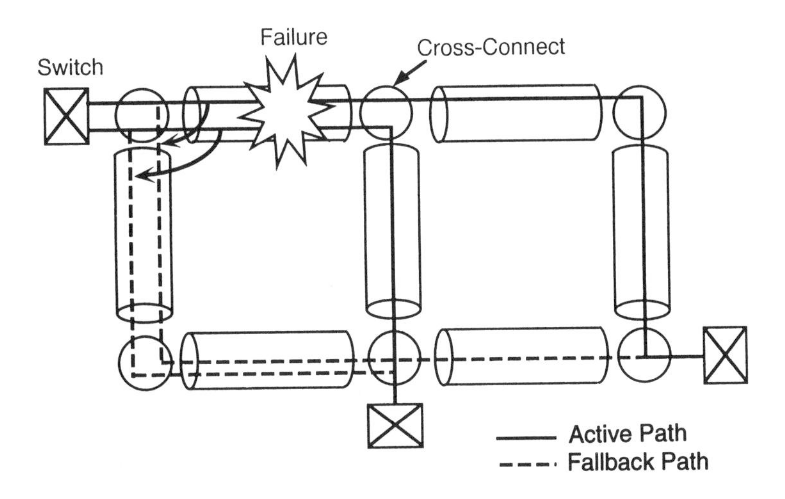

Figure 6.22 Fallback path application (*After*: [25]).

connects at both of the failed link ends (in the case of path route switching at failed link ends) are involved in the process of routing alteration. No routing table renewal or other processing at cross-connect nodes along the protection path is required and this enables rapid restoration.

Effective Multiservice Network Construction

The VPs are effective in creating multiservice networks. This is because VPs provide logical direct links between VPTs and hence provide the means to manage virtual channels according to traffic type. For example, different services, such as public switched services, leased line services, and center-to-end video distribution services, each of which may need service nodes in different locations, can be provided through a single physical network in an integrated fashion. This is made possible by virtue of the logical service separation capability of VPs on common physical links. This is true for either constant bit rate (CBR) or variable bit rate (VBR) services, and for services requiring high reliability (network OAM information is one example) or other attributes.

6.4.3.3 ATM Network Traffic Management Principle

In ATM, the channel and path are logically realized and more effective information transfer is possible than STM; ATM supports various speeds (bandwidths) and burstiness. However, more advanced network resource management is necessary to design and operate ATM networks effectively.

In an STM network, resource allocation or accommodation design for circuits and paths is straightforward and requires no special techniques. This is because the bandwidth for each circuit and path is deterministic and hierarchical. Allocation is physical by assigning the time slots within 125-μs TDM frames to the required bandwidth. On the other hand, in ATM networks, VP/VC bandwidth is nonhierarchical, and cells on each VC/VP are multiplexed asynchronously. Bandwidth allocation, therefore, is done logically (no time slot or cell slot is physically allocated or reserved). For example, in ATM networks, cell loss can occur when multiplexing constant bit rate channels/paths, even if the total capacity of incoming channels/paths is less than outgoing link capacity, if the multiplexing cell buffer memory size for output buffer scheme is less than the number of incoming channels/paths. This occurs, for example, if cells from all incoming ports arrive at the same time and their outgoing port is the same, although the probability of this may be small. Thus the minimum required buffer size must be determined according to the cell loss specification at multiplexing considering incoming and outgoing channel/path capacity and the target outgoing link utilization, even for constant bit rate channels/paths. The difficulty of designing the necessary cell buffer size is greatly worsened when the incoming information traffic is bursty (variable bit rate).

Network Resource Management Principle

The ATM network resource management principle is depicted in Figure 6.23. The basic idea is the same as that in STM networks. Resource management can be divided into three levels. In the first level, VC traffic demands in the network and a specific call (VC) blocking probability (in case of connection-oriented communication) determine switching facility demands and the neces-

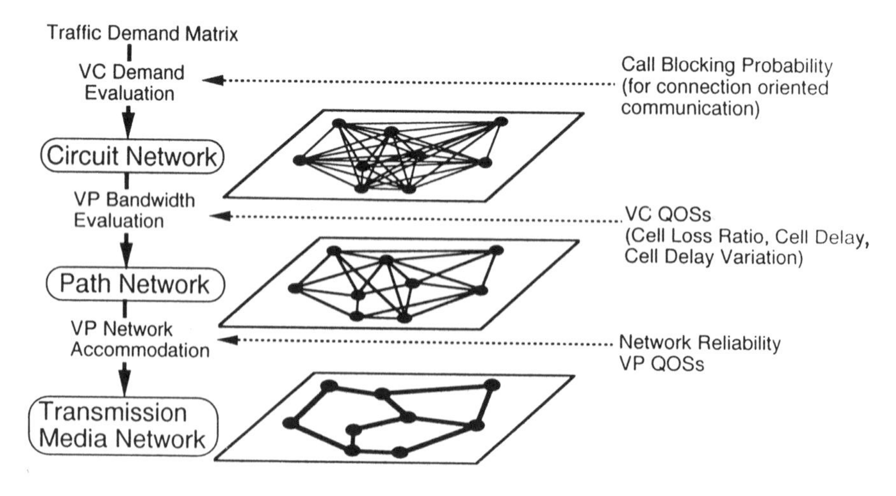

Figure 6.23 ATM network resource management principle.

sary number of circuit demands between switching system units; these requirements determine the circuit network. VP bandwidths between switching units are determined so that the necessary number of VCs can be accommodated and the necessary VC QOSs (cell loss rate, cell transfer delay, etc.) can be satisfied. The created VP network is accommodated within a transmission facility network so that VP QOSs and VP reliability requirements (VP restoration) can be satisfied.

In connection-oriented communication, VC/VP traffic management [27] will be divided in two phases. In the connection setup phase, a connection admission control mechanism decides whether the new VC/VP connection should be accepted or not, based on the VC/VP bandwidth and the QoS requirements and the network's available bandwidth. In this process, the bandwidth requirement of the VC/VP is indicated in the form of traffic source descriptors at T_B reference point, or can be implicitly declared by the service type, which may include traffic descriptors and QoS requirements. On acceptance, the network reserves the necessary bandwidth along the end-to-end VC/VP connection so that the QoS to be provided by the network is satisfied. In the information transfer phase, traffic generated by the source is monitored and controlled (cells sent to the network are accepted or rejected according to violation of the negotiated traffic descriptor values) by the network UPC function [usage parameter control; when this function is performed at internetwork interfaces, it is called NPC (network parameter control)], so that the QoSs of other connections are not degraded. In this process cell priority control can be performed. The important point is that the value of the UPC parameters and the network resource allocation based on the parameters can be determined by the network on the basis of the network operator's policy. For example, for variable bit rate traffic, the network may allocate the network resources and control the user traffic based only on the peak rate, thus minimizing the processing necessary at the cost of lowering the network resource utilization efficiency.

6.4.3.4 ATM Transport Network Element

The organic architecture of the ATM-based transport network is depicted in Figure 6.24. The ATM cross-connect systems can be interlinked directly by the SDH transmission systems of STM-N (cells are mapped into VC-4-Nc (where N = 4, 16)) or by STM-1 according to the traffic demand. ATM-to-STM conversion will be performed if necessary to transmit ATM signals utilizing STM LT-MUX in conjunction with STM signals. In the regional networks, ATM ADM systems forming a physical ring structure will be applied depending on customer distribution and density.

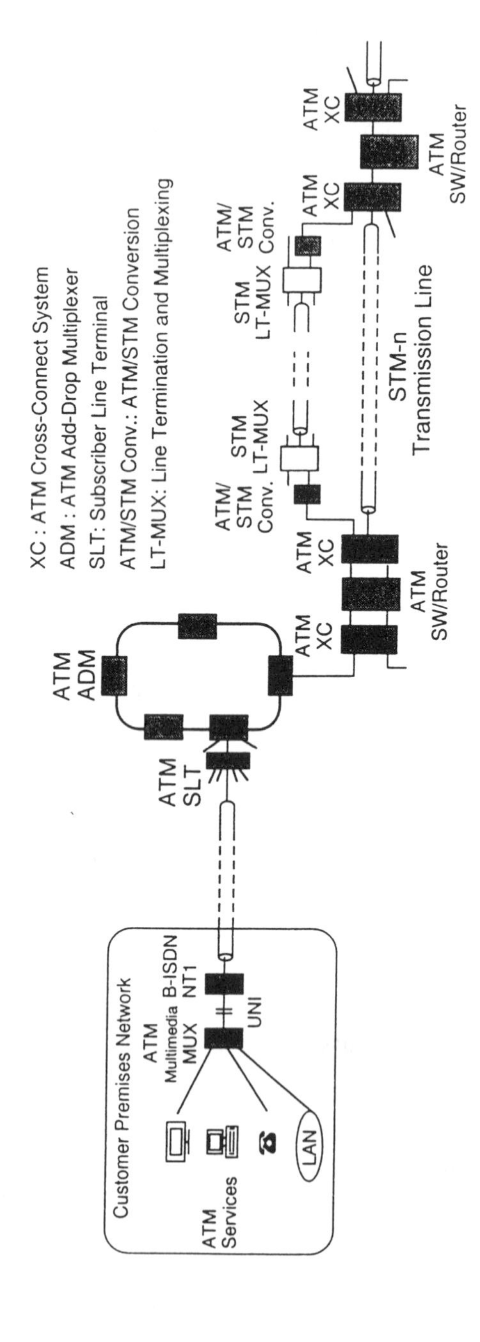

Figure 6.24 Organic architecture of ATM-based network.

Cross-Connect/ADM System

For cross-connection, paths/channels in different lines have to be interchanged. In the process, the multiplexing data width impacts system implementation issues. Existing PDH cross-connect systems are based on the bit multiplexing technique. SDH employs byte multiplexing. ATM utilizes cell (53 bytes or more as defined only within the set) multiplexing. To attain large bandwidth cross-connect/ADM systems, the operating speed within the ATM cross-connect system should be enhanced by utilizing parallel processing up to the cell length. The SDH cross-connect system can employ only byte-parallel techniques. The PDH cross-connect system is basically serial, where the higher speed line is demultiplexed into lower speed lines, and the lower speed line is interchanged by manual wiring or by using a space-division switch. Thus, it is difficult to realize a large-capacity economical PDH cross-connect network.

The requirements of the ATM cross-connect system are as follows: different interface speeds must be efficiently accommodated, modular growth capability must be guaranteed to allow the economical introduction of systems to large or small offices, and large throughput and minimum delay are required. An example of such a cross-connect system architecture [28] is shown in Figure 6.25. The system utilizes a 2.5-Gbps operating speed switch fabric. This is because the cross-connect system is designed to accommodate high-speed interfaces effectively and to minimize cell buffer delay. The hardware operation speed is reduced by utilizing parallel processing techniques in which the number of parallel lines can be up to the cell length. The requirements for the ATM cross-connect switch fabric are nonblocking characteristics in a strict sense, no restrictions on the usage condition of the middle link in multistage switching networks, and cell sequence integrity.

The first requirement is necessary to enable effective VP restoration that does not require any rearrangement if output port bandwidths are available for restoration paths. The second requirement means that the switch connectivity should be determined only by input and output port connection relationships, and should not be restricted by middle link conditions. This allows the blocking stemming from the utilization conditions of the inner link to be avoided even when VP capacity is dynamically altered in the network as explained in a later section. Examples of switch fabrics that satisfy the above requirements are one-stage-switching networks and self-routing multistage networks.

STM-ATM Signal Converter

ATM network facilities will be interfaced to SDH ones via the STM-ATM signal converter [29], which connects SDH paths and VPs, to transmit VPs over STM networks utilizing STM LT-MUX in conjunction with STM signals, and to transmit SDH paths between existing STM-based switching systems over ATM transport networks.

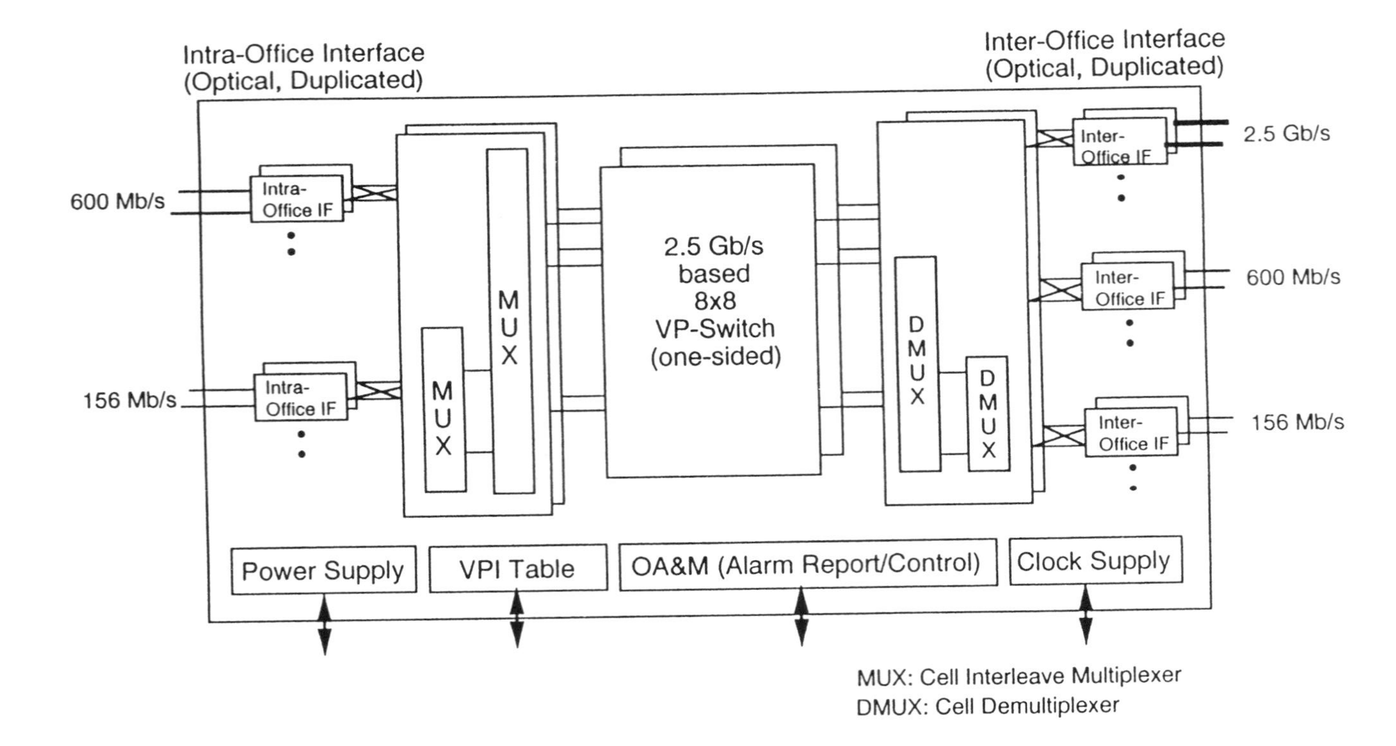

Figure 6.25 Generic cross-connect system architecture (*After:* [28]).

6.4.4 Transport Network Technology Evolution

6.4.4.1 Photonic Transport Network Technologies

Various data communication services such as the frame relay service, SMDS, and cell relay service are being introduced in many countries in keeping pace with the increasing demands of business sectors. Penetration is, however, limited to business subscribers, and the best way to realize the nationwide deployment of broadband multimedia services, such that they become as ubiquitous as the existing telephone service, should be identified.

One of the major requirements for multimedia networks is the flexibility that enables effective handling of a wide range of bit rates, from kilobits to hundreds of megabits per second, and a variety of traffic patterns such as continuous bit rates and variable bit rates. The ATM technology, as described in Section 6.4.3, provides the solution to this requirement. Another requirement is the transport (transmission and cross-connect/ADM) capability enhancement that enables end-to-end broadband (several megabits per second) capacity to be provisioned at, say, twice the cost of the existing telephone service. The point-to-point transmission capability has been increased by more than one order of magnitude in the 1990s through the introduction of optical fiber transmission, and 10-Gb transmission systems are now being introduced. In this context, it is necessary to realize cost-effective trans-access and effective grooming capability at the nodes.

ATM hardware can be realized with existing or reasonable enhancements of existing silicon-based electrical technologies. This, however, limits the increase in cross-connect node throughput to advances in semiconductor process technology, which cannot achieve the quantum leaps in transport capability needed to create cost-effective large-capacity networks. Furthermore, the technologies cannot effectively provide large-capacity paths (trans-access paths) between two nodes, since ATM cross-connection entails cell-by-cell switching at every cross-connect. Another step in technology evolution, therefore, is needed to attain this.

Photonic network technologies are expected to resolve this problem. The potential and effectiveness of photonic network technologies are elucidated in [20]. In this chapter, the network architecture and technologies that will be necessary to realize the said network are described.

6.4.4.2 Optical Paths

The optical path concept was proposed to resolve the above-mentioned problems [5,30]. It can make quantum leaps in both transmission capacity and cross-connect throughput simultaneously by exploiting WDM transmission and the wavelength routing capabilities of paths. The transmission capacity expansion due to WDM and the commensurate increase in cross-connect throughput with wavelength routing enable flexible and cost-effective broadband networking, which allows network-wide deployment of end-to-end high-speed communication capability at low cost.

Optical paths are identified by their wavelengths and accommodate electrical paths. Wavelength paths (WPs) and virtual wavelength paths (VWPs) have been proposed [5,30]. In the WP scheme, an optical path is established between two nodes by allocating one wavelength for the path (see Figure 6.26(a)). The intermediate nodes along the WP perform WP routing according to the wavelength. WPs can be used, for example, for the internode paths that accommodate a number of digital paths or VPs between switching units.

In the VWP scheme, VWP wavelength is allocated link by link and thus the wavelength of each VWP on a link has local significance instead of global significance as in the case of WPs (see Figure 6.26(b)). This is similar to the VPI assignment principle in ATM networks. For this reason, this scheme is called *virtual wavelength path.* At intermediate nodes, wavelength conversion may be necessary but without path termination. The VWP scheme has various advantages over the WP scheme [20]. They include a simple path accommodation design, higher flexibility in network expansion, and fewer network resources (wavelengths or fibers). The single and major disadvantage of the VWP scheme is that it entails wavelength conversion at cross-connects. This will be the key in determining the effectiveness of the VWP scheme.

The application of optical paths is schematically illustrated in Figure 6.27. They are utilized to create large-bandwidth trans-access paths between nodes (or node systems). VPs will be utilized as service access paths to attain

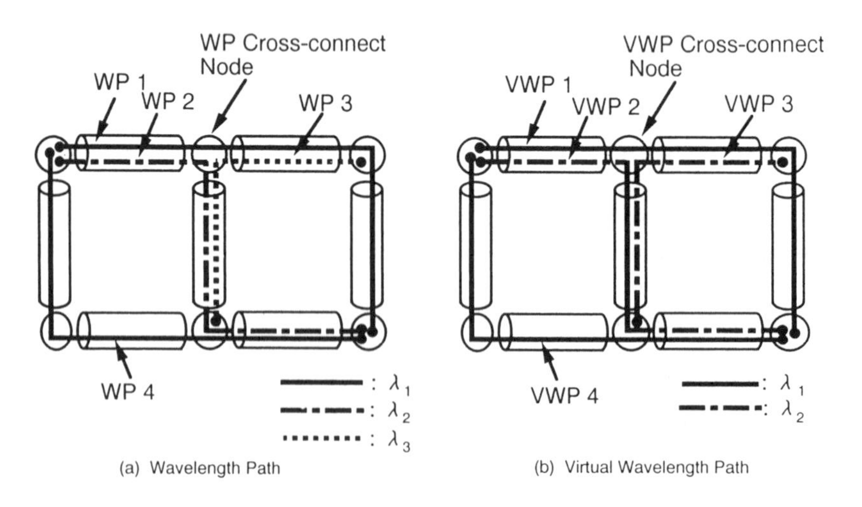

Figure 6.26 Comparison of WP and VWP (*After*: [5]).

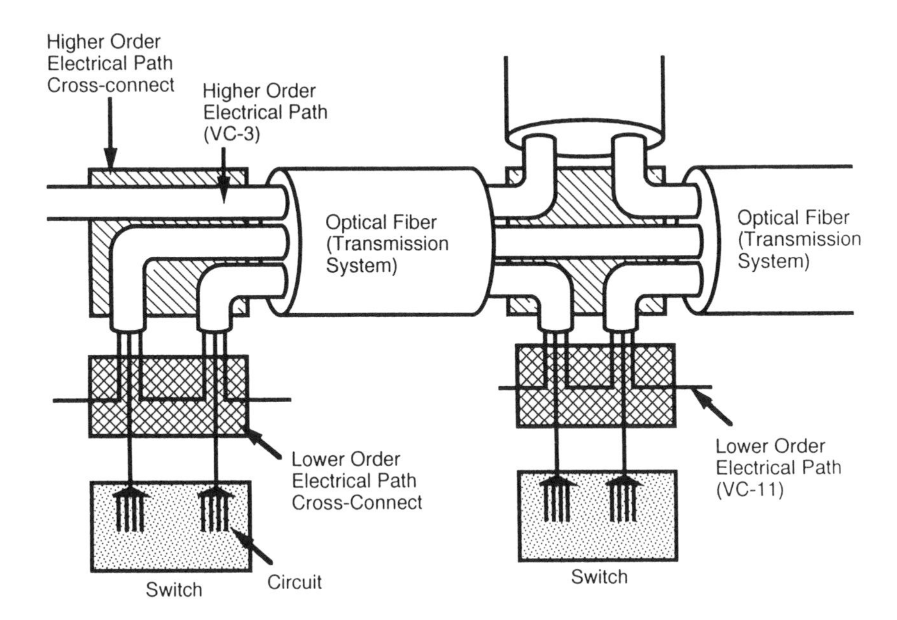

Figure 6.27 Schematic illustration of hierarchical path structure in optical path network (*After*: [31]).

the flexibility needed. The optical path network exploits the combination of WDM transmission and wavelength routing. Optical paths do not replace all electrical paths, but will complement them as trans-access paths, as shown in Figure 6.27. Thus, photonic transport networks will be introduced by overlaying existing electrical path-based networks.

Figure 6.28 compares existing telephone-based networks and future broadband multimedia networks. These broadband networks will utilize different bandwidth VCs instead of B-channels (64 Kbps), and the maximum channel bandwidth will be increased. VPs will be utilized for service access by replacing existing VC-1X paths in SDH networks so that maximum flexibility is realized; bandwidth will be increased in accordance with the channel speed increase. Optical paths of the order of 2.5 Gbps will be utilized considering available cost-effective optical techniques. They will replace the VC-3 electrical trans-access path, and WDM transmission will be utilized instead of TDM. 20-Gb transmission comprised of eight WDM paths per fiber is assumed in the figure; the number will be increased as the technology matures.

Figure 6.29 compares the transport node architectures of ATM networks and optical path networks. For ATM networks that utilize VPs for service access and trans-access, the entire line rate needs to be terminated to access the transmission link and then the entire signal capacity on each link has to

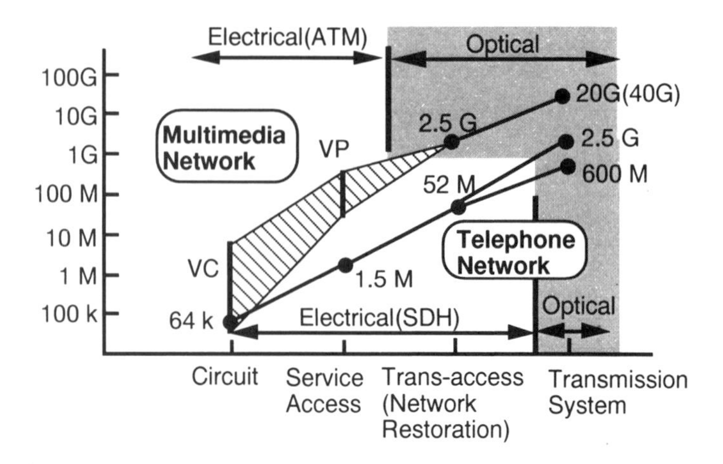

Figure 6.28 Relationship of each functional level (*After*: [31]).

be electrically (cell-by-cell) cross-connected node by node. When optical paths are utilized, on the other hand, the traffic is first cross-connected at the optical level with wavelength routing, so only the necessary portion of total link capacity is terminated at each node; the remaining traffic goes through the node without termination. This is a remarkable advantage when the total link capacity is very large, since it minimizes serial-to-parallel conversion of the bit stream and subsequent necessary electrical processing. The network restoration function can be performed with optical paths. The total throughput of an optical path cross-connect system can be much larger than that of an electrical TDM cross-connect system, and the hardware is simplified when the traffic volume is very large, since no synchronization is needed among optical paths (each WDM channel does not need to be synchronized). This nonsynchronous nature enables us to devise a cross-connect system with high modular growth capability that cannot be achieved with TDM. This is an important point for the economical introduction of the optical cross-connect system, given its vast throughput potential, to transport nodes whose needed throughput ranges from small to large.

6.4.4.3 Optical Cross-Connect/ADM

To realize optical paths, the major network element to be developed is the optical cross-connect/ADM system. In this section, recent technological advances are briefly introduced. The optical cross-connect must be nonblocking in a strict sense to support network restoration. Here, blocking does not include the blocking caused by wavelength collision between input and output ports, which can occur in the WP scheme. The cross-connect architecture is explained taking a recently developed large-scale optical cross-connect as an example. Figure

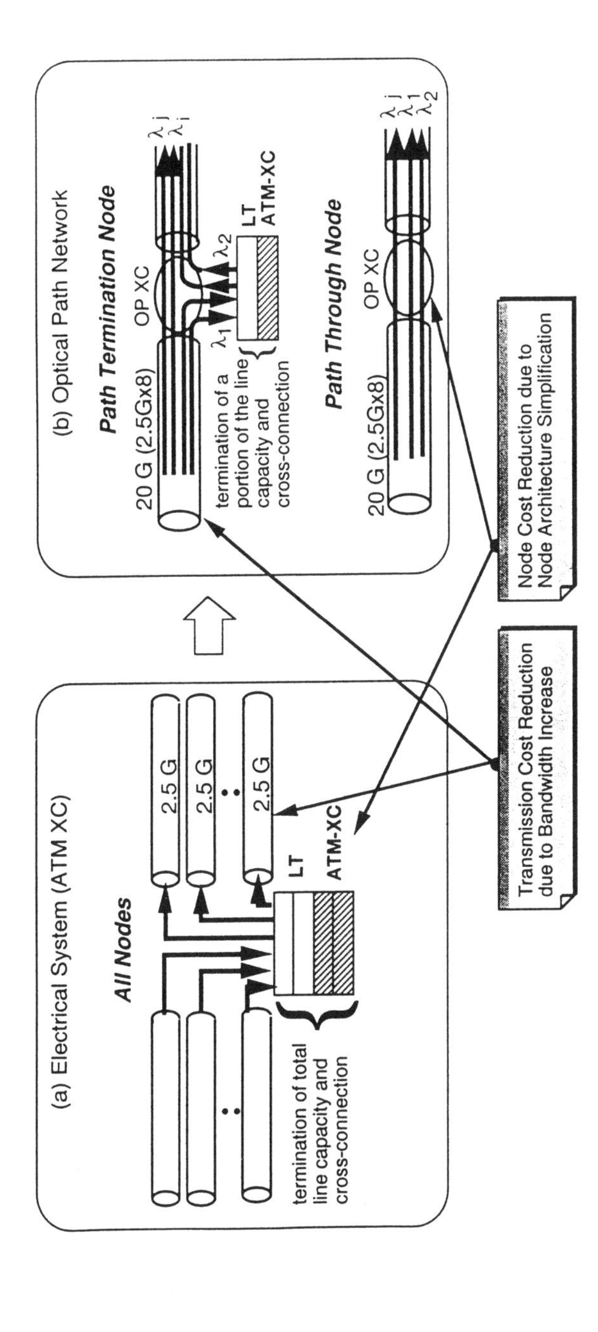

Figure 6.29 Comparison of electrical path and optical path node structure.

6.30 shows the schematic configuration of the WP and VWP cross-connect switch architecture [32]. The cross-connect switch allows any of the *M* incoming optical signals on an input link (*M* wavelengths are multiplexed) to be connected to any of the *N* outgoing ports. The generic configuration of an $M \times N$ junction-type matrix switch, which is a modular component switch of the cross-connect, is also depicted in Figure 6.30.

Optical regeneration at the electrical level (OR/OSs in Figure 6.30) can be eliminated from WP cross-connects that do not need optical regeneration. The WP cross-connect switch with optical regeneration uses fixed-wavelength LDs in OSs, whereas tunable wavelength LDs are used for VWPs. The architecture holds various advantages, some of which are explained next.

This architecture allows easy evolution from a WP cross-connect to a VWP cross-connect by simply replacing the fixed-wavelength LDs in the OS modules with tunable wavelength LDs. This switch architecture thus provides the maximum commonality between WP and VWP cross-connects. Furthermore, the bit rate adaptability of each WP/VWP can be easily attained by upgrading the OR/OS modules while leaving the other major parts unchanged.

Another critical design goal is to maximize the modular growth capability. This permits minimum initial investment while allowing investment to be incrementally increased as traffic demand increases. This is critical for the economical introduction of optical path cross-connects at the outset of service provisioning. This architecture enables very effective modular growth in terms of input and output fiber pairs, as explained in Figure 6.30; the unit of addition/deletion (building block) is delimited by a broken line.

The cross-connect switch hardware based on the said architecture is being developed utilizing planar lightwave circuit (PLC) technologies [33]. Figure 6.30 shows the optical insertion loss of four sets of an 8×16 junction-type matrix switch. This switch was designed for a cross-connect system with the total throughput of 320 Gbps (= 2.5 Gbps $\times$ 8 $\times$ 16), and each junction-type matrix switch board has 20-Gbps throughput. Each board has 8 input ports and any input signal can be delivered to any of the 16 output ports, so $8 \times 16 = 128$ connections are possible. Figure 6.30 shows the histogram of fiber-to-fiber insertion-loss for four sets of the switch boards in the *on* state [34]. The average loss was 12.6 dB. This switch loss includes 8×1 optical coupler loss, so the total average loss of the waveguide splice loss between PLC chip and fibers and connector loss is 3.6 dB. The average on/off ratio attained was 41.2 dB for each path. The on/off ratio is equivalent to the crosstalk when all signal levels are the same.

Optical path cross-connect experiments were successfully performed using the developed cross-connect switches [35]. The experiments modeled a five-node network and eight input optical paths with wavelengths ranging from

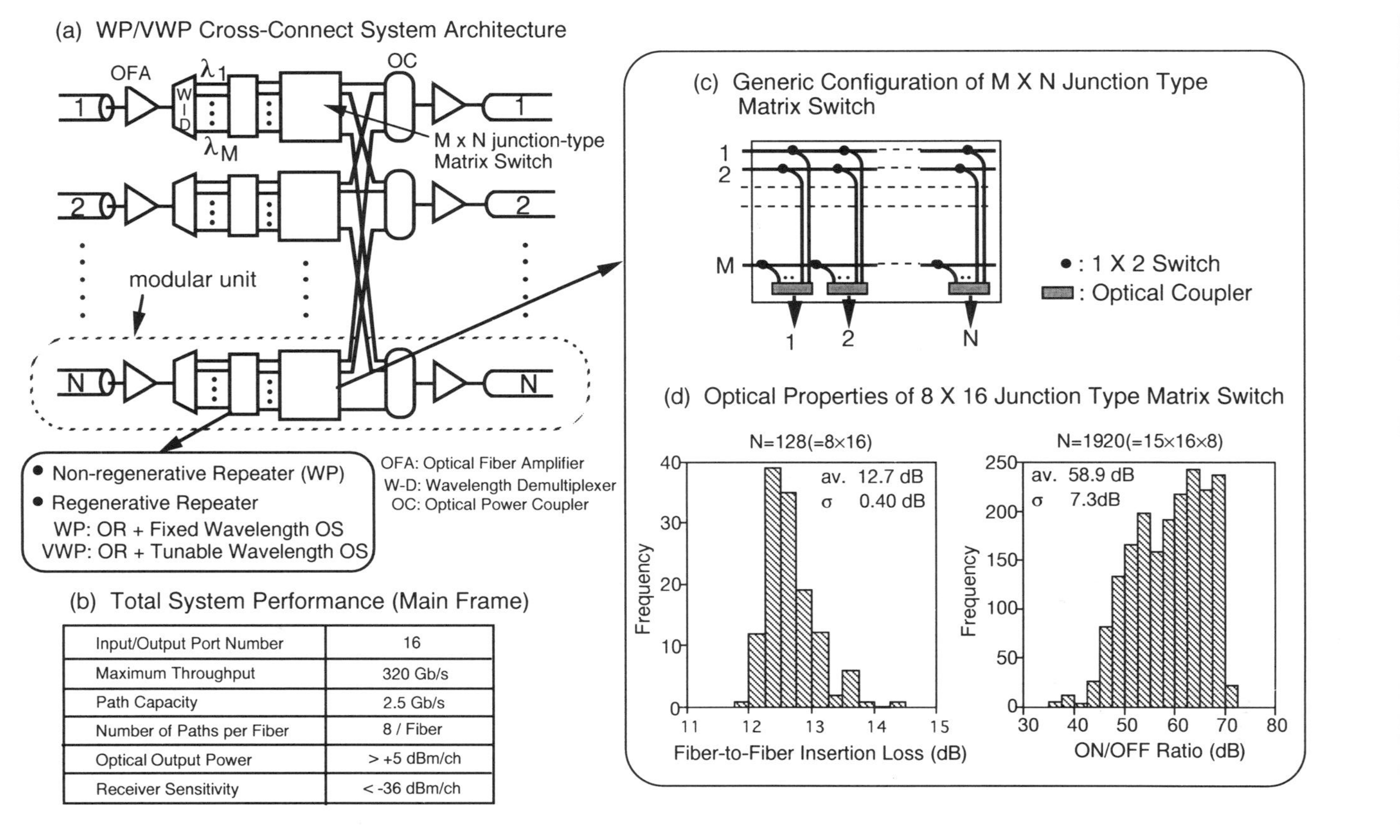

Input/Output Port Number	16
Maximum Throughput	320 Gb/s
Path Capacity	2.5 Gb/s
Number of Paths per Fiber	8 / Fiber
Optical Output Power	> +5 dBm/ch
Receiver Sensitivity	< -36 dBm/ch

Figure 6.30 WP/VWP cross-connect switch architecture and measured optical performances.

1,547 to 1,555 nm with 1-nm channel spacing were transported. The cross-connect hardware performance has been verified.

6.4.4.4 Challenges

We have discussed photonic transport network technologies based on optical paths. Photonic technologies will have a tremendous impact in creating the future transport network on which the bandwidth-abundant B-ISDN will be created. For network realization, technical advances are also needed regarding photonic network OA&M in keeping with the rapid advances seen in hardware development. The network design issues, including wavelength assignment problems and optical network architecture related issues that are not discussed herein, are of great importance [20]. International standardization is critical for the economical introduction of the system and to reap the benefits of the new technologies without delay.

References

[1] Salloum, H. R., and W. T. Anderson, "Bellcore Requirements for Fiber in the Loop Systems," *Proc. IEEE Int. Communications Conf.,* May 1993.

[2] "Generic Requirements and Objectives for Fiber in the Loop Systems," TA-NWT-000909, Bellcore, Issue 2, Dec. 1992.

[3] Bartlett, E. R., *Cable Television Technology and Operations,* New York: McGraw Hill, 1990.

[4] ITU-T Recommendation G.709, "Synchronous Multiplexing Structure," Mar. 1993.

[5] Sato, K., S. Okamoto, and H. Hadama, "Network Performance and Integrity Enhancement with Optical Path Layer Technologies," *IEEE J. Sel. Areas Commun.,* Vol. 12, No. 1, Jan. 1994, pp. 159–170.

[6] Ishio, H., "Progress of Fiber Optic Technologies and Their Impact on Future Telecommunication Networks," *Proc. FORUM '91* (Technical Symposium of TELECOM 91), Geneva, Switzerland, Oct. 10–15, 1991, Session 2.2, pp. 117–120.

[7] Shirakawa, H., K. Maki, and H. Miura, "Japan's Network Evolution Relies on SDH-Based Systems," *IEEE LTS,* Nov. 1991, pp. 14–18.

[8] CCITT I Series Recommendations (B-ISDN), Nov. 1990.

[9] "General Aspects of Digital Transmission Systems; Terminal Equipments," CCITT Blue Book, Vol. III—Fascicle III.4., Recommendations G.700-G.795, Nov. 1988.

[10] ITU-T Recommendation G.707, "Synchronous Digital Hierarchy Bit Rates," Mar. 1993.

[11] ITU-T Recommendation G.708, "Network Node Interface for the Synchronous Digital Hierarchy," Mar. 1993.

[12] ITU-T Recommendation G.803, "Architectures of Transport Networks Based on the Synchronous Digital Hierarchy (SDH)," Mar. 1993.

[13] Sexton, M., and A. Reid, *Transmission Networking: SONET and the Synchronous Digital Hierarchy,* Norwood, MA: Artech House, 1992.

[14] ITU-T Recommendation G.707, "Network Node Interface for the Synchronous Digital Hierarchy (SDH)," Mar. 1996.

[15] Boehm, R. J., Y. C. Ching, and R. C. Sherman, "SONET (Synchronous Optical Network)," *Proc. GLOBECOM '85,* 1985, pp. 46.8.1.

[16] ITU-T Recommendation I.311, "B-ISDN General Network Aspects," Mar. 1993.

[17] ITU-T Recommendation I.432, "B-ISDN User-Network Interface—Physical Layer Specification," Mar. 1993.

[18] Wu, T.-H. *Fiber Network Service Survivability,* Norwood, MA: Artech House, 1992.

[19] Lee, W. S., and D. C. Brown, *Advances in Telecommunications Networks,* Norwood, MA: Artech House, 1995.

[20] Sato, K., *Advances in Transport Network Technologies—Photonic Networks, ATM, and SDH,* Norwood, MA: Artech House, 1996.

[21] Sato, K., and H. Hoshi, "Digital Integrated Transport Network-II," *NTT R&D Document,* No. 697, May 1987.

[22] "Some Key Features on UNI and NNI Considering ATM (A Framework for Discussion)," CCITT SG XVIII, Seoul Meeting, Jan. 25–Feb. 5, 1988, Delayed Contribution D. 1566/XVIII. (Source: NTT.)

[23] Aoyama, T., I. Tokizawa, and K. Sato, "Introduction Strategy and Technologies for ATM VP-Based Broadband Network," *IEEE J. Sel. Areas. Commun.,* Vol. 10, No. 9, Dec. 1992, pp. 1434–1447.

[24] Sato, K., R. Kawamura, and I. Tokizawa, "Introduction Strategy of B-ISDN Based on Virtual Paths," *Proc. FORUM '91* (Technical Symposium of TELECOM 91), Geneva, Switzerland, Oct. 10–15, 1991, Session 2.5, pp. 225–229.

[25] Sato, K., and I. Tokizawa, "Flexible Asynchronous Transfer Mode Networks Utilizing Virtual Paths," *Proc. ICC '90,* Atlanta, GA, Apr. 16–19, 1990, pp. 318.4.1–318.4.8.

[26] Kawamura, R., K. Sato, and I. Tokizawa, "Self-Healing ATM Networks Based on Virtual Path Concept," *IEEE J. Sel. Areas Commun.,* Vol. 12, No. 1, Jan. 1994, pp. 120–127.

[27] ITU-T Recommendation I.371, "Traffic Control and Congestion Control in B-ISDN," Mar. 1993.

[28] Ueda, H., et al., "ATM Cross-Connect System Technology," *NTT R&D,* Vol. 42, No. 3, 1993, pp. 357–366.

[29] Uematsu, H., and H. Ueda, "STM Signal Transfer Techniques in ATM Networks," *Proc. ICC '92,* Chicago, IL, June 14–18, 1992, pp. 311.6.1–311.6.5.

[30] Sato, K., S. Okamoto, and A. Watanabe, "Photonic Transport Networks Based on Optical Paths," *Int. J. Commun. Syst.,* Vol. 8, No. 6, Nov.–Dec. 1995, pp. 377–389.

[31] Sato, K., "Recent Advances in Photonic Transport Network Technologies," *Proc. ECOC '96,* Oslo, Norway, Sept. 15–19, 1996, pp. 2.27–2.34.

[32] Okamoto, S., A. Watanabe, and K. Sato, "A New Optical Path Cross-Connect System Architecture Utilizing Delivery and Coupling Matrix Switch," *IEICE Trans. Commun. Japan,* Vol. E77-B, No. 10, Oct. 1994, pp. 1272–1274.

[33] Kawachi, M., "Integrated Silica Waveguide Technologies," *Proc. OFC'96,* San Jose, CA, Feb. 25–Mar. 1, 1996, pp. 261–287.

[34] Koga, M., et al., "8×16 Delivery and Coupling Type Optical Switches for a 320 Gigabit/s Throughput Optical Path Cross-Connect System," *Proc. OFC'96,* San Jose, CA, Feb. 25–Mar. 1, 1996, pp. 259–261.

[35] Koga, M., et al., "Design and Performance of Optical Path Cross-Connect system Based on Wavelength Path Concept," to appear in Special Joint Issue of *IEEE Journal of Lightwave Technology and IEEE Journal on Selected Areas in Communications.*

7

Multimedia Operations, Administration, and Management and Services

7.1 Operations, Administration, and Management in the Multimedia Era

7.1.1 Introduction

Information communication services in the multimedia era will be integrated with operations, administration, and management (OA&M), allowing customers to confirm, select, and change service conditions, attributes, and media to be used, etc., at any time before, during, or after service use. Moreover, service providers (SPs) and information providers (IPs) will provide OA&M services such as specific information provision, information navigation, and customized service design. For example, specific information provision will include the provision of service performance information, trouble reports, detailed billing information, customer activity logs, etc. In these cases, OA&M capabilities will influence the customer satisfaction of the multimedia communication services.

Customer satisfaction is basically decided based on whether a customer has enjoyed a service at the level that he or she expected; that is, the decision is not made based on the absolute value of any specific parameters, but instead is based on negotiation with the customer. The result of this customer negotiation is called a *service level agreement* (SLA), which is used as a measure of each service provided as well as an object of information communication service management. The quality of service (QoS) is evaluated with this SLA in an integrated manner. In the multimedia era, in which a variety of services are provided by using a variety of elements, SPs and IPs will need to ensure that this negotiation process is quick and efficient and done with the cooperation of customers, that the agreed-on SLAs are accurate, and that the customer understands how they provide attractive value-added services to information communication services.

In this chapter, we discuss the OAM technology at the network element, network, and service levels in the multimedia era. Also, OAM services, such as QoS management and multimedia navigation service, are suggested. In addition, the OAM system technologies and standards-related activities, such as TMN, ATMF, TINA and DAVIC, are discussed.

7.1.2 Multimedia Services and OA&M [1–3]

7.1.2.1 Service Elements and OA&M Elements

Multimedia communication is often referred to as integrated voice, video, and data communication, especially as provided over the broadband integrated services digital network (B-ISDN); however, this is a theoretical, average image and not a typical usage application. The most effective and/or attractive services in some application areas will be used first, and then application areas will be broadened to include the following:

- High-speed computer communication used for a back-end office;
- Applications to provide advancement and increased flexibility in public communication and in the work of a front-end office in a corporate organization;
- Applications to provide advancement and increased flexibility for personal communication such as varied media mail, electronic formalities, information searching/retrieval, etc.;
- Home (entertainment) applications such as both-way, on-demand type image communication and teleshopping.

These are to some extent possible now using narrowband ISDN (N-ISDN) and/or the current Internet; however, multimedia communication transferring real-time video image, large-volume files, etc., is in fact realized via the high-speed B-ISDN communication capability. OA&M in the multimedia era must ensure the availability and progress of the above applications.

Table 7.1 summarizes the multimedia services categories and OA&M factors. From the users' point of view, communication is a "means" to the

Table 7.1
B-ISDN/Multimedia Services

		Service Examples						OA&M Factors		
		Visual	Text	Sound	Data	Comm. Information	Control and OA&M	SLA Factors	Interface	Viewpoints
	Entertainment	CATV		Large Volume Sound		One Way Distribution	One Way (Downward)	Video Quality	Human-Machine	Multichannel
		VOD				One Way (Downward)	Both Way	Video Quality Response Time A Variety of Handling/ Selecting Methods	Human-Machine	Both Way Multichannel Networking Varied Media
Personal Use	Personal Communication	Video Retrieval/ Searching Video Mail	Text Retrieval/ Searching Text Mail	Voice Mail	Data Retrieval/ Searching	Both Way One Way (Downward)	Both Way	Response Time Information Quality (Data Error Rate, etc.)	Human-Machine Human-(Machine)-Human	Varied Media
		*	*		*	Both Way	Both Way		Human-Machine	Personalization
		Video Phone		Telephone		Both Way	Both Way	Realtimeness (Voice-Video Synchron-ization)	Human-Human	Home Processing
		Group Broadcasting	Document Distribution Retrieval	Group Broadcasting		One Way Distribution	One Way (Downward)	Information Quality (Clearness)	Human-Machine	Advancement of OA
Corporate, Public, Governmental Use	Front Office	Video Retrieval/ Searching Video Mail	Text Retrieval/ Searching Text Mail	Voice Mail	Data Retrieval/ Searching Data Transfer	Both Way One Way (Downward)	Both Way	Response Time Information Quality (Data Error Rate, etc.)	Human-Machine Human-(Machine)-Human	Varied Media Networking Virtual Organization
		**	**		**	Both Way	Both Way		Human-Machine	
		Video Phone		Telephone					Human-Human	
	Back-end Office	Live Video Monitoring			High Speed Data Transfer High Speed File Transfer High Speed Remote Control	One Way (Upward) Both Way	One Way (Downward) Both Way	Information Quality (Data Error Rate, etc.) Realtimeness	Machine-Machine	High Speed Processing Large Volume Processing

*Submission/retrieval of forms, etc. from home
**Submission/retrieval of forms, etc. from office

ultimate "end" for a user of transferring information. In conventional types of communication such as facsimile and telephone conversations, a user uses the media to achieve his or her objective such as notifying employees of a meeting date, sending a purchase order, etc. In the multimedia communication environment, this objective becomes much more explicit. For example, interactive terminals will be used not for transferring data but for enjoying a game in a virtual community. It makes progress from the mere presentation/usage of information to the flexible usages by selecting, combining, and/or changing to the most suitable media from time to time, and further creating new values. That is, information communication services must provide an "objective-oriented and nondeterministic" environment that realizes a transparent infrastructure of which the users are not conscious of as "using communication." Conventional communication styles are restricted by the system to allow users select keys/buttons or preset menus, but future communication styles should allow much more flexible activities, such as collecting information from anywhere and further creating new products. From these points of view, a new concept for OA&M that is concerned with the services and their objectives needs to be established in addition to the conventional OA&M concepts for network and network elements.

7.1.2.2 Communication Party Types [1–5]

In multimedia communication, various service types are combined as shown in Table 7.1. The satisfaction of individual users or communication managers is assessed not only by their experiences during the use of a service but by considering a variety of factors affecting the integrity of the total service before, during, or after use. The communication originator and terminator (i.e., machine or human) is an important factor in determining the SLA, evaluating QoS, and making the OA&M evaluation/design policies/standards varied, as shown in Table 7.1.

1. *Human-human communication and human-oriented communication:* Video-on-demand (VOD) movies and games, high-fidelity music, etc., are meant to allow a user to enjoy the application itself with the human senses, so that the attractiveness of the content, fidelity of the transmission, user-friendliness of the interface, etc., will be evaluation factors. In particular, varied high-level human-machine interfaces (HMIs) will be needed according to the user's generation and any handicap(s). For example, using Web browsers and a 3D/virtual reality interface will significantly improve or enhance the human interfaces.
2. *Machine-machine communication:* This type of communication occurs when machines exchange data such as files and transactions for further

processing at the remote location. Human uses the final results of the processing provided as output data. In many cases, communication between machines is automatically and periodically done as a background routine job without human intervention. In this case, the integrity and quality of the transferred data and their transfer process are factors for evaluation of communication.

3. *Human-involved communication:* Notifying employees of a meeting agenda, sending a regular routine report, etc., are processes in which ideas and facts need to be communicated to humans. In these cases, the quality of transmission is not necessarily as important as is making sure the recipient understands the content.

From the OA&M point of view, one of the big issues is how to implement and carry out the OA&M of integrated services while making it possible to operate and manage individual services. This must be realized on the basis of integrated service, network, and OA&M functionalities.

7.1.3 OA&M Services

7.1.3.1 OA&M Categories [6–8]

Modified from the management layers studied by ITU-T, Figure 7.1 shows OA&M areas that are layered and comprise mutually related tasks. According to ITU-T Recommendation M.3010, each area is described as follows:

1. *Business-concerned OA&M:* This area is responsible for the total enterprise to conduct its business. Strategic planning and executive actions

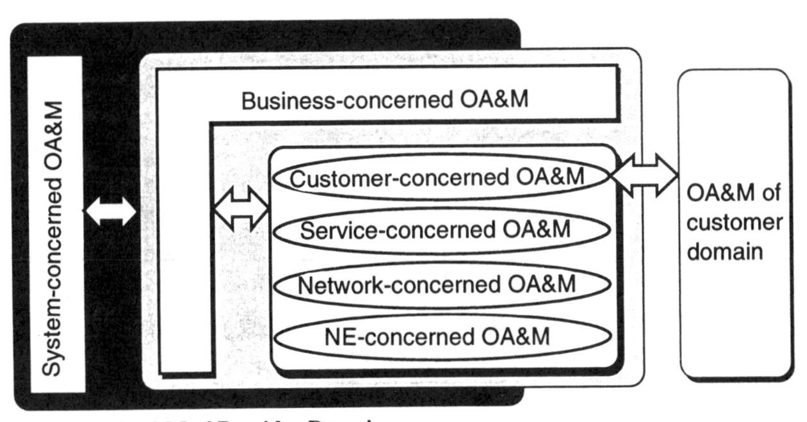

Figure 7.1 The service provider's OA&M tasks.

are supported through or sometimes considered to be included in this layer.

2. *Service-concerned OA&M:* This layer is responsible for the management of services that are provided for customers including those of other service providers. This layer is not concerned with the management of physical entities. It provides the basic point of contact with customers for all service transactions, including opening a new account, service provisioning, service creation, providing information about QoS, service contracts, etc.
3. *Network-concerned OA&M:* This layer provides support functions such as those required to provide an end-to-end view of a network. It receives aggregated or summarized data from the network-element-concerned OA&M layer and creates a global view within its scope or domain. It communicates with other layers via standard interfaces.
4. *Network-element-concerned OA&M:* This layer manages a collection of network elements (NEs), such as node, link, outside plant, etc.

The OA&M customer is concerned with all of the activities of OA&M regarding the customer owned facilities, services, and businesses. This comprises a somewhat similar but reduced set of provider domain OA&M and customer business-specific ones.

System OA&M is concerned with all OA&M activities regarding systems used by a service provider including systems used for OA&M as well as for the company's principal business. Emphasis has so far been placed on the network and NE layers, and thus OA&M has been developed based on what could be done from the bottom-up point of view, rather than what should be done from the service point of view. However, as explained in Section 7.1.2, future multimedia/B-ISDN services will be composed of a variety of mutually related factors, which require OA&M to take a top-down approach from the viewpoint of the customer/user and service. The OA&M has to be implemented in a very flexible manner and be supported by the network and NEs, which are defined as well-structured managed objects.

A look at the history of system development will show that automation of OA&M has progressed from the level of network/NE monitoring and testing to that of control and judgment. Furthermore, the following OA&M functions will be necessary for the progress of multimedia communication:

1. *Automated design, function modification and provisioning of NEs:* With these, the network itself and its OA&M will become fully automated and free running.

2. *Flow-through service provisioning/analysis/development and customizable user interaction functions:* These options will allow the realization of automated service management and user-friendly customer care.
3. *Automated business support function:* This will allow faster decision making and strategy formation.

Historically OA&M has grown according to individual business industry areas such as telecom OA&M, system management, private network (NW) management, and corporate business management. Not only because service providers have computers and local-area networks (LANs) for their internal use but because they have to provide total services to corporate customers by linking their elements with customers' elements, the SPs need all of these types of OA&M. Customers will also need more OA&M capabilities if they are to enjoy a variety of multimedia services as mentioned in the previous section. That is, the activities of the above areas are becoming more and more mutually linked and are converging, resulting in integration of technologies and close cooperation among users, providers and suppliers.

7.1.3.2 OA&M Services [1]

Service OA&M includes all activities required to provide information communication services at the SLA level promised to customers. OA&M services are provided in order to offer additional value to customers on top of or combined with service OA&M, as shown in Figure 7.2. Conventionally, information communication services are directly exposed to a customer, and he or she has to first decide on what to do if problems arise to cooperate with operators/maintenance technicians. OA&M services will free customers from this kind of annoyance so that they realize comfortable services. Embedded-type OA&M service makes the information communication much more user-friendly, and another type of OA&M service is a, so to speak, stand-alone type that can be offered as a value-added service.

This OA&M service is defined as a service in which a customer agrees on the contents and makes a contract with a service provider in the same manner as an information communication service itself. The service is provided at any point of time; that is, before, during, or after information communication, achieved by the cooperation of the customer, OA&M systems, and operator(s).

OA&M services have become recognized as important elements of advanced information services as the intelligent services have been widely introduced and customer demands for high-quality service have increased. In particular, multimedia services will make information communication services much more varied in terms of time, space, and size and combined with the

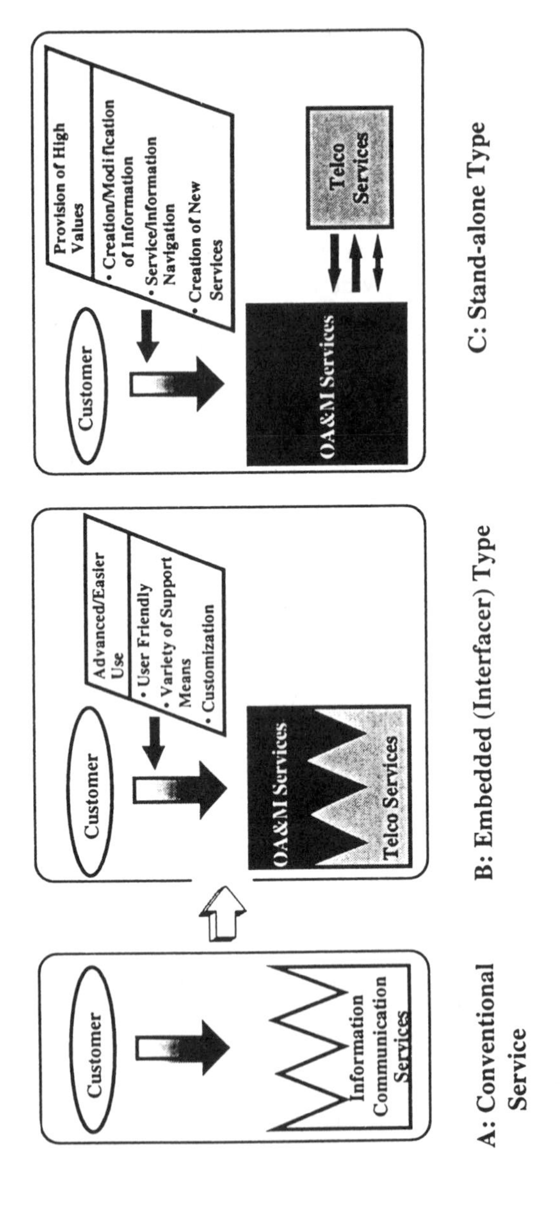

Figure 7.2 OA&M service.

variety of elements explained in the previous section. Thus, information communication services and OA&M services will merge into flexible, objective-oriented ones. That is, in addition to embedded OA&M services that directly or indirectly support the information communications, stand-alone type OA&M services will be provided in such a manner that they should provide a variety of forms of processed/created information, marketing/advertisement information, information/service navigation.

7.1.4 Reference Models and Technologies [1–2]

7.1.4.1 Reference Models

Offering a variety of multimedia services in the near future will be dependent on the development of a platform of a high-speed transport network and access network (including LAN, wireless, cable, etc.), for which research and development activities have recently been vital in many respects [9–12]. In this multimedia service environment, many factors are calling for the establishment of much more sophisticated OA&M than ever before in the conventional telecom service environment as follows.

First, there are many service and network factors. As explained in Section 7.1.2, multimedia services will be roughly grouped into these classifications: high-speed computer communication used mainly for business operations and management of companies (so-called back-end office use); home-use interactive television (TV) services; VOD services; and a means to improve the efficiency and flexibility of personal communications in home/office (OA) use. All of these services are expected to be conveyed on backbone networks based on asynchronous transfer mode (ATM) technology. However, at the startup stage, various methods and paths will be developed according to the characteristics of specific demands since both services and networks will progress at the same time, influencing each other in terms of technology and business, such as frame relay, Fast Ethernet, Gigabit Ethernet, and Internet cable. This will increase the complexity when studying the OA&M functions and system technology, requiring ad hoc solutions and migration methods while pursuing a systematic framework. The proliferation of different types of service terminals including entertainment/game machines, network terminals, and TV sets will accelerate this trend.

Second, there are many business players and management domain rulers. Multimedia communication services cannot be provided by a single provider, but many different types of businesses have to cooperate including transport capability providers, service providers, content providers, equipment providers, and of course customers. The cooperation required will be varied according to the above-mentioned service and network types, and further this cooperation and formation will be constantly changing according to the evolution of multimedia communication. Each player composing a multimedia service delivery chain is required to prepare the OA&M capability for its responsible part, and furthermore that part is usually composed of a number of components in terms of both service and OA&M. That is, when a customer enjoys a service, the successful provision of the service with its OA&M is achieved as a result of the cooperative work of the above players.

The environment explained in the preceding paragraphs suggests that there will be a number of unforeseen changes and emergence of new paradigms in the course of multimedia service progress, such as diverse customer demands, varied business operations including reform of regulation, increased service components (communication, terminal, center, server, OA&M, etc.), and so on. To cope with this complexity, it is important to define a simple model having the minimum number of elements and a clear interrelationship (interfaces) among these elements, including all aspects recognized at least at this moment and allowing new elements to join flexibly. Figure 7.3 shows a simple multimedia/B-ISDN reference model. This model is just one that shows the types of players involved in the services and OA&M activities. Further study is necessary to establish in the industry a shared multimedia/B-ISDN business reference model, logical/physical configuration models based on this, and then concrete implementation methods for these.

7.1.4.2 Elements of the Models

Establishing models to be shared among every type of players is a key study item as mentioned earlier. Together with this, it is also important to define management information, which is a fundamental component for model description, to present a method for its database implementation, and to establish an easy-to-use, flexible application construction method.

With regard to system integration and standardization, the following will be major items:

- *Terminal level:* data format, framework on the handling method, home platform, etc.;
- *Network level:* service handling procedures, high-level functional layer, etc.;
- *Center/server level:* data representation, access interface, etc.;
- *Operations and management level.*

Security is a common issue for all levels.

Another factor of modeling is the one based on a description of Internet technology, such as the Internet user interface, Internet service providers (ISPs),

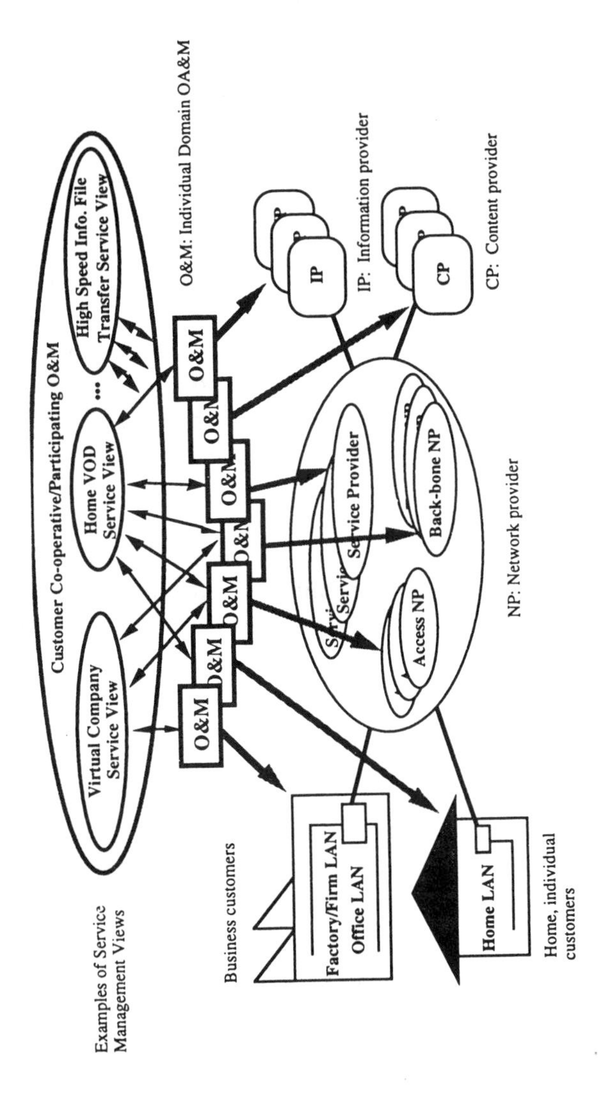

Figure 7.3 Multimedia/B-ISDN reference model.

Internet search mechanism, and a variety of Internet site types. Regarding the OA&M level, it is necessary to establish a systematic method for providing integrated service OA&M crossing over multiple domains of different management agencies/rulers. The cooperation among players involved in the service OA&M is key to solving problems in multimedia services since problems are triggered from customers, service providers, network providers, information providers, sometimes even content providers and equipment suppliers. For the so-called backyard OA&M to support the service OA&M, one of the major issues will be full automation of the OA&M of network and NEs constructed based on the ATM technology [13].

As briefly explained in Section 7.1.3, a unique feature in the multimedia environment is that services and OA&M have become more merged, which has blurred the border between them. That is, selection/change/addition of service conditions, those of media, acquisition of charging/usage information, etc., which are conventionally carried out by the staff of the operating company will be done by customers at any time (before, during, or after) of service according to service scenarios or new scenarios being created during the service. Customer-cooperative OA&M, in which customers themselves can decide on the OA&M activities cooperating with service/network providers, is an important concept from this point of view also. The above-mentioned integrated service OA&M should be constructed so as to realize this concept, with each of the players providing the required information and executing requested actions. Examples of study items will be mechanisms on customer access management, customer negotiation management, service scenario creation management, distributed management cooperating with other (OA&M and service) functions, implementation method as platforms, etc. These aspects are discussed later in this chapter.

7.2 Network Element and Network-Level Operations, Administration, and Management

7.2.1 OA&M Technologies for B-ISDN, ATM, and SDH

Among the many OA&M technologies for B-ISDN, ATM, and synchronous digital hierarchy (SDH), the most important one is information modeling technology, which enables flexible and quick development, deployment, and operation of various types of NEs, transport networks, and access networks. All the services supported by different types of underlying transport networks and access networks are well managed and controlled in terms of management information provided by network management systems (NMSs). A managed object (MO) is an abstract type of management information that spreads over wide range of NEs and networks. Currently, various types of management information have been studied such as alarms for virtual path (VP) or virtual channel (VC), alarms for NE such as power failure, network configuration information such as routing information, trouble ticket information, billing information and tariff, customer contact information, customer-oriented naming, and ownership.

If we are to realize multimedia services effectively, ATM is most suitable as an underlying transport and access network technology. For example, ATM link speed can be scaled from megabits through gigabits. This aspect ensures the longevity of technology by allowing network performance to stay in step with application requirements in the multimedia service era. Figure 7.4 summarizes B-ISDN-related recommendations studied by ITU-T. These recommendations encompass a wide range of I-, Q-, and G-series recommendations [14–16].

For OA&M technologies, SDH has emerged as the first stage of network and NE digitalization. It realizes a unique digital transmission hierarchy in connecting each other across the different carriers' networks globally. Synchronous Optical Network (SONET) is a standard in North America. Its technology

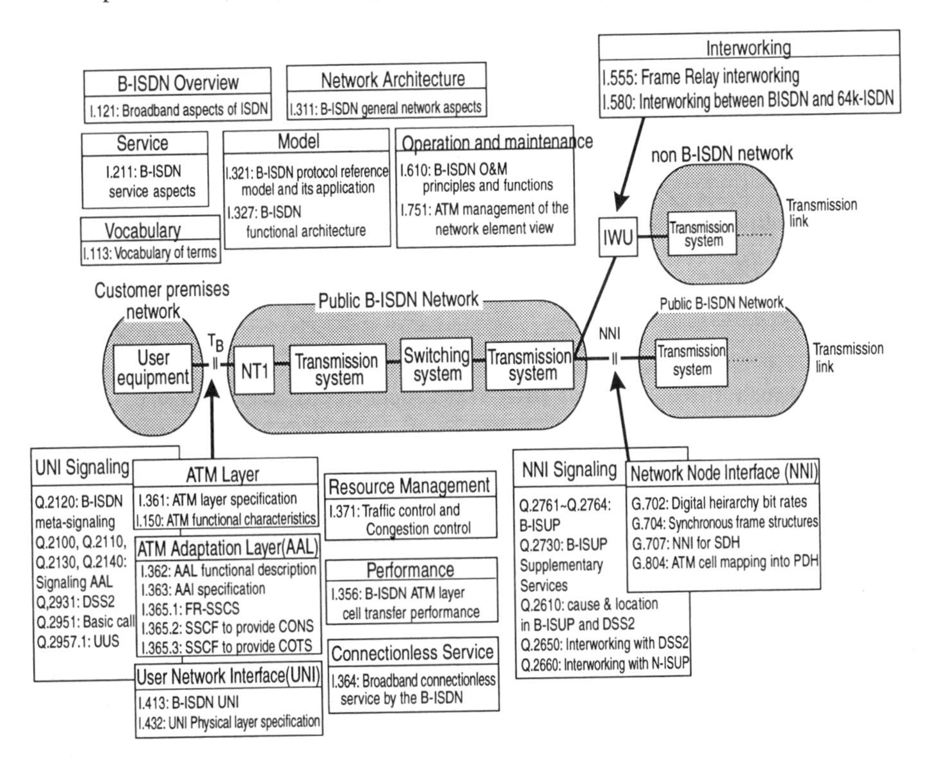

Figure 7.4 B-ISDN-related recommendations.

is very similar to that of SDH. From the OA&M viewpoint, both SDH and SONET technologies can realize real network management systems and NEs through telecommunications management network (TMN) concepts and specifications. ATM technology is the newest one in the telecommunication industry to support emerging multimedia services with the benefits of its high-speed transport capability. ATM technologies concerning ATM networks and ATM NEs are described in the following subsections.

7.2.2 NE Management for ATM Node Systems

This section describes the basics of ATM management technologies for multimedia transport systems.

ATM NE view MOs are being studied by the ATM Forum and European Telecommunications Standards Institute (ETSI) as well as ITU-T [17–19]. ITU-T Study Group (SG)15 has been studying Recommendation I.751 since September 1993. In 1994, ATM Forum and ETSI submitted their recommendations to SG15. Therefore, the present I.751 is based on the contents of these recommendations. Figure 7.5 is the inheritance tree related to configuration and fault management. Shaded MOs are defined in I.751, whereas white MOs are defined in the existing recommendations such as M.3100. Brief explanations of each follow.

ATM cells flow over paths such as SDH VC4. "tcAdaptorTTPBidirectional" represents this type of ATM cell flow. The relationship between this MO and the underlying SDH path TTP (trial termination point) is represented by the "supportedByObjectList" attribute within "tcAdaptorTTPBidirectional." Therefore, the existing SDH path MO can be utilized without modification. If the LCD (loss of cell delineation) is detected, this MO emits an alarm to the NMS.

Three MOs, such as "uni," "interNNI," and "intraNNI," are defined to distinguish three kinds of interfaces, such as the user network interface (UNI), inter-NNI (network node interface), and intra-NNI. These MOs are related to "tcAdaptorTTPBidirectional" through the "underlyingTTPPointer" attribute. Moreover, these MOs have the "loopbackLocationIdentifier" attribute, which represents loopback location. This is a fault management aspect of these MOs. If NMS sets the value of "loopbackLocationIdentification" to true, the loopback OAM cells will be loop-backed at this point.

The "atmAccessProfile" attribute represents the features between the transmission path and VP or between VP and VC. It has two kinds of features: one for numbering and the other for bandwidth. For example, 12 bits are assigned for VPI in the NNI interface. However, some vendors may not support

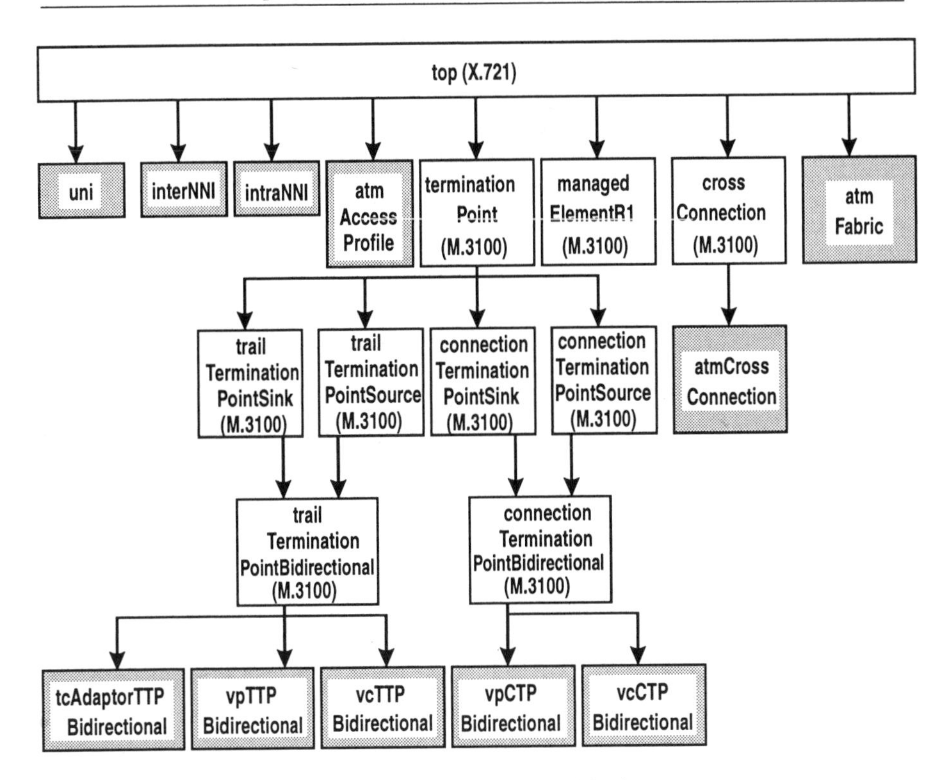

Figure 7.5 Configuration and fault management-related inheritance tree.

12 bits for economic reasons. Therefore, this number may differ between vendors and it will cause some trouble when interconnecting different ATM NEs. To overcome this situation, this MO is defined.

The feature of vpCTP's or vcCTP's configuration management is that this MO has traffic descriptors, and, optionally, QoS class. If the "segmentEndpoint" attribute in these MOs is true, it represents a segment endpoint in the network. When NMS receives a VP-AIS (alarm indication signal) from this MO, NMS may send "loopbackOAMCell" ACTION to the opposite-side NE. This NE will send the loopback OAM cell, which will be loopbacked at the former NE.

"atmFabric" and "atmCrossConnection" represent the cross-connects for VPs as well as VCs.

Figure 7.6 is the naming tree related to performance management. Brief explanations are described next.

When ATM cells enter an NE, HEC (header error control) is executed for each cell. "tcAdaptorCurrentData" has some attributes for performance management related to this HEC. The "erroredCellsHECViolation" attribute

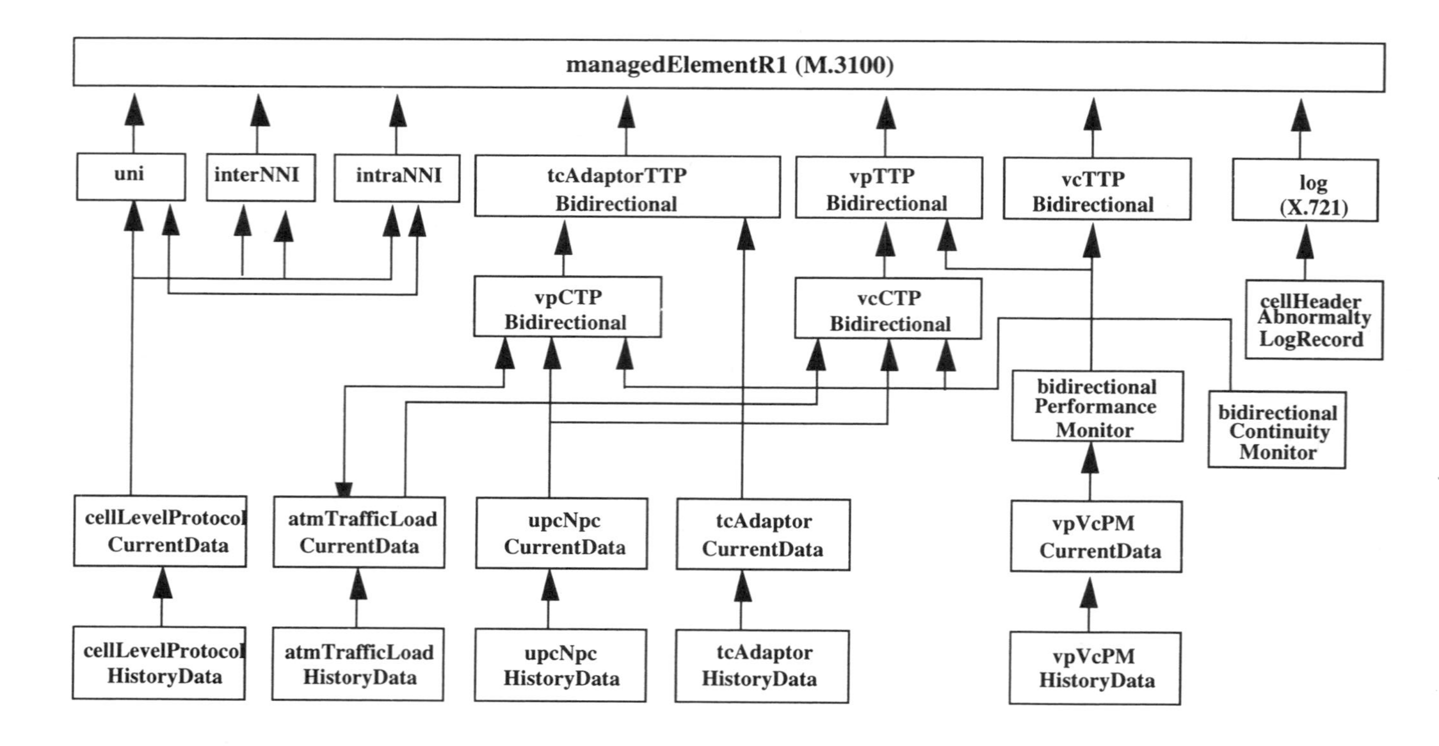

Figure 7.6 Naming tree related to performance management.

represents the number of cells with header bit errors, while the "discardedCells-HECViolation" attribute represents the number of cells discarded due to uncorrectable header bit errors.

On the other hand, the "discardedCellsInvalidHeader" attribute within "cellLevelProtocolCurrentData" represents the number of ATM cells discarded due to header content errors, such as an unassigned VPI/VCI (virtual channel identifier) value or an out-of-range VPI/VCI value.

After this processing, each VP or VC is identified and usage parameter control (UPC) or network parameter control (NPC) is executed. The "discardedCell" attribute within "upcNpcCurrentData" represents a count of ATM cells that were discarded due to UPC/NPC policing, while the "successfullyPassedCells" attribute within it represents a count of ATM cells that were successfully passed by UPC/NPC function.

By using "atmTrafficLoadCurrentData," the number of incoming or outgoing cells can be counted per transmission path or VP or VC.

"bidirectionalPerformanceMonitor" and "vpVcCurrentData" execute performance monitoring of VPs or VCs using performance monitor cells.

"bidirectionalContinuityMonitor" emits a loss of continuity (LOC) alarm to NMS.

7.2.3 Network Management for ATM Networks [20–24]

This section explains the management of ATM network level. Figure 7.7 shows two service management systems (SMSs) and two NMSs as an example of network management system configurations. Because each pair of SMSs and NMSs belongs to a different carrier, this interface between SMSs or NMSs is the X interface. Moreover, each NMS has two sub-NMSs for load distribution or something else.

Some standardization bodies are actively studying these interfaces. ITU-T SG4 and ETSI are responsible for the generic Q and X network view interfaces. On the other hand, ATM Forum is responsible for the Q and X network view interfaces for ATM, which are called the M4 interface and M5 interface, respectively. Figure 7.8 shows the M4 network view of ATM Forum, which is briefly explained as follows. Note that since this ATM Forum management information base (MIB) specifies several options, the following explanation is just one example.

The MOs in bold boxes, which are already defined in the M4 NE view, can be used and referenced by the M4 Network View MOs. "atmNetworkCTP" and "atmNetworkTTP" are used when these M4 NE View MOs are not provided. "atmNetworkTrafficDescriptorProfile" is introduced to allow easy introduction of new kinds of traffic descriptors, such as ABR and VBR. "atm-

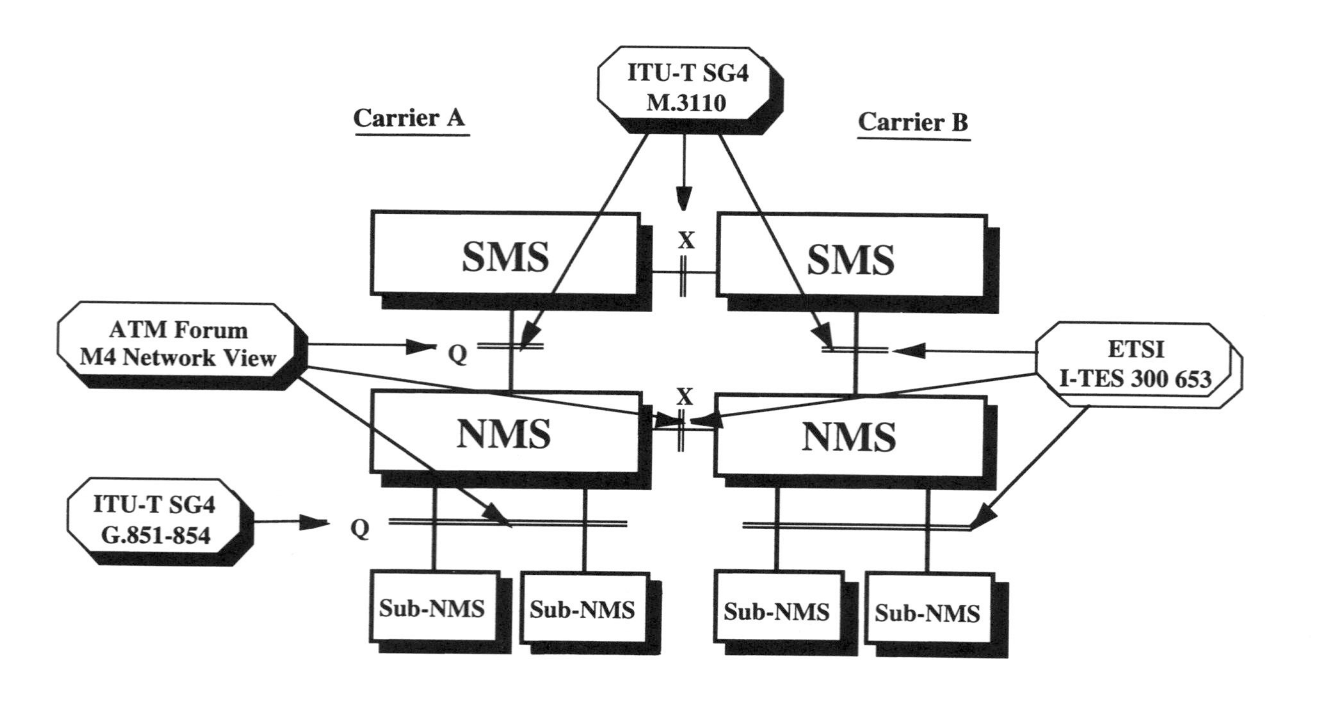

Figure 7.7 Network view recommendations.

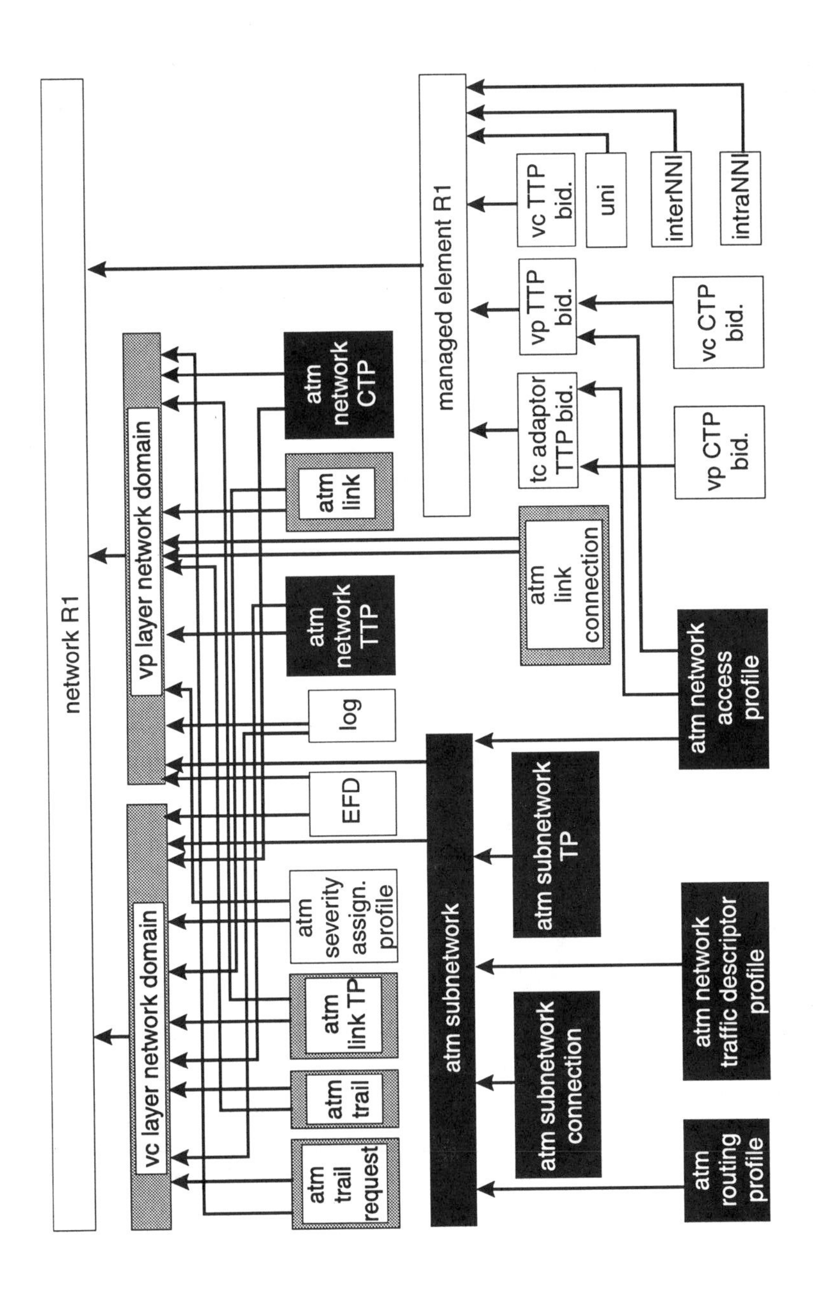

Figure 7.8 Naming tree of ATM Forum network view MIB.

NetworkCTP" may point to "atmNetworkTrafficDescriptorProfile" at points where needed.

"vcLayerNetworkDomain" or "vpLayerNetworkDomain," which is called "LayerNetworkDomain" here, represents part of the VC or VP layer, respectively. Therefore, there may be several "LayerNetworkDomains" within a single network. "LayerNetworkDomain" contains only one top "atmSubnetwork," which may be further decomposed of the lower level "atmSubnetworks" and "atmLinks."

One "atmLink" is terminated by two "atmLinkTPs." "atmLinkTP" is associated with its server TTP by its pointer attribute. Server TTP means "vpTTPBid." at the VC layer, but "tcAdaptorTTPBid." is used at the VP layer. Note that server TTP means "atmNetworkTTP" when these M4 NE View MOs are not provided. Both "atmLinkTP" and server TTP are associated with "atmNetworkAccessProfile" for the purpose of representing the profile of itself. "atmLinkTP" is associated with its "atmNetworkAccessProfile" by pointer attribute, while "atmNetworkAccessProfile" is named via name binding with "LayerNetworkDomain." On the other hand, server TTPs, such as "vpTTPBid.," "tcAdaptorTTPBid.," and "atmNetworkTTP" are associated with their "atmNetworkAccessProfile" by name binding.

In the case where we need to set up "atmTrail" between "atmNetworkTTPs" in "LayerNetworkDomain" by using ACTION, two "atmNetworkTTPs" are first created. Then, a related two "atmNetworkCTPs" and "atmSubnetworkTPs" are also created. After "atmSubnetworkConnection" is created between two "atmSubnetworkTPs," each "atmSubnetworkTP" is bound to its reflected "atmNetworkCTP." Then, "atmTrail" is created after "atmNetworkTTPs" and "atmNetworkCTPs" are bound.

In a manner similar to the way in which we set up "atmTrail," we set up "atmSubnetworkConnection" between "atmSubnetworkTPs" in "atmSubnetwork" by using ACTION, two "atmSubnetworkTPs" are first created. Then, reflected "atmNetworkCTPs" (and "atmNetworkTTPs" if the "atmSubnetworkTP" terminates the "atmTrail." (After "atmSubnetworkConnection" is created between two "atmSubnetworkTPs," each "atmSubnetworkTP" is bound to its reflected "atmNetworkCTP." (Then, "atmNetworkTTP" and "atmNetworkCTP" is bound if the "atmSubnetworkTP" terminates the "atmTrail.") "atmRoutingProfile," which is contained in "atmSubnetwork," represents a set of topological routing constraints that can be applied to a new "atmTrail" or "atmSubnetworkConnection" during its setup. To provide a deferred "atmTrail" setup service, "atmTrailRequest" is defined. It is created as a result of ACTION on "LayerNetworkDomain."

Existing "alarmSeverityAssignmentProfile," "EFD," and "log" are used for fault management.

7.3 Service Operations, Administration, and Management

7.3.1 Introduction

Multimedia services such as VOD, home shopping, and network games are expected to be provided in the near future via wide-area networks (WANs) such as the Internet and high-speed, broadband networks. From the customers' viewpoint, it will be important for them to be able to easily select, using only a TV-type home terminal set, a service from among many providers and to subscribe to multimedia services, check the service charges, and make inquiries/complaints about the service.

This section explains a concept and requirements for multimedia service OA&M and OA&M service, and discusses service OA&M functions and service navigation functions to meet these requirements. The OA&M data and functions for supporting the businesses of content and service providers and for realizing well-managed services must be developed for network and system management as well.

7.3.2 Service and Business OA&M

7.3.2.1 Objectives for OA&M [7,8,25]

The key goals of service OA&M are to improve customer service, reduce cost, and shorten time to the market. In other words, service OA&M is the efforts of service managers to find ways of achieving total profit improvement through the improvement of operational processes and creative application of networking and information processing. Service OA&M includes all kinds of processes and systems employed to deliver services to the customers. Service OA&M is concerned with service creation, order handling, customer administration, marketing, problem handling, billing, and so on.

On the other hand, the purpose of business OA&M is to achieve business goals in terms of return on investment, market share, employee satisfaction, etc. Success of business OA&M will be based on the successful service OA&M.

7.3.2.2 Layered Management Concept [26,27]

The layered management concept has been introduced in a standards arena where management interface specifications are developed for the integration of network and service OA&M systems. The concept is described in ITU-T Recommendation M.3010, "Principles for Telecommunication Management Network." The concept is illustrated in Figure 7.9, where the upper management layer is dependent on a lower layer. Reference points are focal points for

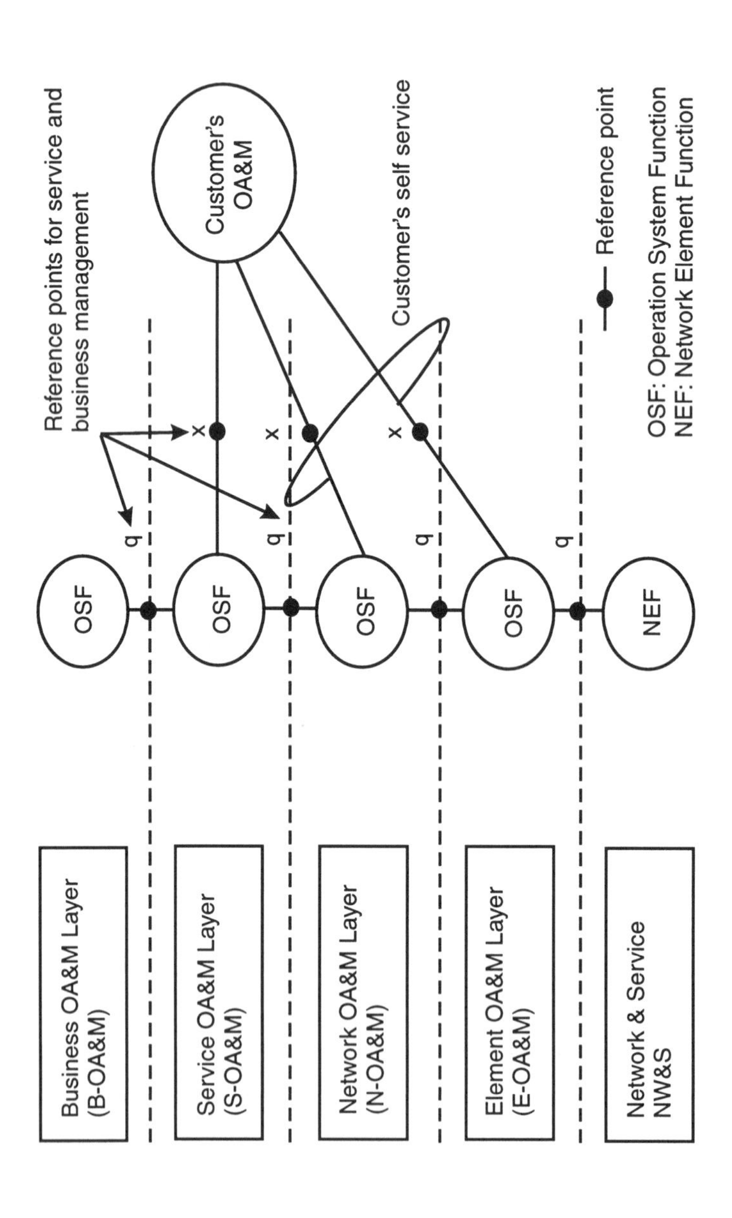

Figure 7.9 Layered management concept [7–25].

the exchange of management information, and the base for the development of standards interface specification.

Business and service OA&M can better be recognized in this concept. Service OA&M views have to be well established to be shared with the customer's management side. Lower layers do not necessarily have to be shared with the customer side. Some skillful customers will share lower layer management information for their own specific management strategy and/or mechanisms. Across the reference point x shown in Figure 7.9, management information is exchanged between customers and service providers. For the definition of management information, the layered management concept is useful to classify information in a well-organized manner among service providers and customers, which leads to stable management information modeling.

7.3.2.3 Service Delivery Chain [27,28]

The service delivery chain is another useful concept for identifying the objectives of OA&M. Figure 7.10 shows a typical example of a service delivery chain. In this figure, the layered management concept is added on the delivery chain concept to explain precisely the service offering mechanisms. The telecommunications market depends on given political and economical situations of society. However, the delivery chain of telecommunication services to end customers is another principle that gives us another base with which to identify OA&M goals. The service delivery chain will provide ideas for making decisions about OA&M targets based on the well-recognized mechanisms of offering multimedia telecommunication services, identifying clearly who is the customer and what is the target of OA&M, etc.

7.3.3 Multimedia Service OA&M

7.3.3.1 Multimedia Service OA&M [1,2,29–31]

The typical classifications for multimedia services are given in Table 7.2. The concept of a network configuration for providing multimedia services is illustrated in Figure 7.11. Those players who are involved in multimedia services such as VOD and electronic shopping are content providers, service providers, network providers, and end customers. The content providers provide video sources or commodities and their catalog. The service providers connect a service providing system (SPS), such as head-ends, contents servers, and information databases having various content, to a network and end customers can access the SPSs to watch video or to purchase goods through a service consumer system (SCS) such as a set-top box or a personal computer/network terminal (PC/NT), etc.

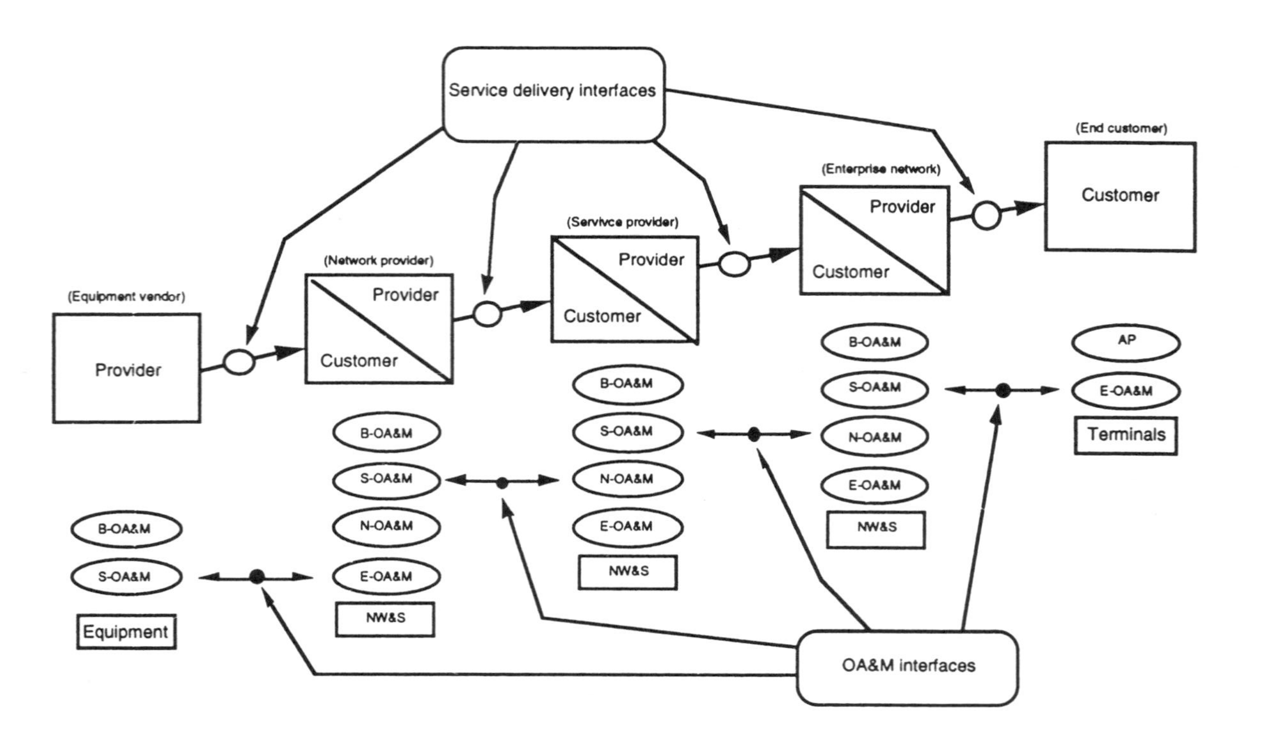

Figure 7.10 Service delivery chain.

Table 7.2
Sample Classification of B-ISDN Services

Classification	Examples	Major Customers
Interactive service Center-access Messaging service	VOD, information retrieval (e.g., newspaper article)	Home customer (personal use)
High-speed communication service among systems	High-speed file transfer	Business customer (enterprise)
Distribution service	CATV	Home customer (personal use)

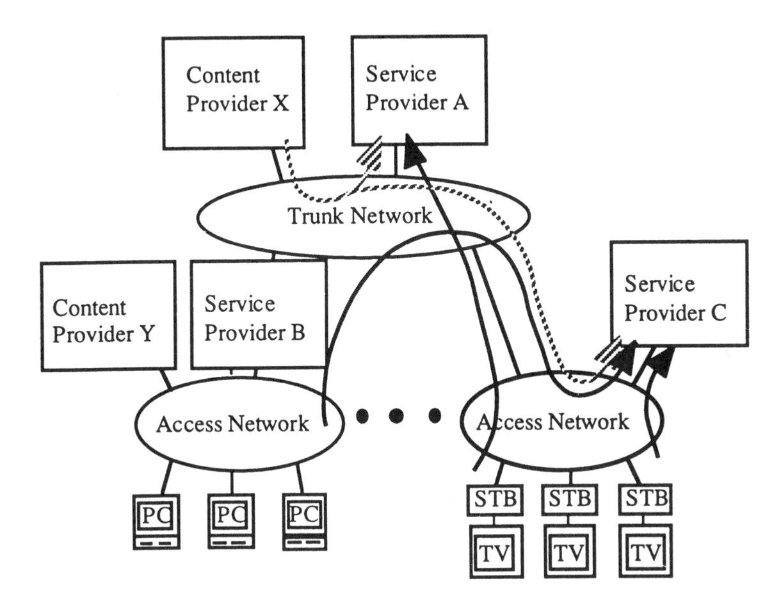

Figure 7.11 Service network configuration example.

Continuing service OA&M by the network providers involves operation tasks, such as measurement to ensure QoS, rapid response to recover from system failures, the calculation of service charges, the creation of billing information, and the monitoring of service use status.

The functions that the service OA&M system should provide in order to satisfy the requirements given are listed in Table 7.3.

An example of the multimedia service OA&Ms is described next. It will be easier for customers to use their own screen display to make complaints to the network provider about any service problems. Communication with the

Table 7.3
Service OA&M Function Examples

Function User	Function
End customer	• Applying for a service from the terminal • Setting locks, preferences, identification codes, etc., from the terminal • Sending complaints and receiving responses via the terminal
Network and service providers	• Usage charge proxy surcharging • Collection of references and use charges from the end customer • Distribution of use charges to the service providers • Restriction of service use by end customers as indicated • Registration of end customer's use to the server

customer is displayed on the terminal of the service operation center, which then needs to determine the cause of the problem. Possible screen layouts are shown in Figure 7.12. It can be seen that service OA&M requires detailed information on the state of the network. This makes it important for the service OA&M systems to work closely with the network OA&M systems to collect and manage the information required and with the OA&M systems controlling SPS and SCS.

7.3.3.2 Customer Cooperation/Participating OA&M [1,2]

The ultimate goal of service provider's OA&M is to achieve customer satisfaction. Particularly in the multimedia era, fundamental requirements for future OA&M will be responsiveness and speed and accuracy in every elemental task that is common to all types of services, such as customer inquiry, service provisioning, fault repair, and service subscription profile changes.

Customer satisfaction is further increased by "customer cooperative OA&M," in which a customer can decide on the OA&M activities with his or her service provider. As OA&M service interfaces become increasingly mechanized and then automated, OA&M will become fully responsive to customer needs. However, mechanization or automation does not always lead to customer satisfaction. Therefore, service providers must collect precise information to determine each customer's needs in his or her specific environment, and then develop human–machine customer interfaces that are adaptable to the varied customer requirements. Furthermore, to ensure fair competition allowing differentiated service offerings by various types of service providers, all players must develop and share a framework that establishes what kinds of information should be common among service providers and what should be left to each provider.

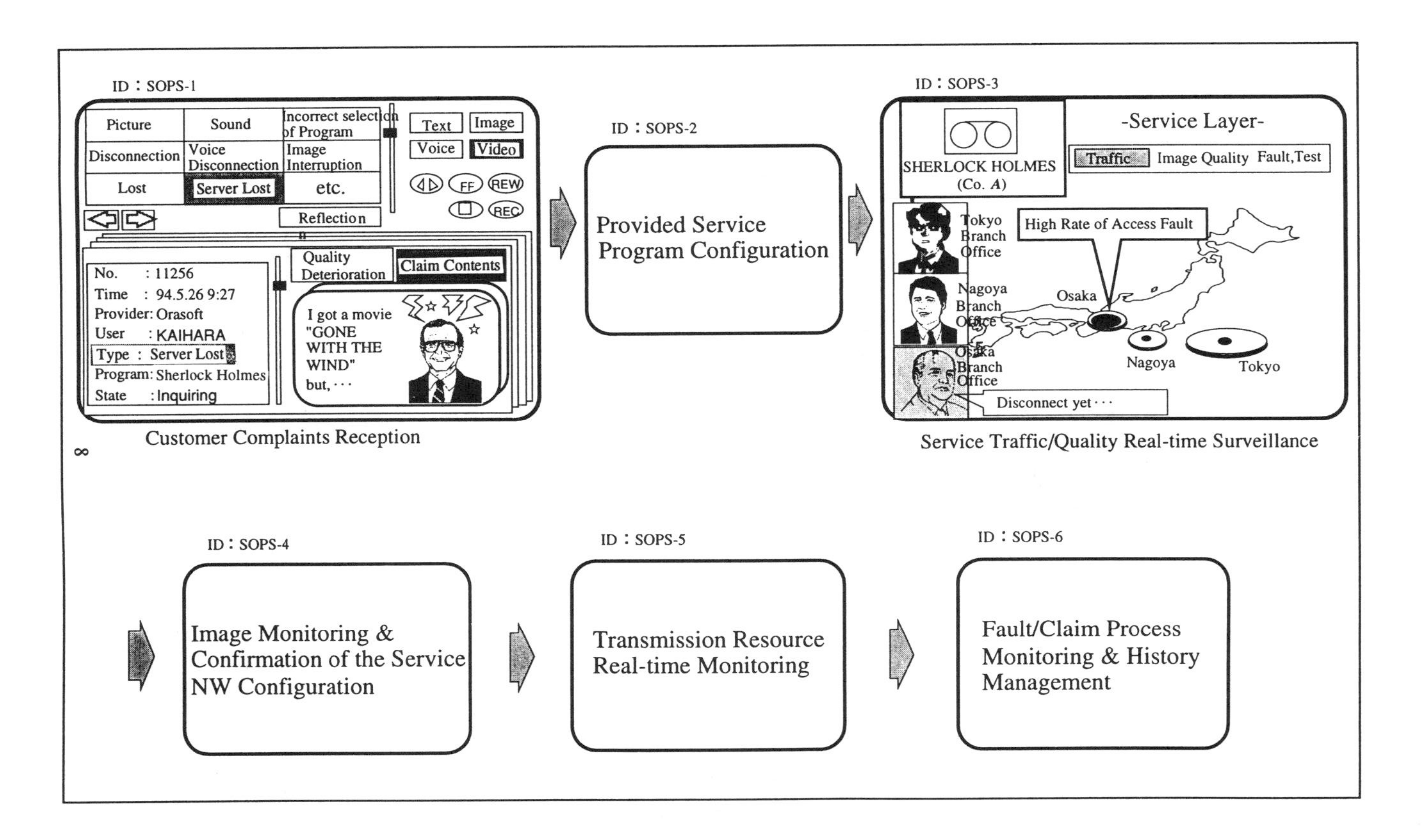

Figure 7.12 Service operation functions for carrier's operator (example of claim handling).

7.3.4 OA&M Service Requirements

7.3.4.1 Definition and Examples [1,2,28–31]

Telecommunications OA&M supports provision of telecommunications services to customers by customizing, monitoring, and controlling networks and network systems. The degree of achievement of OA&M objectives can be assessed by customer satisfaction, as mentioned in Section 7.3.3.

The parameters used in defining the agreed level of QoS are gathered into the service level agreement (SLA). The SLA is also used as a reference for assessing services offered to customers, and further is itself a target of OA&M. In line with this idea, "OA&M service" is defined as another type of customer service (see Section 7.13).

OA&M services evolve by considering what response should be made to customer demands. When offering a new service, the timeliness, precision, and quality of the response can be incorporated into one service field. OA&M services and telecommunications services are provided in parallel in all service providing processes for customers. Operators and OA&M systems cooperate in providing OA&M services. Thus, OA&M services should be based on contracts between customers and service providers as are telecommunications services.

Telecommunications history has shown that OA&M services did not have to be explicitly recognized. This was because the primary customer demand was for subscription to the telephone service itself rather than for total services, including customer service. However, OA&M services become essential when

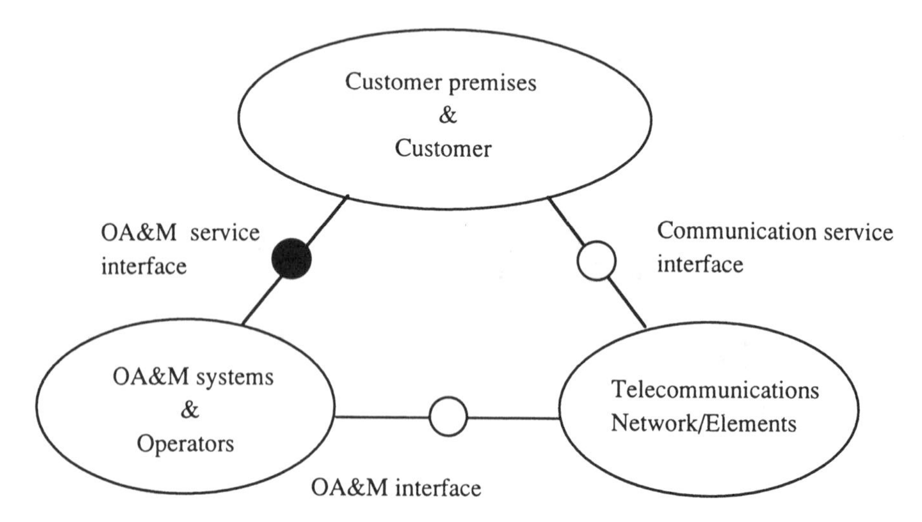

Figure 7.13 OA&M service interface.

customers expect higher QoS or more sophisticated telecommunications services in the multimedia era. Besides, interoperable common interface specifications and standardization will be essential for OA&M services when these services are provided through human-to-machine and/or machine-to-machine interfaces in multicarrier/multivendor environments, since customers will want to be able to freely select services from among providers through a single interface. Thus, providing products and/or services having interoperable interfaces is an essential requirement for service providers and suppliers in competitive markets.

Four types of people and organizations—content provider, service provider, end customer, and network provider—are involved in the configuration of the type of service network for multimedia services including OA&M services. They all have different perspectives on the provision and use of services.

The foremost matter is to offer multimedia services to the end customers and at the same time provide profits to the content providers and service providers. Also important are OA&M services that ensure smooth and efficient service use and provision, such as the collection of service charges from the end users, the recording of content use, and the number of potential users (latent demand) (see Figure 7.14). Examples of multimedia OA&M services are described next.

Consider this example for customers: When a new customer connects his or her SCS to the network terminal, screens such as those in Figure 7.15 will appear so he or she can register his or her service order (SO). As illustrated in Figure 7.15, it is important that customers can easily send messages and requests to the carrier via their screen display in a PC, TV set, or equivalent type of terminal.

Here is an example for service providers: Figure 7.16 shows an example of OA&M services for service providers. One such service will determine to which of a provider's multiple SPSs a customer will be connected, based on that customer's location. This will help evenly load the SPSs of service providers with multiple same-service SPSs on the network. Another example is that of providing data on service use of the service and programs.

7.3.4.2 OA&M Service Data [29]

The provision of OA&M services to end customers, content and service providers, and network providers is specifically a matter of providing the various data and functions required for service use and provision. OA&M services that need to be provided for each type of user are described next.

1. *End customer:* The customer can apply for the use of services via an access terminal at the time of its installation in the home and begin to use the service immediately on a plug-and-play basis. When some

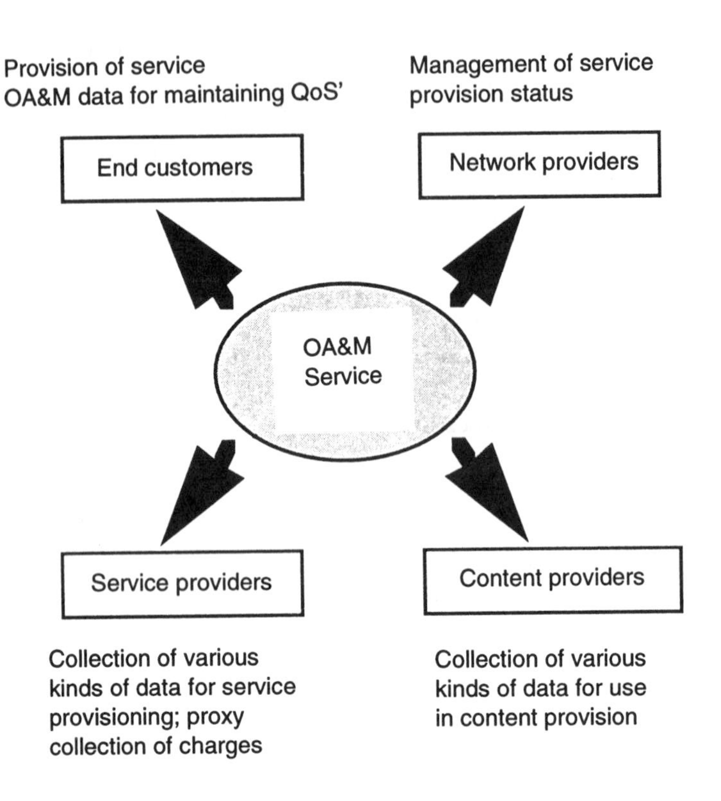

Figure 7.14 Users of the OA&M service.

service abnormality occurs, the customer can use the access terminal to complain of the problem and to receive a response concerning it. The user can also use the terminal to get detailed service charge information and the registered navigated service programs on a list at any time.

2. *SPSs:* Service charges can be collected from the end customer on behalf of the service provider. Service can be suspended for customers whose payments of the service charge are delinquent. Data on service use such as the service and program titles and times used is usually available. Also available are data on the number of end customers who could not access the service because of busy SPSs or another reason (latent demand) as well as time periods in which system failures occurred, the affected areas, and the time of recovery.
3. *Content provider:* Data indicating the degree of end customers' interest in the content being provided and temporal trends in use are available,

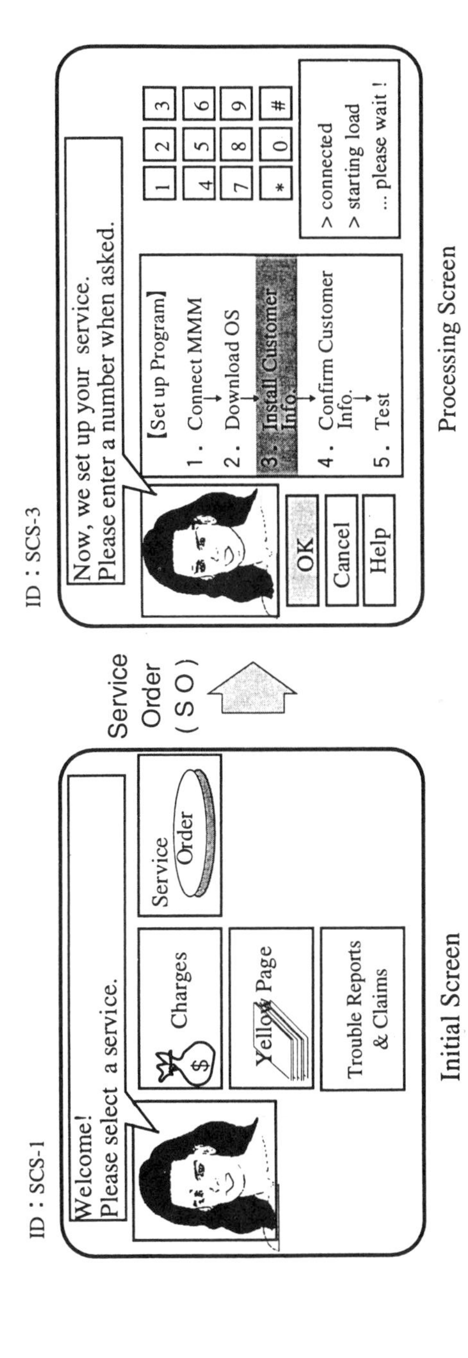

Figure 7.15 Customer screens for the OA&M service (example of a service order).

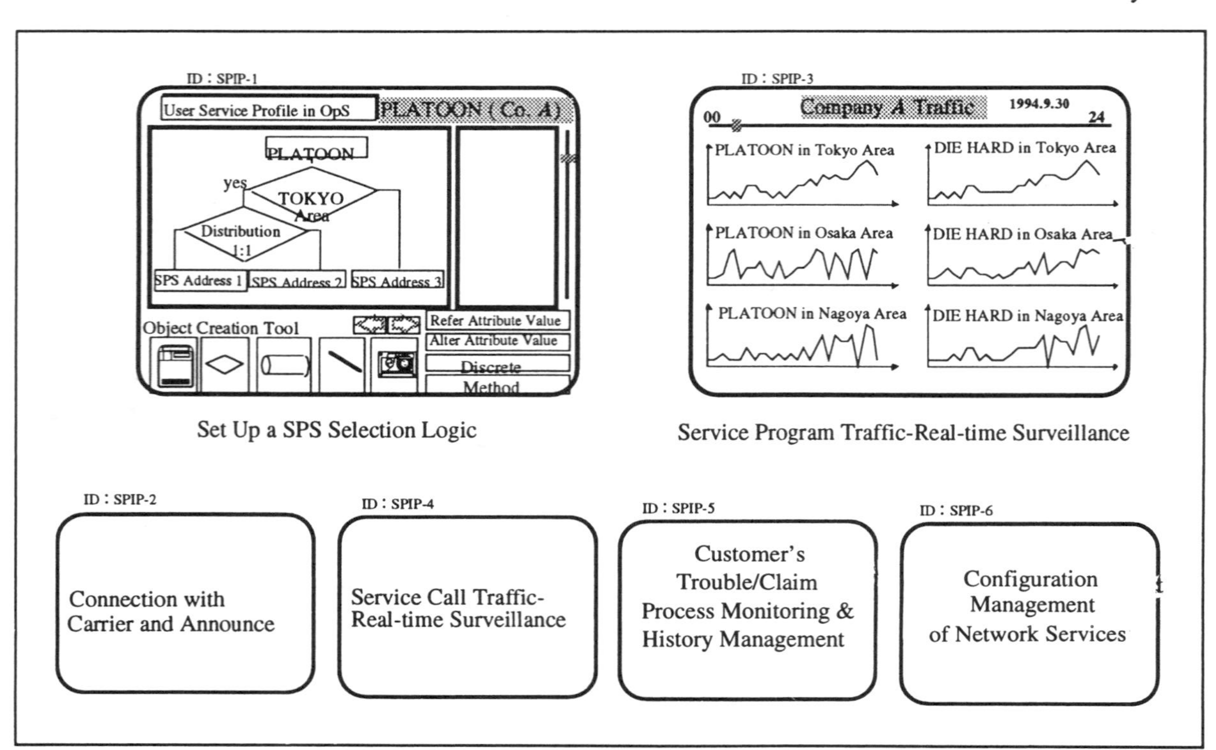

Figure 7.16 OA&M services for service providers.

including latent demand. For each service and program, the characteristics of the end customer and the patterns of use can be known as well as the degree of interest in the service provider's servers.

4. *Network provider:* Available services are the provision of the situation of the use of proxy services for content providers and service providers and the situation of the use of programs, etc. The occurrence of conditions that affect service provision, such as service access conditions, network abnormalities, and server abnormalities, can be known promptly.

Various kinds of data on end customers and the history of complaints are managed and can be used in handling inquiries and complaints. The OA&M service data that the service OA&M systems should provide in order to satisfy the requirements given are listed in Table 7.4.

Table 7.4
OA&M Service Data Examples

Data User	Required Data
Customers	• Service charge information • List of services offered • Service abnormality situation and time of restoration
Service providers	• Sales figures for services and programs • Information on customers that used the service • Use of services and programs • Number of end customers that could not access services and programs and reasons for the failures • Abnormalities in the server, the network and the terminals, the cause of the abnormalities, and the time of restoration
Content providers	• Information on customers that used the content • Use of services and programs • Number of end customers that could not access services and programs and reasons for the failures
Network providers	• Frequency of use for each service server • Complete set of the above information • Services and programs on each server, and their historical records • End customer complaints and their historical records • Setting made by the end users, such as lock and preference information, and identification codes

7.3.4.3 Intelligent Dynamic Service Provisioning (IDSP) [32,33]

Service provisioning falls under the field of OA&M services. The need to bring about new services and modify the existing ones without delay becomes increasingly important as telecommunication services become more diversified and complex in the multimedia/B-ISDN era. Customers, meanwhile, have essential preexisting objectives such as supporting business and home uses, and telecommunication services are the means to achieve these objectives. By employing these services, the demand to achieve these essential objectives can be met quickly and easily, whenever, wherever, and in whatever form is required. In progressing toward this goal, it is becoming increasingly important for customers to be able to provide their own telecommunication service simply and flexibly for their own use, based on their own knowledge and understanding of the network specifics. There are three major requirements:

1. It should be possible to utilize telecommunication services based on the customers' own concepts.
2. It should be possible to freely interconnect telecommunication services to any customer terminal.
3. Network and terminal sides should achieve a degree of cooperation that makes the network seem invisible.

Example services that will be available in the multimedia/B-ISDN era include remote distributed education, remote medical diagnosis, and electronic shopping. All of these services can be perceived as common physical facilities, for example, a school, a hospital, or a department store, that are mapped onto the network as virtual sites, which we refer to here as *communities.* Service provisioning is defined here as the construction and management of a telecommunication environment for the purpose of achieving online communities.

Intelligent dynamic service provisioning (IDSP) is a new way to provide OA&M services in the multimedia/B-ISDN era. IDSP has three distinguishing attributes that enable this service provisioning approach to satisfy the preceding requirements:

1. *Intelligent:* IDSP is intelligent, thus giving customers the control to freely set up and modify telecommunication services in any way they desire. It generates parameters that best suit the desired telecommunication services.
2. *Dynamic:* IDSP is dynamic, meaning that new telecommunication services can be set up and customized at any time and in real time.

IDSP can create or modify service plans and can control network/NEs and terminals in real time on a per-call basis or during calls.

3. *Customer participation:* IDSP is also built on this notion of customer cooperative participation as explained earlier [2]. IDSP capabilities are dispersed on both the network and terminal sides, and service provisioning involves close cooperation between them.

7.3.4.4 Intelligent Trouble Management [34]

The major desire of customers is that they receive trouble-free services and adequate service OA&M against network/service problems. However, problems, including natural disasters, cannot always be avoided. Therefore, trouble management is a key function in solving the problems and maintaining the high communications capability of a network when multimedia service users encounter problems with QoS.

The intelligent trouble management (ITM) process explained here is designed to support public network service and corporate service providers. Its goals are to provide OA&M services on investigating causes (e.g., interface mismatching and disconnection) of NW/NE trouble (e.g., inability to originate a call or receive calls), to track the progress of trouble resolution, and to estimate service restoration/recovery time.

The ITM action results offer (1) an OA&M scenario for recommending exact or precise operational instructions to operators when managing trouble, (2) accurate and timely OA&M irrespective of operator skill, and (3) OA&M services to improve services for customers.

A realistic approach to achieving successful ITM is to use a computerized trouble management system (ITMS) to integrate and harmonize computing power with the processing ability of operators.

Resolution of two major issues will lead to realization of ITMS. First, the operator's management knowledge should be translated into the knowledge that the system can easily recognize, and it should combine this knowledge and generate various OA&M scenarios that operators can easily understand. Second, in order to achieve accurate troubleshooting when there is no history of trouble at the introduction stage of multimedia services and future new network technologies, model-based knowledge, that is, communication and management knowledge in network and NEs, should be translated into knowledge that the ITMS can easily recognize.

The fundamental mechanism of the ITM is illustrated in Figure 7.17. The scenario generation mechanism and scenario execution mechanism provide a solution for the first issue.

Service, network, and NE troubleshooting based on model-based reasoning is a method for solving the second issue. This method, characterized by

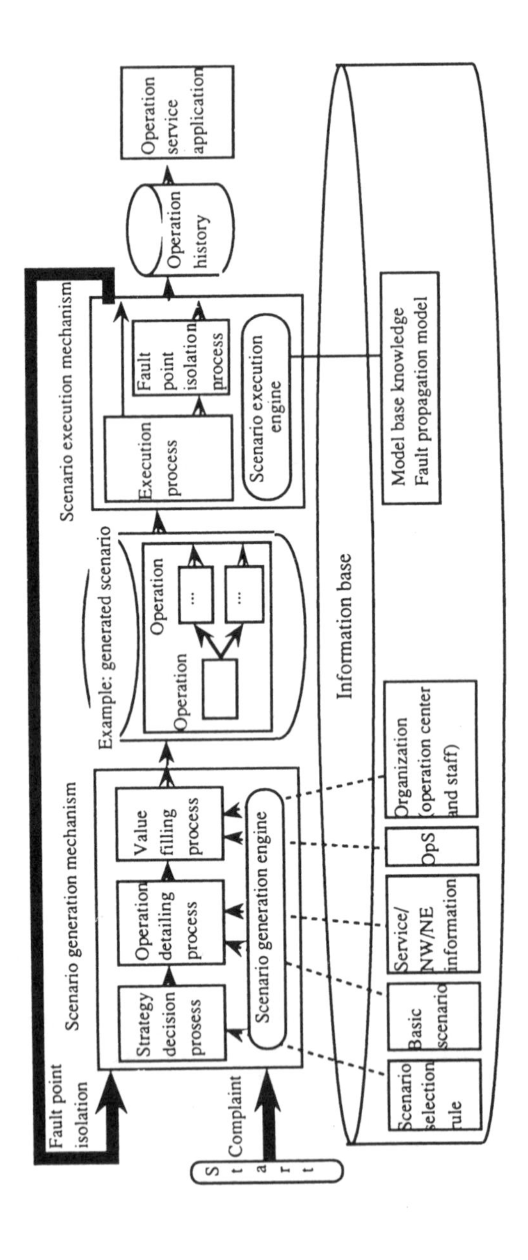

Figure 7.17 Fundamental mechanism of ITM.

use of the state transition knowledge of each network and NEs, traces the process of state changes, compares all states with symptoms, and infers what fault has occurred.

7.3.5 Multimedia Navigation Service

7.3.5.1 Definition and Examples [29–31]

From the end customers' viewpoint, it will be important for them to be able to easily select a service among many service providers by simple operation of his or her terminal of daily use, such as a TV set, PC, etc. From the service providers' viewpoint, they want to access as many end customers as possible all over the world.

It would inconvenience the end customer who wants to watch a movie using VOD, if he or she had first to find out which service provider was showing the movie before accessing an SPS. A good solution to this would be to display a list of all movies from all suppliers on the display. The act of selecting one of them would then automatically connect the end customer to the appropriate SPS.

Navigation service is defined as a service that finds the SPS needed for the required end customer service, connects his or her SCS to the SPS, and provides the end customer with information about charges. Figure 7.18 shows the concept of this service. Selecting a service menu (Yellow Page) from the initial menu list screen on a TV set will display available service categories. For example, selecting "movie" will list a variety of movie titles, and further choosing one of them may sometimes show a clip from the movie. When the customer enters his selection via SCS, the connection will be made toward an appropriate SPS. Navigation service generally offers to end customers the ability to access any of a variety of services that are available by simply specifying the type of service they desire.

As the number of SPSs connected to the network increases, the importance of the navigation service will increase and become essential. Search engines becoming popular in Internet services are also examples of navigation services.

7.3.5.2 Service Navigation Functions [30,31]

In the case that a given SPS offers some service program such as games and VOD, for example, a screen for selecting VOD or games is presented when the end customer accesses the SPS. If the end customer selects VOD, then another screen listing movie titles is presented. The end customer then selects the title of his or her desired movie, and the selected movie is presented. This process, up to the point when the movie has been selected, is called the *service*

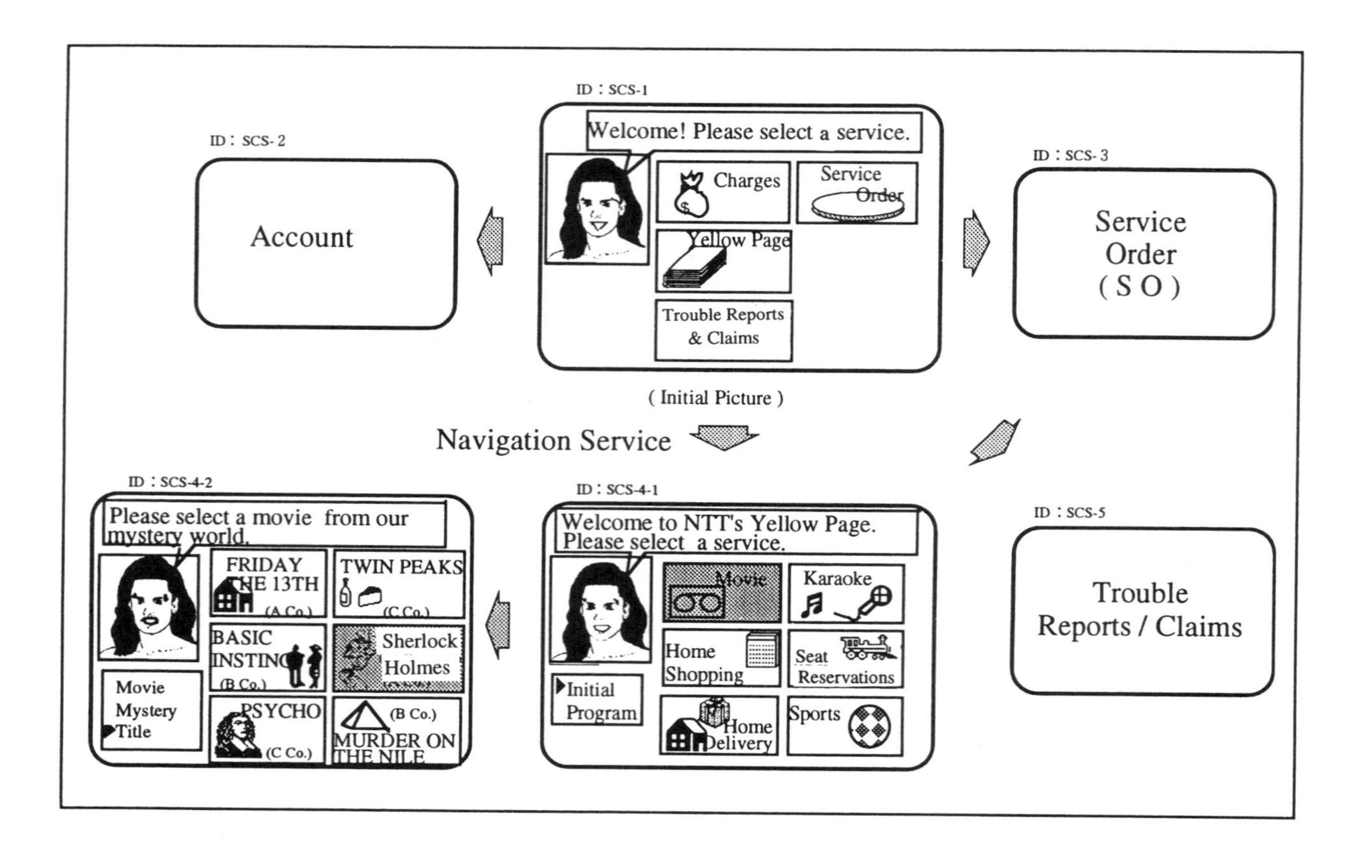

Figure 7.18 End customer screens for navigation services.

navigation function. For the end customers, service navigation functions are very important. They help end customers to select a service and access the service without wondering what SPS is supporting the service.

There are several unique functional requirements for multimedia navigation service as discussed in the following:

1. *Storage of, and access to, image data:* An image-media SPS for video services including VOD should accommodate up to 1,000 connections at a time at current levels of technology. However, several tens of thousands of connections must be accommodated at a time since it is necessary to navigate and connect a number of such SPSs on the network. Its average holding time is estimated to be only several minutes long since each connection is just to show the service content. Therefore, an SPS is needed that can handle a large number of connections with a short holding time at any one time.
2. *Automated updating of image data:* The service menus that SPSs provide have to be updated from time to time. SPSs send updated service navigation screens and introductory images whenever the menus are updated. This calls for the establishment of a mechanism that manages title images and introductory images as a managed object to avoid troublesome work of updating menus.
3. *SPS selection algorithm:* Service providers may want to deploy many SPSs that provide the same service to achieve wide-area coverage of the network throughout the country or even other countries. In this case, connections must be controlled so as to avoid traffic concentration to one SPS and to achieve uniform traffic distribution. To achieve this, an algorithm to decide destination SPSs must be provided, and service providers must be given a means to freely change these distribution rates.
4. *Function to handle SPS failures:* By using the function described in item 3, alternative SPSs can be assigned to allow for any service outage due to failures or congestion. However, when there is only one SPS providing a certain service and it fails, it will be a good customer service to prevent unsuccessful access to it and a good saving of network resources as well. For this purpose, failure of an SPS must be detected through the control system of the SPS and then display a message informing end customers of the SPS's service outage on the service navigation screen.
5. *Detecting excessive simultaneous connections to an SPS:* Frequent incompletion of call attempts due to excessive simultaneous connection

requests to a certain SPS has to be detected and controlled. The SPSs themselves cannot detect incomplete calls that occur in the network. Therefore, the connection state of each SPS must constantly be watched, and the function discussed in item 4 must be performed when the number of simultaneous connections reaches a maximum, causing frequent call incompletions.

7.4 OA&M System Technologies

7.4.1 Network-Level OA&M System Construction [35–40]

As a practical example to develop a network-level OA&M system according to TMN concepts, this section introduces the ATM Transport Networks Management and Operations System (ATMOS), which uniformly manages a nationwide ATM transport network to offer multimedia services. This system achieves its configuration, fault and performance management functions using 100 managed object classes.

7.4.1.1 ATM Transport Network and ATMOS System Overview

Figure 7.19 depicts the ATM transport network configuration composed of several NEs, such as NT, subscriber line terminal (SLT), and cross-connect system (XC), in order to provide ATM-VPs to users. This figure also shows ATMOS, which manages NE and SDH networks in addition to this ATM network.

ATMOS uses a multilayered operation system architecture. The uppermost layer comprises workstations with human–machine interface functions. The SHS handles operation scenarios such as VP provisioning management and trouble ticketing. To establish an ATM VP between two endpoints, the workstation invokes operations to the SHS, which starts to manage the PMS (path management system) or NE. Next, there is a path management layer that represents communication paths connecting multiple NEs. As each component is geographically distributed, the data communication network (DCN) interconnects all components. A combination of leased lines and LAN routers forms this DCN. The transmission control protocol/Internet protocol (TCP/IP) protocol is selected to facilitate the remote maintenance of operation programs. International Organization for Standardization (ISO) protocols, such as common management information protocol (CMIP), are mapped to this network using RFC1006. In addition, a directory service agent (DSA) is installed to provide network and presentation addresses for management association establishment between two components. When an operator needs configuration

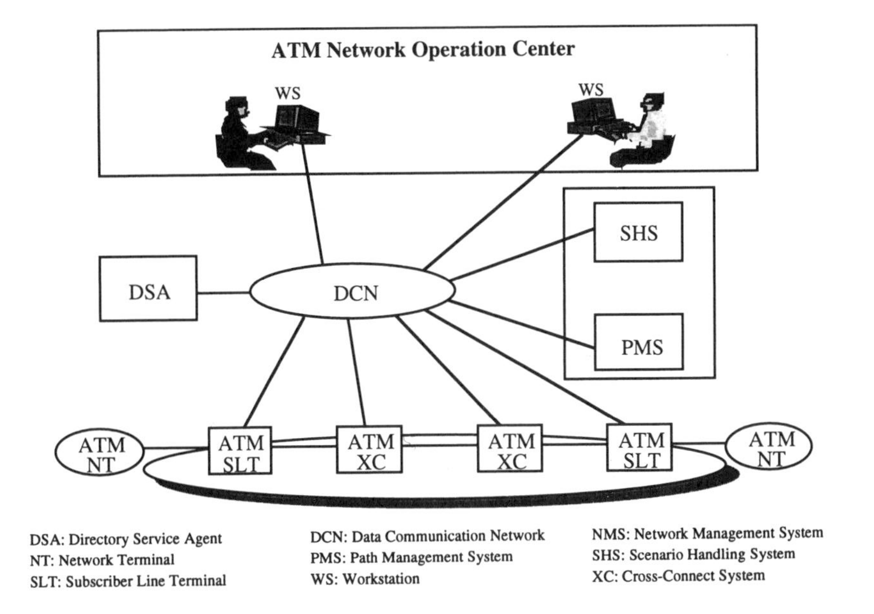

Figure 7.19 ATM transport network configuration and its OA&M system (ATMOS) configuration.

information on a certain NE, the workstation invokes operations directly to the NE after obtaining the NE's communication address from DSA.

7.4.1.2 ATMOS's Detailed OA&M Specifications

ATMOS has several management functions: NE/network provisioning as configuration management, NE/network failure reporting and testing as fault management, and NE/network performance data collection as performance management.

Since the CMIP protocol is widely adopted for subsystem interconnection, management functions in subsystems, SHS, PMS, and NE are modeled using Guidelines for the Definition of Managed Objects (GDMO) managed objects. By the platform technology mentioned in the next section, the corresponding operation program is easily produced. For transport network operations, ISO X.700 series, "Generic Network Model (M.3100)" and "Synchronous Digital Hierarchy (SDH) Management Objects (G.774)" from the ITU-T recommendations are the minimum applicable. For ATM network operation, new managed object classes are newly defined since object definitions were undergoing standardization in ITU-T and ATM Forum. MO features are as follows:

1. *MOs within NE:* SDH-related managed objects were specified as subclasses derived from G.774, while ATM-related ones were newly

defined. Because physical configurations of line cards are different from vendor to vendor, functional block MOs are newly defined in order to hide the differences between vendors.

2. *MOs within PMS:* ATM specific network MOs are newly defined. These objects are derived from the network; connection and trail defined in M.3100.
3. *MOs within SHS:* Operation scenario-related managed objects, such as construction ticket and trouble ticket, were newly defined. The construction ticket manages the operation scenarios for the control of the PMS and NEs concerned, while the trouble ticket manages alarm logs and messages to be sent to appropriate operators.

One example for construction ticket behavior is shown later. Management requests invoked by an operator through the human–machine interface are converted to some ACTIONs on a construction subsystem object in SHS. If "search-and-create-ticket" ACTION is received by a construction subsystem object, it spawns a ticket object, such as a "vpTrailConstructionTicket," if there is no corresponding ticket object. This ticket object has the ticket ID (identification) attribute, the user Label, the create date, and the event destination attribute. Event destination indicates a system to which the ticket object is to send its notification.

7.4.1.3 Use of Platform Technologies

The requirements for the OA&M system are usually changed in parallel with development activities. Accordingly, the MO specification, which is a basis of the OA&M system specification, may be changed. Specification changes affect the development period and have high costs associated with them. The development of ATMOS was effectively done utilizing the Telecommunication Network Management Systems (TNMS) engine, which translates GDMO-defined MO class definitions into OA&M system software. ATMOS was developed by a small number of people in a short period using this development method. It was confirmed that high software productivity was obtained by using platform technologies.

The upper part of Figure 7.20 shows the development method using the TNMS engine, which is composed of GDMO Translator, TNMS Kernel, and Super Manager.

GDMO Translator

The GDMO Translator analyzes MO definitions written in GDMO and generates C++ source files on TNMS Kernel. MO definitions consists of two

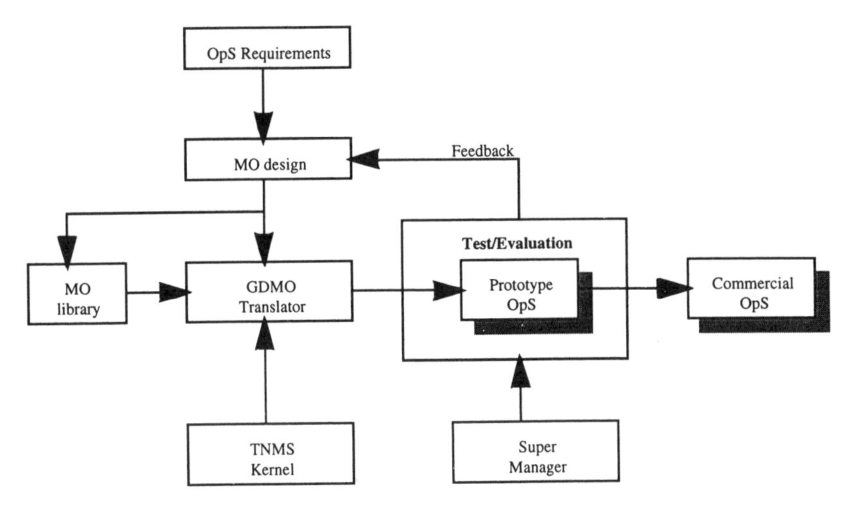

(1) OA&M System Development method

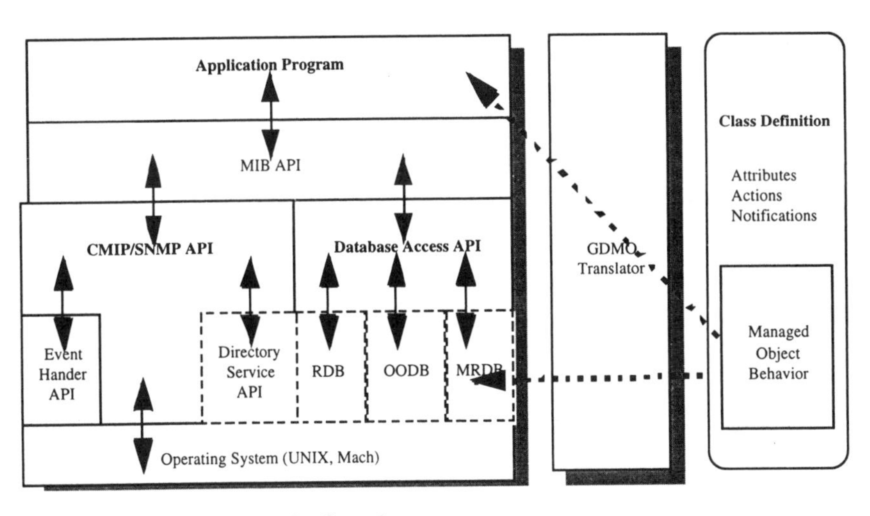

(2) TNMS Kernel Configuration

Figure 7.20 The TNMS engine.

parts. One is the formally described definition, such as class, attribute definitions, and ASN.1 (Abstract Syntax Notation 1) descriptions. The other is programs that a programmer produces in accordance with MO behavior. By using the GDMO Translator, the former is incorporated onto DB, while the latter is incorporated onto MIB application program interface (API) as application programs.

TNMS Kernel

The functional configuration of TNMS Kernel is shown in the lower part of Figure 7.20. There are two ways to access MOs. If MOs are located at another management system, MO access is executed using CMIP. On the other hand, if MOs are located at its own management system, MO access is executed via some database management systems (DBMS), such as relational DB (RDB), object-oriented DB (OODB), and memory resident DB (MRDB). Considering these system environments, TNMS Kernel has the following features:

- Database access API provides a common access method to any DBMS.
- MIB API provides a common access method to both CMIP API and Database Access API.
- Directory service API provides location-free MO definitions.
- Event handling API provides multithreaded service to concurrently execute application program.

Super Manager

Super Manager is a kind of debugging tool that provides necessary debugging commands in addition to some browsing functions, such as ASN.1 data and management information tree (MIT) structure. Since the same GDMO definitions used in the OA&M system are also used in this debugging tool, it saves many debugging costs and much time.

7.4.2 Service OA&M and OpS Construction

7.4.2.1 Requirements for Service OA&M and Functions of OpS

Four types of people and organizations as communication parties are involved in the multimedia services. They all have different perspectives on the provision and use of the services, as explained next.

- *End customer (EC):* ECs use services, such as video/movies and games, through home access equipment such as set-top boxes (STBs) or personal computers.
- *Service provider (SP):* SPs provide the services for ECs.
- *Content provider (CP):* CPs provide the contents of services, such as video source materials and game application software, for SPs. Here, CP includes an information provider (IP).
- *Network provider (NP):* NPs provide the networks to convey information communication to realize services.

They would like to do the tasks relating service OA&M as follows:

- ECs want to know how much services will cost and want to ask SPs about problems.
- SPs want to register users, to charge ECs for their usage, to solve claims and problems, to survey QoS, and to survey usage trends of services.
- CPs want to survey usage trends of contents.
- NPs want to manage the quality of network and solve network problems.

To meet the preceding requirements, the following functions of service OA&M are needed:

- End customer management for SPs and ECs;
- Content management for SPs and CPs;
- Service traffic management for SPs and CPs;
- Problem handling for SPs and NPs;
- Account management for SPs and ECs;
- Fault management for SPs and NPs.

The specific functions of the multimedia service OA&M compared to the existing communications service OA&M are service traffic management and contents management.

7.4.2.2 Implementation of the Service Traffic Management Function

Although operation information includes many details such as service usage information, customer information, and network fault information, service usage information is the most important for multimedia service OA&M because the information regarding each EC's usage patterns is useful for not only a service surveillance but also for SP and CP business support. It is important that service usage information can be monitored in a real-time manner by SPs or CPs on demand.

The important issues of service usage information are how to gather and process the information, and what viewpoint (management view) should be used when presenting it to the providers.

Utilizing the most effective management views yields efficient service operation, smooth service provision, and good business support. Management views are:

- *EC view:* Extract information according to the individual EC to know his or her own service usage and charging information.
- *SP view:* Extract information according to the individual services to know the popular services and service status, or to individual CPs to choose CPs offering attractive contents.
- *CP view:* Extract information according to the individual content title to incorporate this information into their content creation plan, or to provide individual service to choose attractive SPs and services.
- *NP view:* Extract information according to (1) the individual EC to manage the access networks of ECs, (2) to the individual SP to survey the access networks of SPs or (3) to the individual NP to survey the network, handle network service faults, discern network bottlenecks, and finally reflect this information in the network construction plan using the NP's network OA&M.

To realize management from the preceding viewpoints, a new service traffic generation method needs to be introduced. The "call and traffic" model of existing network services can be extended to provide a consistent model of service usage information and service management views. The new "service call and service traffic" generation model created is illustrated in Figure 7.21. "Service call" corresponds to service usage information, and "service traffic" corresponds to the service management views in this figure.

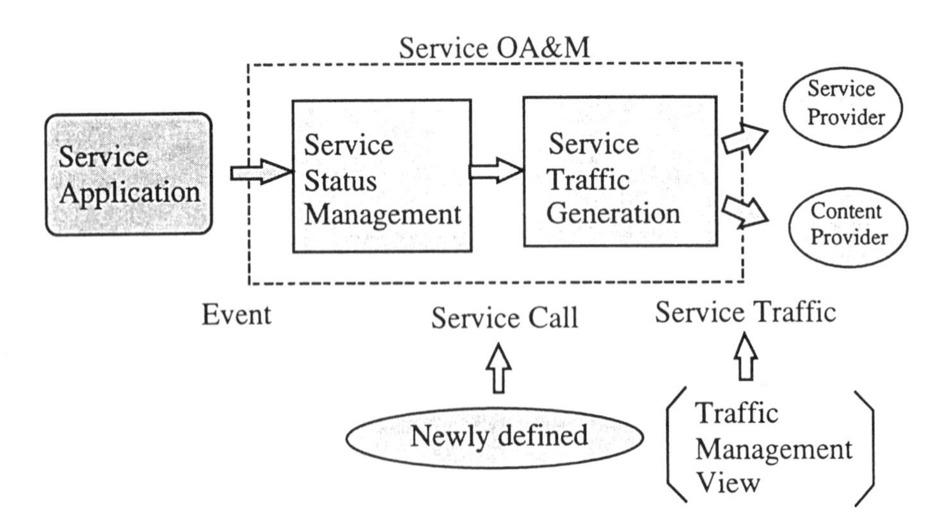

Figure 7.21 Service call/traffic generation model.

Here the service calls for multimedia communication that are newly defined are shown in Figure 7.22.

It is assumed that the EC uses the service from time port *ts* to *te*.

In the case of the telephone service, we have defined this period from *ts* to *te* of service use duration a *call*. In the case of the multimedia services, at first the EC selects a favorite movie he or she wants to watch in the navigation phase. This is defined as a *navigation call*. Then the EC rents the movie from an SP and watches it. During the rental phase the EC can play, pause, and stop the movie at any time and as many times as desired. Each "play" is defined as a *play call*, and a rental phase is defined as a *rental call*.

7.4.2.3 Linkage Between Service OA&M and Network OA&M

As described earlier, as one of the OA&M functions, a fault management function is provided for SP and NP. The SP and NP each has a fault management view. It may be enough for a SP to know that a faulty point should be distinguished at the level of SCS, network, service server, or community antenna television (CATV) server. On the other hand, more detailed information related to each piece of equipment may be necessary for NPs. Therefore, an effective link between service-level OA&M and network-level OA&M is needed, for which the layered management architecture shown in Figure 7.23 should be applied. This figure illustrates the configuration of the service OA&M and network OA&M link, where the service OA&M and network OA&M are layered.

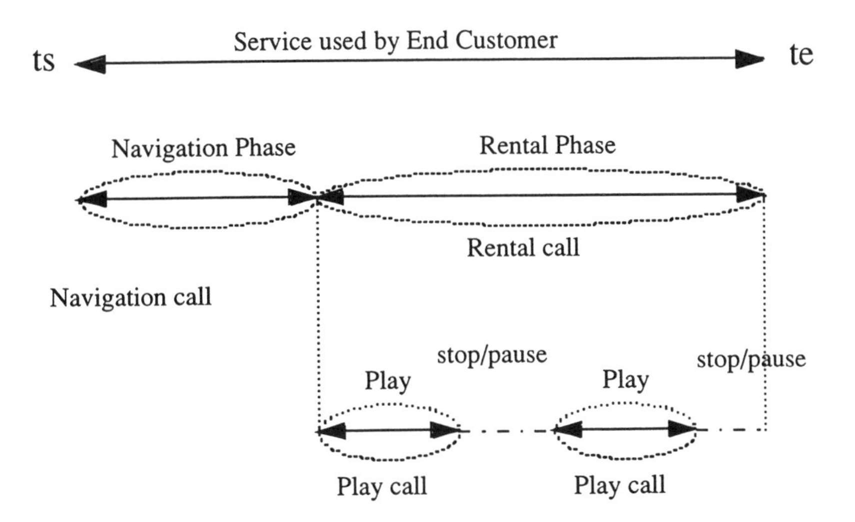

Figure 7.22 New definitions of multimedia service calls.

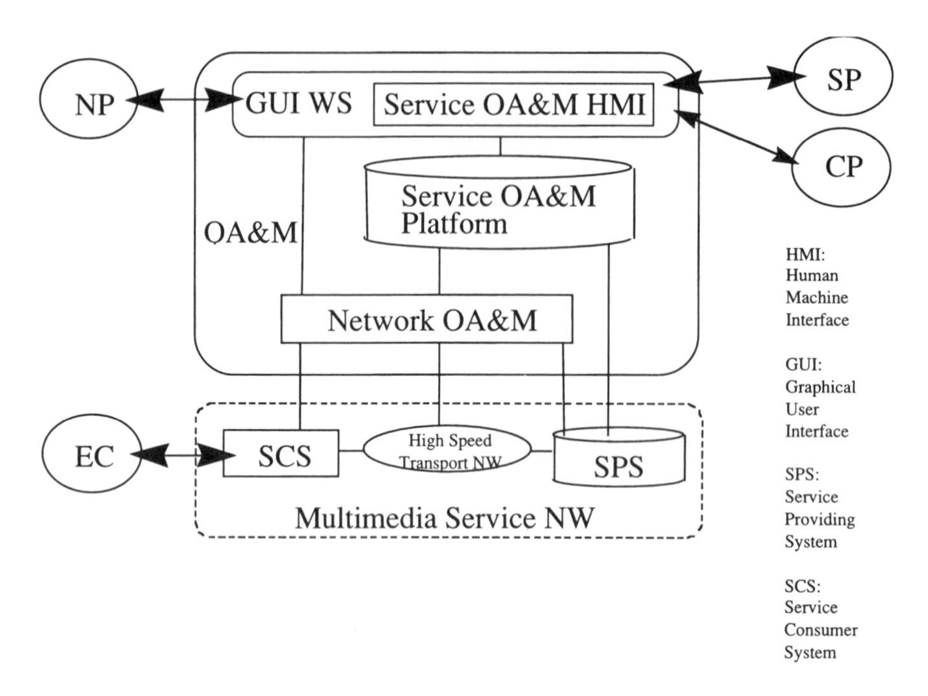

Figure 7.23 Layered management structure of service and network OA&M.

The multimedia service network includes SCSs, service servers, and high-speed transport networks. An EC uses services only through his or her SCS.

Network OA&M monitors the status of the multimedia service network and reports the alarms of the network. The service OA&M platform collects the information concerning usage of services and billing from the service servers to generate traffic data and charging data and then provides them to a human-machine interface (HMI) at the service OA&M center.

SPs and CPs use the service OA&M through the network OA&M HMI in the graphical user interface-workstation (GUI-WS). NPs use the network OA&M through the GUI-WS.

7.4.3 Configuration of Service Navigation System

7.4.3.1 The Navigation Process

Navigation is a process of reaching a service by means of making successive choices, for example, selection of a service category, a service provider, and an offer within a particular service. A navigation process consists of two cascaded phases: the service selection phase and the service brokering phase.

The service selection phase is a process by which a user chooses a target service by referring to the catalogs offered by a navigation provider. The service brokering phase is a process by which a session is established between the user and the service/service provider, which is selected by the service selection phase. In actual navigation, these processes may be spirally chained; however, for simplicity, it is assumed that a navigation process has just two phases.

In this section, a provider or such an object offering a navigation service is called a navigation provider (NP) to distinguish it from a service provider offering a target service for users.

7.4.3.2 Meta-Information for Navigation

Meta-information of services is necessary for NPs to offer a catalog that is used for users to browse and select services in a service selection phase. Meta-information is knowledge with a target service attribute or characteristics. It is important to separate the knowledge from both navigation scenarios and content used in a catalog.

This policy ensures that meta-information is reused by various navigation applications and maintains the knowledge itself. This is an example of a model–control–view separation design policy, which is popular in the GUI design area. Control and view are, however, closely coupled in some situations. If a walk-around style navigation scenario for a shopping catalog is taken, it might need content that represents the position and direction in a virtual shopping mall to show users where they are and where they can go.

In the case of a popular hierarchical menu style scenario, however, another content might be needed.

Meta-information for navigation is categorized as follows:

Service-Meta Information

- *Service category:* Information for grouping services;
- *Service description:* Information on what the service is;
- *SP description:* Information on who provides the service;
- *Service schedule:* Information on when the service is up;
- *Session profile:* Information on necessary resources/features to establish a session;
- *Service access point (SAP):* Information on the network address of the SP offering the service.

Catalog Meta-Information

- *Content for NP:* Content information used in a catalog such as video, photo, text, etc.

Protocol Meta-Information

- *Query protocol:* Information on protocols used in a service selection phase between NP and SP/NP, and between NP and SCS;
- *Operation protocol:* Information on protocols used in registration/maintenance of the meta-information.

Usage Data

- Information on the navigation activity, such as service selection frequency and popularity, which is collected by NP, used in catalogs and submitted to outer systems

7.4.3.3 Configuration of Navigation System

Figure 7.24 depicts a flow of the meta-information described earlier, where the navigation repository is the container of the meta-information. In this figure, there are three processes: navigation, catalog generation, and usage data analysis. These are independent of each other, so that a navigation system is divided into at least three subsystems or servers. Figure 7.25 shows the sample architecture of the navigation server. The navigation server consists of the session manager, session, stream manager, and utilities.

The session manager creates sessions, monitors activity, and submits usage data to the navigation repository and usage data analysis process.

The session is an instance of the object offering a service. There are two types of sessions from the navigation viewpoint. One is a navigation session (N-session), which offers the two phased-navigation mentioned earlier. The other is a target-service session (S-session), which is established in a service

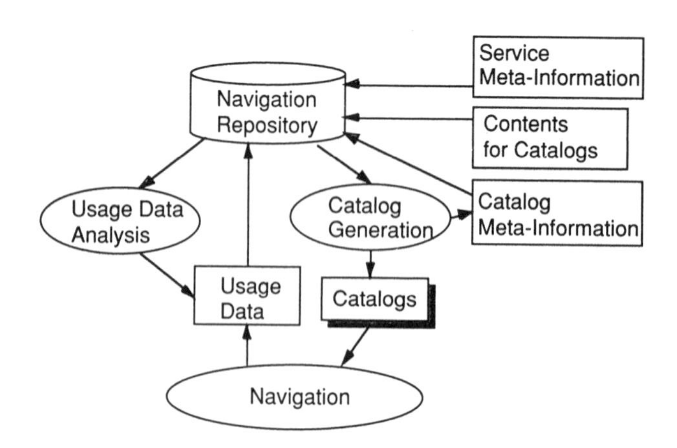

Figure 7.24 Navigation information flow.

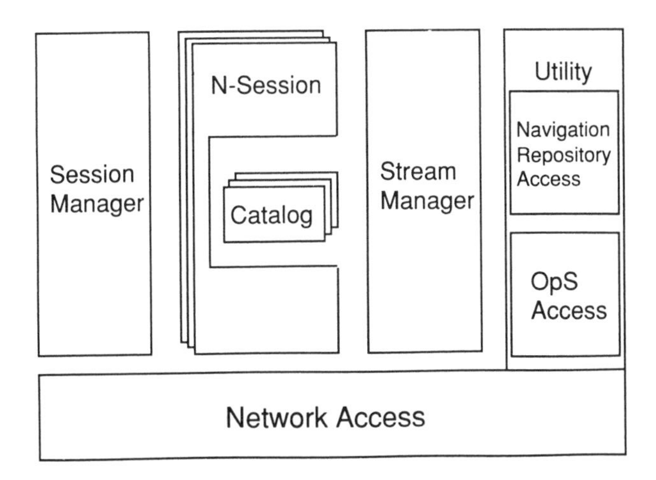

Figure 7.25 Navigation server configuration.

brokering phase by NP and is provided by SP. As is understood, session manager and N-session are the heart of the navigation system.

The stream manager deals with video stream transfer, video play control, and downloading of application-specific data/programs.

Utilities are support functions to access meta-information and content in the navigation repository, and to cooperate with service OA&M systems (outer systems).

7.5 Standards-Related Activities

7.5.1 Telecommunications Management Network

7.5.1.1 Introduction

The concept of TMN was proposed in 1986 within the ITU-T. The original purpose of the TMN was the management of ISDN. The standards for management networks, both human's and machine's, are essential to exchange management information. For a human's management network, standards have been produced within ITU-T. The standardization of the machine's network is a new effort in the standards arena. The study of the TMN and its standards will be one of the most important in the forthcoming multimedia communications era of the 21st century.

ITU-T recommendations do not actually offer everything necessary for the implementation of the TMN. TMN recommendations are the base for various implementations over the world. Therefore, they are called *base stan-*

dards. In addition to these base standards, a number of local issues need to be determined as well as the technologies to be used for implementation. All of these aspects have to be shared among the TMN developers and the users.

It might be difficult to say which areas the ITU-T should have responsibility for in these standardization activities. In reality, ITU-T recommendations do not cover all aspects for implementation. Therefore, additional activities are necessary to develop specifications required for the TMN implementation. NMF, described in Section 7.5.2 is the typical example of forums/consortia to complement the ITU-T TMN.

This section describes the TMN overview and future trends.

7.5.1.2 Principles for TMN [25]

There are three aspects to the principles for TMN. They can be considered separately when planning and designing a TMN:

1. TMN functional architecture;
2. TMN information architecture;
3. TMN physical architecture.

The functional architecture describes the appropriate distributions of functionality within TMN to allow for the creation of functional blocks from which a TMN of any complexity can be implemented.

The information architecture, based on an object-oriented approach, gives the rationale for the application of Open Systems Interconnection (OSI) systems management principles to the TMN principles.

The physical architecture describes realizable interfaces and examples of physical components that make up TMN.

TMN Functional Architecture

The TMN functional architecture is based on a number of TMN functional blocks, as shown in Figure 7.26. The functional blocks provide the TMN with functions that enable it to perform the TMN management services. A data communication function (DCF) is used for the transfer of information between the TMN functional blocks. Pairs of TMN functional blocks communicating with each other are separated by reference points. The TMN functional blocks are listed below.

1. *Operation Systems Functional (OSF) block:* The OSF block processes information related to telecommunications management for the purpose of monitoring and/or controlling telecommunications functions.

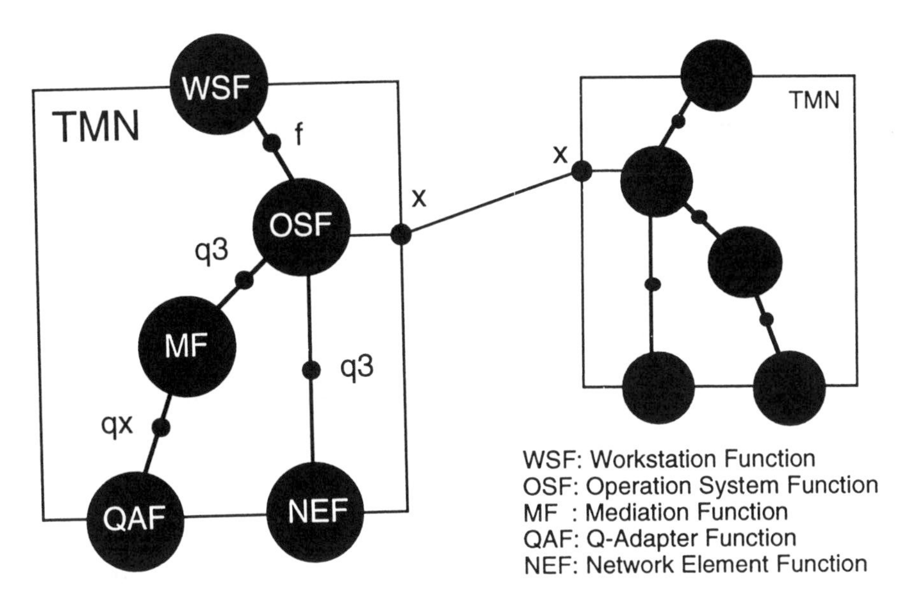

Figure 7.26 TMN functional architecture [7–25].

2. *Network Element Functional (NEF) block:* The NEF block communicates with TMN for the purpose of being monitored and/or controlled. The NEF block provides the telecommunications and support functions required by the telecommunications network being managed.
3. *Workstation Functional (WSF) block:* The WSF block provides the means to interpret TMN information for the human user, and vice versa.
4. *Mediation Functional (MF) block:* The MF block acts on information passing between an OSF and NEF (or QAF described later) to ensure that the information conforms to the expectations of the function blocks attached to the MF. Mediation function blocks may store, adapt, filter, threshold, and condense information.
5. *Q-Adapter Functional (QAF) block:* The QAF block is used to connect non-TMN entities. The responsibility of QAF is to translate between a TMN reference point and a non-TMN (e.g., proprietary) reference point.

TMN Information Architecture

The TMN information architecture describes an object-oriented approach for transaction-oriented information exchanges. Other approaches may be required but have not yet been determined. This architecture mainly intends to use the OSI systems management standards.

The TMN methodology makes use of OSI Systems management principles and is based on an object-oriented paradigm. Within TMN, management systems exchange information modeled in terms of managed objects (MOs). Managed objects are conceptual views of the resources that are being managed. Thus, a managed object is the abstraction of such a resource that represents its properties as seen by management. A managed object may also represent a relationship between resources or a combination of resources.

The manager/agent concept is imported from the OSI systems management concept in Recommendation X.701[41]. Within TMN, the management processes will take on one of two possible roles, manager role or agent role, for a specific management association. The manager role is the part of the distributed application that issues management directives and receives notifications. The agent role is the part of the application process that manages managed objects. The role of the agent is to respond to directives issued by a manager. It will also reflect to the manager a view of these objects and emit notifications reflecting the behavior of these objects. A manager is the part of the distributed application for which a particular exchange of information has taken the manager role. Similarly, an agent is the part that has taken the agent role.

To interwork, communicating management systems must share a common view or understanding of at least the following information: supported protocol capabilities, supported management functions, supported managed object classes, available MO instances, authorized capabilities, and containment relationship between objects. These information pieces are defined as *shared management knowledge.*

TMN Physical Architecture [25]

TMN functions can be implemented in a variety of physical configurations. The TMN physical architecture consists of several TMN building blocks as shown in Figure 7.27, according to the set of functional blocks that it is allowed to contain. They are discussed in the following list.

- *Operation systems (OS):* A system that performs OSFs. The OS may optionally provide MFs, QAFs, and WSFs.
- *Mediation device (MD):* A device that performs MFs. The MD may also optionally provide OSFs, QAFs, and WSFs. MD can be implemented as hierarchies of cascaded devices.
- *Q-adapter (QA):* A device that connects NE-likes or OS-likes with non-TMN compatible interfaces to Qx or Q3 interfaces.
- *Data communication network (DCN):* A communications network within a TMN that supports the DCF. The DCN represents an

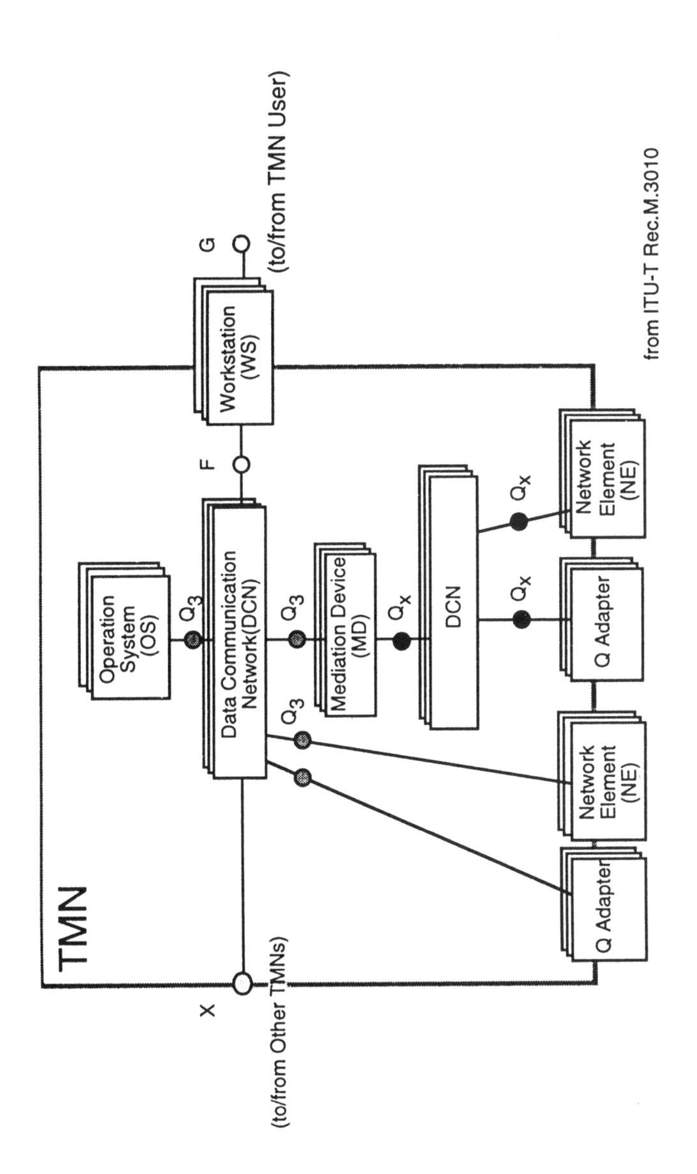

Figure 7.27 TMN physical architecture [7–25].

implementation of OSI layers 1 to 3 and provides no functionality at layers 4 to 7. The DCN consists of a number of individual interconnected subnetworks of different types. For example, the DCN may have a backbone subnetwork that provides TMN-wide connectivity between a variety of subnetworks providing local access to the DCN.

- *Network element (NE):* Comprised of telecommunication equipment and support equipment considered to belong to the telecommunications environment that performs NEFs. The NE may optionally contain any of the other TMN functional blocks according to its implementation requirements.
- *Workstation (WS):* A system that performs WSFs. The workstation function translates information to a displayable format and vice versa.
- *Interoperable interfaces:* If two or more TMN building blocks are to exchange information they must be connected by a communication path and each element must support the same interface on that communication path. The interoperable interface defines the protocol suite and the messages carried by the protocol. Transaction-oriented interoperable interfaces are based on an object-oriented view of the communication and therefore, all the messages carried deal with object manipulations.

TMN standards interfaces (Q3, Qx, X, and F) are defined to interconnect TMN building blocks as also shown in Figure 7.27.

7.5.1.3 OSI Systems Management

As previously stated, OSI systems management technology was selected as the basis for the TMN interface specification. Although the OSI systems management standards are not intrinsically involved in TMN, the concepts of OSI systems management have become so intimately associated with TMN that it is impossible to understand TMN without understanding OSI systems management. This section describes OSI systems management from the perspective of TMN to help readers understand the TMN interface specification easily.

OSI systems management standards are being developed under the cooperation of International Standards Organization (ISO) and ITU. Although OSI systems management and TMN have evolved almost in parallel, they have quite different features. The original purpose of the development of OSI systems management was the management of computer/data networks. In reality, their evolution has been considerably influenced by the TMN requirements.

The following subsections review OSI systems management concepts, presents the organization of the TMN interface specifications, and discusses the benefits of using the OSI systems management as the basis for the specification of the TMN interfaces.

OSI Systems Management Overview [41]

Figure 7.28 illustrates the key concepts of OSI systems management. An MO is the conceptual view of the real target (physical or logical) of management. The real targets of management are network resources that need to be monitored and/or controlled to achieve management objectives, for example, removal of failure or performance degradation in the network. Real resources such as line card (physical) or circuits (logical) can commonly be represented by an MO. MOs with same properties are instances of an MO class. A line card could be an example of an MO class. It groups all instances of line cards that implement circuits. The management information base (MIB) is the conceptual repository of the MOs instances. An MO class is defined by attributes, the management operations, the behavior, and the notifications.

The essential part of OSI systems management is the definition of two roles, the manager and the agent roles. The manager is the specific entity in the managing system that exerts the control, the coordination, and the monitoring. It issues the requests to perform operations against the agent. It also receives the notifications emitted by the MOs and sent by the agent. The agent is the specific entity in the managed system to which the control, the coordination, and the monitoring are directed. It receives and executes the requests sent by the manager, and sends the notifications to the manager.

The manager and agent may communicate using the seven-layer OSI protocol suite. A key element for the suite is the common management information service element (CMISE), which is one of the building blocks used in the application layer. CMISE consists of a service definition, the common

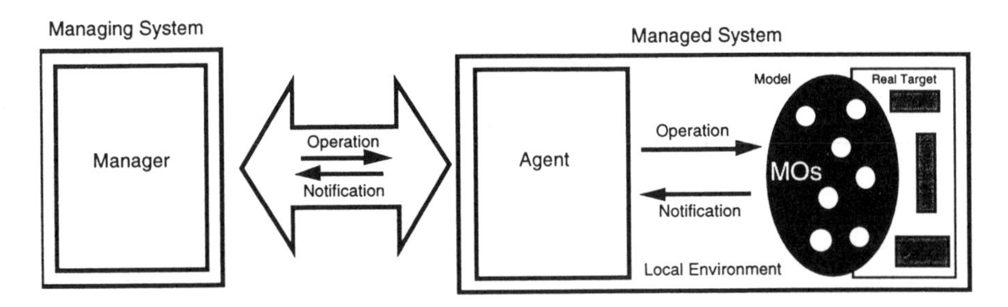

Figure 7.28 OSI systems management [7–41].

service information service (CMIS); and a protocol specification, the common management information protocol (CMIP).

The agent receiving the message is responsible for carrying out request(s). It maps the request(s) on the MO(s) into request(s) on real resources. The mechanisms used for the mapping are implementation specific and not subject to standardization. It is understood that the OSI concept gives one of the concrete examples of TMN information architecture. Using this concept, the resources are modeled so that the manager and agent have a common view. All messages exchanged between manager and agent have a basic form of either requesting something of one or more objects or an object informing another system of some event. The request may be as simple as returning the value of a parameter, or as complicated as asking the NE to reconfigure itself.

Management Information Modeling

The specification of object-oriented information is called *information modeling.* The majority of the effort spent in defining TMN interfaces has gone into the development of these information models. OSI systems standards gives tools for the modeling. The development of the model itself is the task to be done by the TMN experts.

TMN standards intend to develop two kinds of interface standards: generic and technology-specific management information model. Recommendation M.3100 [22] contains generic network management information models, which are generic enough to describe information exchanged across all TMN interfaces, independent of the telecommunications technology. The objects specified in the technology-specific standards are often imported from the generic standards or are subclass of generic objects.

Inheritance (also called subclassing) is the procedure of specifying the new object class based on a previously defined object class. Thus, the new object class has all the characteristics of the base object class (superclass) with some new characteristics. Generic- and technology-specific modeling is the policy to ensure some level of similarity between different technology-specific information models.

Allomorphism is a capability that can be used to manage the telecommunications technologies in a generic manner. It is the procedure of specifying a subclass that masquerades as a superclass. One use of this is to allow a technology-specific object to be treated as a more generic object. Thus, a technology-specific object can be managed as a generic object.

The disadvantage of the approach described is that technology-specific management capabilities are inaccessible. A related use of allomorphism is to provide a generic set of management capabilities in certain situations, while providing vendor-specific enhancements in other situations. More experience in the use of TMN standards will be needed to determine if allomorphism is truly a useful concept.

The coordination will be as important as the mechanisms in the development of information models for providing the generality and consistency desirable in TMN interfaces.

Benefits of the Use of OSI

After discussion among the TMN experts, the application of OSI systems management standards was decided. The following advantages have been considered for the decision:

- OSI protocols are the standards that have been accepted as real international standards among ITU as well as ISO members.
- Object-oriented modeling technologies have already been developed to be used as an appropriate tool within TMN interface specifications.
- Important tools are also becoming available such as naming rules and a data-specification language (ASN.1).

Although implementation technology was not mature enough to introduce the TMN interface specification in all kinds of environments, benefits will gradually be recognized to facilitate its introduction.

7.5.1.4 Implementation of TMN [42–46]

No example exists for the complete implementation of the TMN interface specification. This does not mean that there has been no effort to implement TMN. First, two examples described in the following subsections explain why the complete implementation of TMN cannot be achieved at one time. We also point out that ITU-T recommendations do not give every kind of information necessary for the implementation. Second, an attempt to develop a complete set of specifications for TMN implementation is introduced.

Examples of TMN Implementation

The following are collected from the experiences of one SP in the development of management systems for its telecommunications network.

Proprietary Qx: One SP developed specifications for the management interface between the mediation system and telecommunications equipment. According to the TMN principles, it could be called the *proprietary Qx interface specification.* The development was completed in the pre-TMN era, when there were no TMN recommendations for Qx interface specifications. Similar cases were found over the world in the same period of time, pre-TMN era.

Figure 7.29 shows the message example of that specification. The management network architecture was very close to the TMN architecture. Therefore, the requirements were almost the same as far as the protocols were concerned. When it comes to the information model, it was not really a model. It was a real, not abstract, view of the managed resources. This kind of model can be used only in one SP, and could easily become obsolete as a result of changes in equipment.

At the planning stage of the system, there was no strong demand for generic information modeling. Equipment in a real environment was manufactured following the specification of the SP. Only a standardized protocol was required to construct a management network that was common among various types of equipment.

This proprietary Qx follows a TMN-like architecture, uses proprietary protocols and messages, and has been successfully used for a decade. This fact explains that the TMN architecture or concept itself is worth applying to the management environment. However, multivendor environment and object-oriented information modeling will be essential.

Proprietary Q3: Figure 7.30 shows an example of the so-called *proprietary Q3* scenario developed for SDH management by the same SP. In this case, the specification followed the TMN recommendations: TMN architecture, principles, protocol specifications, modeling procedure based on OSI systems management, etc. It might still have to be called proprietary because interoperability with management systems that were developed independently could not be ensured. Besides, this was inevitable because the management information

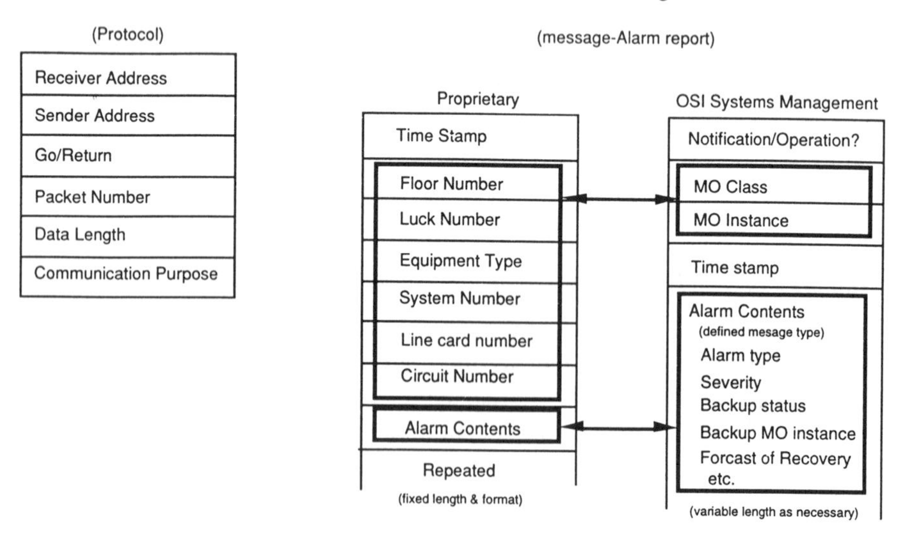

Figure 7.29 Example of proprietary Qx.

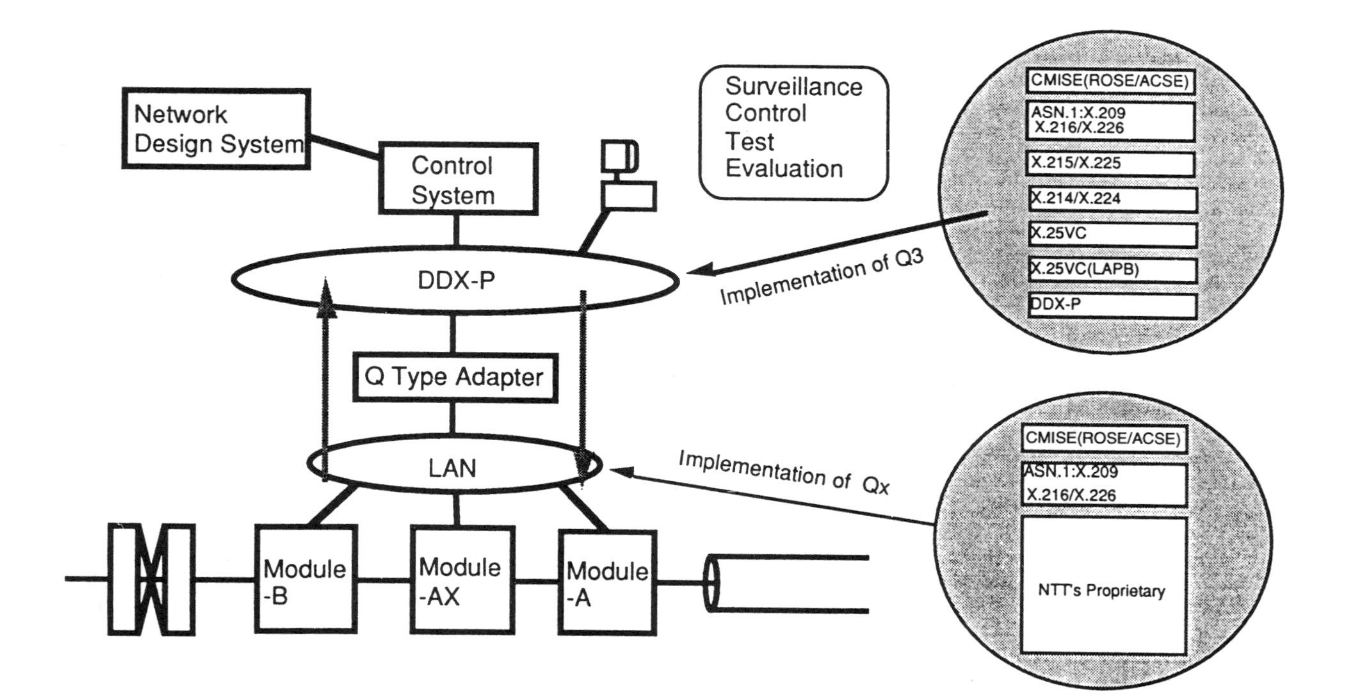

Figure 7.30 Example of proprietary Q3 for SDH based on TMN.

model in the recommendation had not been mature enough for unified implementation.

This example tells us that management information modeling is the most important for the future of TMN. Also, the TMN recommendations do not give all specifications necessary for implementation such as the decision of optional parameters for protocol implementation.

Set of Specifications for Implementation

From the preceding two examples, we can see that further specifications were required to implement TMN. Figure 7.31 illustrates the analogy of the situation.

Recommendations indicate fundamental, generic, or common aspects of TMN specifications. These aspects can be recognized as a mandate for TMN. In the implementation of TMN, local matters as well as implementation technologies will not necessarily be common among all of the TMN users of the world. Specifications for these implementation issues could change more rapidly than the fundamental issues. This means that another development mechanism would be very necessary for TMN implementation specification. Current forums/consortia type activities give one solution for the requirements.

7.5.1.5 Future Trends [47]

Figure 7.32 shows a summary of the TMN history within the ITU-T. The study of O&M interfaces for intelligent transmission terminals triggered the TMN study whose original intention was the specification of protocols for the O&M interface. During the course of the study, experts pointed out the

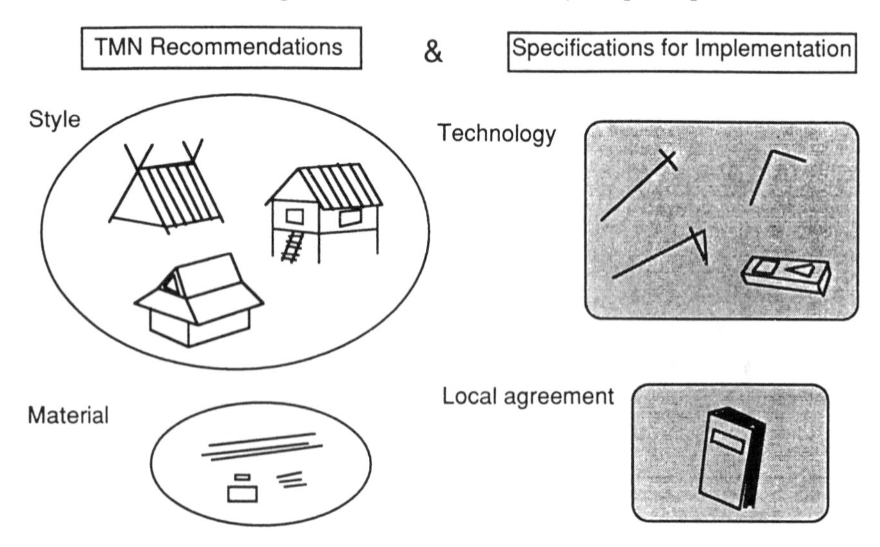

Figure 7.31. Analogy of TMN implementation.

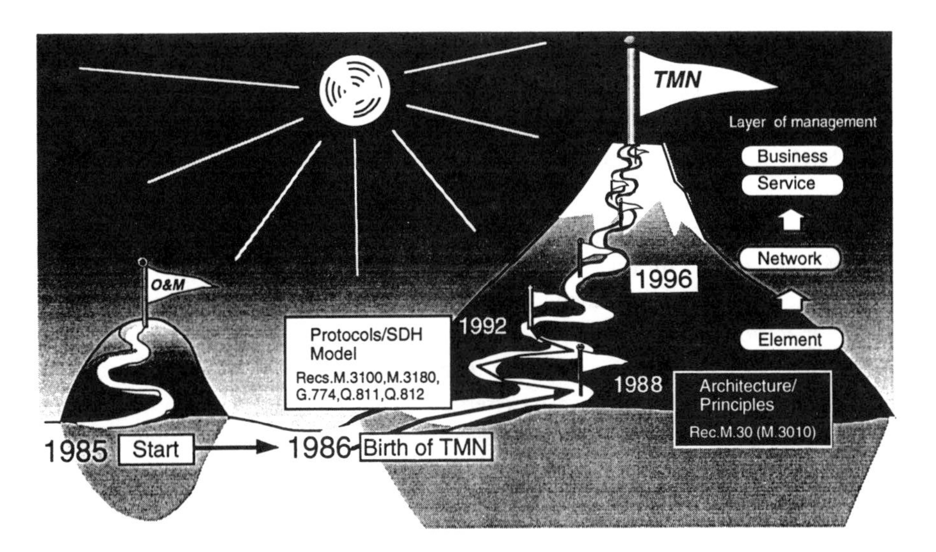

Figure 7.32 Future of TMN.

necessity for a more fundamental investigation for the management of telecommunications.

The O&M interface specification looked rather simple, that is, like a rather low mountain or hill to be attacked as shown in Figure 7.32. The original goal was to fulfill the requirement for a part of telecommunications management, and the final destination of the TMN development was to achieve the full integration of the management activities including human and machine.

To climb up the high mountain of TMN, it is very necessary to develop a sufficient plan or strategy. The plan has actually been set up and carried out within the TMN experts group. As shown in Figure 7.32, the first step was the TMN architecture, the second one was the protocol specifications, and so on. However, precise steps for the future are not necessarily clear enough. Significant activities are still required to list requirements for future TMN and to prioritize them. Then, the development of technologies for the implementation of the TMN will be essential.

The precise future plan needs to be finalized to be shared among TMN users as well as suppliers. No one can climb up the mountain of TMN at once. A step-by-step approach is the only way to achieve the TMN objectives.

7.5.2 NMF

NMF is a nonprofit consortium, funded entirely by its members—almost 200 of the world's leading telecommunications service providers, suppliers, and users. Its members identify the service and network management problems that are most critical to solve—the problems that no company can afford to solve alone. The service management for the multimedia services are within its scope to be studied in cooperation with other relevant organizations.

NMF has developed a reputation for delivering business-focused specifications that are easy to work with, yet detailed enough to serve as the function for products that can really integrate, automate, and interoperate across business. At NMF, they focus on the business-critical tasks that matter to their members. They issued the Service Management Business Process Model, on which they deliver specific solutions called NMF solution sets and NMF component sets.

NMF solution sets are packages of standards and specifications that meet a specific business need. They are extremely precise, in that they do not permit options in implementation. For this reason, when any two management systems implement a solution set, they should interoperate. Solution sets can be applied either horizontally and vertically, as shown in Figure 7.33.

NMF component sets specify "pieces" of the management infrastructure that enables the many solution sets to be implemented. They are the building blocks to be used in constructing management systems. Component sets will be of most interest to developers of management systems who wish to buy underlying platforms and tools that will interoperate with other suppliers' products.

NMF had issued six solution sets and five component sets as of October 1995 followed by some other additional solution sets and component sets.

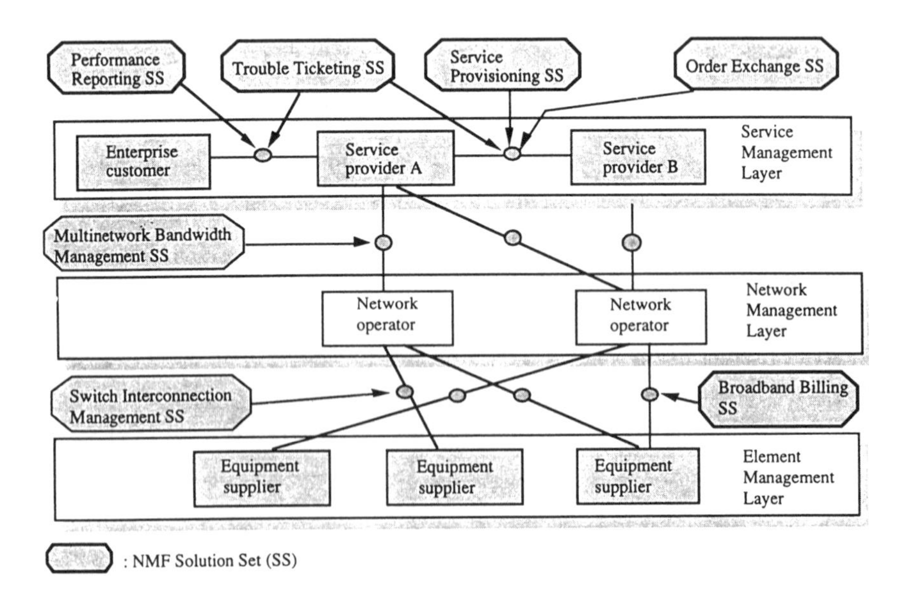

Figure 7.33 Examples of NMF solutions sets.

Requirements for computing systems used for SPs are specified as SPIRIT (Service Provider Integrated Requirements for Information Technology) specifications.

NMF solutions are comprehensive, encompassing not only processes and systems, but the underlying infrastructure as well. NMF's work programs are organized into service management, network management, and platforms and technology to enable them to focus talented resources on specific areas while maintaining an end-to-end view of the problem to be solved.

Each of the programs is focused on critical, competitive issues such as the following:

- Improving customer care systems and processes;
- Making multivendor operations a reality;
- Delivering scalable distributed computing;
- Smoothly integrating legacy systems;
- Making TMN practical and cost effective.

NMF has focused on the tasks of the communication SPs, which naturally comprise any type of new SPs, such as multimedia SPs, Internet SPs, and cable SPs, in the multimedia era. Solution sets and component sets for multimedia service OA&M need to be studied.

NMF provides some useful information on its WWW server: http://www.nmf.org/.

7.5.3 ATMF

ATM is designed to carry multiple types of traffic simultaneously—voice, data, and video. It addresses today's needs now, while enabling completely new applications that will emerge in the future. ATM is a universal technology, applicable from inside workstations, through LANs, and both private and public WANs. The ultimate deployment of one technology, ATM, will allow seamless integration of networks, with a reduced support and operational cost.

The ATM Forum (ATMF) is an international nonprofit organization formed with the objective of accelerating the use of ATM products and services through a rapid convergence of interoperability specifications. In addition, the ATMF promotes industry cooperation and awareness. Among many subworking groups (SWG) in the Technical Committee, Network Management SWG, which is a focal point of network management activity under the umbrella of ATMF. The goal of ATMF is to develop requirements and interface specifications for network management information flows between management systems and the network, and between management systems across administrative boundaries such as private-to-public and public-to-public networks.

To enable a consistent network management framework in an ATM era, an ATM network management model has been defined as shown in Figure 7.34. In this figure, five management interfaces are specifically defined as follows:

1. *M(1) interface:* A management interface between an end user and private network management system (NMS);
2. *M(2) interface:* A management interface between a private ATM switch and private NMS;
3. *M(3) interface:* A customer network management (CNM) interface between a private NMS and local exchange carrier (LEC) NMS;
4. *M(4) interface:* A management interface between a LEC ATM switch and LEC NMS;
5. *M(5) interface:* A management interface between a LEC NMS and an international carrier (IC) NMS.

In addition to these management interfaces, the private user network interface (UNI) and public UNI are defined as communication interfaces.

Major work areas are finished as follows:

- Requirements and interface specification (M3) for management information flow between public network and private network;
- Requirements (M4) for management information flow between a public NMS and the public network; and

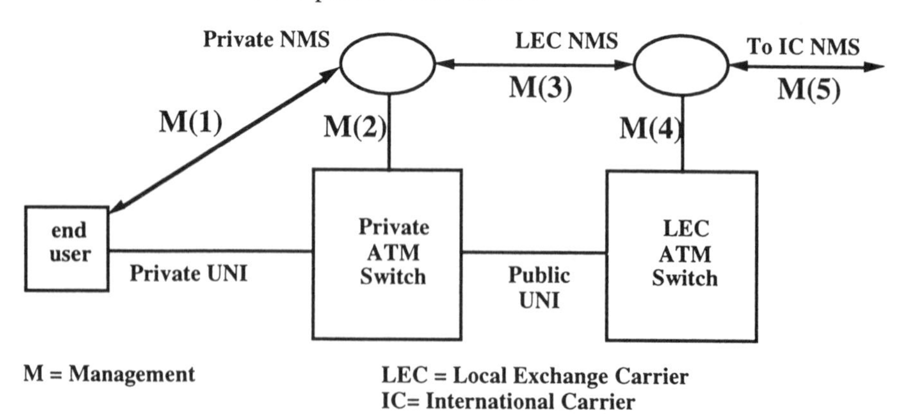

Figure 7.34 ATM network management model.

- MIB specification for management information flow between a public NMS and the public network.

Information on ATMF and its activities can be found at http://www.atmforum.com.

7.5.4 TINA and OMG

7.5.4.1 Introduction

To enable flexible, easy, and fast software deployment in the multimedia era, an emerging IT-based computing and networking architecture that has been studied by the Telecommunication Information Networking Architecture-Consortium (TINA-C) and the Object Management Group (OMG) becomes more important. OMG provides a wide range of distributed processing environment (DPE) products based on its Common Object Request Broker Architecture (CORBA) specifications. TINA provides a consistent information networking architecture based on DPE technologies supported by OMG.

7.5.4.2 TINA

TINA is studied by TINA-C, which is an international collaboration aimed at defining and validating an open architecture for telecommunications systems for the broadband, multimedia, and information era.

TINA Architecture

TINA architecture defines concepts and principles of the structuring of software, and for the constrains that should be applied during the specification, implementation, execution, and operation of software.

A structure of telecommunications software is defined as shown in Figure 7.35. At the bottom are the hardware resources, such as processors, memory, and communication devices. Above this is a software layer that contains the operating systems, communications, and other support software found in computing systems. This layer is called native computing and communications environment (NCCE). Above this is the DPE layer, followed by the telecommunications applications layer.

TINA Applications Model

Management layering, as defined by TMN, consists of NEs, an element management layer (EML), network management layer (NML), service management layer (SML), and business management layer (BML). The TINA architecture defines three layers with service, resources, and element. The management

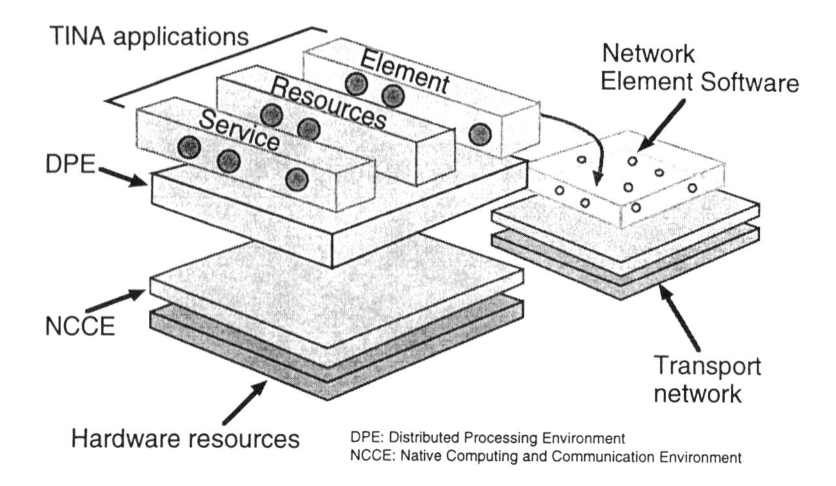

Figure 7.35 TINA reference model.

layering principles are used to partition the software in the application layer as follows:

- *Element layer:* The element layer is populated by objects that represent atomic units of physical or logical resources, defined for allocation control, usage, and management purposes. Objects in the element layer are called *elements.* The element is responsible for communication with the actual device (network element software), which may follow proprietary or standards based protocols and interfaces.
- *Resources layer:* The resources layer contains objects that maintain views and manipulate collections of elements and their relationships. It also provides the service layer with abstract representation of elements. From a TMN perspective, the resources layer corresponds to the EML and the NML.
- *Service layer:* The service layer is populated by objects involved in the provision of the services to stakeholders. Objects in this layer are either specific to a given service or are service independent. The first embodies logic, data, and management capabilities specific to a service. The second provides generic service access, control and management capabilities.

Information on TINA-C and its activities can be found at http//:www. tinac.com.

7.5.4.3 OMG

Object management architecture (OMA) is a center of all the activity undertaken by OMG. OMG was formed to help reduce the complexity, lower the costs, and accelerate the introduction of new software applications.

OMA Reference Model

The OMA reference model identifies and characterizes the components, interfaces, and protocols. The OMA can be viewed as two major segments consisting of four critical components, as shown in Figure 7.36:

1. *Application oriented:* The OMA characterizes interfaces and common facilities as solution-specific components that rest closest to the end user.
2. *System oriented:* Object request brokers and object services are more concerned with the "system" or infrastructure aspects of distributed object computing and management.
3. *Vertical market oriented:* Domain interfaces are vertical application or domain-specific interfaces. These coupled with combinations of inherited common facility (CF) and OS interfaces will provide critical application frameworks to a wide variety of industries.

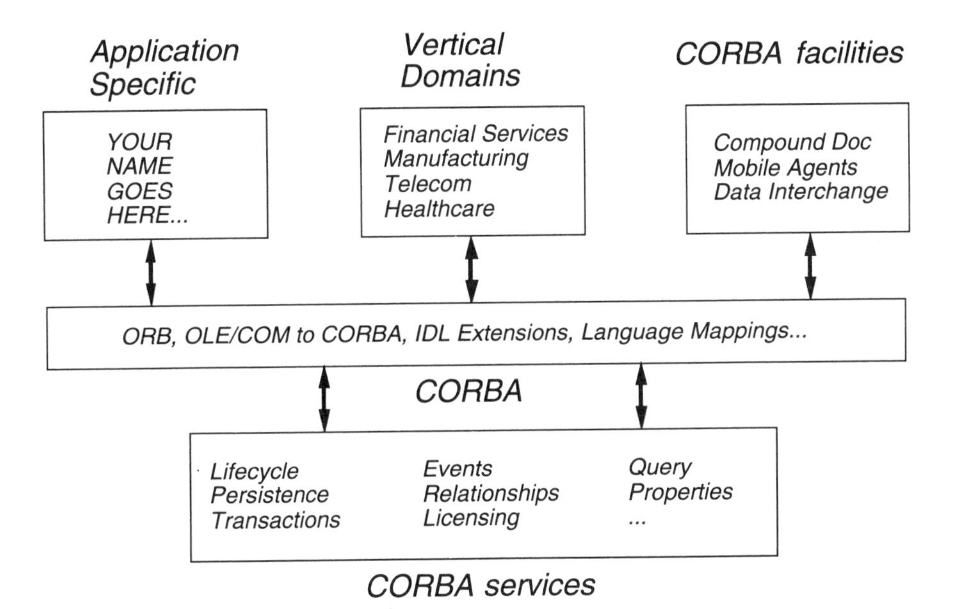

Figure 7.36 OMA reference model.

OMA Component Definitions

OMA components are defined as follows:

- *Object request broker:* CORBA is the commercially known communications heart of the standard.
- *Object services:* These components standardize the life-cycle management of objects. Object services provide for application consistency and help to increase programmer productivity.
- *Common facilities:* These provide a set of generic application functions that can be configured to the specific requirements of a particular configuration.
- *Domain interfaces:* These represent vertical areas that provide functionality of direct interest to end users in particular application domains.
- *Application objects:* These represent component-based applications performing particular tasks for a user.

Information on OMG and its activities can be found at http://www.omg.org.

7.5.5 DAVIC [48]

7.5.5.1 Introduction

The Digital Audio-Visual Council (DAVIC) is a nonprofit association established in 1994. Its purpose is to promote the success of emerging digital audiovisual applications and services such as VOD, broadcast, and games. In December 1995, DAVIC announced their first specification, DAVIC1.0, which defines the first set of technical tools enabling the development of systems that support basic applications such as VOD.

The specification was defined by the policy of only one tool per function. It provides a system reference model, protocols between entities in the model, and APIs for service applications.

Some functions specified by the DAVIC1.0 are not DAVIC original ones, but come from other bodies. For example, B-ISDN and ATM specifications are imported from ITU-T/ISO and ATM Forum; the Moving Picture Experts Group (MPEG) and the Multimedia and Hypermedia Information Coding Experts Group (MHEG) are from ISO/IEC (International Electrotechnical Commission); Internet-related protocols are from IETF; and CORBA-related specifications are from OMG.

From the OA&M viewpoint, DAVIC1.0 specifies only a basic framework. This may be why DAVIC would not provide systems or services but tools. The last call for proposal, CFP6, addresses DAVIC system management. Improvements for OA&M function might be made in a near future release: DAVIC1.2 or later.

A few service providers have announced they would use DAVIC-compliant systems for their services in 1996. This section outlines the DAVIC system reference model (DSRM) and protocols.

7.5.5.2 DAVIC System Reference Model

The DAVIC system consists of four entities: content provider system (CPS), service provider system (SPS),[1] service consumer system (SCS), and delivery system (DS).

The DS has a very broad meaning in DAVIC, but it may be interpreted as a network that delivers information to other DAVIC entities.

The CPS offers information content to SPS via DS. The SPS is a so-called server, and provides the access operation interface for a content or value-added content-related service to its client, SCS.

The SCS refers to the functionality of a set-top box and provides a user interface, SPS access via DS, incoming content decoding, and presentation function.

The DS consists of a core network, an access network, and an in-house network. The core network provides a switched connection from/to CPS and SPS, via an access network to SCS. The access network is a collection of equipment performing transmission, multiplexing, concentration, and broadcasting of services information flows between end users and the core network. It has no switching capability. The in-house network can range from a simple wire to a so-called LAN.

The DSRM classifies the communication between the DAVIC entities in three categories, called *planes:*

1. *User plane:* Transparent user information flow between user information objects;
2. *Control plane:* Information flow between control information objects to control user plane;

---

1. In Sections 7.3 and 7.4 we used the term *service providing system* (SPS) instead of DAVIC's terminology *service provider system* (SPS) because the latter may suggest a broader sense of a service provider system to conventional telecommunication people, thus causing some confusion.

3. *Management plane:* Information flow between management information objects to perform system management functions, such as fault, configuration, accounting, performance, and security.

In each plane, four service layers are defined between the entities (objects) to specify all logical peer-to-peer service interactions as follows:

1. *SL0:* Principal service layer. Service itself, for example, sending or receiving and presenting content information such as movies. Interactions in this layer are transparent to DS.
2. *SL1:* Application service layer. Resources needed by principal service users and providers. Interactions in this layer are transparent to DS.
3. *SL2:* Session and transport service layer. Resource control and management to establish end-to-end communication for SL1 clients.
4. *SL3:* Network service layer. Basic lower network services such as addressing/routing, connection services, and physical layer services to SL2 clients.

DSRM defines five information flows from a source object to a destination object as follows:

1. *S1:* One-way content-information flow on the user plane (SL0 and SL1) from SPS to SCS. A typical S1 example is video represented in MPEG2 TS stream over ATM.
2. *S2:* Bidirectional control information flow on the control plane (SL0 and SL1) between SPS and SCS. This is typically used for controlling video play and downloading application program/data from SPS. The S2 protocol is ISO/IEC DIS13818-6 MPEG2 digital storage media command and control user-to-user interface (DSM-CC UU) over OMG CORBA2.0.
3. *S3:* Bidirectional control-information flow on control plane (SL2) between users (SPS/SCS) and DS. This is used to establish, modify, and terminate a session among SPS, DS, and SCS. The protocol for S3 is DSM-CC UN (user-network signaling). This flow and S2 are heart of DAVIC systems.
4. *S4:* Bidirectional control-information flow on the control plane (SL2 and SL3) between users (SPS and SCS) and DS. This is for call/connection control and the resource control in B-ISDN network defined by ITU-T Q.2931,Q.2130 and Q.2110.

5. *S5:* Bidirectional control-information flow on the management plane (SL0-SL3). This category includes the information flow for layer management and system management. The S5 protocol is SNMP for users (SPS/SCS) and CMIP for DS.

Figure 7.37 depicts a DSRM layer structure, primary objects in each entity, and information flow between entities. Figure 7.38 shows the DAVIC1.0 protocol stack over ATM network.

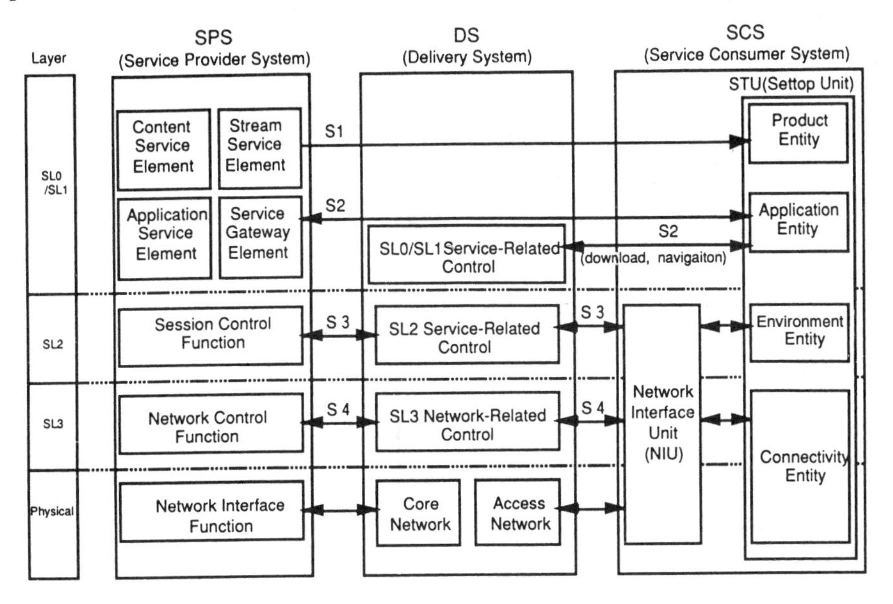

Figure 7.37 DAVIC system reference model.

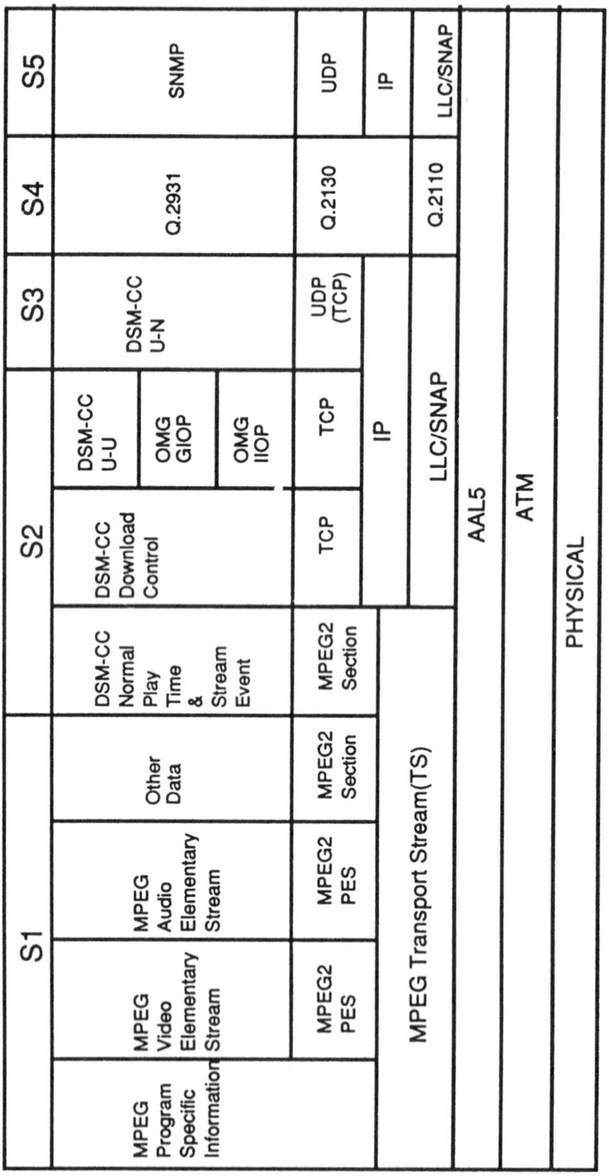

S1
S2
S3
S4
S5
MPEG Program Specific Information
MPEG Video Elementary Stream
MPEG Audio Elementary Stream
Other Data
DSM-CC Normal Play Time & Stream Event
DSM-CC Download Control
DSM-CC U-U
OMG GIOP
OMG IIOP
DSM-CC U-N
Q.2931
SNMP
MPEG2 PES
MPEG2 PES
MPEG2 Section
MPEG2 Section
TCP
TCP
UDP (TCP)
Q.2130
UDP
IP
IP
MPEG Transport Stream(TS)
LLC/SNAP
Q.2110
LLC/SNAP
AAL5
ATM
PHYSICAL

DSM-CC : Digital Storage Media - Command and Control
PES : Packetized Elementary Stream
GIOP : Generic Inter-ORB Protocol
IIOP : Internet Inter-ORB Protocol
U-N : User - Network
U-U : User - User

Figure 7.38 DAVIC 1.0 protocol stack (DAVIC 1.0, Part 7).

References

[1] Yoshida, M., "Service Operation and Management Towards the Multimedia Era," *NTT R&D,* Vol. 44, No. 6, 1996 (in Japanese).

[2] Ejiri, M., and M. Yoshida, "Progress in Telecommunications Services and Management," *IECIE Trans. Commun.,* Vol. E-78-B, No. 1, Jan. 1995.

[3] Eldering, C. A., "Customer Premises Equipment for Residential Broadband Networks," *IEEE Commun. Mag.,* June 1997, p. 114.

[4] ITU-T, Recommendation I.211, 1991.

[5] Shimazaki, N., "Current Trends and the Near Future in the Information Networking Service Industries," *IEEE Commun. Mag.,* July 1996.

[6] ITU-T, Recommendation M.3010, 1992.

[7] Adams, E. K., and K. J. Willetts, *The Lean Communications Provider,* New York: McGraw-Hill, 1996.

[8] Aidarous, S., and T. Plevyak (Eds.), *Telecommunications Network Management into the 21st Century,* New York: IEEE Press and IEE, 1994.

[9] Deloddere, D., et al., "Interactive Video on Demand," *IEEE Commun. Mag.,* Vol. 32, No. 5, 1994, pp. 82–88.

[10] DAVIC, "Contributions to DAVIC Opening Forum," Forum in San Jose, CA, June 1994.

[11] Murase, T., "Overview and Future Direction of the VI&P Comprehensive Experiments," *NTT Rev.,* Vol. 8, No. 6, 1996, pp. 46–52.

[12] Shimizu, A., "Present Status of Joint Utilization Tests of Multimedia Communications," *NTT Rev.,* Vol. 8, No. 2, 1996, pp. 8–13.

[13] See, for example, "Special Issue on Self-Healing Networks for SDH and ATM," *IEEE Commun. Mag.,* 1995.

[14] ITU-T, I-Series Recommendations I.113, 121, 150, 211, 311, 321, 327, 356, 361-365.3, 371, 413, 432, 555, 520, 610, 751.

[15] ITU-T, Q-Series Recommendations Q.2100, 2110, 2120, 2130, 2140, 2610, 2650, 2660, 2730, 2761-2764, 2931, 2951, 2957.1.

[16] ITU-T, G-Series Recommendations G.702, 704, 707, 804, 826.

[17] ITU-T, I-Series Recommendations I.311, 321, 326, 353, 356, 371, 610, 751.

[18] ATM-Forum af-nm 0020-000, af-nm-0027.001.

[19] ETSI, ETS DE/NA5-2210 Ver.3.

[20] ITU-T, I-Series Recommendations I.326.610.751.

[21] ITU-T, G-Series Recommendations G.805, 852-01, 853-01, 854-01.

[22] ITU-T, Recommendation M.3100.

[23] ITU-T, X-Series Recommendations X.720-722.

[24] ETSI, I-ETS 300 653.

[25] ITU-T, Recommendation M.3010.

[26] Mistry, R., et al., "OA&M for Full Services Access Networks," *IEEE Commun. Mag.,* March 1997, pp. 70–77.

[27] Matsusita, M., T. Okazaki, and K. Fujimoto, "A Telecommunications Management Integration Network," *IECIE Trans. Commun.*, Vol. E-78-B, No. 1, Jan. 1995.

[28] Fry, M., et al., "Multimedia Service Delivery with Guaranteed Quality of Service," *Proc. NOMS'96,* Vol. 3, 1996, pp. 476–481.

[29] Mori, T., et al., "Multimedia Service Operation System and Service Navigation System," *NTT Rev.,* Vol. 8, No. 6, 1996, pp. 88–95.

[30] Kaihara, M., and H. Tokunaga, "A Study of Multimedia Service Operation and Management Features," *Proc. 1995 Pacific Workshop on Distributed Multimedia Systems,* Mar. 1995, pp. 96–100.

[31] Kaihara, M., et al., "Multimedia Service Operation: Management and Navigation," *NTT R&D,* Vol. 44, No. 6, 1995 (in Japanese).

[32] Tsuchida, H., and K. Fujimoto, "Intelligent Dynamic Service Provisioning Architecture in the Multimedia Era," *Proc. Int. Conf. Communications (ICC'96),* June 1996, Vol. 2, pp. 1117–1122.

[33] Nose, M., et al., "Intelligent Dynamic Service Provisioning (IDSP)," *Proc. Network Operations and Management Symp. (NOMS'96),* Kyoto, Japan, Apr. 1996, Vol. 3, pp. 711–722.

[34] Fujimoto, K., et al., "Intelligent Trouble Management System Based on Operation Scenario and Fault Simulation," *IECIE Trans. Commun.,* Vol. E-78-B, No. 1, Jan. 1995.

[35] Capell, R., et al., "Evolution of the North Carolina Information Highway," *IEEE Network,* Vol. 8, No. 6, Nov./Dec. 1994, pp. 64–70.

[36] Alexander, P., et al., "ATM Net Management: A Status Report," *Data Commun.,* Sept. 1995, pp. 110–116.

[37] Shimizu, S., et al., "Implementing and Deploying MIB in ATM Transport Network Operations System," *Proc. ISINM'95,* Santa Barbara, CA, 1995.

[38] Yoda, I., et al., "Object Oriented TMN based Operations Systems Development Platform," *Proc. ICC'94,* New Orleans, LA, 1994.

[39] Fujii, N., et al., "ATM Transport Network Operation System in Japan," *IEEE Commun. Mag.,* Sept. 1996.

[40] Anderson, J., et al., "Operations Standards for Global ATM Networks," *IEEE Commun. Mag.,* Dec. 1996, pp. 72–85.

[41] ITU-T, X Series Recommendations X.700-701.

[42] Network Management Forum, "A Business Strategy: Implementing TMN Using OMNI-Point," 1994.

[43] Network Management Forum, "A Technical Strategy: Implementing TMN Using OMNI-Point," 1994.

[44] Glitho, R. H., and S. Hayes, "Telecommunications Management Network—Vision vs. Reality," *IEEE Commun. Mag.,* Mar. 1995, pp. 47–52.

[45] Sidor, D., "Managing Telecommunication Network Using TMN Interface Standards," *IEEE Commun. Mag.,* Mar. 1995, pp. 54–60.

[46] Yamagishi, K., et al., "An Implementation of a TMN-Based SDH Management System in Japan," *IEEE Commun. Mag.,* Mar. 1995, pp. 80–85.

[47] Matsushita, M., O. Miyagishi, and M. Yoshida, "NTT's approach to open telecommunications management, Part 1 and Part 2," *NTT Rev.*, Vol. 6, Nos. 2 and 3, Mar. 1994.

[48] http://www.davic.org/.

8

Wireless and Mobile Multimedia Networks

8.1 Introduction to Wireless and Mobile Services and Networks

In this chapter, we examine wireless/mobile services and networks. "Wireless" and "mobile" are used synonymously, and asynchronous transfer mode (ATM) means "multimedia." This is not always the case. Untethered network connectivity may or may not support user and node mobility. ATM networks can transport diverse information, such as data, voice, and video, in the same network. Yet its methods to provide performance guarantees to achieve quality of service are not well matched to the vagrancies of wireless links. An approach to multimedia applications that adapts to the varying performance of the network may be a more reasonable approach. In this chapter, our primary focus is on mobile access via wireless networks to multimedia data, including World Wide Web access and the delivery of real-time audio and video streams.

Cellular phone and paging service subscriptions are expanding rapidly. The Yankee Group, a market forecasting firm, has estimated that the number of cellular telephone subscribers in the United States will exceed 40 million during 1997, 60 million by 2000, and 100 million by 2004. More than one in three Americans will be carrying cellular telephones! The Scandinavian countries already exceed this. The May 4, 1996, issue of the *Economist Magazine,* predicted that by the year 2000, the number of cellular telephones would equal 40% of the fixed-line telephone lines in the United States (compared with 20% in 1995). Such predictions have fueled the frenzy for personal communications system (PCS) licenses. More than $20 billion has been spent before even the first PCS telephone has been turned on.

PCS is exciting because it marks the total digitization of the airwaves, offering higher voice capacity and improved wireless data services. Unlike the North American analog standard, there is no single PCS standard. The most prevalent are CDMA, PCS 1900, and IS-54. Code-division multiple access (CDMA), developed by Qualcomm, is based on spread spectrum, a technique that spreads a user's waveform over a wide spectrum with the ability to superimpose multiple waveforms to achieve high capacity. Several operators, such as Sprint Spectrum and PrimeCo PCS (Airtouch, Bell Atlantic, Nynex, and US West), have chosen this system. PCS 1900 is a version of the GSM (Global System for Mobile Communications) digital cellular system widely used in Europe. Pacific Bell Mobile Services has chosen this scheme, based on time-division multiple access (TDMA). IS-54 is the PCS version of the North American digital cellular standard, and while it is also based on TDMA, it differs in many details from PCS 1900. AT&T Wireless has selected this system for deployment. None of these schemes will interoperate.

Relative to cellular telephony, wireless data has developed slowly, even though portable computers are the fastest growing component of the PC market. In 1997, more than $25 billion was spent on portable computers worldwide. This will double by 2000. While wireless data services have been available for almost a decade, there are only a few hundred thousand subscribers in the United States. Wireless local-area networks (LANs) represent such a small market that some major companies, like IBM, have left the market.

Why has wireless data developed so slowly? The reasons are complex:

- *Complex pricing:* Some services charge a fixed rate for airtime, others by the amount of data sent, or a flat rate. It has been argued that only usage-based pricing will prove to be a good incentive for infrastructure providers to improve their service, yet users desire predictable charges. Wireless environments are characterized by high error rates, and users will be overcharged if they pay for packets transmitted rather than received.
- *Poor performance:* Wide-area data has achieved transfer rates below 20 Kbps (thousand bits per second), inadequate for multimedia applications, and coverage is often limited. While wireless LANs achieve better data rates, perhaps a megabit per second, this still falls short of standard Ethernet rates. High errors on wireless links also inhibit performance. Noninteroperability among different wireless LANs or wide-area services is a further impediment.
- *Killer application:* Compelling applications for wireless data have yet to be found. Mobile video conferencing might be the next killer

application, though needed bandwidth is high. Paging, short messages, and electronic mail are also popular, and are becoming well integrated with new digital telephony services. Web access has potential, but requires higher bandwidths than those now provided in the wide area.

- *Form factor:* Today's laptop computers are too large to use on the move. Most places in which you are likely to use a laptop, like an airport, could be equipped with wireline access. Personal digital assistants (PDAs) are smaller pocket-sized devices, but these are also difficult to use while moving. Some of the larger PDAs can connect wirelessly via a PCMCIA card for one of the wide-area wireless networks. A third class of device, the communicator/smart phone, is an intriguing addition. It combines a cellular telephone, a data modem, and a PDA in a single pocket-sized device. These are the first devices that really can be used while on the move.

Wireless communications broadly consist of amateur, industrial, consumer, business, military/aerospace, and long-haul applications. Amateur refers to the use of the radio spectrum by amateur operators for personal voice and data transmission. Industrial uses include automotive, such as intelligent highway and vehicle navigation systems based on the Global Position System (GPS). Consumer usage includes residential cordless telephony, cellular telephony, and paging.

Business applications overlap with consumer usage, spanning cellular, paging, wireless public branch exchange (WPABX), wireless local-area networks (WLANs), private/special mode radio (PMR/SMR), and mobile data. WPABX is a more advanced form of cordless telephony supporting mobility within a premises. Examples include the European DECT system, the Japanese Portable Handyphone System (PHS), and Bellcore's WACS and PACS systems [1]. There are a number of wireless LAN products on the market, most operating in the unlicensed industrial–science–measurement (ISM) band. IEEE Standard 802.11 offers a standardized specification for a wide range of WLAN technologies that will enhance interoperability for the future. The Europeans have been working on the HiperLAN specification, supporting data rates up to 20 Mbps over short distances. The U.S. Federal Communications Commission (FCC) has recently opened up the 59- to 64-GHz band for unlicensed WLAN development; this could support data rates up to 100 Mbps and beyond. Private/special mode radio systems are systems used by dispatch operators, such as taxi cabs and delivery services. Increasingly, they are being positioned as an alternative to digital cellular. Mobile data are the wide-area packet radio services used primarily for e-mail access to mobile users. PCS is not a well-defined concept, but encompasses a combination of residential cordless phones, cellular telephony, paging systems, and wireless PBXs [2].

Wireless communications are used by the military and the aerospace industry for communications to vehicles, ships, and planes, using satellites as well as terrestrial methods. Long-haul services are point-to-point microwave links following railroads and utility pipelines, and dot the ridgelines and high-rise buildings in urban areas.

This chapter is organized as follows. In Section 8.2, we review the technical challenges faced by multimedia wireless and mobile systems. Section 8.3 addresses wireless multimedia devices: design issues and the state of the art in portable computers, PDAs, and wireless communicators. Section 8.4 examines network architectures for wireless and mobile systems: the mechanisms for tracking users as they move through the network, directing packets to their current locations, and managing the low quality of wireless links. Section 8.5 describes the quality-of-service issues faced when implementing multimedia applications over wireless networks. Section 8.6 discusses the integration of wireless and wireline networks. We predict future developments in Section 8.7 and provide a glossary of terms in Section 8.8.

8.2 Multimedia Challenges in Wireless and Mobile Systems

8.2.1 Historical Perspectives

8.2.1.1 Radio Basics

James Clerk Maxwell developed his equations in the 1850s that showed that all radiation is the same phenomenon, differing only in wavelength and frequency, the product of which is the speed of light (300,000,000m per second). Henrich Hertz, working in the 1880s, provided the first demonstration of electromagnetic radiation that was not light yet could operate at a distance. Electrical sparks on one side of his lab induced a spark several meters away. These "Hertzian waves" would become known as *radio* [3].

Radio is a form of electromagnetic radiation whose wavelengths are in the range of hundreds of meters (frequencies of 100s of kilohertz) down to millimeters (frequencies up to 100 GHz) [4] (see Figure 8.1). The last 100 years has seen the harnessing of ever shorter wavelengths for communications. Infrared radiation has also been used, and it exists in the range of hundreds of nanometers, just below the wavelengths of visible light.

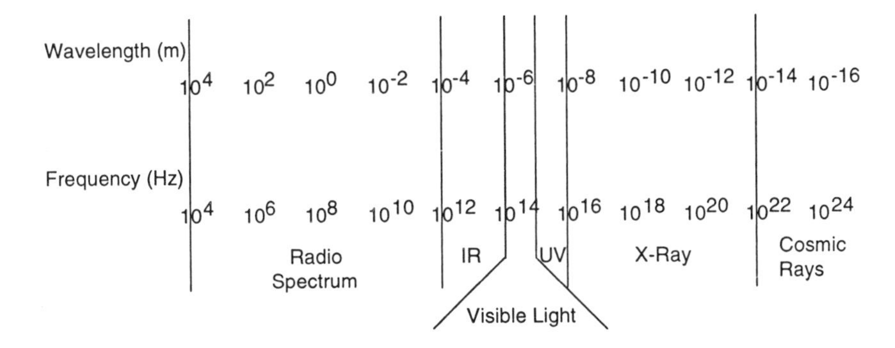

Figure 8.1 The electromagnetic spectrum.

An AM radio operates at a frequency of approximately 1 MHz and a wavelength of 300m. A single wave is about 1,000 ft. FM radio and television stations operate around 100 MHz with wavelengths of about 3m (10 ft). Shorter wavelengths imply smaller transmitting and receiving antennas, an advantage for mobile communications. Cellular telephones operate in the spectrum below 1 GHz with wavelengths of 30 cm (1 ft). Half a wavelength is standard: a cellular antenna is a few inches while a television antenna is several feet. Satellites operate in the 10-GHz region, with wavelengths of only 3 cm (1 in). While this implies a very small antenna, satellite signals are dramatically degraded as they travel over very great distances from the orbital position of the satellite to the earth. An antenna dish (18 to 24 in.) is needed to receive these faint signals.

Radio signals are identified by their wavelengths. Below 30 kHz, they are known as very low frequency (VLF). Between 30 and 300 kHz, they are called low frequency (LF). At 300 kHz to 3 MHz, the signals are called MF or medium frequency. The most sought-after frequencies for mobile communications are found in the next three regions. The high frequencies, or HF, are between 3 and 30 MHz. The very high frequencies, or VHF, are located between 30 and 300 MHz. This is where the lower channels of TV and the FM radio stations are. The ultra high frequencies, UHF, are in the range of 300 MHz to 3 GHz. Here are the higher TV channels, cellular telephones, wireless LANs, and paging frequencies. The 3- to 30-GHz range is called super high frequency, or SHF, and above 30 GHz, it is called EHF, or extremely high frequency. These latter frequencies, with their very short wavelengths, are the domain of satellite communications and some experimental forms of wide-area and in-building wireless LAN technologies.

Radio waves pass through and around physical objects by propagation: reflection, diffraction, and scattering. Reflections occur when a propagating wave impinges on an object that is large compared to its wavelength. For mobile communications wavelengths, reflecting objects include the surface of the Earth, buildings, and walls. Diffraction occurs when the radio path is obstructed by a surface with sharp irregular edges. Thus, radio waves can bend around obstacles, even when a line-of-sight (LOS) does not exist. Scattering occurs when a wave hits an object smaller than the wavelength of the propagating wave. Examples include foliage, street signs, and lamp posts [5].

Radio stations can be picked up at distances far beyond the visible horizon. At these frequencies, the radio waves reflect off the ionosphere (see Figure 8.2(a)). Short-wave radio bounces between the ionosphere and the ground, making it possible to hear transmissions around the world. At the higher frequencies, the waves behave more like light, and follow a LOS beam that may be blocked by certain materials, especially metal (see Figure 8.2(b)). Infrared waves, for example, cannot pass through walls.

Mobility affects radio propagation. The communications path, or channel, varies with location and time. Radio waves are scattered by nearby objects, causing interference with the stronger direct radio signal path. This is called *multipath.* Objects can shadow receivers and transmitters, yielding lost connectivity. Radio signal strength will attenuate with distance. These effects result in rapid fluctuations in the received power, changing across orders of magnitude in a time frame measured in microseconds.

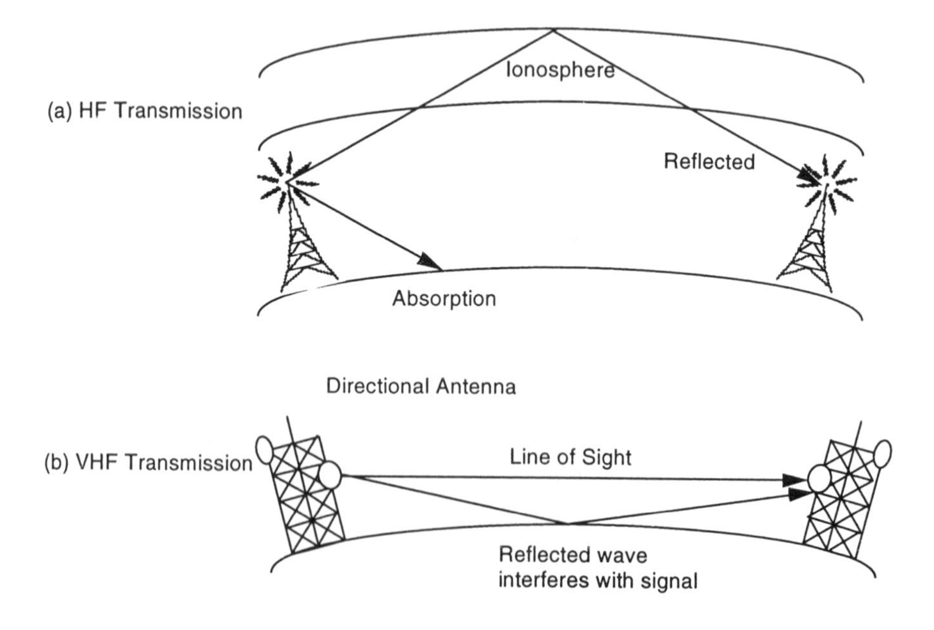

Figure 8.2 Absorption and reflection of radio waves at different frequencies.

8.2.1.2 Evolution of Cellular Telephony

The ability to communicate anywhere, anyplace, and anytime has long been a dream. Its possession, and the timely information it could provide, would ensure the power of kings, nation states, and business cartels.

We review the development of telegraphy, that is, "writing at a distance," and telephony, that is, "speaking at a distance." In the last few years, radio has merged with telephony to yield today's cellular and personal communications systems.

Communications Before the Industrial Age

Examples can be found in antiquity of "muscle-powered" communications methods: runners, homing pigeons, or horse relays. Additional methods operated within visible distances, extended by the telescope: smoke signals, torch signaling, heliographs (flashing mirrors), signal flares, and semaphore flags [6]. However, these were too dependent on weather conditions, were easy to intercept, and it could still take months to communicate over very long distances.

Napoleon's Secret Weapon: The Optical Telegraph

The late 18th century yielded the first comprehensive infrastructure that transmitted messages faster than transportation. This was the optical telegraph, a mechanical construction of articulated arms that could encode several hundred symbols by their positions. It provided news to the capital while allowing the revolutionary government to control armies at its borders. By the mid-1800s, this system had grown to cover over 5,000 km with 556 stations in operation [6].

What Hath God Wrought: The Electrical Telegraph

This was soon superseded by development of the electrical telegraph. Samuel Morse developed his variable-length dot and dash code in 1838. Now information could move without significant mechanical delays and at well beyond visible distances. Telegraph lines soon spanned the globe to meet the needs of the Imperial powers [7]. New technologies were developed to bridge water. The transoceanic telegraph cables were among the highest risk private sector endeavors of the nineteenth century. After several unsuccessful attempts, Cyrus Field laid the first transatlantic cable in 1858. President Buchanan and Queen Victoria exchanged the first telegrams. But the line failed within a few months. After the American Civil War, a new cable was laid in 1866, operating continuously thereafter [3].

Marconi and Radio

In 1895, Guglielmo Marconi demonstrated that electromagnetic radiation, created by spark gap, could be detected over several miles [8]. After the Italian government rebuffed him, Marconi traveled to England where he demonstrated his system, winning the interest of the Royal Navy. In 1898, he installed radio sets on the Royal yacht, reporting on ship regattas from sea. In 1901, Marconi received the first long-wave transmission across the Atlantic. After continuous experimentation, using ever more powerful transmitters and longer wavelengths, he achieved regular transatlantic radio cable service in 1907. Only in the 1920s was it discovered that short waves, reflected off of the ionosphere, offered a better communications method.

Broadcast Radio and the Rise of Shortwave

After World War I, Marconi's spark gap was replaced by vacuum-tube amplifiers combined with sensitive tuning circuits. Thus began the convergence of radio and telephony. In 1915, the first transcontinental wireless voice transmission was accomplished. In 1920, the first commercial radio station began operating in Pittsburgh. Short-wave reached maturity, replacing long-wave for global communications. Marconi discovered the phenomenon of ionospheric "skip." By 1924, Marconi used shortwave to communicate by voice from England to Australia.

The Telephone Network

Alexander Graham Bell demonstrated his early system in 1876. The early days of the phone system were characterized by many independent operators. Subscribers to different providers could not call each other even if they were in the same town. No rules of interconnection existed and long-distance "toll" calls were rare. In 1913, AT&T and the government reached an agreement: AT&T would build a national network in return for becoming a regulated monopoly [1].

Development of Mobile Radio

In 1921, Detroit was the first city to experiment with police dispatch via radio. Initially the system was one way, with transmission from a centralized transmitter to the cars. Transportable, low-power transmitters were yet to be developed. The first such system was deployed in Bayonne, New Jersey, in the early 1930s, using "push to talk" communications. In 1935, Edwin Armstrong invented frequency modulation (FM). Compared with amplitude modulation (AM), FM virtually eliminated background static. World War II played a critical role in stimulating the development of a commercial FM manufacturing ability.

In 1946, AT&T received the first mobile telephone license for St. Louis. The initial car phone systems were based on a single centralized transmitter covering a metropolitan area. Meanwhile, Bell Labs was refining the cellular system. The concept is based on many smaller transmitters, spread across a geographical area. System capacity is improved by reusing the allocated spectrum on a cell-by-cell basis.

In 1947, AT&T proposed an allocation for mobile telephony of 150 two-way channels, with 100-kHz spacing. The FCC did not act. In 1949, AT&T proposed mobile telephony in the UHF bands. The FCC allocated the spectrum to UHF TV stations. In 1957, AT&T proposed 75 MHz for mobile telephony near 800 MHz. The request was once again rejected. Finally, in 1970, the FCC allocated 115 MHz for mobile services. This was eventually reduced to an allocation of 40 MHz for mobile telephony in 1974 [4].

Evolution of Mobile Telephony in the United States and Europe

Figure 8.3 tracks the evolution of these technologies in the United States and Europe. The U.S. technology was deployed by 1979 as North American AMPS (Advanced Mobile Phone System). Europe developed region-specific and incompatible analog cellular systems in mid-1980s: for example, TCAS in Great Britain and NMT (Nordic Mobile Telephone) in the Nordic countries [4].

Cellular telephony was an immediate success. Capacity became strained. Capacity can be increased by introducing additional cell sites. Another way to increase capacity is to pack more channels into the same bandwidth. Nar-

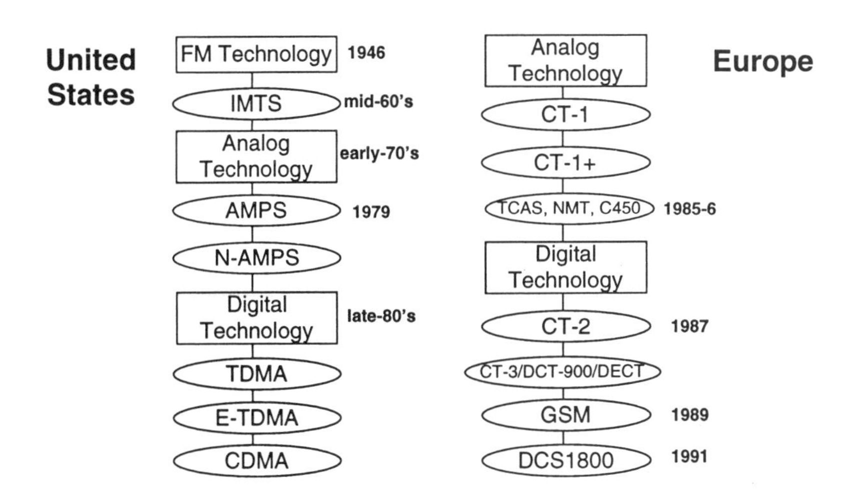

Figure 8.3 Evolution of mobile telephony in the United States and Europe.

rowband AMPS (N-AMPS) was introduced to triple the capacity of the analog system through more aggressive voice compression.

The United States and Europe began developing digital techniques in the mid-to-late 1980s, to achieve higher system capacities. The Europeans made rapid progress on digital cordless telephone standards. These offer lightweight handsets, long battery life, low-cost base stations, and 32- to 64-kbps data rates mixed with telephony. In addition, they focused on a single standard for the continent: GSM (Global System for Mobile Communications) [9]. In the United States, two alternatives have been adopted: TDMA (the North American Digital Standard, IS-54) [10] and CDMA (IS-95) [11]. Europe has emerged with a single digital cellular system, while the United States is pursuing deployment with multiple incompatible systems.

8.2.1.3 Evolution of Wireless Data

The Internet

The Advanced Research Projects Agency (ARPA) of the U.S. Department of Defense provided substantial early funding for networking research. During the 1960s, it developed the ARPANet. During the 1970s, it led the development of internetworking and extended the packet-switching concept to radio and satellite networks [12,13].

Packet switching arose from the desire to build highly survivable systems. The telephone system's centralized switches, and its rigid concept of connection, represent vulnerabilities. If a switch is broken during a connection, the connection fails. Packet switching replaces switches with distributed routers. Messages are divided into independently routed packets. This allows traffic to be statistically multiplexed over the available communications paths, gracefully adapting to the traffic demands and better utilizing existing link capacity without reserved bandwidth. Local information is used to choose the next hop; there is no centralized point of control in the system.

In 1967, ARPA proposed an ambitious program to connect computers at its research sites using point-to-point lines and specially designed message processors. UCLA became the first node on the ARPANet, attaching the first interface message processor (IMP) in 1969. At the same time, the Aloha Project at the University of Hawaii built a packet-switched network based on fixed-site radio links. They developed contention-based media access protocols called "Aloha," and applied these to satellites as well as radio systems. They became the foundation of Ethernet's media access technique.

In 1972, the ARPA Packet Radio Program applied packet techniques for the battlefield. A packet-switched satellite network (SATNet) was also created. A need arose to link these into a true "network of networks." In 1973, the specifications for the transmission control protocol (TCP) were developed. It shifted reliability from the network to the end hosts, thus allowing the protocol to operate over unreliable links [14].

ARPA's Packet Radio Program

The ARPA packet radio program demonstrated how packet-switching concepts could be extended to broadcast radios. Three technical developments made the early 1970s ripe for "highly integrated" packet radios: (1) the microprocessor, enabling sophisticated digital control in a small package, (2) surface acoustic wave (SAW) devices from which small-area matched filters could be built, and (3) communications protocols, like Aloha, able to achieve intelligent management of the shared radio channel through dynamic scheduling [15]. In 1983, ARPA launched the Survivable Radio Networks (SURAN) program. Its goal was to demonstrate scaling in packet radio networks able to operating in harsher environments [16].

Wireless LANs

In the late 1970s, IBM Research Labs in Switzerland pioneered diffuse infrared (IR) technology for factory environments. HP Labs investigated radio-frequency (RF) spread-spectrum technology, seeking to link together distributed terminals. Similar work was under way at Motorola using the 1.7-GHz band [17,18]. In 1985, the industrial-scientific-measurement (ISM) bands were made available for LANs using spread spectrum. Since FCC licenses were not required, many small companies entered the market. As the underlying integrated circuits technology advanced, it is now possible to purchase radio interfaces suitable for use by virtually all laptop and PDA devices on the market today. The major drawback is that each manufacturer has defined its own waveforms, making interoperability impossible.

In 1990, the IEEE 802.11 committee was formed to standardize the media access protocols for the 2.4-GHz ISM band. This efforts offers the possibility of a new generation of interoperable devices. In 1992, Winforum was established to define the etiquette rules for the unlicensed PCS bands [19], allowing independent devices to share available bandwidth. The technique is also being used for recently allocated data bands. At the same time, HiperLAN standardization began in Europe. Their goal is to achieve data rates in the 20-Mbps range, albeit within a room sized cell.

Wide-Area Wireless Data Systems

In 1983, IBM and Motorola designed a wide-area data system, ARDIS, to support IBM's National Service Division [20]. By 1990, the system was deployed in more than 400 metropolitan areas. In 1993, it was extended to provide nationwide roaming. IBM sold out to Motorola in 1994. By then the number of users had grown to approximately 35,000. The system was recently upgraded to 19.2 Kbps.

The RAM mobile data system, Mobitex [21], is a joint venture in the United States by Bell South, Ericsson, and Motorola. The first system was deployed in Sweden in 1986 and in the United States in 1989. The U.S. system covers 100 metropolitan areas. It supports national roaming, and will eventually support international roaming. Data rates are modest at 8.0 Kbps. In 1994, the system had approximately 12,000 U.S. subscribers.

Cellular digital packet data (CDPD) provides data over analog cellular. IBM developed the concept in 1991. An effort was undertaken in 1993 to draft the specifications, with deployment beginning in 1995. The service provides a data rate of 19.2 Kbps, and uses idle channels in the analog cellular system, by judiciously hopping among the available channels. It supports the TCP/IP protocols, including a version of Mobile IP.

Summary

Table 8.1 shows spectrum allocations for existing and proposed wireless services. It shows the trend toward higher frequencies for newer services. Another trend is the recent spectrum allocations for wireless data, such as the unlicensed PCS band (1910 to 1930 MHz), the unlicensed national information infrastructure band (5.15 to 5.35 GHz and 5.725 to 5.875 GHz), and the unlicensed millimeter-wave band (59 to 64 GHz).

8.2.2 Overarching Technical Challenges

8.2.2.1 Untethered Access

Wireless communications provides freedom from traditional network tethers. There is no need to physically connect the node, though there must be radio or infrared coverage if the node needs access wireline resources. This makes it possible to provide network access to users on the move or working without a prewired infrastructure. An increasingly important use is the ability for collaborators to come together to form a network around a conference table, to share notes, drawings, or other data without setting up a network in advance.

With this capability comes numerous challenges [22]. Foremost is the hash communications environment that manifests itself as lower bandwidth, higher latencies, and higher burst error rates than wireline technologies. Since the user depends on the wireless channel for network connectivity, the probability of disconnection rises dramatically. Thus, end nodes must be able to operate autonomously when the network fails.

Table 8.1
Spectrum Allocations in the United States

Service	Frequencies	Comments
AM radio	535–1605 kHz	Broadcast audio entertainment
Cordless telephony	46.6–47 MHz base station 47–50 MHz handsets	Early generation of dubious quality; better systems available in the 915 MHz ISM band
Broadcast television	54–72 MHz VHF Ch 2–4 76–88 MHz VHF Ch 5,6 174–216 MHz VHF Ch 7–13 470–608 MHz VHF Ch 14–36 614–806 MHz VHF Ch 38–69	Analog video broadcast entertainment; 6 MHz per analog TV channel
FM radio	88–108 MHz	Analog stereo audio broadcast
Digital television	174–216 MHz VHF Ch 7–13 470–698 MHz UHF Ch 14–51	Proposed reallocation; 6 digital channels will be carried in old 6-MHz analog channel
Specialized mobile radio	806–821 BS/851–866 HS MHz 896–901 BS/935–940 HS MHz	Dispatch radio; emerging digital services similar to cellular with no roaming costs
Cellular phones	824–849 MHz 869–894 MHz	Advanced Mobile Phone Service (AMPS); being replaced by TDMA and CDMA
Narrowband PCS	901–902/930–931/940–941 MHz	New allocations for advanced 2-way paging and messaging; auctioned for $617 million
Industrial-scientific-measurement	902–928 MHz 2358.25–2441.75 MHz 5737.5–5862.5 MHz	Unlicensed bands used for many applications; 915 and 2400 MHz primary bands for today's wireless local area networks
Broadband PCS	1850–1990 MHz broadband 1910–1930 MHz unlicensed	New allocations for cellular services and wireless local loop; raised $20 billion
Digital audio	2320–2345 MHz	New satellite-based digital audio services
Wireless cable	2500–2655 MHz 2655–2690 MHz	Multichannel multipoint Distribution service; for 2-way voice, video, and data
Unlicensed natl info infrastructure	5.15–5.35 GHz 5.725–5.875 GHz	To be used for wireless LAN applications; 20 Mbps will be possible
Direct broadcast satellite	12.2–12.7 GHz downlink 17.3–17.8 GHz uplink	Primarily for broadcast entertainment; data capabilities may also be possible

Table 8.1
Spectrum Allocations in the United States (continued)

Service	Frequencies	Comments
Teledesic satellite constellation	18.8–19.3/28.6–29.1 GHz	Global satellite network delivering voice, video, and interactive multimedia services
Mobile satellites	1525–1544/1626.5–1645.5 MHz 1545–1559/1646.5–1660.5 MHz	Geostationary; voice, video, data services
Big LEOs	1610–1626.5/2483.5–2500 MHz	Low earth orbit; voice, low speed data Intenet-like services;
Fixed satellite services	17.7–18.8/19.7–20.2 GHz 28.35–28.6/29.5–30 GHz	e-mail, videoconferencing, data access, data distribution
Millimeter wave	59–64 GHz	Unlicensed, potential for 100-Mbps WLAN

Wireless nodes must handle wide variation in connectivity. Wide-area networks offer 10-Kbps bandwidths while WLANs achieve 1 Mbps. As a node moves between inside and outside connectivity, its bandwidth and latency undergo significant changes. Applications must be designed to adapt to these radical changes in connectivity.

Coverage and proximity to base stations determine the connection quality. But nearby users also affect network performance. Users closer to a base station are likely to obtain better performance (the "near-far problem"), while they may interfere with each other attempting to gain access to the wireless media (the "hidden terminal problem").

The final challenge is security. In wired networks, one must gain access to a physical port to attach. Mechanisms limiting access through physical security no longer work. A fixed assignment between a physical attachment and a network address is no longer appropriate. Wireless nodes attach anywhere or at anytime. This implies that stronger mechanisms for authentication of roamers in a wireless environment are needed. Mechanisms to circumvent over-the-air wiretapping (especially of the authentication handshake!), such as encryption, are strongly needed in a wireless networking environment.

8.2.2.2 Mobility

Wireless devices need not be mobile: consider a computer attached to a 5m satellite dish. When devices are both wireless and mobile, the challenge is to support network access as the devices move around. To achieve "all the time" connectivity, the routing functions must allow end nodes to change their network attachment on the fly. Packets must be forwarded to the node as it moves. This becomes more challenging when the routing algorithms include multicast protocols for multipoint collaborative applications [23]. This functionality is called *location transparency,* since independent of location, a node will receive its packets.

Services can also depend on the node's location. These include resource discovery mechanisms that make it possible to find local services, like the nearest printer or file server. "Follow me" services are also possible. Phone calls could be routed to the nearest phone, based on the user's current point of attachment. Some services may migrate as the user moves, for example, a special format translating "proxy" may follow the user into the roamed in subnetwork, to reduce the latency to perform its translation.

This raises issues of privacy and user tracking. While tracking user location may be desirable from a resource allocation viewpoint (e.g., spectrum can sometimes be borrowed from adjacent cells that are not as heavily loaded), it is possible to have too much information about the location of individuals. It is necessary to treat users in an anonymous fashion for anything but the internal processing demands of the network.

8.2.2.3 Ad Hoc Multihop Connectivity

The wireless channel usually connects the mobile terminal to a base station in a single hop. The base station itself is connected to a wired infrastructure. Consider now a system where there is no wired infrastructure, and connectivity must be maintained to mobile users. An example of such a system is a multihop packet radio network [24]. Multihop communications present new challenges. Due to constraints on transmission power, not all radios are within range, and packets may need to be relayed from one radio to another on their way to their intended destinations.

These so-called "instant" infrastructure systems are more complex to manage than traditional single-hop centrally controlled systems. The salient features may be described as the "3Ms": real-time multimedia, multihop, and mobile. Real-time multimedia plus multihop has been investigated in certain satellite systems and terrestrial packet radio networks. Real-time multimedia plus mobile is pervasive in cellular radio systems [25]. Multihop plus mobile was well studied in the ARPA Packet Radio Program. While implementations have combined any two of these, none support all three.

8.2.2.4 Portability

Providing quality interaction with small devices is a challenge. They are more constrained in their computing performance and have lower quality displays than desktop machines. Furthermore, the data you carry may be at risk through theft or breakage.

Small form factors constrain the user interface. Truly portable devices have limited real estate for keyboards. The user interface must take advantage of icons, handwriting, or speech-based inputs. For certain users, a head-mounted display may be advantageous.

Since the devices are small, their ability to store data is limited. Flash memory can be used for small form factor and low-power storage, but it remains expensive.

A challenge is the power limitations imposed by the energy you can carry. This limits the display technology, the amount of processing done locally (power management techniques place the system in a sleep mode), and the amount of storage you can carry (aggressive disk spin down techniques will be necessary) [26].

8.2.2.5 Adaptability

A key attribute of wireless and mobile systems is their need to adapt at all system levels. At the physical level, we see adaptation in terms of choice of channel, power control, level of error correction, and antenna beam forming. As we move up the network layers, the router layer must adapt in response to user movement. The transport layer must adapt to the higher error rates of wireless links. Applications must adapt to the bandwidth and latency changes seen as a mobile node moves among wireless subnetworks.

8.2.2.6 Metrics

Traditional network quality is determined by metrics like bandwidth, latency, and performance as a function of the number of users supported. Good end-to-end performance over wired and wireless subnetworks must be achieved. Wireless and mobile networks must also be evaluated in terms of capacity, such as the number of users supported per megahertz of available spectrum and the number of users supported per base station. The range of a wireless network is another metric. The cell size and transmission/power range of devices determine coverage and error rates. Smaller cells, lower power, and greater reuse of frequency support more users, but at a greater cost. Economic issues also affect the complexity of the network and the associated network interface devices. Handoff is a new function in mobile networks, and the latency with which it is accomplished will have implications for the performance of multimedia traffic.

8.2.3 Definitions and Terms

Wireless information systems are systems with the ability to compute, communicate, and collaborate anywhere, anytime. A number of terms have been used to describe these, including ubiquitous computing, nomadic computing, decoupled computing, and wireless computing. While no clear-cut consensus exists for these definitions, for the purposes of this chapter, we suggest the following definitions.

8.2.3.1 Ubiquitous Computing

Ubiquitous computing [27] envisions an environment where devices are so inexpensive that there are hundreds in an office. With such numbers of devices, any attempt to interconnect them would be physically impossible; thus wireless communications is necessary. Ubiquitous devices demand modest bandwidth, for example, 64 Kbps or less, although the number of bits per second per cubic foot may be very large. These ideas, first presented more than 5 years ago, have so far had modest effect. The issue is not computers everywhere, but information everywhere. It is the network which has become ubiquitous, through Internet access and the World Wide Web.

8.2.3.2 Nomadic Computing

Nomads are always on the move. Nomadic computing refers to the ability to compute in any support environment [28]. This is the wireless analog of nonlocal users using local machines to do computation. In this model, individual organizations will construct their own wireless infrastructures, linked by wired internetworks. A user should be able to attach to the network even within another's wireless infrastructure. Support for nomadic computing must address security and privacy as nomads move between organizational infrastructures. The host infrastructure must retain security against the nomad, while its operations must be kept private from the owners of the host infrastructure.

8.2.3.3 Decoupled Computing

Decoupled computing supports operation when detached from the infrastructure. It enables operations like file access even when disconnected from the server [29]. Using file caching, it is possible to update a file even when disconnected. When connection is reestablished, the cached files are resynchronized with the server. It is also possible to perform decoupled computing for applications operating as a spool. Results are trickled back and forth as bandwidth becomes available. Eudora e-mail is an example that allows users to read and compose mail on a detached computer, connect to a network server over phone lines, transfer mail, and then disconnect. Some distributed file systems support disconnected operation [30,31]. This is an important element of wireless computing because the support infrastructure is not universally available.

8.2.3.4 Wireless Computing

Wireless computing refers to computing systems that are connected to their working environment via wireless links. This term is applied to the computing devices participating in a wireless LAN, with gateways to wired networks [32]. The key is the ability to work within a collection of computing devices and servers for sharing data and information. This implies symmetric bandwidth between the wireless node and the network, and a desire for as a high bandwidth as possible, approaching Ethernet speeds. Under this definition, wireless e-mail is a particularly primitive form of wireless computing!

8.2.3.5 Mobile Computing

Mobile computing, as a discipline and as a set of technologies, is a union of all of the above. The overarching technical issues fall into five broad categories: object identification, registration with the infrastructure, establishment of connectivity, mobility within the network, and privacy and security. Objects in a mobile environment must be named and identified, even as their network point of attachment changes. The issue is what forms of identification must remain invariant, and what needs to change to reflect user and terminal mobility. In the telephone network, a user is identified by a telephone number. Mobility is associated with the terminal, not the subscriber. In GSM, users are assigned "smart cards," and can be reached at any handset containing their card. Subscriber mobility is different from terminal mobility. Computers in the Internet are uniquely identified by their IP address.

The second issue is registration. Mobile users and their equipment must be authenticated, authorized to perform certain functions, and allocated resources. The bootstrap is registration: the user makes herself known to the local system, it communicates with her home system to verify that the user is a subscriber in good standing and that the home system vouches for her ability to pay for the services that will be used.

The third issue is connection establishment in the face of mobility. After registration, a mobile node may be identified by a locally assigned IP address, but how does the rest of the Internet discover this? One solution is to use an extra level of indirection through the invariant home IP address. This is how the cellular telephone system operates. A user is given a cellular telephone number in her home region. When she roams, the home system keeps track of where she is so calls can be forwarded to her current location. Whenever she moves, her current location is made known to the home system. Connections establish the communications path in advance of information transfer while provisioning the path for the performance the user expects. As users reposition, the connections must be extended, adding latency associated with forwarding traffic from one system to the next, or torn up and reestablished, which has its own latency implications.

The fourth issue is mobility and the handoff process. Handoff introduces errors, dropped connections, and high latency. In mobile telephone systems, there may not be available channels in the moved-to cell, so a connection could be dropped. For mobile computers, adding a node to the cell may reduce the performance for the mobiles already there. Further, as devices move through the network, services may follow it. This form of process migration helps to reduce the latency of application execution.

The final issue is privacy and security. Mobile nodes and users must be authenticated to obtain services. Encryption protects users from eavesdropping. It is important that the registration process not be overheard and replayed by intruders. Some systems never exchange passwords in the clear, and keep a history of challenge-response interactions. A user may not wish her identity to be fully disclosed to the visited system; the acceptance of the home system to pay her debts should be sufficient.

8.3 Emerging Mobile Multimedia Technologies and Devices

8.3.1 Devices

Table 8.2 captures the diversity of today's computing devices, spanning high-end desktop personal computers to pocket-sized PDAs. At their heart is the microprocessor, improving in performance at an astonishing rate. Portable devices must consider battery life, with performance metrics shifting from MIPS to MIPS per watt. The table summarizes network connectivity, CPU power, memory and disk resources, screen size, and bit depth. As devices get smaller, their processing performance, connection speeds, memory, storage and screen size, and screen capabilities, are reduced.

The following subsections describe portable computing devices in more detail: the laptop computer, the PDA, and the network computer/"thin client." The latter does limited local processing, and depends on wireline resources for processing and storage. This are also known as communicators.

8.3.1.1 Laptop Computers

Laptop computers are efficiently packaged desktop machines. The form factor is largely determined by the full-sized keyboard. Increasing capacities of memory and disk have led to high capability in a 7-pound package about the size of a loose-leaf binder.

Table 8.2
Wireless Computing Devices

Device	Bandwidth bits/sec	CPU	Memory & Disk	Screen Size	Bits/Pixel
High-end PC	Ethernet (10–100 Mbps) ISDN (128 Kbps)	266-MHz Pentium Pro	64/4G	1280 × 1024	16–24 color
Low-end PC	"	150-MHz Pentium	16/1G	1024 × 768	8–16 color
High-end notebook	Cellular (9.6–19.2 Kbps) or Wireline (28.8–33.6 Kbps modem)	"	"	800 × 600	8 color
Low-end notebook	"	100-MHz 486	"	640 × 480	4 gray-scale
PDA	2400–14.4 Kbps modem	20-MHz RISC or 486	2/0	320 × 200	1–2 gray-scale

The major limiting factor in these devices is power consumption. Table 8.3 shows the typical laptop computer power consumption for a 386-based laptop computer of several generations ago [22]. Battery weight is a major consideration; batteries often weigh more than a pound. The main power consumers are the display backlights and the disk drive. Nonreflective LCD displays require their own power-consuming light source. Sophisticated power

Table 8.3
Typical Laptop Power Consumption

Component	Power (W)	Component	Power (W)
Base system (2 MB, 25 MHz)	3.650	IC card slot	0.100
Base system (2 MB, 10 MHz)	3.150	Additional memory (per MB)	0.050
Base system (2 MB, 5 MHz)	2.800	Parallel port	0.035
Screen backlight	1.425	Serial port	0.030
Hard drive motor	1.100	1.8-in PCMCIA hard drive	0.7–1.3
Math coprocessor	0.650	Infrared network	0.250
Floppy drive	0.500	PCMCIA modem, 14.4 Kbps	1.365
External keyboard	0.490	PCMCIA modem, 9.6 Kbps	0.625
LCD screen	0.315	PCMCIA modem, 2.4 Kbps	0.565
Hard drive active	0.125	GPS receiver	0.670

management techniques in laptop computers help reduce the power demands. User activity is monitored, and when it falls below a threshold, the processor is slowed, the display backlight is dimmed or turned off, and the disk drive is spun down.

Processors for portable devices demand new metrics. MIPS is no longer adequate. Better metrics include MIPS per watt, a measure of efficiency with direct impact on battery life and heat dissipation issues in highly integrated systems, MIPS per square millimeter, which measures the silicon costs of the processor core and the ability to surround it with specialized functions for a more highly integrated design, and bytes per task, a measure of code density that captures the space premium of RAM/ROM space.

8.3.1.2 Personal Digital Assistants

Personal digital assistants (PDAs) are handy computing devices that perform fixed personal information management (PIM) functions, such as address book, calendar, notepad, and to-do lists [33,34]. They are attractive because of their small form factor and portability. But they present some design challenges dictated by the small package.

Small devices contain low-quality displays due to constrained size, resolution, and pixel depth. Display area is severely limited by device size, which must fit in a palm. Typical sizes are 240 by 320 pixels, bilevel, or four grayscale levels. This means that the designer must create a user interface within a small screen area and without colors. This is a significant impediment for video or high-quality images.

Displays are reflective since the backlight represents the single most significant power drain. This implies that they are not visible without direct sunlight. PDAs are similarly limited in their inputs; a conventional keyboard is unwieldy. Some incorporate a calculator-style keyboard, but these are too small for serious typing. They have been replaced with stylist-based touch-sensitive displays with keyboards that pop up on the display, and which support entry by tapping the stylist on the display. Alternatives include capture of digital ink—the drawn strokes are captured and stored in an uninterpreted form—or handwriting recognition either with a user-trained system or with an easy-to-recognize alphabet such as Grafitti as used in the US Robotics Pilot.

Table 8.4 tabulates attributes of several PDAs. Memory size and processor speeds are limited, necessitated by low sales prices and long battery life. The majority of PDAs are based on slow Intel x86 processors, although several incorporate the ARM RISC processor or derivatives of the Motorola 68000 family. Most of the embedded functionality is implemented by software in ROM. Writable memory is limited to a few megabytes, and is used to hold data records for the various PIM applications. A distinct trend has been toward

Table 8.4
Comparative Capabilities of Laptops and Personal Digital Assistants

Device	Mem Size	Processor	Batteries	Weight	Display
Armstad Pen Pad PDA 600	128 KByte	20-MHz Z-80	40 hrs 3AAs	0.9 lbs	240 × 320 10.4 sq. in.
Apple Newton Message Pad	640 KByte	20-MHz ARM	6–8 hrs 4 AAAs	0.9 lbs	240 × 336 11.2 sq. in.
Apple Newton 110 Pad	1 MByte	20-MHz ARM	50 hrs 4 AAs	1.25 lbs	240 × 320 11.8 sq. in.
Apple Message Pad 2000	5 MByte	160-MHz ARM	3–6 weeks 4 AA	1.5 lbs	480 × 320 16 gray lvl
Casio Z-7000	1 MByte	7.4-MHz 8086	100 hrs 3 AAs	1.0 lbs	320 × 256 12.4 sq. in.
Tandy Z-550 Zoomer	1 MByte	8-MHz 8086	100 hrs 3 AAs	1.0 lbs	320 × 256 12.4 sq. in.
Sharp R-3000 Zaurus			70 hrs 2 AA	11 oz	320 × 240 touch sense
AT&T EO 440 Pers Comm	4–12 MByte	20-MHz Hobbit	1–6 hrs NiCd	2.2 lbs	640 × 480 25.7 sq. in.
Sony MagicLink PIC-1000	1 MByte	15-MHz Dragon Proc	13 hrs 6 AAs	1.25 lbs	480 × 320 13.5 sq. in.
Sony MagicLink PIC-2000	2 MByte	15-MHz Dragon Proc	13 hrs 6 AAs	1.25 lbs	As above backlight
HP OmniGo 100	1 MByte RAM 3 MByte ROM	16-MHz 80186	2 AA	11.6 oz	240 × 240 touch sense
HP OmniGo 200	As above	As above	As above	As above	"Holo-graphics design"
HP OmniGo 700LX	2 MByte	7.9-MHz 80186	2 AA	14 oz	640 × 200 4 gray lvls
Portable PC circa 1995	4–16 MByte	33-MHz 486	1–6 hrs NiCd	5–10 lbs	640 × 640 40 sq. in.
High-end laptop PC circa 1996	16–32 MByte	133-MHz Pentium	1–6 hrs Li Ion	5–10 lbs	800 × 600 256 colors
High-end desktop PC	32–64 MByte	200-MHz Pentium Pro	NA	50 lbs w/ display	1280 × 1024 16 million

improved operating life with fewer batteries, indicating successful reduction of power demands. Fifty to 100 hr of operation is common for the latest PDAs.

8.3.1.3 Communicators

Communicators are an emerging class of devices between the extremes of a PDA and a cellular telephone. They integrate the PIM applications common to PDAs with a digital modem and cellular telephone. The next generation of two-way pagers, which provide Internet-based electronic mail functions, may evolve into this class.

An example is the Nokia 9000 communicator. When closed, it looks like a pocket cellular telephone. When opened, it looks like a calculator PDA. One of its features is its "Internet" button. Pressing this will automatically connect you with an Internet service provider. A simple World Wide Web browser is built into the device.

These devices can execute "built-in" applications independent of network connectivity. There are also devices which are little more than displays. An example is the Berkeley InfoPad. The Pad is a low-power, portable multimedia terminal capable of displaying text and graphics, playing audio and compressed video, recording audio, and capturing pen inputs [26]. The InfoPad depends critically on wireline computing resources to achieve basic functionality [35].

8.3.1.4 Hardware Component Technologies

Processors

A desire for low power and long battery life places limits on the performance of an embedded processor, since clock cycle directly influences power consumption. Several vendors support processor cores for PDAs. Intel's Polar chip set is based on the 386. The Draco chip set is a two-chip implementation of a low-power 486 architecture. The AMD Elan chip set is based on the 386, incorporating the logic for a PDA on a single chip. NEC's V800 architecture contains 16-bit and 32-bit instructions to improve code density and reduce memory accesses. The V820 integrates system logic to reduce chip count for a full PDA. The Sony MagicLink is built around the Motorola Dragon 68349 processor, consisting of a 68020, 6K bytes of memory, and system logic.

The Motorola PowerPC offers a case study spanning many markets. The original 601 was designed for desktop workstations. A low-voltage version, the 601v, operating at 2.5V, has also been introduced. The 602 is a low-power version, specifically oriented toward consumer devices. It operates at 3.3V, and supports a variety of power saving modes. At full 66-MHz operation, it consumes 1.2W. Performance without an external cache is rated at 40 SpecInt92s (one SpecInt92 is equal to the performance a Digital VAX 11/780 on a well-defined set of benchmark programs). The 603 is designed for portable applications, but at higher performance. At 80 MHz, the 603 consumes 3W and achieves 75 SpecInt92. The 603e-200 is a 200-MHz version that operates with a 2.5V power supply. The 603e-240 operates at 240 MHz. It consumes 3.5W at 100 MHz, 4W at 166 MHz, and 5.5W at 240 MHz. The 604 and 620 chips are designed for higher performance desktop configurations, and consume substantially more power, reaching 30W at 133 MHz for the 620 for a predicted performance of 225 SpecInt92s.

Another case study is the MIPS R4100 family. Its 64-bit architecture supports (1) a standby mode that suspends internal operation, but continues to cycle the processor's external bus, yielding a 90% power reduction; (2) a suspend mode that halts all activity, resulting in a 95% cut in processor power demands; and (3) a hibernation mode that shuts down the whole processor, requiring a hard reset to wake up.

Our last case study is the ARM processor. It is a RISC-based architecture, operating at 33 MHz at 1W, able to achieve 30 Dhrystone MIPS (Dhrystone is a synthetic workload of typical instructions found in job mixes). A highly integrated chip, it includes support for 8-bit audio, 256-bit color video selected from a 16 million color palette, and 16-gray level support. It supports the direct attachment of dynamic random access memories (DRAM). It integrates support for PCMCIA devices. A typical system consists of the 7500, 8-DRAM chips and a PCMCIA support chip. At 33 MHz, with a 5V power supply, the ARM 7500 consumes 0.5W when driving an LCD display.

Disk Drives

Disk drive capacity has increased while their physical sizes have decreased. Yet the 2.5-in disk form factor is still too large for handheld devices. Because disk drives represent the second most power-consuming component, today's PDAs come with a limited amount of battery-backed up SRAM—typically under 4 MBytes—which can be extended with expensive flash RAM on PCMCIA cards. These cost approximately $100 for 1 to 2 Mbytes. The market is waiting for small devices to take off, so that this technology can start following the traditional semiconductor decline in prices.

LCD Displays

Color LCD technology is poised to become ubiquitous. The most common technology is transmissive color twisted-nematic (TN) or super-twisted nematic (STN) LCD with internal backlighting. Backlighting consumes significant power, and transmissive devices do not work well in direct sunlight.

Reflective displays are progressing rapidly. Holographically formed polymer-dispersed liquid crystals (H-PDLCs) and polymer-stabilized cholesteric texture liquid crystals (PSCT) appear viable [36]. The latter reproduces color by stacking red, green, and blue panels, reflecting up to 50% (compared with 3% to 5% in transmissive LCDs). The H-PDLCs have the potential to reflect even more incident lumination. The technology uses optical interference techniques to set up planes of modulated liquid crystal droplets at predesignated positions. These reflect light of a certain wavelength when in the off state (the droplets are misaligned) and transmit the light when on. The effective reflective index can be varied with voltage, thus controlling the intensity of the reflected light.

Batteries

Battery weight will continue to be a significant factor in system portability. There are two major approaches: (1) reduce power demands through low power design and (2) exploit new battery technologies. Modern systems exploit both strategies.

The main issue is battery weight and volume. These determine the battery's energy storage. Figures of merit include watt-hours per pound (battery carry weight) and watt-hours per cubic centimeter (battery size). Consider a portable device with a power demand of 2W. If its battery is rated at 20 W-hr per pound, and it weighs 1 pound, then it should be able to power the system for 10 hr.

In metal-oxide-silicon (MOS)-based processors and associated electronics, power consumption is dominated by switching power. This is described by the following equation:

$$\text{Power} = \text{Capacitance} \times \text{Voltage}^2 \times \text{Frequency}$$

From this, it is clear that power demands can be reduced by one of three approaches:

- Reduce capacitance by increasing VLSI integration, reducing chip counts, and increasing advanced packaging technology, such as multichip modules.
- Reduce voltage by dropping from 5V to 3.3V to 2.5V and below. Savings here dramatically reduce power demands because of the squared term.
- Reduce frequency by operating at lower speeds, especially when the system is not actively used. This implies intelligent, power-sensitive system operation, exploiting spin-down strategies for the elements that consume the most power.

The four major battery technologies in use in today's portable systems, nickel-cadmium (NiCd), nickel-metal-hydride (NiMH), lithium ion (LiIon), and lithium poly (LiPoly), are plotted in Figure 8.4. The figure shows watt-hour per kilogram versus watt-hour per liter. Improved W-hr/kg yields lighter

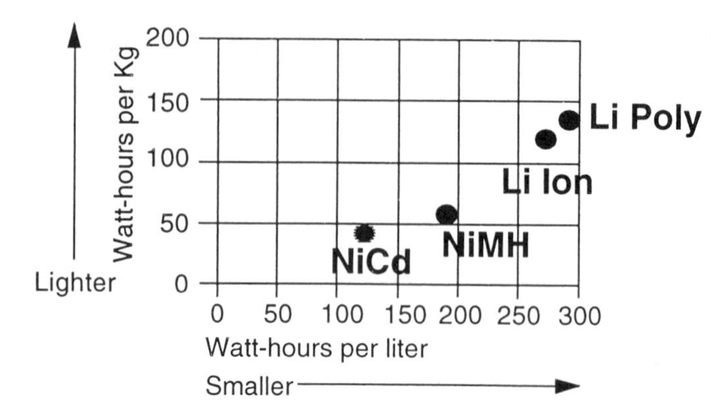

Figure 8.4 Battery technology.

weight batteries for same power demand. Improved W-hr/liter yields a smaller battery for same power demand.

NiCd batteries are the most widely used. Most laptops have migrated to NiMH, yielding better energy storage for carry weight and substantially improved energy storage per volume. Batteries in these technologies are constructed from steel cylinders containing hazardous liquid electrolytes. NiCd and NiMH batteries suffer from "memory effect": They operate best when fully discharged before being recharged.

LiIon batteries have begun to appear in the newest laptops and cellular phones. Their energy storage capability is more than twice that of NiCd technology. LiPoly batteries are about 10% more efficient than LiIon. Both use solid electrolytes, making it possible build the batteries as "sandwiches." This allows the batteries to be formed into arbitrary space-filling shapes, a significant advantage versus other battery technologies.

8.3.1.5 Video and Speech Encoders/Decoders

A key requirement for mobile wireless multimedia systems is the ability for source coders to make efficient and flexible use of the available bandwidth. As the topology and other network attributes change, the bandwidth and reliability of individual links also change. Voice and video source coding algorithms support rate-adaptive compression. The source must adjust its rate based on the network's QoS. In addition, source coding permits "on-line" rate adjustment based on changing conditions. For example, multiresolution decompositions of the video stream assign different portions of the encoded bitstream to different priority levels. The priority information is used at lower network layers in decisions regarding how packets containing voice and video information are handled (e.g., if bandwidth becomes scarce, low-priority packets are dropped).

Video Coding Algorithms for Wireless Environments

To reduce complexity and increase robustness, algorithms for wireless video place less emphasis on motion compensation than on intraframe coding. Furthermore these algorithms employ low-complexity transforms and quantization schemes. Coding and synchronization techniques are not only for protection from burst errors, but also for recovery from deep fades in which the signal is lost for several seconds. They adapt to changing bandwidth allocations from the network, and to changing error characteristics of the channel.

The transform constitutes one of the most important elements of a compression algorithm. While transforms do not perform compression, they furnish an alternative representation that can be compressed in the subsequent quantization step. The block discrete cosine transform (DCT) is the transform employed in JPEG [37] and MPEG [38], as well as in video conferencing standards such as H.261. The DCT performs well when low compression ratios are used, but at high compression ratios severe blocking artifacts result in the reconstructed image. Wavelet transforms offer an alternative and can lead to efficient still-image compression if the right filters are used. The principle behind the wavelet transform is to hierarchically decompose an input signal into a series of successively lower resolution reference signals and their associated detail signals [39,40]. This representation is very useful to lower networking layers which use the range of available priorities to decide which layers to forward in case of scarce bandwidth.

An example of a discrete wavelet transform (DWT) algorithm specifically developed for a wireless application is described in [41]. The coding scheme consists of a six-level DWT followed by scalar quantization, run-length coding (RLC), and Huffman coding, with error protection provided by convolutional codes. Each step was modified for the wireless system. Scalar quantization is performed adaptively on a subband-specific basis. Coding rate adaptivity in response to changing network bandwidth allocation is achieved by varying the quantization factor uniformly over all subbands. Rate control to within 0.02 bpp (bits per pixel) is achieved by feedback from the destination. Quantization rate adjustment is the most effective way to reduce the rate at the source if network feedback is available. For a more drastic rate reduction, one or more of the subband streams can be dropped if bandwidth becomes scarce. This requires a local decision at the network node or gateway, based on local network measurements and on packet priorities, without source intervention or end-to-end feedback.

Speech Coding

The design of high-quality speech coders for wireless networks is a challenging task; good quality must be maintained with low power consumption under time-varying channel conditions and limited bandwidth. The design must account for parameters such as bit rate, delay, power consumption, complexity, and quality of coded speech. Noise robustness, hierarchical or embedded encoding, and low delay are critical. Noise robustness is important because of fading and interference in the wireless channels. Embedded encoding allows for rate adaptation; as the bit rate is reduced, the degradation in speech quality is graceful. Low delay is important since delays of 30 ms or higher are noticeable to human listeners.

In the past, speech codec design was mostly driven by bandwidth-efficiency considerations, the target application being telephonic where the channel does not vary and the signal-to-noise ratio (SNR) is high. For example, code excited linear prediction (CELP) based coders are popular because of their low bit rates. Their performance deteriorates in the presence of background noise, and the coders' complexity is high.

A speech codec that overcomes these limitations was reported in [42,43]. It has the characteristics of low delay, adaptive bit allocation, and quantization, and is noise-robust. The coder consists of analysis/synthesis filter banks, a perceptual metric, an adaptive bit-allocation scheme, and a quantizer. The subband coder processes input frames of 20 ms at 8 kHz. A significant aspect of this design is rate adaptivity achieved with an embedded code. The quantizer was made embedded by enabling truncation of the least significant bits (LSB) of the index. The network takes advantage of this by dropping the least significant bits first when congestion or reduction of bandwidth is detected.

8.3.2 Mobile Applications

8.3.2.1 Collaboration Technologies

Mobile systems are motivated by their ability to provide time-critical information to users on the move. A common scenario is emergency response operations after a natural disaster. Conventional communications infrastructures may not exist or have been disabled. Rapidly deployed wireless infrastructures can restore communications quickly. In such scenarios, group collaboration must be supported.

The collaboration tools developed for the multicast backbone, or *MBone* [44], are becoming popular on the Internet: *vic* for video conferencing, *vat* for audio conferencing, and *wb* for a shared whiteboard and writing surface [23]. The tools process real-time packet video and audio streams in a variety of standard digital formats. These streams are delivered without a requirement that they be transmitted reliably. A shared whiteboard enables the exchange of annotations or the presentation of image and graphic information among collaborators. Whiteboard data must be delivered reliably.

A key enabler for group communications is multicast protocols [45]. IP multicast extends the best effort unicast model of the Internet to enable efficient multipoint, multiparty communications. This is achieved by delivering packets along a spanning tree rooted at the source host, duplicating packets to recipients only at the branch points in the tree. This ensures that no packet is sent more than once over any physical link.

Multiparty communications introduces two problems: the possibility of congestion and the need to manage heterogeneous receivers. It does not make sense to forward to a given receiver more data than can be handled by the bottleneck link between it and the source. Doing so only introduces additional congestion, causing packets to be dropped, and reducing the quality seen by participants. Higher level protocols have been developed that allow the sender and receiver to communicate about the successful delivery of packets. This is the real-time transport protocol (RTP) [46]. Should the number of delivered packets fall below a threshold, the receiver alerts the sender, who adapts by sending at a lower rate. The sender occasionally probes the network by sending at a higher rate. This approach is an example of so-called *application-level framing* [47].

The whiteboard application represents a different challenge. Because data must be delivered reliably, scalable reliable multicast [48] is needed. The application determines how to deal with data received out of sequence. For the whiteboard, it is not so crucial that a drawn line be delivered in strict segment-by-segment sequence, as long as all of the data are eventually received and rendered. The whiteboard application can buffer received segments and explicitly request repair retransmissions for those not successfully delivered, while simultaneously rendering those that have been successfully received.

Wireless communications adds extra complexities when delivering these real-time streams over the lossy, bandwidth-constrained wireless channel. The schemes mentioned work well because they require little explicit support from the network. The applications gracefully adapt to the bandwidth and error rates exhibited by the network.

8.3.2.2 File Systems that Support Mobility

Ficus and Coda are file systems designed especially to insulate applications from variations in network connectivity. Disconnected operation may be desirable even when network connectivity is good, because this can conserve power or maintain radio silence.

We focus on Coda [29]. A client's application uses a local file cache to allow continued operation even when disconnected from its file servers. When connected, a local agent hoards files of interest. The cache services requests for file accesses without touching the network when the client is disconnected.

Cache misses appear as errors to applications programs. Once the client is reconnected, updated cached files are propagated to other users, and any conflicts are resolved.

Conflict detection and resolution must be implemented carefully. For structures like file directories, Coda resolves many kinds of concurrent updates. For example, the concurrent insertion of files with different names is straightforward. Less obvious resolutions are necessary if a user updates a file that has been deleted by another.

Coda supports special mechanisms for when the client is connected to its servers via a low-bandwidth link. It provides "trickle" reintegration that asynchronously and opportunistically spools changes in the local cache to the servers. It also provides a "patience" model that allows users to specify a service time threshold for cache misses, since waiting for a file to be downloaded over a slow-speed network could be intolerable.

8.3.2.3 Mobile Database Systems

Individuals will use their information appliances to obtain information, the latest news, or directions to a restaurant [49]. Frequently requested information could be filtered from broadcast information channels, so-called "push" communications. Broadcasting involves careful scheduling, balancing bandwidth against worst-case latency. If many subscribers are interested in the latest baseball scores, these should be scheduled frequently. If a substantially smaller user community is interested in cricket scores, we would expect this to be scheduled for broadcast at a much lower frequency, utilizing less bandwidth but also introducing increased latencies to obtain up-to-date information.

Users may not be willing to wait for the next broadcast of infrequently asked for information, or the information they need may be too individualized to be aggregated into a reasonable subscriber community. Such information will need to be explicitly "pulled" by the user, perhaps through explicit queries against databases.

With mobility, location becomes an important attribute in database queries. A user is interested not only in the number of taxis within 1 mile of 57th Street, but also the number of taxis within 1 mile of where they are standing now.

8.3.2.4 User Interfaces

Because portable devices have limited real estate for keyboards, there has been interest in icon- or pen-based interfaces. Some devices eliminate the keyboard, replacing it with function-specific buttons. These devices still support functionally specific "virtual" keyboards, which are rendered on the touch-sensitive display and selected with a stylist.

For devices with pen input, some support handwriting recognition and others capture pen strokes. Recognition has proven to be difficult. Some devices support a modified alphabet that is easier to recognize, but requires the user to modify her handwriting.

Video and image capture in small devices is becoming ubiquitous. Charged coupled devices (CCDs) are becoming ubiquitous because of video camcorders. Highly integrated CCD cameras have been declining in price and are being integrated into PDAs.

8.4 Wireless Communications and Mobile Networking Technology

8.4.1 Physical-Layer Considerations

8.4.1.1 Basics of Radio Propagation

The basics of radio propagation can be described in terms of three mechanisms: *reflection, diffraction,* and *scattering* [5]. Through these, it is possible for a radio wave to be received despite lacking a line of sight to the transmitter. Reflection occurs when a propagating wave impinges on an object that is large compared to the signal wavelength. Real-world objects like the surface of the Earth, buildings, walls inside buildings, and so on will cause waves to reflect. With the trend toward shorter wavelengths, smaller objects become reflectors. Diffraction occurs when the radio path between transmitter and receiver is obstructed by a surface with sharp irregular edges. Thus, waves bend around the obstacle, such as around the corners of buildings. Finally, scattering occurs when the incident wave is reflected in many directions simultaneously when hitting an object that is smaller than the wavelength. For the cellular and PCS frequencies, such objects include foliage, street signs, lamp posts, and even humans!

Mobility has a profound effect on the radio channel. The received signal power varies significantly with user location and time. Over short time periods or distances, the receiver sees rapid fluctuations in the received power. Contributors to this variability include *multipath* scattering from nearby objects, *shadowing* from dominant objects, and *attenuation* effects due to the distance between transmitter and receiver. Even when stationary, the received signal power may drop due to nearby moving objects. These are called *fades.* Short-duration fades arise because of destructive cancellation. This is also called *Rayleigh fading* or *frequency-selective fading.* Long-duration fades are caused by shadowing—an object obstructs the receiver from the transmitter—or attenuation. In free space, received power falls off inversely with the square of the distance, although other effects may cause a more accelerated reduction in received power.

This leads to the concept of *delay spread*, also known as *multipath dispersion*. Multipath propagation yields signal paths that are of varying lengths, with different arrival times at the receiver. This "smears" the received signal, causing intersymbol interference if old signals are delayed long enough. Delay spread limits how fast symbols can be sent. A rule of thumb sets the transmission rate at less than the inverse of twice the delay spread. Typical values in the wide area are 0.2 μs in an open field, 0.5 μs in a suburban environment, and 3.0 μs in a dense urban environment. Such numbers indicate a maximum transmission rate of 160,000 bps for urban environments. GSM uses additional techniques, described later, to sustain a signaling rate of 270,000 bps per channel. Since wireless LANs propagate over shorter distances, delay spreads are substantially shorter, permitting higher data rates.

How do radio designers mitigate these impairments to achieve higher data rates? The first technique is *antenna diversity:* the use of two antennas separated by a quarter wavelength to combine received signals. If one antenna experiences a short-duration fade, the second is likely to be outside of the fade. A second strategy is *equalization.* It works by correcting the received signal through a complex process that subtracts a delayed and attenuated image of the direct signal from the actual signal. The weightings for the subtraction are determined by occasionally transmitting an agreed-on training sequence. Its variation from the expected signal indicates how the receiver should correct receptions in the near future. GSM uses this technique to achieve its high data rate. The final technique is *forward error correction,* which allows the receiver to reconstruct lost data. Alternatively, the link can support automatic retransmission protocols, called automatic repeat request (ARQ).

8.4.1.2 Digital Modulation

In radio transmission, information is encoded by modulating a carrier wave. The carrier is modified with amplitude changes such as AM, frequency changes such as FM, and phase changes. AM modifies the signal's transmission power based on the input data symbol. FM modifies the carrier frequency as a function of this input. Noise has a greater effect on amplitude rather than frequency, so FM is preferred. FM is also a *constant envelope technique:* the carrier wave's amplitude does not vary with changes in the input. It is desirable because it leads to more power efficient amplifiers, an important concern for portable, battery-operated devices. Phase modification can be accomplished through frequency modification, and vice versa, so these approaches are interlinked.

Demodulation removes the carrier signal to obtain the original baseband waveform. Detection extracts the symbols from this waveform. *Coherent detection* is a technique that requires a replica carrier wave at the receiver. The receiver uses it to detect the signal, and cross-correlates the detected signal with the replica. This is called *synchronous detection.* Noncoherent detection does not exploit reference phases. It is called *envelope detection.* This yields a less complex receiver, but results in worse performance.

One of the modulation scheme figures of merit is *power efficiency.* This refers to its ability to preserve the fidelity of the digital message at low power levels, such as at the fringes of signal coverage. Noise immunity can be improved by increasing signal power. Power efficiency measures how much increased signal power is needed to achieve a particular bit error rate (BER). A second figure of merit is *bandwidth efficiency.* This measures the scheme's ability to accommodate data within a limited bandwidth.

There are numerous ways to encode binary information in a modulated carrier wave. *Amplitude shift keying* (ASK) turns the carrier on or off to represent 1 or 0. Because the information is encoded in the signal's amplitude, the approach is sensitive to noise. *Frequency shift keying* (FSK) represents 1 or 0 by two different frequencies slightly offset from the carrier frequency. *Binary phase shift keying* (BPSK) uses alternative sine wave phases to encode 1 or 0. Such a scheme is simple to implement, makes relatively inefficient use of bandwidth, but is very robust to noise.

To improve bandwidth efficiency, schemes have been developed for multilevel modulation. *Quadrature phase shift keying* (QPSK) encodes two bits per symbol by modulating a carrier wave by shifts of $\pi/2$ radians. While exhibiting twice the bandwidth efficiency of BPSK, it requires a more complex receiver and suffers from some noise problems. A strategy to overcome this is *minimum shift keying* (MSK), a form of frequency shift keying that selects the minimum spacing between alternative frequency states that allows the modulated carrier to be smoothly meshed. *Gaussian minimum shift keying* (GMSK) is a variation that passes the waveform through a low-pass filter. It is extensively used in modern digital cellular and cordless telephones. Another widely used approach is *$\pi/4$-shifted QPSK.* The scheme superimposes two QPSK phase shift constellations, and limits phase transitions to smaller amounts. It is more spectrally efficient than GMSK, but less power efficient.

8.4.2 Media Access

8.4.2.1 General Issues in Media Access in Wireless Systems

Media access is the way a network node gains access to the physical media to transmit or receive data. The media are shared among several nodes, and a contention mechanism is needed so that only one transmitter is active at a time. Most MACs are *random access:* the mechanism allows nodes to access the media at any time. Thus simultaneous transmitters will result in collisions and failed packet delivery. Alternative schemes are based on scheduled or orthogonal assignments to the channel, through the allocation of a time slot or orthogonal code, made statically or dynamically on a demand basis.

Media access is complicated by two problems. The first is *hidden terminals.* As a sender transmits its packets to a receiver, the hidden terminal transmits, causing a collision and packet loss at the receiver. The hidden terminal might go undetected by the sender. This makes it impossible to use collision detection mechanisms like those in standard Ethernet algorithms. The second problem is the *near-far problem.* Radios are designed to capture the nearest strong signal. Those closest to a base station are more likely to gain access to the channel than those further away, leading to unfairness.

8.4.2.2 Contention-Based Schemes

The *Aloha protocol* was the first contention-based scheme. A node transmits whenever it desires. Because of hidden terminals, a positive ACK from the receiver is required to indicate successful reception. If the ACK is not received within an interval, the transmitter backs off and retransmits later. If transmitters adhere to transmissions during specified slot times, called *slotted Aloha,* the chance of collisions are reduced and channel utilization is improved.

A variation is the *carrier sense multiple access/collision avoidance scheme* (CSMA/CA). The transmitter listens before transmitting. If nothing is heard, the station gains access. Since hidden terminals remain a problem, explicit ACKs are still necessary. The sender backs off and retransmits if a collision is detected, usually because of a time out.

Inhibit sense multiple access (ISMA), or *digital sense multiple access* (DSMA), works well for cellular wireless networks. The base station transmits a busy tone whenever the channel is being used. A sender waits until the busy tone is removed before it can attempt to send. If a collision is detected, the sender backs off and retransmits later.

The Distributed Foundation Wireless MAC (DFW MAC) of the IEEE 802.11 specification uses explicit signaling to reduce hidden terminal problems. It is based on request-to-send (RTS) and clear-to-send (CTS) signaling (see Figure 8.5). An RTS from sender to receiver indicates that the sender wishes to communicate. CTS signaling, from receiver to transmitter, indicates that the receiver is ready and available. Without CTS, the sender will not transmit, since the receiver is probably busy. Data frames are acknowledged frame-by-frame. A negative acknowledgment (NAK) indicates that a transmitted frame was corrupted and needs to be retransmitted. The receiver may also send RxBUSY. This indicates that the receiver is currently busy, and the transmitter should try again later.

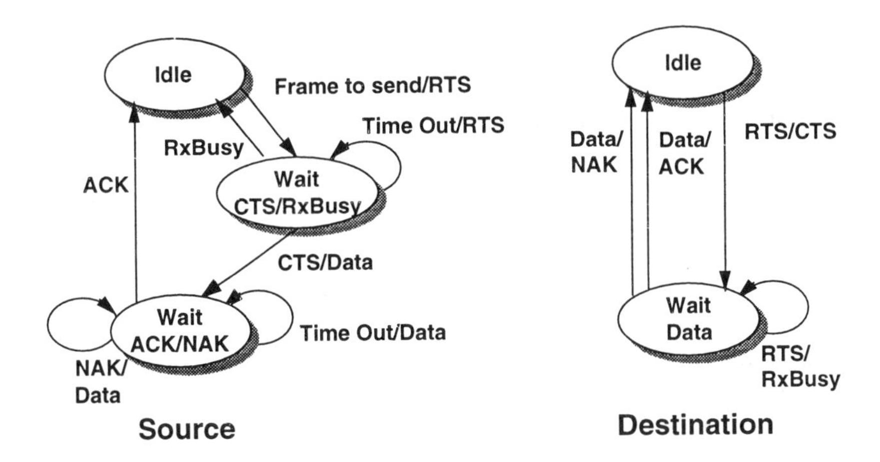

Figure 8.5 IEEE 802.11 MAC-layer protocol signaling.

8.4.2.3 Time-Division Multiple Access

Time division allows multiple users to share a channel through a time allocation scheme. Access to the channel is arranged into time slots, and these are allocated in a static fashion to individual users. It has several advantages. TDMA allows the channel to be shared among users without the contention inherent in CSMA schemes. Variable bit rates can be supported by ganging slots. It requires less stringent power control because of reduced interference. Mobile assisted handoff is possible because the terminal uses available receive slots to measure base station signal strength.

The technique also has disadvantages. It exhibits a pulsating power envelope that causes interference with devices like hearing aids. There is inherent complexity in the slot and frequency allocations. Finally, the high data rates imply a need for equalization.

Packet reservation multiple access (PRMA) combines TDMA with slotted Aloha [50]. Transmissions are organized into slots, a packet per slot. Frames are a logical grouping of slots. A slot within a frame may be available or reserved. The information is broadcast from the base station. Reservations are obtained when terminals contend for available slots. The base station indicates when data are received correctly; if not, the packet collision causes contending senders to try again later. The terminal succeeding in obtaining the slot reserves it subsequently. The base station detects when there are no data being sent in a reserved slot, thus indicating that the terminal is releasing its reservation.

8.4.2.4 Code-Division Multiple Access (CDMA)

Spread-spectrum techniques "spread" the radio signal over a wide band by modulating it with a unique code word. The receiver's correlator distinguishes a sender's signal from others by examining the wide band with a time-synchronized duplicate of the code word. The signal is recovered by a despreading process at the receiver. The spread spectrum waveform is more resistant to multipath effects and more tolerant of interference.

Direct sequence spread spectrum (DSSS) operates by sampling or "chipping" bits at a higher frequency than the data rate, thus spreading the signal's energy over wider frequencies. This permits diversity recombination at the receiver. CDMA techniques choose the codes so that the correct code reconstitutes the transmitted waveform; using the wrong code will yield a waveform that appears to be all zeros (see Figure 8.6).

Frequency hopping spread spectrum (FHSS) transmits data across a sequence of frequency channels (see Figure 8.7). Slow FHSS transmits multiple bits before hopping to a new frequency. Fast FHSS spreads a bit's transmission across frequencies. The ability to span a wide range of frequencies makes this system able to tolerate interference.

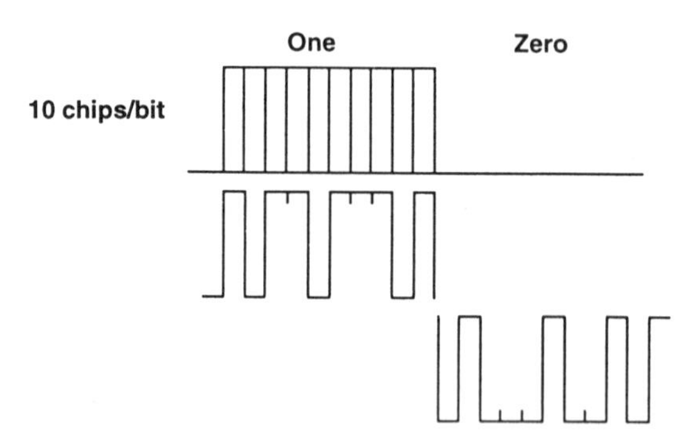

Figure 8.6 Direct sequence chipping sequence.

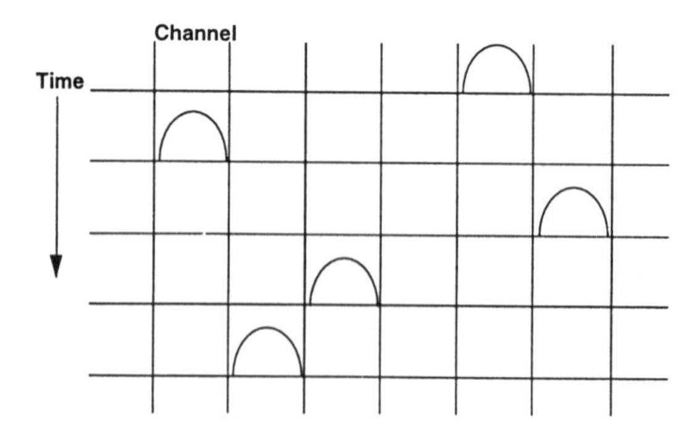

Figure 8.7 Frequency hopped spread spectrum.

8.4.3 Mobile Networking

8.4.3.1 Cellular Concepts

The attenuation of radio waves can be exploited for high capacity using the concept of frequency reuse. A given set of channels is used within a cell. If the same channels are used in an adjacent cell, interference will arise. If the users of the same frequencies are kept apart, the communications link can be designed to tolerate a low level of interference.

A common reuse pattern is shown in Figure 8.8. Each hexagon partitions the channels into one of seven sets. A *cluster* of seven cells covers the available frequencies. Cell diameters are chosen so that the cochannel interference is sufficiently low.

A system design chooses a reuse factor to maximize the capacity per area, subject to interference. If cell size is constant and spectrum allocation per cluster is fixed, then more cells per cluster yield fewer channels per cell, less capacity, and less cochannel interference because the cochannel cells are farther apart. Less cells per cluster implies more channels per cell, more capacity, and more cochannel interference.

The GSM reference terminology is shown in Figure 8.9. A mobile switching center (MSC) handles call routing for a region. It interfaces the mobile system with the public switched telephone network (PSTN). The home location register (HLR) organizes call redirection/routing information for local subscribers. The visitor location register (VLR) locally stores subscriber information for visitors. It participates in the registration of subscribers moving into the region. The authentication center (AuC) and the equipment identification register (EIR) are used to grant service to mobile stations (MS). The EIR maintains a list of legitimate, fraudulent, and faulty MSs.

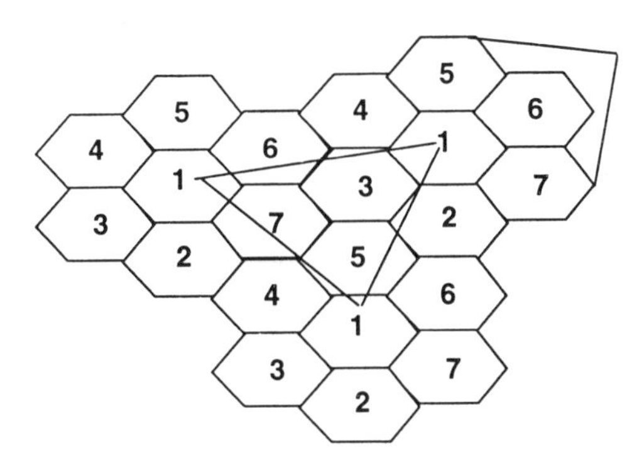

Figure 8.8 Cellular reuse concept.

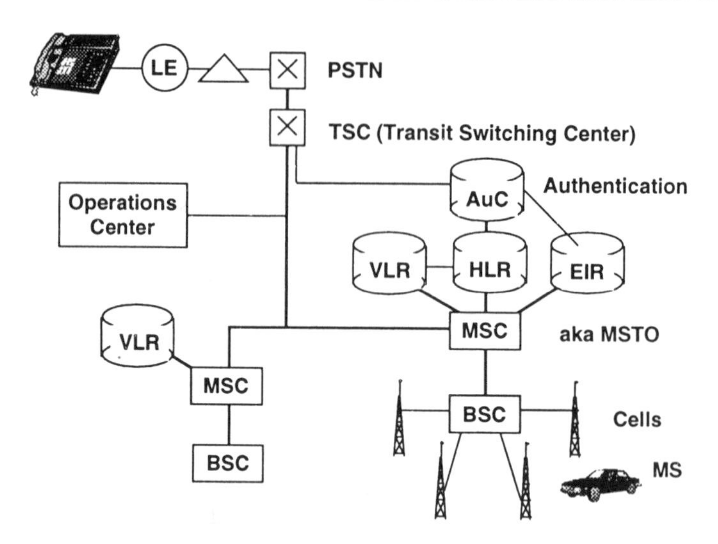

Figure 8.9 GSM terminology.

For AMPS, a call to a mobile subscriber goes as follows. The MS is located within the region. The MSC dispatches a request to all of the BSs it controls. The mobile equipment ID is broadcast as a message over all forward control channels. The MS responds on a reverse channel. This is detected by a BS, which relays the ACK to the MSC. Then the MSC instructs the BS to assign the call to an unused voice channel. The BS communicates the frequencies to the MS, which tunes to the channel. An alert is sent to the MS to cause it to ring. The call is now "in progress." The MSC modifies the transmit power as the MS moves around the BS, to minimize the MS's power drain. As the MS moves out of range of its old BS in the direction of a new one, the MSC chooses to assign it to a frequency of the new BS. This is handoff, and its purpose is to maintain call quality as users move. The control signals to instruct the MS to tune to new frequencies or to adjust its power are transmitted in-band, at a frequency that is not noticed by the human ear.

IS-54 offers an alternative: mobile assisted handoff (MAHO). Signal measurements at the MS determine when to hand off. These occur during unused time slots when the MS tunes to adjacent BSs to take signal strength measurements. The MS stores signal strength and bit error rates for several BSs. The MSC obtains these measurements on demand. The MSC is still responsible for causing the handoff to take place.

8.4.3.2 Mobile Internetworking Routing Protocols

Mobile IP directs traffic to mobile users through indirection [51]. A flow takes place between a correspondent host (CH) and the mobile host (MH). Every node has an identifying address: its IP address. Packets contain header fields for the sender (IP_{CH}) and the receiver (IP_{MH}). Mobile IP circumvents the association of IP addresses with specific networks. MHs can attach and detach themselves from networks as they roam the Internet. Changing the IP address is insufficient. It could take days for the mapping between the original node name and its new IP address to propagate throughout the network.

In the Mobile IP specifications [52,53], every MH has a home network and a global IP_{home} address. In the MH's home network resides the home agent (HA), intercepting packets destined for IP_{home}. These are encapsulated inside another packet. The sender address is that of the original CH. The care-of address is that of another agent in the roamed in network: the foreign agent (FA). The packet is sent via IP routing to the FA, where it is deencapsulated and delivered to the MH. The MH can communicate directly to the CH, but the CH must always indirect through the HA. The MH can request a local IP address by invoking the dynamic host configuration protocol (DHCP). This address is used by the HA directly, obviating the need for a foreign agent.

When an MH enters a new subnetwork, it first discovers an FA willing to provide it with redirection. This is built on the existing Internet control message protocol (ICMP) for router discovery. Once accepted locally, the MH registers its new care-of address with its HA. Registration attempts must be authenticated. A malicious user could hijack an MH's packets by furnishing its own care-of address. Mobile IP uses a challenge/response method based on a secret key shared by the MH and the HA. Only an MH that knows the secret can compute the authentication response given the challenge.

The major performance challenge is circumventing the dogleg routing between CH, HA, and FA. This is eliminated if the CH caches bindings between IP_{home} and the MH's care-of address. Maintaining these bindings is called *route optimization.* For example, when a CH first sends a packet to an MH via its HA, the HA sends a binding update to the original sender. Until the binding expires, the CH use's the care-of address. If the MH moves to a new subnetwork, it requests its old FA to forward packets to its new address, while alerting CHs to the new address. To avoid long forwarding chains, the bindings time out, forcing the CH to flush its cache and reestablish the binding.

8.4.3.3 Handoff

The easiest way to increase network capacity is to decrease cell size (cell splitting). In addition to increased capacity, transmission power is also reduced, increasing the battery life in mobile devices. However, the number of cell crossings increases per-unit time, placing a greater demand on the network's ability to support handoffs. Frequent crossings also impact the quality of service. Network connections may be extended from one base station to the next, requiring forwarding. This increases the latency by which data are delivered to end devices. It is also advantageous if the network's spectrum resources can be allocated to where they can be best be used; that is, where the most users are located.

To make handoffs seamless, it is desirable to minimize handoff latency, the frequency of handoff and its effects on QoS, the probability of dropping connections across handoffs, and "call blocking" and the effects of admission control. Several strategies have been suggested for addressing these. Multicast routing can reduce handoff latency [54]. The system tracks the MS as it moves through the network, predicting the next BS. It then uses multicast routing to transmit packets both to the current BS (the forwarding BS) and to a small set of potential next BSs (the buffering BSs), which form the multicast group. Signal strength measurements at the MS determine to which BS to hand over. The one with the strongest beacon becomes the new forwarding BS, while the old forwarding BS becomes a buffering BS. Since packets have been sent to both, handover takes place with minimal delay. As the signal strength from a given BS goes below threshold, the BS is pruned from the multicast group. As a new BS is detected, it may be joined to the multicast group. Note that the mobile device makes all of these decisions.

Handoffs can be reduced by superimposing wide-area "macrocells" on smaller area "microcells," partitioning the spectrum among them. If the mobile is crossing microcell boundaries frequently, it is assigned to the covering macrocell. If a mobile remains in a given macrocell too long, then it is reassigned to a microcells.

When a mobile user moves into a new cell, there is no guarantee that spectrum will be available. Channel allocation schemes have been devised to accept new mobiles as they enter the cell [55]. Any resources thus reserved may limit the ability to support new connections from MSs in the cell. Thus, call blocking is traded against connection dropping. The techniques span from reserving a fixed number of channels (*guard channels*), to determining the number of reserved channels based on the density of users in nearby cells, to borrowing channels from nearby cells with few users, to scheduling channels to cells based on prediction of user motion or time of day.

8.4.3.4 Packet Radio Networks

Packet radios (PRs) provide communications to fixed or mobile network nodes using radios for the physical links. A packet radio network is fast to deploy and provides connectivity to moving units. PRs also introduce new complexities: the dynamic nature of the network topology and the half-duplex operation of the radios [56]. Packet radio networks can be single hop or multihop. The network can be constructed from fixed infrastructure, or can form its connectivity in an ad hoc manner [16].

Physical/Link Layer Considerations

Physical connectivity depends on local RF propagation. Cell sites cannot be surveyed in advance. It is not possible to take advantage of larger antennas or greater antenna processing. PRs are identical, highly portable, and always moving. Terrain effects, distance between nodes, and antenna height directionality are critical issues for maintaining physical connectivity.

Military systems make extensive use of spread-spectrum methods. They demand more flexibility in the coding schemes. Preamble spreading codes may be different from those used for the data portion. Codes could also be changed bit by bit to reduce the probability of interference. Transmitter power is chosen not only to ensure successful reception at the receiver; it must also be selected to hide the network from adversaries.

For broadcast reception, all transmitters use a common preamble code. A transmitter can also use a receiver-directed preamble, directing its transmission to a node that has "tuned" to that code. Several packets can be in the air, yet be unlikely to interfere. A transmitter can also send information in the preamble instructing the receiver about the waveform used for the data portion. Codes can be changed on a packet-by-packet basis.

For the data link, automatic repeat request (ARQ) and forward error correction (FEC) techniques are used. These must be designed to cooperate with spread-spectrum systems; the possibility of correlated codes for at least some part of the packet is high.

PRs are designed to operate in half-duplex mode. A radio transmits a packet, transitions to receive, receives an ACK, receives the next packet for forwarding, and transitions back to transmit. It cannot receive a new packet while it is transmitting a packet.

There are two ways to implement hop-by-hop ACK. The first makes ACK signaling explicit, giving them higher priority than data. A second scheme uses echo or passive ACKs. When a downstream radio forwards its packet, it can be heard upstream. This can be interpreted as an implicit ACK that the receiver obtained the original packet.

This reduces the number of explicit messages between nodes. However, the latency from when a packet is sent to when it is ACK'd will be longer, since forwarded packets are placed at the bottom of the receiver's transmission queue. The node cannot send the next packet until the preceding one has been received. Because longer packets have greater interference probability, using shorter explicit ACKs may improve performance. Note that passive ACKs cannot be used with receiver-directed spreading codes.

Multihop Multimedia Networks

The major challenge is to account for resources so that bandwidth reservations can be made. In single-hop networks such accounting is simple; MHs learn of each other's requirements, directly or through a BS. This solution can be extended to multihop networks by creating clusters of radios. Access is controlled and bandwidth is allocated in each cluster. Clusters have been used before in PR nets, but mainly for hierarchical routing rather than for resource allocation [24].

Several architectures are reported in Figure 8.10. Cluster and Token-CDMA make use of clusters. SWAN and Virtual Net rely more heavily on CDMA for channel sharing and define clusters of two nodes communicating with a proper code. While in Cluster and Token-CDMA the clusters are spatially separate, in SWAN and Virtual Nets the clusters consist of two nodes and are overlapped. Regardless of the method, a common requirement is dynamically reconfiguring clusters despite mobility and failures.

Another challenge is to reroute VCs and reallocate bandwidth when radios move. This is critical in multihop networks where not only sources and/or destinations move but any node along the path can move. To overcome this, it is necessary to introduce a "soft state" VC scheme coupled with fast reservations a la PRMA [50]. Power control is essential in multihop networks, to

Layer \ Alg	Clustering	SWAN	Virtual Net	Token CDMA
Network	call management with QOS acceptance, fast VC, datagram			
	QOS routing (loopfree B-F)	probing, power adj. min. power routing	source routing	QoS routing (loop-free B-F)
Link Control	ACKs backpressure, priorities, voice/video rate adjustment			
	low priority drop	adaptive power adj.	power control	low priority drop
MAC	TDMA, CDMA PRMA (voice/video) Slotted CSMA (data)	TDMA, CDMA with channel probing	TDMA, CDMA	Token, CSMA CDMA
Connectivity Management	lowest ID clusterhead election	link probing, adaptive power control	virtual topology maintenance	highest degree clusterhead election

Figure 8.10 Alternative network architectures.

achieve efficient spatial reuse. This is particularly critical in CDMA channels in order to mitigate the near-far effects [57].

Network Routing

For routing, link existence and quality must be determined. The best route is the one with the smallest number of hops with "good" connectivity. Link quality can be determined by measuring signal strength, SNR, or bit error rates [24]. Link quality can be improved through higher transmission power, or more aggressive error correction or retransmission schemes. Link capacity is not only a function of its own traffic; traffic on nearby links also affects the capacity. Network conditions vary more rapidly in PR nets, with links being lost and new ones established. Network management must disseminate connectivity information more rapidly than in the wireline case. Network partitions are also more likely.

Routing algorithms choose a good path based on link connectivity. Three schemes are possible: *flooding, connection-oriented,* or *connectionless.* Flooding transmits a packet on all links from the source to immediately connected nodes, which do the same. More sophisticated schemes generate additional routing table updates, which are likely to be out of date with rapidly changing topologies.

Connection-oriented routing (point-to-point routing) maintains a sequence of hops between any source and destination. Keeping this information up to date, despite topology changes and network partitions makes this strategy unusable for many networks.

The final scheme is connectionless routing. This requires no knowledge of end-to-end connections. Packets are forwarded toward their destination, with local nodes adapting to changes in network topology.

Routing information must be distributed through the network. For small networks, this is accomplished by a centralized routing server. Distributed algorithms behave better. Each node determines the best next hop in the direction of the destination. To keep this information up to date, routing tables are periodically exchanged among neighbors. In the ARPA PR net, *packet radio organization packets* (PROPs) are broadcast by each node every 7.5 sec. Nodes thus identify their one-hop neighbors. A tier table is constructed as the messages flood the network: a node that receives a PROP indicating that the originating node is n hops away and stores the node as being $n + 1$ hops in its routing table (unless it already knows of a shorter route). When this completes, each radio knows how many hops it is from each other radio as well as the neighbor that sent it the message indicating the shortest distance. When a node discovers a new next node with better link quality for a given destination, it passes this information on to its neighbors.

There are schemes combining centralized and distributed approaches. For very large packet radio networks, it is useful to impose hierarchy on the network topology, thus hiding far away changes. The next hop routes to these far flung nodes are unlikely to change as rapidly as routes within a cluster. The *cluster head* maintains hop-by-hop routes for each node within a cluster and routes in the direction of "border radios" for intercluster packets. An added complexity is that clusters are born and die as nodes move. Sophisticated algorithms are needed to manage these processes.

A final issue is packet forwarding. If several transmission attempts fail to deliver a message to the next node along the route, the message is broadcast to any node that can complete the route. To avoid loops, the forwarding node must be closer to the destination than the transmitting node. Thus localized rerouting techniques can fix broken routes.

Packet forwarding causes flooding if multiple nodes hear the forwarding request and choose to forward. An optimization to reduce this is "filtering based on overheard traffic." If a packet is queued at a node, and it hears it being forwarded from a different node, it assumes this to be a passive ACK, and removes the packet from its send queue.

Network Management Considerations

Additional algorithms are needed for congestion and flow control to ensure that resources are not oversubscribed, so that packets are likely to reach their destination. One approach is to reserve the needed resources on the source to destination path; if the request cannot be satisfied, the connection is refused. These techniques are described in more detail in Section 8.5. Advanced reservation is difficult in a PR nets because of the potential for rapid topology changes.

It is possible to control the rate at which packets are being forwarded. A sender cannot deliver packets because the downstream is congested. A pacing protocol slows the retransmission attempts by increasing the interval between successive attempts.

8.4.3.5 Wireless Local-Area Networks

The two technologies used for implementing wireless LANs are *infrared* (IR) and *radio-frequency* (RF) [58]. Each has strengths and weaknesses. IR transmitters and receivers are inexpensive. Virtually every laptop computer and PDA is now being sold with a direct IR interface. IR obviates the need to deal with spectrum regulation. Since walls and windows are opaque to IR, it is possible to achieve high spectrum reuse. But IR is prone to interference from the sun and incandescent light. It cannot be used outside or inside buildings with substantial sunlight. IR is susceptible to shadowing effects from objects moving near the receiver/transmitters.

RF uses more expensive components than IR. RF covers areas inside buildings that are too large for IR, such as large corridors or classrooms. RF disadvantages include the potential for expensive licensed operation, except in the unlicensed bands, and the need to design for the complexity of radio propagation and interference.

For IR, fading is not a problem because the wavelengths are very short, and the transmission distances are short. The detectors are large with respect to the wavelengths being detected. These issues represent a major problem for radio waves.

Direct IR has been demonstrated to operate over a kilometer with little difficulty. Diffused IR has a practical propagation limit of 20m. Given a theoretical limit of 260 Mb-m/s, it should be possible to achieve 13 Mbps in a room with a dimension of 20m. A comparable RF picocellular system contains cells in the range of 50m to 100m in diameter, though the effective coverage depends on local geometric details, like corridors. For both technologies, intersymbol interference limits effective bandwidth.

8.5 Supporting Quality of Service for Multimedia Applications

8.5.1 Connection-Based Versus Connectionless Approaches

Quality of service (QoS) refers to the traffic-dependent performance metrics associated with a connection, such as the bandwidth, latency, or likelihood of message loss that a connection can tolerate for the data it is transporting. QoS requirements become guarantees through admission control mechanisms invoked when a new connection is initiated. It will be aborted if its QoS requirements cannot be guaranteed by the network [59].

The telephone system has been designed to meet certain quality levels. These include metrics such as the perceived audio/speech quality, worst-case end-to-end switching delays, and likelihood of call blocking. Certain values for these are determined, and the network is provisioned to meet the desired number of connections. Performance guarantees are thought to be important for real-time, interactive applications.

The Internet is based on a best effort model, without guarantees. This is a good match for data that comes in interactive bursts (e.g., paging, messaging, e-mail, or "user in the loop" data entry), interactive (e.g., file transfer or Web access, where the user waits for completion) or asynchronous bulk transfer (e.g., net news).

Best effort service allows the network components to be simple. This explains the rapid diffusion of TCP/IP, operating over many link technologies, including twisted-pair, coax, fiber optic, wide-area wireless, local-area wireless, packet radio, and satellite.

The best effort model makes real-time implementation more challenging. Existing Internet applications are sensitive to losses but tolerant to latency. Real-time applications, on the other hand, are tolerant of losses but sensitive to latency. Despite this, real-time point-to-point and multipoint conferencing has been successfully developed. Examples include RealAudio audio broadcast, Internet telephony, CUSeeMe conferencing, and vic and vat conferencing applications. The latter achieves good real-time performance by using application level protocols such as RTP [46].

RTP is appropriate for real-time data streams that can tolerate some losses. Video and audio streams are examples of loss-tolerating data types. They can adapt to jitter by allocating buffering at the receiver. An RTP source moderates its transmission rate based on reports of successfully received data. If the sender's rate exceeds that being received, the sender reduces its rate. Periodically, it attempts to increase its rate to discover if a greater data rate can be supported. In this way, the sender and receiver adapt to the available bandwidth in the network without requiring any support from the network.

8.5.2 ATM's Approach for QoS

ATM networks form the foundation of integrated voice, video, and data services, combining the flexibility of the Internet with the telephone system's ability to guarantee performance [59]. ATM networks have some advantages. ATM scales to very high data rates because of fast switching and data multiplexing based on fixed-format, small-sized "cells" containing virtual circuit identifiers rather than full network address (using this approach requires that connections be established before the first data can flow between the connection's endpoints). ATM implements sophisticated queuing that enables priority-based schemes. ATM's inherent connection orientation, or virtual circuits (VC), is critical for QoS. Connection setup allows the network to perform resource allocations and balance the traffic demands across links.

ATM's VCs are unlike conventional circuit-switched networks. Rather than allocating exclusive resources to connections, VCs are statistically multiplexed over shared physical links. The connections share resources and can interfere with each other. ATM's VCs do not guarantee reliable, in-order delivery. Out-of-order delivery is unlikely since the cells follow same paths. Reliable transport is left to higher level protocols.

ATM's connection orientation presents some problems for mobility. Movement between cells requires existing connections to be torn up and reestablished, a time-consuming task that changes the existing semantics of connections [60]. Even short data transmissions must invoke the full latency of setup and negotiation for resources. Finally, ATM is not a good match for lossy wireless links. ATM has no agreed-on mechanisms for recovery or retransmission at the link layer. There is a question whether ATM will be an end-to-end technology or a link or subnetwork technology. Unless native ATM connectivity exists between the ends of the connection, service guarantees cannot be enforced. There are already millions of nodes on the Internet that are not attached to ATM networks. To ensure interoperability, it will be necessary to run TCP/IP over ATM. Interfacing ATM's guarantees with the emerging Internet capabilities for flows and reservations remains a topic of lively debate.

QoS can be categorized into *guaranteed service* and *best effort service.* The former is achieved if the connection obeys its traffic descriptor. If the network determines that it can support this, it allows the connection to be established and guarantees that its requirements will be met. The requirements may consist of specifications related to bandwidth, delay, or the probability of packet loss. Guarantees of this kind are required when the application must have tight coupling between the connection end points.

ATM offers two kinds of guaranteed service classes: *constant bit rate* (CBR) and *variable bit rate* (VBR). CBR is appropriate for constant rate source data streams that demand a small delay jitter. It is motivated by constant bit rate encodings for audio and which places limits on latency. VBR is appropriate for traffic patterns that have a sustained rate but are bursty. The peak rate may occur only for short bursts. These classes tolerate higher delays and variations in delay. QoS is specified as worst-case delay, delay-jitter, and tolerance for lost packets. Compressed audio and video fit here.

Best effort traffic falls into two categories: *available bit rate* (ABR) and *undefined bit rate* (UBR) services. ABR guarantees zero loss rate if the source follows the network's traffic management signals. The network provides resource management (RM) cells that indicate what the available bandwidth is on the connection's bottleneck link. No other guarantees are provided other than lossless delivery, as long as the source does not exceed a specified maximum bandwidth. ABR is only available for those sources that are able to process RM messages. UBR is like datagrams: The network provides no guarantees. Best effort services are appropriate for applications the do not demand real-time performance, and which can gracefully adapt to the bandwidth available in the network: file transfers (FTP), World Wide Web access, and remote teletype access (telnet).

8.5.3 The Internet's Approach for QoS

The Internet Engineering Task Force (IETF) has been extending the Internet's best effort services with guaranteed services. These are more like promises than strong guarantees. The IETF designers believe that it is too difficult to enforce such strong guarantees [45]. IETF guaranteed services are *delay tolerant* and *delay intolerant*. The former are able to tolerate some degree of packet losses and delays. For these, a predictive service is adequate, that is, the network predicts a maximum delay that can "occasionally" be exceeded.

The set of specifications for providing performance promises to applications is still being developed within the IETF. These consist of traffic and network service specifications, admission control procedures, resource reservation protocols, and packet scheduling mechanisms. They are associated with IPv6's flows, which have some commonalities with ATM connections. It identifies packets with common traffic characteristics.

The service specifications request classes of service. These include controlled delay service and the level of delay that can be tolerated, and predictive service and a target delay. The specification is complex, because it specifies minimum, maximum, and statistical distributions of the flow's expected bandwidths and latencies.

Given a service specification, the network chooses to admit a new flow or deny access for those whose specifications exceed what can be provided. The network can police a flow to ensure that it meets its traffic specification.

The mechanism for resource reservation is the resource reservation protocol (RSVP) [61]. It operates quite differently from ATM's connection establishment and signaling. RSVP is based on soft state in routers, trading off hard service guarantees for greater robustness. It is closely integrated with multicast, which supports dynamic path establishment, as well as receiver-oriented control mechanisms and received rate reports. Resource allocation is separated from route establishment. In multicast routing, receivers determine a path through the network to the sender. Senders distribute traffic specifications while receivers distribute their desired network service. These sender-directed path messages and receiver-directed reservation messages are built on existing protocols. They periodically signal the network about their changing traffic and reservation specifications. Should a receiver fail, its reservations time out and thus become available for allocation to others. If a router fails, the multicast tree is reconstructed using existing routing algorithms, and a new set of reservations is made.

RSVP allows QoS to change dynamically and permits receivers to specify different QoS requirements. In ATM, QoS is fixed for the life of a connection.

Once the necessary resource reservations have been made, the final step is to implement these in every router on the path. This is achieved through packet classification and scheduling within routers. Classification maps packets on flows into their reservations. Scheduling ensures that the queued packets obtain the service they requested.

8.5.4 QoS and Wireless LANs

Wireless links complicate issues of network QoS. In fiber optic links, error rates are extremely low, link bandwidths are very easy to predict, and QoS parameters are largely determined by how the queues are managed within the switches or routers. Losses are due almost entirely to congestion-related queue overflows rather than link losses. Wireless links, on the other hand, sustain high bit error rates, potentially high latencies, and unpredictable link bandwidths. Wireless links are inherently time varying. Two wireless end nodes sharing the same link could see widely different link bandwidths, depending on their relative proximity to the base station, their location in a radio fade, or a multipath environment that causes the receiver to lose synchronization. Cochannel interference from a nearby transmitter could degrade link quality. Hidden terminals could cause time-consuming backoffs and reaccess attempts that further degrade network performance.

Since link quality varies on small time scales, improving a wireless link through error coding or higher transmit power is difficult. Improving a link for one user can hurt others, for example, increasing the coding for one receiver could steal available link bandwidth from others. Increasing transmit power in one cell increases cochannel interference, thus reducing the quality of other users' links. In general, it is difficult to achieve "guarantees" between senders and receivers in such a complex environment [62]. Given the more constrained bandwidth of wireless links and their higher error rates, it seems likely that these will represent the QoS bottleneck.

QoS can be achieved for wireless users through packet scheduling at the base station. This is a natural place for centralizing admission and bandwidth allocation decisions for a wireless cell. Packets within the cell must pass through the base station and the base station decides whether to accept a new connection. These can be scheduled to give precedence to high-priority traffic or to achieve minimum delays [63].

8.5.5 QoS Support in Ad Hoc, Multihop Networks

Ad hoc, multihop, multimedia networks can be selfstanding, or may be connected to a wired backbone. The multihop combined with reconfiguration due to mobility poses novel challenges. There are several issues: clustering, VC maintenance, mobility management, and power control. One essential requirement is the ability to adjust to continuously changing network conditions. The users and applications must be made aware of the available resources (e.g., bandwidth) so that they can make the best use of them. The user QoS is dynamically adjusted as conditions change rather than being "contractually" frozen at connection setup time, like in ATM.

8.5.5.1 Clustering and Virtual Circuits for Multimedia Support

To support multimedia traffic, the wireless network layer must guarantee QoS to real-time traffic. A possible approach consists of the following [64]: (1) partitioning of the multihop network into clusters, so that controlled, accountable bandwidth sharing can be accomplished within each; (2) establishment of VCs with QoS guarantee; and (3) use of QoS routing to track QoS metrics along each path.

The clustering algorithm partitions the network into clusters [64]. Optimal size is dictated by spatial reuse of the channel and delay minimization. Within each, nodes directly communicate with the cluster head and communicate with each other in at most two hops. The clustering is based on a distributed, lowest ID cluster head election scheme. Each node is assigned a distinct ID. Periodically, each node broadcasts the list of IDs it can hear and elects the lowest ID node as its cluster head. As a result, a node becomes either a cluster head n, or a gateway l, or an ordinary node m. In Figure 8.11, nodes 1, 2 and 4 are cluster heads; 8 and 9 are gateways. The clustering algorithm converges

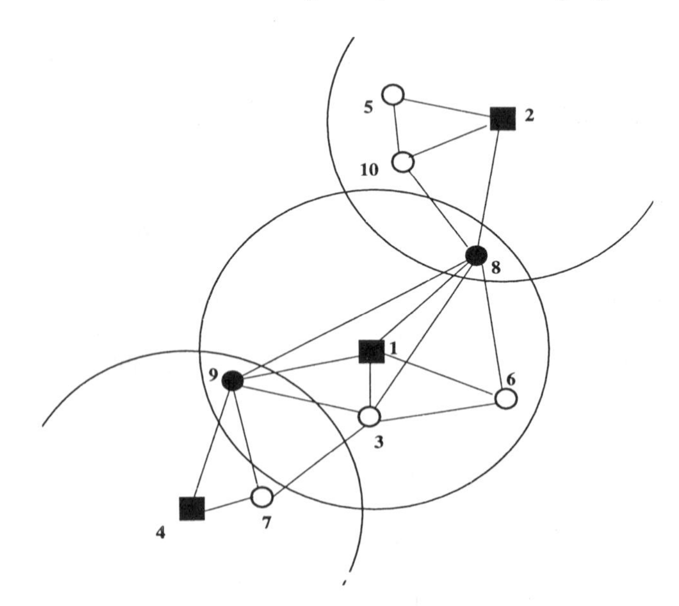

Figure 8.11 Examples of clustering.

rapidly. In the worst case, the convergence is linear in the total number of nodes.

Within each cluster, the MAC is implemented using a time slotted frame. The transmission time scale is organized in frames, each containing a fixed number of slots. The network is synchronized using an elastic synchronization scheme [65]. The frame is divided into two phases: control phase and information phase (see Figure 8.12).

The control phase supports slot and frame synchronization, clustering, routing, power measurement and control code assignment, VC reservation, and ACKs. By exchanging connectivity information between neighbors, each node elects its cluster head and updates its routing tables. Cluster heads broadcast ACKs as well as slot reservations and code assignment for each VC within their covering area. Each real time connection is assigned to a VC, an end-to-end path along which slots have been reserved. The number of slots per frame is determined by the bandwidth required by the VC. The path and slots of a VC may change dynamically during a connection.

The information phase must support both virtual circuits and datagrams. Since real-time traffic needs guaranteed bandwidth when active, slots must be allocated to the VC at setup time (see Figure 8.13). The remaining slots (free slots) can be accessed by datagram traffic using a random access scheme such as slotted Aloha.

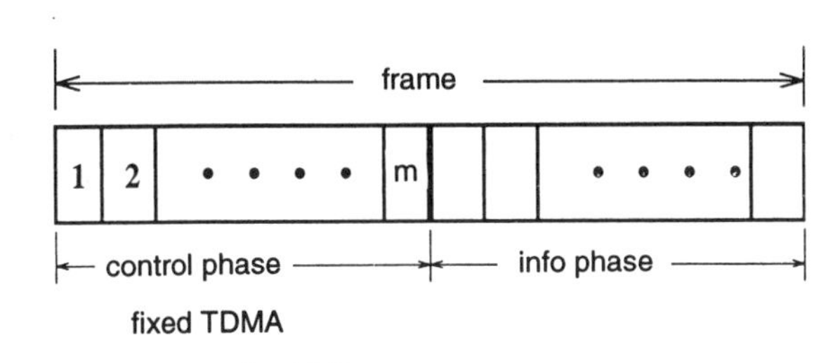

Figure 8.12 Channel access frame.

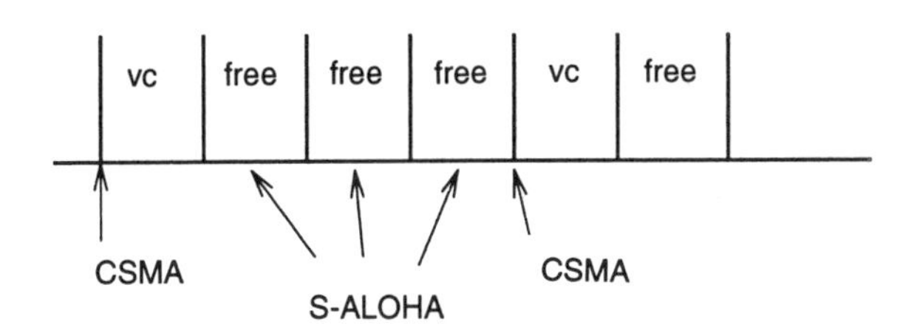

Figure 8.13 Slot allocation and channel access in the information phase subframe.

To reduce interference between neighbors a different spreading code is used in clusters. Gateway nodes retune their receivers to the codes of the neighboring clusters to maintain connectivity. Within a cluster, CDMA is used to increase channel capacity: two or more VCs share the same time slot using different codes. Transmit power is carefully adjusted slot by slot to yield an acceptable signal-to-interference ratio (SIR). To support multimedia, bandwidth availability and various other channel attributes (e.g., SRI, packet drop rate, CRC failure) are propagated along with minimum hop information. The source node can make use of the advertised QoS on the path by blocking calls with infeasible rate requests, by adjusting the rates to the advertised bandwidth, and by optimizing the video rate/channel coding strategy.

8.5.5.2 Multimedia Support in Presence of Mobility

As nodes move, routes must be updated and clusters must be reconfigured. Dynamic route updating is intrinsic in the distance vector scheme and is automatically supported. Some extra care has been devoted to suppress temporary loops occurring during reconfiguration [66]. The distributed cluster algorithm quickly reconfigures clusters [64].

A conventional VC setup scheme is suitable only for a static network. In a mobile network, the time to set up a new VC may be comparable to the interval between path changes. The conventional reconfiguration scheme cannot keep up with MH movements. A VC setup based on fast reservations is a better approach.

With fast reservations, each packet in the stream is routed individually, based on its destination. The first packet in the stream, upon successfully capturing a slot in the subframe, reserves it subsequently. If the slot is unused for a certain number of frames, it is declared free by the cluster head and is returned to the free pool. This was inspired by PRMA [50], and allows the VC stream to select a new path to the destination when the old path fails. The fast reservation VC is akin to the "soft state" connection and flow ID concept proposed in RSVP [61]. Each path change can cause some disruption: out-of-order, delay to acquire a slot on the new path, lack of free slots, possible looping, etc.

The disruption caused by rerouting can be mitigated by exploiting rate-adaptive, hierarchically encoded voice/video compression schemes. Least significant packets within a hierarchically encoded video or voice stream can be dropped during rerouting. This alleviates congestion and reduces reservation delay if bandwidth on the new path is scarce, albeit at the expense of temporary signal quality degradation. Quality is restored after rerouting, to the degree that there is sufficient bandwidth on the new path.

8.5.5.3 Transmit Power Control

By controlling power, we mitigate interference and improve spatial reuse, thus increasing network capacity. Transmitted signals reach their intended receivers, generating minimal interference on others sharing the same channel and minimizing the power needed to maintain transmission quality. During the operation of the clustering algorithm, power control allows for reuse of bandwidth by multiple transmission links. The distributed power control with active link protection (DPC/ALP) algorithm updates transmitter power based on measurements of the SIR at the receivers [57,67,68]. The algorithm exploits spatial reuse, allowing more links to coexist in the channel satisfying their SIR requirements. It is fully distributed. It provides active link protection (ALP), maintaining the SIRs of operational links above their required threshold values, while new interfering links try to access the channel and establish communication by achieving their SIR target levels. However, if this is not possible, new links are suppressed without hurting operational links. Guaranteed QoS through active link protection is important in multimedia networks where traffic has stringent QoS requirements.

DPC and ALP also help with mobility management. Since the nodes continuously move, their mutual interference changes dynamically. By adjusting the powers, we maintain the operational links and sustain the essential structure of the network, despite the fact that its actual topology is gradually stretched and twisted due to mobility.

8.5.5.4 Dynamic QoS Adjustment in a Wireless Network

One of the challenges in multihop wireless networks is to make the best use of available network resources. Mechanisms must be put in place to inform the source of the available QoS in the networks. There must be agents to dynamically adjust the characteristics of the data stream enroute, to match the changing network conditions.

Three main options are available: (1) end-to-end QoS reporting, (2) network QoS reporting, and (3) network QoS delegation. In end-to-end QoS reporting, the receiver in a session measures the QoS parameters relevant to the application and reports them to the source. For example, for video, the destination measures the bit errors and feeds back that information to the source, which then adjusts frame rate, encoding scheme etc.

This is the approach used in video codecs to adjust quantization. It is also the scheme used by TCP to adjust the window (based on destination feedback). The advantages include customized QoS reporting, network protocol independence, and true end-to-end performance. The disadvantages include inaccurate reporting of network resources, an inability to separate destination congestion from network problems, difficult destination feedback in multicast, and potential unfairness.

An alternative is to use QoS reporting: the network layer propagates measurements to the application. For example, the routing algorithm reports to the source the bandwidth available to a given destination. The source selects a proper video encoding rate.

A potential problem is the network traffic overhead caused by the reports. Different applications require different QoS data, with varying refresh frequency. An interesting compromise is QoS delegation: the application delegates QoS renegotiation to the network. For example, in a video conference over heterogeneous media, agents at the wired/wireless boundary transcode video, or drop the high-resolution substreams. "Smart" packets may carry instructions on what to do (e.g., transcoding method) to the payload when specific network conditions are met.

To make network QoS reporting and network delegation practical, a set of APIs must be developed between network and applications layer. This set defines the syntax and semantics of the messages/commands. Furthermore, a careful integration of these schemes in the same integrated services network (ISN) must be accomplished, lest the schemes interfere and become counterproductive.

8.5.6 Reliable Transport

The most widely used reliable transport layer protocol is the transmission control protocol (TCP). It is a byte stream connection-oriented protocol that combines congestion control with flow control. It uses a sliding window flow control scheme at the sender and cumulative ACKs from the receiver. Congestion avoidance and control is implemented by how the sender grows or shrinks the congestion window. It is initially grown in response to each received acknowledgment by successive doubling from a starting size of one segment (the transfer unit). The window doubles for each round-trip time (RTT). This is the slow start phase. Once an implementation-dependent threshold is crossed, the sender enters *congestion avoidance.* Here the window grows in increments of a single segment for each received acknowledgment. Using these mechanisms, the sender probes the network to discover just how much data can be flowing between sender and receiver.

As the receiver receives each segment, it generates an ACK to the sender indicating the number of bytes that have been successfully received so far. If received out of sequence, the receiver generates a *duplicate ACK.* This indicates that a segment was received, but it acknowledges only the last segment received in sequence, thus duplicating a previous ACK. One way that a lost segment is discovered is when a threshold number of dup ACKs, usually three, has been received. This causes the sender to retransmit the lost segment and to halve its congestion window (*fast retransmission and recovery*).

Since this method depends on segments actually arriving at the receiver, an alternative is needed if none are generating ACKs. This is provided by coarse timeouts. TCP implementations set the timeout interval to the mean RTT plus four times the standard deviation. If no ACK is received within a time out, the sender retransmits the segment, sets its window size to one segment, and re-enters slow start.

Fast retransmission works well when the window size is large. It is not effective when connections are short lived or involve small amounts of data. If the RTT is highly variable, then the coarse timeouts will wait too long before expiration.

Alternative schemes have been proposed based on receiver-directed rate control. Rather than using ACKs, they probe the network to determine the rate which it can clock out the data stream. The receiver recovers by explicitly indicating to the sender what has not been received, requesting its explicit retransmission. Further, the receiver determines the rate at which it is actually receiving the data stream. It reports this to the sender to control its rate. The receiver builds its own estimate of packet interarrival times. When these are exceeded by a threshold, the receiver triggers its retransmission request.

TCP was designed expecting the links to exhibit low error rates. The dominating loss phenomenon is congestion. This is why retransmission mechanisms are coupled with reductions in window size. However, if the bottleneck is due to a lossy link, these approaches reduce throughput unnecessarily. It has been widely observed that wireless links suffer from burst losses, due to fading effects that can multiply their impact by causing the receiver to lose synchronization at the level of the physical channel.

One way to reduce link errors is to implement mechanisms at the link layer such as FEC or link layer retransmission (ARQ, automatic repeat request). FEC reduces throughput even when there are no link errors. It does not help much with fades caused by shadows, because these may persist for a significant number of bit times. To combat these problems, the error correction scheme must also be implemented with interleaving, but this increases latency. An entire interleave block must be received and verified before it is forwarded. This places a limit on how much data it is feasible to interleave, and fades may still persist for longer than the feasible interleave block.

Link layer retransmission still causes reordered data delivery to the receiver, that is, a block is sent to the receiver but is lost on the link due to an error burst. A subsequent block might be sent and received before the original is resent. This has been observed to trigger dup ACKs and end-to-end retransmissions at the transport layer. AIRMAIL [69] reorders buffers at the link layer to avoid out-of-order delivery to higher protocol layers. But this increases delays and variability in the interpacket delivery times. Link layer design must be carefully examined for their interactions with the transport layer.

One method for fixing these problems is to split the transport layer connection into two connections: one between the sender and the base station, and a second between the base station and the mobile host [70]. Timers are separately tuned for the appropriate behavior for the wireline and the wireless connection. It allows special mechanisms to be used for loss recovery over the wireless link without distributing the conventional mechanisms for congestion control on the wireline connection.

However, the approach breaks the end-to-end semantics of TCP. Data are ACK'd that never arrived at the end node. A BS failure could compromise application behavior. Strange interactions across the connections are possible. A wireline sender can time out because of errors on the wireless connection that slow down connection progress.

An alternative approach is to develop a TCP-aware link layer scheme [71]. The BS maintains soft state of TCP segments enroute to MHs within its cell. A layer of software above routing intercepts the TCP ACKs, and uses a threshold number of dup ACKs to trigger local retransmissions. A key aspect of this approach is that it intercepts the dup ACKs without forwarding them to the sender. This shields the sender from the effects of local losses, which would cause it to reduce its congestion window and experience reduced throughput. The technique can be combined with *selective ACKs* to reduce the complexity of determining which segments must be retransmitted. A selective ACK indicates the byte range that generated the ACK.

This assumes that the limiting phenomenon is losses on the wireless link. Other congestion effects may dominate. The technique can be generalized by using explicit loss notification to signal the sender that the connection has suffered error losses. This can be combined with other mechanisms, such as random early detection (RED) gateways [72], which detect congestion on the connection path. The sender determines how to deal with the connection: Slow down to avoid congestion or keep the rate unaffected because link layer schemes will repair the localized error-related problems.

Interruptions in the connection across cell-to-cell handoffs present problems for transport. Packets are directed to a mobile host's old BS as it moves towards its new BS. If these are simply dropped by the old BS, the sender is likely to time out and retransmit them. Not only does this waste network bandwidth, it reduces the sender throughput because of TCP slow start. If they are forwarded from the old to the new BS, while the underlying mobile routing now directs packets to the new BS, packet reordering becomes likely.

This can cause sender retransmissions and/or timeouts. Reference [73] proposes a scheme in which the new BS automatically generates dup ACKs to trigger fast retransmission and recovery from the sender more quickly. Alternatively a multicast-based handoff scheme, as described in [54], can minimize the effects of handoff on TCP connections by ensuring that the stream is delivered to the new base station in advance of handoff.

8.5.7 Issues in Asymmetric Connections

Asymmetric connections are those in which the bandwidth in one direction of the connection far exceeds the bandwidth available for the opposite path. An extreme example of this is direct broadcast satellite (DBS), which sends data to the user at several hundred kilobits per second, but use much conventional telephone modems at tens of kilobits per second, for the return path. Even cellular data overlays like CDPD involve an element of asymmetry. The BS schedules the forward link, and transmits at high power to the mobile hosts with reduced likelihood of losses. The MHs share the return link, must contend for it with other users, and have an increased likelihood of losses because of their more limited transmit power.

Asymmetries also arise because of the nature of the traffic patterns. Web access pulls data from servers to clients, which generate ACKs to the sender. More data flows toward clients than they source. Consider a half-duplex PRnet. These receive a sequence of frames, changing from receive to transmit mode, forwarding on the frames, then returning to receive mode to obtain an ACK that has been successfully received downstream. Switching between receive and transmit mode involves high latency. Throughput can be optimized by remaining in one mode for as long as possible. Yet if the flow of data is primarily in one direction, throughput could be wasted on the short ACKs going against this flow. Furthermore, ACK traffic can contend with the downstream data traffic, causing collisions and backoffs that reduce connection throughput.

Poor performance on the ACK path limits the performance on even a high- bandwidth forward path. This is becoming more problematical as IPv6 moves toward larger addresses. One solution is to implement header compression on the return path for the ACK packets. A second solution is to delay ACKs so that each one acknowledges a larger range of received data. For example, a simple strategy is to acknowledge every other packet received. One must be careful to avoid triggering retransmissions from the sender because an ACK has been delayed too long.

This raises a second issue. TCP does a good job of adapting its coarse timeouts as long as the RTT estimate does not exhibit a high variance. Asymmetries in the connection bandwidth, differential loss rates, and congestion effects on the paths lead to higher variability in the RTT estimate. This means that when losses do occur, the retransmission timeouts can become very large, thus significantly degrading connection if losses occur reasonably often. Again, this observation points out the weaknesses in the current TCP design when faced with the complexities of reliable transport over wireless links.

8.6 Integration of Wireless and Wireline Networks

8.6.1 An Application Scenario

Consider the following scenario, drawn from the domain of emergency response. A police helicopter observing a serious auto accident on Golden Gate Bridge immediately calls for an emergency medical team to be dispatched. The helicopter, using a wide-area wireless net, transmits low frame rate digitized video of the crash to the medical team, to assist them in assessing the severity and the need for support units. They decide to request backup. The helicopter crew also establishes a two-way audio conversation with the ambulance driver to direct him to the scene via the least congested route, perhaps forwarding the medical team a street map with a hand-annotated route.

At the scene, the medical team and police units free the victims, stabilize their condition, and head them to Mt. Zion Hospital in San Francisco. The victims are identified. One of them has a serious head injury, requiring immediate surgery. The team requests the preparation of a suitable medical facility. Critical medical information—such as patient reactions to various medications—is downloaded from a regional medical Picture Archiving and Communications System (PACS) at the UCSF Medical Center to the teams in the ambulances via the wide-area wireless net. Simultaneously, detailed patient records and archived neuro-images are forwarded to an image file server at Mt. Zion via a high-performance wireline network. An in-hospital wireless network allows the emergency room physicians to retrieve and view the patient's images and records on their PDAs. They decide that a CAT scan should be performed on the patients. The new scan, combined with existing patient information, will support their planning of the proper surgical suite and determination of a course of operating procedure.

After the ambulance arrives at Mt. Zion, the patient is routed to the CAT scanner. PACS automatically appends the new neuro-images with his existing patient information. The surgical team assesses the injury's seriousness, and decides that a 3-D reconstruction of the CAT images is needed. The hospital requests UCSF's high-performance computers to construct a 3-D visualization of the head trauma and a computer-assisted guide for the surgical procedure. These are forwarded to Mt. Zion via a wireline network, where they are wirelessly rerouted to the surgeon's PDA while he is enroute to the operating room. Its unusual nature prompts him to hold an impromptu video consultation with a distinguished neurosurgeon on his way to a lecture at the UCSF Medical School across town. Just as the two doctors agree on a course of surgery, the stretcher is wheeled into the operating room, and the neurosurgeon steps into his lecture hall.

This scenario exhibits generic requirements for wireless multimedia applications:

- Seamless interoperation of wide-area wireless communications for mobile units, able to exploit higher bandwidth inside buildings, yet still providing useful functionality over wider area, lower bandwidth, higher latency wireless technologies;
- Access by mobile applications to the full computation and information resources of distributed processing, visualization, and digital library access;
- Rapid exchange of media-rich data, including video images, audio communications, annotated maps and graphics, handwriting and text, all in support of decision makers on the move, in real time where possible, and in near real time otherwise;
- Type-sensitive data exchange adapted to the available communications quality, exploiting mechanisms like bandwidth-adaptive compression, latency-adaptive mobile host front-end processing and wireline back-end processing.

8.6.2 Wireless Overlay Networks

Future mobile information systems will be built on heterogeneous wireless overlay networks, extending traditional wired and internetworked processing "islands" to hosts on the move over a wide area. No single technology offers wide-area coverage, high bandwidth, and low latency, with support for vehicular as well as pedestrian mobility. To yield flexible connectivity over wide areas, the wireless internetwork must be formed from multiple wireless overlays interconnected by wired segments.

For many emerging applications, such as those that support collaboration and wireless data access for mobile users, (near) real-time audio/video is an absolute requirement. Table 8.5 identifies the performance characteristics of representative wireless technologies. It also gives the approximate achievable

Table 8.5
Characteristics of Alternative Overlay Technologies

Type of Network	Bandwidth	Latency	Mobility	Typical Video Performance	Typical Audio Performance
In-room/ Building (radio-frequency infrared)	> 1 Mbps RF: 2–20 Mbps IR: 1–50 Mbps	≪ 10 ms	Pedestrian	Two-way, Interactive full frame rate (compressed)	High-quality 16-bit samples 22-kHz rate
Campus-area packet relay	≈ 64 Kbps	≈ 100 ms	Pedestrian	Medium-quality slow scan	Medium-quality Reduced rate
Wide-area (cellular, PCS)	19.2 Kbps	> 100 ms	Pedestrian Vehicular	"Video phone" or freeze frame	Async "voice mail"
Regional-area (LEO/VSAT DBS)	Asym up/down 0.1–4.8 Kbps 12 Mbps	≫ 100 ms	Pedestrian Vehicular Stationary	Seconds/frame async video playback	Async "voice mail"

video and audio performance. To support users and applications roaming across such a heterogeneous collection of networks, the applications must be able to adapt to the available network performance.

It may appear that the concept of multiple overlay networks will be impeded by the MH's need for multiple transmitter/receiver systems. Yet even today it is possible to simultaneously configure a laptop with network adapters for in-room diffuse IR, in-building RF, campus-area packet radio, and wide-area CDPD.

Not only do handoffs take place "horizontally" within a single network, they occur "vertically" between alternative overlays. Mobile IP only provides an initial solution. However, the extension is straightforward. For each overlay network and network interface, the mobile host has a different IP address. The HA maintains not one binding between the IP_{home} and the locally assigned address, but a table of these. The MH participates in the choice of binding. For example, an application might specify that high-priority traffic traverse the available overlay with the lowest latency, regardless of cost. Less critical traffic might be constrained to travel over the lowest cost connections currently available. Because the goal is to get the data through, signal quality, bit error rates, and the resulting probability of packet loss and retransmission must be considered as part of the handoff decision. Under certain conditions, the handoff algorithm might defer switching its connection to a higher latency channel, even though it has better signal quality due to stronger signal strength, from a low latency channel with a weak signal.

The above assumes that multiple interfaces may be powered on. This is an expensive proposition for the MH. To extend battery life, only one network interface is powered on. Thus the MH is connected to only one overlay network at any time, though it will (likely) be registered with every wireless overlay to which it has access. While its primary connection is supported by one of its interfaces, the other interfaces are in a standby mode. They are powered up from time to time to determine the quality of connectivity to their respective overlays. A trade-off exists between the frequency of determining link quality, the power consumed by the operation, and how up to date the information is about the network state to drive handoff decision making. When the mobile is low on power, a good strategy is to connect to the network with the lowest transmitter power demands, and the handoff algorithm in the mobile may scale back the frequency of probing other overlays. System context may also dictate the frequency of probes. For example, as the mobile moves toward the fringe cells of its current network overlay, a good strategy would be to increase the probes to alternative overlays.

8.6.3 Client-Proxy-Server Applications Model

Client-server processing is a popular way to structure network applications. Mail programs like Eudora or Web browsers fall into this class, as do many database applications. The security concerns associated with the Internet have introduced a new paradigm for structuring applications: client-proxy-server. A proxy is a software intermediary that executes between the client and the server at a well-known location in the network, usually at a machine controlled by the client's organization but outside of its security firewall. Proxies provide a convenient place to perform useful operations on behalf of clients, well beyond the original function of filtering Internet traffic through a corporate firewall. Proxies can mitigate the lossy, constrained bandwidths of wireless links by changing the representations of data enroute to the client, performing data type specific compressions, caching data for rapid reaccess, and prefetching data in anticipation of future access [74]. Thus, proxies can make client applications running on impoverished devices appear to behave as if they were running on well-connected machines.

Mobile clients vary in their available hardware resources, ease of programming, level of software sophistication, and quality of connectivity. Clients span the range from small display and limited computing capability PDAs to full-featured laptops. At the low end, the clients have limited memory, slow processors, and limited run-time environments. These constraints seriously limit the amount of processing it is feasible to do locally on the end device. Thus, an application-level strategy for handling client variability is to provide a system architecture that supports on-demand, data-type-specific, lossy compression based on the semantic meaning of the data. The data being delivered to the client can be tailored to the specific constraints of the end device.

There are several kinds of variations to which a client must adapt. Network variation includes bandwidth, latency, and the error behavior of the network. When plugged into the campus Ethernet, the network bandwidth seen by the client is high, network latency is low, and errors are few. When connected by a wide-area wireless data network, however, the bandwidth is low, latency is high, and errors are potentially high.

Hardware variations include screen size and resolution, whether they support full color or reduced grayscale, the amount of available local memory, and the CPU power. We can expect portable computers to be less capable than desktop machines at any time in the technology evolution. Furthermore, as these migrate toward very low cost consumer items, they will be considerably less capable than typical desktop machines.

Software variations include the application-level data encodings that the client can handle given the processing and display capabilities of the end device. For example, a motion JPEG video decoder can easily execute on a laptop. However, a PDA would be hard pressed to execute such code with a reasonable latency. As another example, the client may not be able to render PostScript to the screen. It may not able to handle nonstandard HTML extensions or even run network protocol extensions like IP multicast.

There are three design principles for adapting client applications to these variations. The first is to exploit data-type-specific lossy compression mechanisms, which have been called *distillation* and *refinement*. The semantic content of the data is used to determine how information can be compressed and prioritized enroute to the client. Distillation is a highly lossy data-type-specific compression algorithms that preserves most of the semantic content of a data object, while adhering to a particular set of constraints. An image can be distilled by reducing its color depth, resolution, and size. These may be determined by the end device's display capabilities. Similarly, video can be distilled by reducing its frame rate as well as its frame-by-frame image characteristics. A PostScript file can be distilled by extracting its alphanumeric text for display on an end device that does not have the software to render postscript directly. A map data type can be distilled by sending landmarks first, followed by more detailed topographic data.

Refinement is the process of fetching a part of the source object at increased quality, including possibly the full quality of the original representation. For example, a portion of a distilled image could be refined into its full-resolution original representation.

For example, consider a 10-Kbps wide-area wireless data link. Suppose the user downloads an 800-Kbyte image. This would take 800 sec—15 min!

If a compression algorithm took 10 sec to distill this into a 16-Kbyte lower resolution image, it would take 26 sec to complete this transfer. This shows the importance of considering the overall latencies in determining the effectiveness of adaptation, not simply the additional processing time to perform the compression functions.

A second principle is "on-the-fly" adaptation of data types. The input representation is transcoded into the desired output representation that can be handled by the end application. For example, the operational software on some PDAs can efficiently render PICT images to its display. Therefore, it is desirable to transcode JPEG and GIF images into PICT for rendering on such a device. Similarly, end devices may be able to handle certain video formats, but not others. The UC Berkeley InfoPad has hardware support for video in VQ format. It can participate in a motion JPEG playback if the appropriate transcoder functions are provided. These schemes introduce additional latency to compute the new target representation, but they can also significantly reduce the end-to-end latency if the final representation can be significantly reduced in size.

The final principle is to push the complexity away from the mobile clients and servers. First, it is desirable for servers to be insulated from client limitations. It certainly simplifies server design if all the possible client variations can be ignored. Second, clients will be running on machines that are battery limited, and it is worthwhile to move processing-intensive tasks from these to the wireline servers whenever possible. Processing-intensive distillation, refinement, and transcoding functions can be performed at an intermediate "proxy" between the client and server. The proxy executes on a well-connected, processor server complex that provides cycles for executing proxy functions.

The basic proxy software architecture consists of two major components: the type-specific distillers to optimize the QoS for the client in real time and a network connection monitor (NCM) that continuously measures the end-to-end bandwidth and quality of connectivity to the proxy's clients. The job of the NCM is to notify the proxy of any changes in bandwidth or connectivity which may affect its transcoding decisions.

8.7 Future Directions in Wireless Multimedia

Wireless data services are still in their relative infancy. There are many successful examples of sophisticated applications-specific wireless systems—such as those in use by car rental companies for convenient check-in on the bus from the terminal or package tracking services used by parcel delivery services. For example, the United Parcel Service's TotalTrak System supports 60,000 widely dispersed mobile units communicating with a central database server. More than 100 different cellular carriers are integrated into a single system, with integrated billing, support for roaming, and redundancy in the communication service. This system is in reliable use on a daily basis, nationwide.

General-purpose wide-area wireless services have not been nearly as successful. Complex pricing, low bandwidth, and poor Internet connectivity have limited their use. There is an increasing understanding that in the wide area, it is important to adopt an Internet service provider model, rather than merely providing the data pipe. CDPD represents one attempt to overlay Internet-capable data services onto analog voice channels in North America. However, the data capabilities are an afterthought to the voice services. The Europeans, on the other hand, are defining a general-purpose packet radio system, or GPRS, on top of the GSM cellular phone standard. The cellular operator will be able to transport data packets over wireless channels, intermixing digitized voice and data, while routing the data onto the Internet at the mobile switching center.

Wireless LANs have supported the Internet protocol suite for quite some time. The issue here is that existing WLANs are considerably slower than commonly deployed wired networking alternatives, like 10- and 100BASE-T Ethernets. Because of their limited bandwidth, scalability to larger numbers of users is a problem. Network interfaces have also been expensive, in part because sales volumes have been so modest. Today, the trends are toward ever higher bandwidth within ever smaller cells to achieve a high level of bits per second per cubic meter. The more widespread adoption of standards like IEEE 802.11 and the use of etiquette rules to allow sharing of spectrum by multiple users, will help make wireless LANs more efficient, more interoperable, and better able to coexist with other wireless users and technologies. In addition, dramatically increased wireless bandwidth appears to be on the horizon. The FCC's recent allocation of the unlicensed NII band at 5.15 GHz (350 MHz) and in the 60-GHz spectrum band (5 GHz of available spectrum) suggests that much higher bandwidth will be available soon.

Users are rapidly discovering that access is the killer application. In other words, a user should be able to access the same network resources from their mobile device as they can from their office machine. This makes it necessary for a wireless user to be continuously connected via the best available network. Such connectivity could be provided through wireless overlays, with in-building networks providing high bandwidth in the local area, and wide-area networks providing reduced bandwidth connectivity elsewhere. Since overlay bandwidth, latency, and error rates vary over several magnitudes, it is critical to provide adaptivity through proxy services. The IP "Dial Tone" will become ubiquitous, with IP to the pager, phone, PDA, and laptop, becoming standard. The Internet provides the means for service integration. But it is important to remember that a phone is not a workstation. With its limited bandwidth, storage, and display capabilities, network support for heterogeneity through the proxy adaptation will become ubiquitous.

8.8 Glossary of Terms

ATM (asynchronous transfer mode): A high-speed method of digital data communications based on rapid switching of small, fixed-sized data packets called *cells.* It is currently being deployed in telephone networks and backbones of high speed data networks.

CDMA (code-division multiple access): A multiple access scheme that uses code sequences as traffic channels in a common radio channel.

cell splitting: A method of increasing capacity by reducing the size of the cell.

FDMA (frequency-division multiple access): A multiple access scheme that divides an allocated spectrum into different radio channels.

frequency reuse factor: A number based on frequency reuse to determine how many channels per cell.

GSM (global system for mobile communications): A digital cellular system based on time-division multiple access that is standard throughout Europe, and which is being deployed around the world.

half-duplex operation: Operating method in which transmission is possible in both directions, but only in one direction at a time.

handoff: A frequency channel will be changed to a new frequency channel as the device moves from one cell to another without the user's intervention.

IS-54: The North American digital cellular standard, it is based on a time-division multiple access scheme.

PCS (personal communications systems): PCS is a new generation of all-digital wireless services, combining features of cellular voice, paging, and data access.

PCS 1900: The implementation of the European GSM System in North America.

PDA (personal digital assistant): A portable computing device containing built-in functions such as to-do lists, address book, calendar, and memo pad.

QoS (quality of service): QoS refers to a network's ability to meet the data delivery latency and bandwidth requirements necessary to support real-time applications such as audio and video conferencing.

TDMA (time-division multiple access): A multiple access scheme in which a user has a periodic time slot assigned for his or her use.

References

[1] Calhoun, G., *Wireless Access and the Local Telephone Network,* Norwood, MA: Artech House Publishers, 1992.

[2] Cox, D., "Wireless Personal Communications: What is It?" *IEEE Personal Commun. Mag,* Apr. 1995, pp. 20–35.

[3] Bray, J., *The Communications Miracle,* New York: Plenum Publishing, 1995.

[4] Calhoun, G., *Digital Cellular Radio,* Norwood, MA: Artech House Publishers, 1988.

[5] Andersen, J. B., T. S. Rappaport, and S. Yoshida, "Propagation Measurements and Models for Wireless Communications Channels," *IEEE Commun. Mag.,* Jan. 1995, pp. 42–49.

[6] Holzmann, G. H., and B. Pehrson, *The Early History of Data Networks,* Los Alamitos, CA: IEEE Computer Society Press, 1995.

[7] Headrick, D. R., *The Invisible Weapon: Telecommunications and International Politics 1851–1945,* New York: Oxford University Press, 1991.

[8] Masini, G., *Marconi,* Marsilio Publishers, New York, 1996.

[9] Rahnema, M., "Overview of the GSM System and Protocol Architecture," *IEEE Commun. Mag.,* Apr. 1993, pp. 92–100.

[10] Falconer, D. D., F. Adachi, and B. Gudmundson, "Time Division Multiple Access Methods for Wireless Personal Communications," *IEEE Commun. Mag.,* Jan. 1995, pp. 50–57.

[11] Kohno, R., R. Meidan, and L. B. Milstein, "Spread Spectrum Access Methods for Wireless Communications," *IEEE Commun. Mag.,* Jan. 1995, pp. 58–67.

[12] Hafner, K., and M. Lyon, *Where Wizards Stay Up Late: The Origins of the Internet,* New York: Simon and Schuster, 1996.

[13] Norberg, A. L., and J. E. O'Neill, *A History of the Information Processing Techniques Office of the Defense Advanced Projects Agency,* Minneapolis, MN: Charles Babbage Institute, 1992.

[14] Tanebaum, A. S., *Computer Networks,* 3rd ed., Englewood Cliffs, NJ: Prentice Hall, 1996.

[15] Kahn, R. E., et al., "Advances in Packet Radio Technology," *Proc. IEEE,* Vol. 66, No. 11, Nov. 1978, pp. 1468–1496.

[16] Lauer, G. S., "Packet-Radio Networks," Chap. 11 in *Routing in Communications Networks,* M. Steenstrup, Ed., Englewood Cliffs, NJ: Prentice Hall, 1995.

[17] Pahlavan, K., "Wireless Intraoffice Networks," *ACM Trans. Office Info. Syst.*, Vol. 6, No. 3, July 1988, pp. 277–302.

[18] Pahlavan, K., T. Probert, and M. Chase, "Trends in Local Wireless Networks," *IEEE Commun. Mag.*, Mar. 1995, pp. 88–95.

[19] Steer, D. G., "Coexistence and Access Etiquette in the United States Unlicensed PCS Band," *IEEE Personal Commun.*, Fourth Quarter 1994, pp. 36–43.

[20] DeSimone, A., and S. Nanda, "Wireless Data: Systems, Standards, Services," *ACM Wireless Networks J.*, Vol. 1, No. 3, Oct. 1995, pp. 241–253.

[21] Khan, M., and J. Kilpatrick, "MOBITEX and Mobile Data Standards," *IEEE Commun. Mag.*, Mar. 1995, pp. 96–101.

[22] Forman, G. H., and J. Zahorjan, "The Challenges of Mobile Computing," *IEEE Computer Mag.*, Vol. 27, No. 4, Apr. 1994, pp. 38–47.

[23] McCanne, S. R., "Scalable Compression and Transmission of Internet Multicast Video," Ph.D. Dissertation, Report No. CSD-96-928, Dec. 1996.

[24] Jubin, J., and J. D. Tornow, "The DARPA Packet Radio Network Protocols," *Proc. IEEE*, Vol. 75, No. 1, Jan. 1987, pp. 21–32.

[25] Keeton, K., et al., "Providing Connection-Oriented Network Services to Mobil Hosts," *Proc. USENIX ASSOCIATION Mobile and Location-Independent Computing System*, Aug. 1993, pp. 83–102.

[26] Sheng, S., A. Chandrakasan, and R. W. Brodersen, "A Portable Multimedia Terminal," *IEEE Commun. Mag.*, Dec. 1992, pp. 64–75.

[27] Weiser, M., "The Computer for the Twenty-First Century," *Sci. Am.*, Vol. 265, No. 3, Sep. 1991, pp. 94–104.

[28] Kleinrock, L., "Nomadicity: Anytime, Anywhere In a Disconnected World," *ACM Mobile Networks Appl.*, Vol. 1, No. 4, Jan. 1997, pp. 351–358.

[29] Satyanarayanan, M., "Mobile Information Access," *IEEE Personal Commun. Mag.*, Vol. 3, No. 1, Feb. 1996, pp. 26–33.

[30] Kistler, J., and M. Satyanarayanan, "Disconnected Operations in the Coda File System," *ACM Trans. Computing Syst.*, Vol. 10, No. 1, 1992.

[31] Kuenning, G., G. Popek, and P. Reiher, "An Analysis of Trace Data for Predictive File Caching in Mobile Computing," *USENIX Conf. Proc.*, June 1994, pp. 291–306.

[32] Katz, R. H., "Adaptation and Mobility in Wireless Information Systems," *IEEE Personal Commun. Mag.*, Vol. 1, No. 1, 1994, pp. 6–17.

[33] Davids, N., "Personal Digital Assistants: Part 1," *IEEE Computer Mag.*, Sept. 1996, pp. 96–100.

[34] N. Davids, "Personal Digital Assistants: Part 2," *IEEE Computer Magazine*, Nov. 1996, pp. 100–104.

[35] Narayanaswamy, S., et al. "Application and Network Support for InfoPad," *IEEE Personal Commun. Mag.*, Vol. 3, No. 2, Apr. 1996, pp. 4–17.

[36] Crawford, G. P., T. G. Fiske, and L. D. Silverstein, "Reflective Color LCDs Based on H-PDLC and PSCT Technologies," *Proc. SID 96 Digest*, pp. 99–102.

[37] Wallace, G. K., "The JPEG Still Picture Compression Standard," *Commun. ACM*, Vol. 34, 1991, pp. 30–45.

[38] Le Gall, D., "MPEG: A Video Compression Standard for Multimedia Applications," *Commun. ACM,* Vol. 34, 1991, pp. 46–58.

[39] Chien, C., et al., "A 12.7 Mchips/sec All-Digital BPSK Direct-Sequence Spread-Spectrum IF Transceiver," *IEEE J. Solid-State Circuits,* Vol. 29, No. 12, Dec. 1994.

[40] Stedman, R., et al., "Transmission of Subband-Coded Images via Mobile Channels," *IEEE Trans. Circuits Syst. Video Technol.,* Vol. 3, Feb. 1993, pp. 15–22.

[41] Villasenor, J., B. Belzer, and J. Liao, "Wavelet Filter Evaluation for Image Compression," *IEEE Trans. Image Proc.,* Vol. 4, Aug. 1995, pp. 1063–1070.

[42] Shen, A., "Perceptually-based Subband Coding of Speech Signals," Master's Thesis, Los Angeles: Electrical Engineering Department, UCLA, June 1994.

[43] Shen, A., et al., "A Robust and Variable-Rate Speech Coder," *Proc. Int. Conf. Acous. Speech Sig. Proc. (ICASSP),* Vol. I, 1995, pp. 249–252.

[44] Macedonia, M. R., and D. P. Brutzman, "MBone Provides Audio and Video across the Internet," *IEEE Computer Mag.,* Apr. 1994, pp. 30–36.

[45] Peterson, L. L., and B. S. Davies, *Computer Networks: A Systems Approach,* San Francisco: Morgan Kaufman, 1996.

[46] Schulzrinne, H., et al., "RTP: A Transport Protocol for Real-Time Applications," IETF, Audio-Video Working Group, Jan. 1996, RFC-1889.

[47] Clark, D., and D. Tennenhouse, "Architectural Considerations for a New Generation of Protocols," *Proc. of ACM SIGCOMM Conf.,* Philadelphia, PA, Sept. 1990.

[48] Floyd, S., et al., "A Reliable Multicast Framework for Light Weight Sessions and Application Level Framing," *Proc. ACM SIGCOMM Conf.,* Boston, Sept. 1995, pp. 342–356.

[49] Imielinski, T., and B. R. Badrinath, "Mobile Wireless Computing: Challenges in Data Management," *Commun. ACM,* Oct. 1994.

[50] Goodman, D. J., and S. X. Wei, "Efficiency of Packet Reservation Multiple Access, *IEEE Trans. Vehicular Technol.,* Vol. 40, No. 1, Jan. 1991, pp. 170–176.

[51] Johnson, D., and D. Maltz, "Protocols for Adaptive Wireless and Mobile Networking," *IEEE Personal Commun. Mag.,* Feb. 1996, pp. 34–42.

[52] Ioannidis, J., D. Duchamp, and G. Maguire, Jr., "IP-based Protocols for Mobile Internetworking," *Proc. SIGCOMM '91,* pp. 235–245.

[53] Myles, A., D. Johnson, and C. Perkins, "A Mobile Host Protocol Supporting Route Optimization and Authentication," *IEEE J. Sel. Areas Commun.,* Vol. 13, No. 5, June 1995, pp. 839–849.

[54] Balakrishnan, H., S. Seshan, and R. H. Katz, "Reliable Transport and Handoff Protocols for Cellular Wireless Networks," *ACM Wireless Networks J.,* Vol. 1, No. 3, Dec. 1995, pp. 469–482.

[55] Tekinay, S., and B. Jabbari, "Handover and Channel Assignment in Mobile Cellular Networks," *IEEE Commun. Mag.,* Vol. 29, No. 11, Nov. 1991, pp. 42–46.

[56] Leiner, B. M., D. L. Nielson, and F. A. Tobagi, "Issues in Packet Radio Network Design," *Proc. IEEE,* Vol. 75, No. 1, Jan. 1987, pp. 6–20.

[57] Bambos, N., and G. Pottie, "On Power Control in High Capacity Cellular Radio Networks," *Proc. IEEE GLOBECOM,* Vol. 2, 1992, pp. 863–867.

[58] Bantz, D. F., and F. J. Bauchot, "Wireless LAN Design Alternatives," *IEEE Network,* Mar./Apr. 1994, pp. 43–53.

[59] Walrand, J., and P. Varaiya, *High Performance Communications Networks,* San Francisco: Morgan Kaufman, 1996.

[60] Acampora, A., and M. Naghshineh, "Control and Quality of Service Provisioning in High Speed Microcellular Networks," *IEEE Personal Commun.,* Vol. 1, No. 2, 1994.

[61] Zhang, L., et al., "RSVP: A New Resource Reservation Protocol," *IEEE Network,* 1993, pp. 8–18.

[62] Acampora, A., "Wireless ATM: A Perspective on Issues and Prospects," *IEEE Personal Commun.,* Vol. 3, No. 4, Aug. 1996, pp. 8–17.

[63] Floyd, S., and V. Jacobson, "Link Sharing and Resource Management Models for Packet Networks," *IEEE/ACM Trans. Networking,* Vol. 3, No. 4, Aug. 1995, pp. 365–386.

[64] Gerla, M., and J. Tsai, "Multicluster, Mobile, Multimedia Radio Network," *ACM Wireless Networks J.,* Vol. 1, No. 3, Oct. 1995, pp. 255–266.

[65] Ofek, Y., "Generating a Fault Tolerant Global Clock using High-speed Control Signals for the MetaNet Architecture," *IEEE Trans. Commun.,* 42:5, May 1994, pp. 2179–2188.

[66] Murthy, S., and J. J. Garcia-Luna-Aceves, "A Routing Protocol For Packet Radio Network," *Proc. ACM MOBICOM'95,* Nov. 1995, pp. 86–95.

[67] Bambos, N., S. Chen, and G. Pottie, "Radio Link Admission Algorithms for Wireless Networks with Power Control and Active Link Quality Protection," *Proc. IEEE INFOCOM'95,* Boston.

[68] Bambos, N., S. Chen, and G. Pottie, "Channel Access Algorithms with Active Link Protection for Wireless Networks with Power Control," Technical Report UCLA-ENG-95-114, Los Angeles: UCLA School of Engineering and Applied Science, 1995.

[69] Ayanoglu, E., et al., "AIRMAIL: A Link-Layer Protocol for Wireless Networks," *ACM Wireless Networks J.,* Vol. 1, No. 1, Feb. 1995, pp. 47–60.

[70] Bakre, A., and B. R. Badrinath, "I-TCP: Indirect TCP for Mobile Hosts," *Proc. 15th Int. Conf. Distributed Computing Systems,* May 1995.

[71] Balakrishnan, H., et al., "A Comparison of Mechanisms for Improving TCP Performance over Wireless Links," presented at ACM SIGCOMM Conference, Stanford, CA, Aug. 1996.

[72] Floyd, S., and Vol. Jacobson, "Random Early Detection Gateways for Congestion Avoidance," *IEEE/ACM Trans. Networking,* Vol. 1, No. 4, Aug. 1993, pp. 397–413.

[73] Caceras, R., and L. Iftode, "Improving the Performance of Reliable Transport Protocols in Mobile Computing Environments," *IEEE J. Sel. Areas Commun.,* Vol. 13, No. 5, June 1995.

[74] Fox, A., "Adapting to Network and Client Variability via On-Demand Dynamic Distillation," presented at ACM Architectural Support for Programming Languages and Operating Systems Conference, Boston, Oct. 1996.

[72] Floyd, S., and Vol. Jacobson, "Random Early Detection Gateways for Congestion Avoidance," *IEEE/ACM Trans. Networking,* Vol. 1, No. 4, Aug. 1993, pp. 397–413.

[73] Caceras, R., and L. Iftode, "Improving the Performance of Reliable Transport Protocols in Mobile Computing Environments," *IEEE J. Sel. Areas Commun.,* Vol. 13, No. 5, June 1995.

[74] Fox, A., "Adapting to Network and Client Variability via On-Demand Dynamic Distillation," presented at ACM Architectural Support for Programming Languages and Operating Systems Conference, Boston, Oct. 1996.

9

Multimedia Communication Services and Security Issues*

9.1 Introduction

As multimedia communication networks (MCN) evolve, they will encompass most aspects of wireline and wireless networks. They will also be predominantly digital networks. The implementation of new technologies (e.g., ATM), increased use of existing technologies (e.g., IP), planning for more distributed network architectures, and the introduction of new services (e.g., desktop multimedia conferencing) are all factors driving multimedia network evolution. Increased interconnection of various networks, such as telcos, interexchange carriers, Internet providers, information service providers, and video service providers are a key aspect of multimedia network evolution. Another important factor is that multimedia networks will have an increased dependency on software to control critical network elements and some will increasingly rely on the use of external service logic (separate from switch fabric) and external databases.

Additionally there will be increased access by customers to management information (e.g., customer network management services). The integration of old networks and new networks, the coexistence of older technology and

Ms. Szot wishes to thank the following individuals for their input, guidance and support during the development of this chapter: Harry A. Hetz, Kathleen F. Jarosinski, James W. Prince, Edward E. Balkovich, Glenn Moore, Sarah Deutsch, Christine Huff, John McCollum, and Stephen J. Szot.

Mr. Lubchenko wishes to thank Prodip Sen for his guidance and support prior to and during the preparation of this chapter, and to James W. Prince for his review comments.

newer technology, and the addition of new network designs all provide significant opportunity for security problems. There will be increased risks in areas such as disclosure of information, fraud, unauthorized usage, loss of service, loss of data, and misdirected information. There is an increased need for protection of Internetwork interfaces, protection and proper use of network elements, and protection of network and customer data. Network planners need to implement the necessary security capabilities and features to protect networks, and to provide security to customers using those networks. Additionally we must consider the legislative and regulatory arena; for example, to meet requirements of the Communications Assistance for Law Enforcement Act (CALEA) legislation industry members are working to ensure the capability to permit approved wiretapping.

Each of the major topics considered for network security (access control, confidentiality, integrity, authentication, nonrepudiation, key management) is introduced in this chapter. Each area is discussed briefly to give the reader an understanding of what it is, the value it provides, and the general ways in which implementation can occur. With any implementation one must consider appropriate protocol reference models (PRM) and important aspects of each PRM plane (e.g., user, control, management) as well as the location in the network that needed security services can be implemented: end-to-end, end-point-to-switch, switch-to-switch (hop-by-hop).

Two of the more widely embraced network protocols supporting multimedia networks are asynchronous transfer mode (ATM) and the Internet protocol (IP). This chapter briefly examines both of these protocols from a high-level view and identifies some current relevant security activities for each. For the area of ATM we review major generic network security services and how they apply in an ATM network. Included in this will be the advantages and disadvantages of implementation at the ATM layer services for the user and control planes, and identification of items that could be done at higher layers.

The next major section covers the aspects of security implementation in IP-based networks: both the "public" Internet and in private networks. This will include items such as IPSEC, SNMPv2, PGP, PEM, SSL, public key infrastructure, and DNS security.

Finally we take a look at what the industry standards world has done to address the security area. We cover traditional standards bodies such as ISO, ITU-T, T1, and ETSI and industry forums such as the ATM Forum, Network Management Forum, and the IETF.

9.2 Security Considerations for Multimedia

Security aspects of multimedia communications networks can vary widely depending on how you define multimedia. Regardless of how you describe multimedia, however, security is a very critical part of networking today. And security is a very complex and broad topic, not one that can be described in any great detail in one chapter. It is important to recognize the significance of security for future and current users and implementers of multimedia networks. In this chapter basic security issues are reviewed, and issues/developments pertaining to ATM- and IP-based multimedia networks are highlighted. There are numerous technology alternatives and security solutions—this material does not attempt to choose or pick among them, but instead highlights some of their well-known advantages and disadvantages. As much as this may seem like a cliché, the old adage of the chain only being as strong as the weakest link is still meaningful and useful for the discussion of security.

Security issues can range widely from establishing IDs and passwords, to protecting files on PCs, establishing policy for Internet access within a corporate network, and replicating video streams at set-top boxes.

Multimedia itself may not bring to mind any new considerations for security, if you simply regard it as the compilation of the voice, data, image, and video networks, areas that have at least some security mechanisms today. With the advent of this consolidation/integration we do see a greater amount of Internetworking taking place. So what is of particular interest is the networking of multimedia—simply stated, when multiple networks are connected (at least two) with multiple users (at least one), a level of Internetworking has been achieved. Also of particular interest is the level of interactivity that you experience with a multimedia service. So two of the key multimedia network characteristics that we focus on are Internetworking and interactivity and their relation to security of multimedia networks.

In the last decade we have seen dramatic advances in technologies that enable and support multimedia services. These advances provide both benefits and challenges when it comes to security. For example, increases in computational power at the desktop will aid in the faster processing of complex mathematical calculations that permit cryptography in a more practical real-time environment. This same advance also provides some additional capability for those intent on cracking (for better or worse) cryptographic algorithms and compromise the confidentiality of the data stream.

9.2.1 Internetworking

Multimedia applications help to increase the demand for more bandwidth, and this in turn motivates users and implementers of multimedia networks to consider alternatives to building or expanding expensive, private networks; public networks can be used to provide this needed bandwidth with lower costs, greater reliability, and in some cases provide a more dynamic means of accessing other locations. While compression techniques can help reduce bandwidth needs and the size of data to be transferred, the need for speed continues to increase. As an example consider the ongoing increase in analog modem speeds—where users and equipment suppliers attempt to squeeze every bit of bandwidth from the current base of analog lines. The integrated services digital network (ISDN) services are also providing digital access at high rates. And there is even more promise of speed with newer digital subscriber line (DSL) technologies such as asymmetrical digital subscriber line (ADSL), rate adaptive DSL (RADSL), and HDSL. In the future the deployment of fiber in the access network will eventually provide a tremendous amount of potential bandwidth to multimedia users.

Users generally do not consider the issue of connectivity until they find a destination they cannot reach. They want as much potential connectivity as possible without having to think about it. Like the telephony network currently in place, if they need to call someone they want to be able to just pick up the phone and dial. From a user perspective the Internetworking aspects may be transparent. A prime example of this can be seen with the Internet and an e-mail service. Most users have no awareness of the potential number of networks their mail may traverse. But with each network, and each component of each network, there may be a risk. This scenario highlights the reality of Internetworking and the need for securing sensitive data traversing networks.

9.2.2 Interactivity

Multimedia services and applications are considered to have high potential for being dynamic—there is some element of change involved. Interactivity is the primary means by which we make that change occur—we have an exchange of information. To have this interactivity one needs to have a bidirectional connection—not necessarily symmetrical, but permitting communication in both directions. Consider the examples of surfing the Web and watching television. Accessing a Web site involves selecting the path and choosing options that are of greatest interest, and the process is fairly interactive. However, watching television only allows the changing of channels and powering on or off.

The degree of interactivity available to the user is, of course, dependent on the amount of control that the user has and the number of options available to him or her. There is some correlation with the degree of control and the amount of information exchanged. For users a greater degree of control may lead them to sharing more sensitive data or exposing themselves to greater risk. Take for example Web sites that ask for personal data before allowing you to proceed. At the other extreme, network elements with less control and intelligence may need to send more data back to other parts of the network. This could lead to issues with remote access vulnerabilities and potential exposure of sensitive data transfers.

Delay tolerances play a big part in interactivity—how much delay the user is willing to tolerate, how long they are willing to wait, or if the delay has a negative impact on the service quality itself (e.g., disrupts a normal conversation in video conferencing) [1]. Delay can come from many sources, one of which is the network. Understanding, managing, and manipulating (and ultimately reducing) this delay is very critical to the successful deployment of key multimedia services. An example is interactive gaming over the Internet—while there is much interest in this type of service, it is not clear if the delays and resulting application impacts have reached satisfactory levels yet. Delay can be impacted by malicious users, and can be an issue of service/network availability and responsiveness.

9.2.3 Establishment of On-Demand Connections

Another important aspect to interactivity is the ability to establish a connection on-demand. But while on-demand, signaled connections provide a more dynamic, flexible, and potentially economical means of setting up paths to providers and other users, this control plane mechanism can also be one of the most vulnerable to security attacks. Signaling is generally an issue with connection-oriented networks and technology and is not required with a connectionless environment.

Looking at embedded systems it is important to highlight how signaling links now connect the network's distributed computing environment and support shared call processing. The interconnection of these signaling links, without proper security measures, could provide the opportunity for improper communications between network elements of the involved networks. This kind of exposure could make networks vulnerable to monitoring of call processing, private customer data, and network data, and possible interference with proper operations of the network [2].

9.3 Service Provider Perspective

The number and type of multimedia providers is almost as numerous and varied as the number of technology and application drivers. The so-called traditional telephony providers no longer see themselves in the traditional sense, and are pursuing opportunities to provide high-speed data services, video services (wireline and wireless), and Internet access services in addition to voice/telephony/low-speed data services (wireline and wireless). Similarly cable TV providers no longer see themselves as being solely in the business of one-way video delivery—they are now exploring the added services of high-speed data, telephony, and two-way video. Finally, many others, such as competitive access providers (CLECs) and Internet service providers (ISPs) are also jumping into the multimedia fray. The multimedia technologies are numerous and generally available, and many companies are looking to implement them in their networks. Customers (are expected to) have many choices of multimedia communication services and providers.

Protection of the underlying network is an important security issue for multimedia service providers, as is the ability to offer customers value added security services as needed or desired. Service or network providers planning for implementation of security mechanisms need to consider:

- The nature of the security threat;
- Strength of security needed;
- Location of security solution(s);
- Cost of available mechanisms;
- Speed and practicality of mechanisms;
- Interoperability.

After performing a security analysis on the network in question, a better appreciation for the amount, or strength, of security should be obtained. Once the threats have been evaluated and solutions for prevention or detection of threats have been identified, then identify the options of where these solutions could be implemented (e.g., in each router or just at the border routers). Follow this with a survey of available security tools and mechanisms that meet your requirements. These tools need to be evaluated for their speed and practicality so that users are not spending all their time encrypting files or trying to manage huge lists of random numbers (keys). The evaluation should also consider how the tools will work in the network, especially in a multivendor environment, because even if that is not an objective it may become a reality.

9.4 User Perspective

Users of multimedia networks can range from a large business all the way down to private consumers. From a business user perspective it seems like those with the most to lose are the ones most intent on obtaining appropriate security measures for multimedia communication networks. Losses could come in the form of lost potential future business if services such as electronic commerce, for example, are not offered. Yet there is also great risk if those services are not offered in a manner that protects both the consumer and the provider.

Some other drivers from a user perspective include protection of sensitive data (intellectual property) for businesses, government/military information from a national security (classified/sensitive data) perspective; medical data (privacy/accuracy), and financial data (privacy, billing/fraud, theft of service, overall national financial security).

Telemedicine is a great application for the consumer when they can gain the insights/invaluable input and experience of a remotely located medical expert/specialist. Issues of privacy and accuracy with these kinds of data must be taken into consideration when transferring files and imaging information over networks.

The ongoing presence of the prank macro virus associated primarily with Word documents is just one example of a contamination that can and does occur with exchange of documents between users. Exchanging files over networks, between users, and retrieval from servers, etc., is promulgating these viruses! Purveyors of these viruses continue to modify them, requiring ongoing updates to antivirus software to permit detection and deletion of the virus.

Electronic commerce is expected to increase rapidly over the next few years. The primary enabler for most consumers is the availability of the graphical user interface (GUI) found with the World Wide Web (WWW). As this relatively new method of doing business grows we expect to see some new approaches to doing business electronically. Consumers and retailers alike must wrestle with privacy and fraud issues in this new commerce environment.

9.5 Legal and Government Perspective

9.5.1 Export Control

In the United States security and security-related issues are greatly impacted by regulatory and legislative activities. A prime example is that of the current (as of this writing) policy on encryption. The following is an excerpt from a November 15, 1996, White House press release:

> First, President Clinton has signed an executive order directing the transfer of jurisdiction for the export control of commercial encryption products from the State Department to the Commerce Department. Second, President Clinton has designated Ambassador

> David L. Aaron as Special Envoy for Cryptography. "We are moving forward to implement the encryption export liberalization plan that I announced in October," said Vice President Gore. "These two actions will help to promote the growth of international electronic commerce and robust secure global communications in a manner that protects the public safety and our national security."

It was also noted in this press release that [3]:

> Specifically, it allows the export of 56-bit key length encryption products under a general license contingent upon industry commitments to build and market future products that support key recovery.

With this action cryptography took a step closer toward commercialization, and a step away from being considered a "munition." The basic rationale behind the government's position is that the they do not want encryption ever to be so strong that they cannot "break" it. This is, as you can imagine, a hotly debated subject, especially for privacy activists. Key size restrictions limit the strength of encryption and undermine confidentiality; they also limit the commercialization of this service, as many suppliers simply remove encryption in order to sell to non-U.S. companies and sites. Key escrow is also viewed as weakening the security as the key is known to another party (the trustworthiness of that other party is debated by some) and nonrepudiation is harder to prove because of the involvement of that other party.

Cryptography Envoy Aaron, in a speech to the industry in early 1997, reiterated the current White House view that key recovery is the means to accomplish strong encryption while protecting the public infrastructure. It is expected that these issues will continue to be debated and may have changed since the writing of this text. Readers interested in learning more on this topic are directed to relevant government and private industry web sites.

9.5.2 Law Enforcement Community

The Communications Assistance for Law Enforcement Act of 1994 (CALEA) was instituted to support law enforcement's ability to perform authorized interception of communications in an environment of ongoing advances in telecommunications technology. An important aspect of CALEA is that it requires, to the extent "reasonably achievable," that the telecommunications industry assist in this process by meeting the government's surveillance requirements. This assistance is accomplished via modification of the existing network and design of new equipment, services, and facilities so law enforcement may conduct all types of electronic surveillance. Key groups impacted by this law include telecommunications common carriers (e.g., telcos), wireless service providers, and telecommunications equipment manufacturers (e.g., switch vendors) [4].

In an effort to support the needs specified in CALEA, the industry has been working on development of a "Lawfully Authorized Electronic Surveillance" (LAES) standard. This standard defines the interfaces between a telecommunication service provider (TSP) and law enforcement when there is a need to conduct lawfully authorized electronic surveillance. In this context electronic surveillance refers to the interception and monitoring of communication content (e.g., voice and data) and/or information pertaining to the communication itself (e.g., sending party address) [5].

9.5.3 Intellectual Property

While intellectual property rights are not the focus of this book, one must be aware that patents and their licensing affect the availability and cost of various technologies. In this case, many, if not all, cryptographic algorithms are covered under patents—some are licensed for no fee, others are not. Our advice: developer beware, buyer take care. This also has some interesting ramifications with respect to standards, because most standards bodies today require that adoption of patented technology within a standard come with a promise of licensing on reasonable and nondiscriminatory terms and conditions. The particulars of "reasonable and nondiscriminatory" are left to the lawyers and courts to sort out.

9.6 Network Security Issues

The following list of threats and countermeasures is conveyed to provide a high-level view on the types of issues that are of concern, and the mechanisms or tools that can be used to prevent, at least detect, these security events.

9.6.1 Threats

Some in the industry segregate threats into passive and active. Passive attacks could involve those acts of wiretapping or analysis of user traffic (to determine meaningful patterns in type, volume, timing, source, and destination), where the data in question are not necessarily modified. Active attacks could take the form of masquerades, replays of old messages, denial of service, or alteration of a message, where the attacker is taking a proactive approach. The following series of threats are examples and are not intended to be an exhaustive list.

9.6.1.1 Disclosure of Information/Interception

This is an attack on data confidentiality, where the unauthorized party gains access to information or resources. This can take two forms: an attack on the content, or an attack where communication patterns are analyzed. Wanting to keep transmissions of information secret is of great concern—whether this is for business reasons, personal privacy reasons, concern for fraud, or for reasons of national defense. And not having communication patterns analyzed is also of concern; for example, if usage is up dramatically some might speculate about impending significant plans, such as a merger. It is not always possible to keep transmissions secure, so a well-established approach is to make the transmission so unintelligible to anyone else that it is of no practical use or value.

9.6.1.2 Unauthorized Modification

Any transfer of data could be subject to modification en route. This could be done in any number of ways, such as accessing a copy of the data at an interim node, or by tapping a facility; either way an unauthorized party has gained access to, and may then modify, the data in question. The classic "man in the middle" scenario can be a form of unauthorized modification. This is an attack on integrity; integrity and access control countermeasures can help with this area.

9.6.1.3 Unauthorized Usage

Gaining access to network resources and services without proper authorization is the focus of this threat. This can range from logging into remote network elements to tapping into a shared medium to steal services (e.g., cable). Access control and authentication countermeasures help mitigate these issues.

9.6.1.4 Denial, Loss of Service, and Interruption

This is an attack on availability; the service or system in question is not performing as designed. In a telephony network this might equate to not being able to get a dial tone when trying to make a phone call; or for cable services this might mean that the cable system is down; and for corporate data services an example could be having the LAN go down. The basic concept here is that some security element has contributed to the interruption (either short- or long-term) of services. Implementing access controls and authentication can be helpful in this area.

9.6.1.5 Repudiation and Nonrepudiation

In many instances it is important to know that the receiving party has indeed received the intended data. A common example of this is with e-mail. Viewed another way this can be the verification that the originating party was indeed who they claim to be—for example, when ordering products or services via the WWW. Proper authentication techniques provide support for this area.

9.6.2 Countermeasures

A number of security services are available that can help prevent attacks like those just noted. And there are additional services that can help detect when either malicious or nonmalicious security problems occur.

9.6.2.1 Data Confidentiality

With networked systems, it is possible that there is not complete confidence in, and control over, the path that your data will be taking. This is especially true for users sending data across a public network (the Internet, for example), or if you are a network provider interconnecting with another network.

Encryption is a primary means of providing confidentiality. Encryption typically involves the use of complex mathematical algorithms and unique values to transform material into an unusable form—hopefully unusable for those without the algorithm and unique value. This value is the key—figuratively and literally. The key can be of any length, but the longer it is, the harder it is to figure out; this greater key length is also more cumbersome for the user to use and manage. So a good balance should be struck—something that is practical to use and manage but is also hard to crack.

The basic concept here is that the original material (plaintext) is transformed by the algorithm and key to a protected state (ciphertext). Here are some key concepts and tools that are important with cryptography:

1. *Secret key cryptography:* Parties involved both know the single secret key, and use this one key to both encrypt and decrypt the material to be kept in confidence. Since only one key is used, and both parties use it, this method is sometimes called symmetric cryptography. It is imperative that secret keys be kept secret, and that users consider changing this key with some frequency.
2. *Public key cryptography:* Involves the use of two keys: the encrypting party can use their private key (known only to them) to encrypt the material such that the decrypting party using the sender's public key (well known) have validation on the identity of the encrypting party to decrypt and are assured that the message was confidential. Additionally the sender can use the public key of the intended recipient so that they are the only party that can decrypt, since the recipient's

private key is the only key able to decrypt the message. Public key cryptography is also called *asymmetrical cryptography.*

3. *Session key:* Used with public key encryption systems, this scheme involves a third key, randomly generated for every message, or session, encrypted. As an example, PGP uses a 128-bit IDEA session key. The flow in general can be described as follows: The session key is generated and is used to encrypt a message (e.g., using the IDEA algorithm for PGP); the public key algorithm encrypts the session key with recipient's public key (e.g., PGP uses RSA); the encrypted message and encrypted session key are put together and the message is sent. What is of special note in this process is that the most processor intensive algorithm (public key) is only encrypting the session key, and not the entire message. The entire message is encrypted with a one-time only session key (that may not even have been disclosed to the user).
4. *Hash function:* Hashing is a simple form of security transformation, used primarily for message authentication and integrity assurance. Hashing takes a message (plaintext) and transforms it into a fixed-length value (e.g., 128 bits for MD5). One of the criteria for a hashing function to be secure is that even if the hashing function, the message and its digest are known, it is very difficult to construct another message that produces the same digest. By using a password in the process, and computing the hash over the message plus a secret password, the receiving end (that knows the secret password) can be assured that the sender is as claimed and that the message has not been modified when they compute the same message digest.
5. *Key management:* Managing keys used with encryption is very important for obvious reasons. But with public key encryption it is important to have a well-known process for sharing public keys. For this process a key certificate is used, which contains the public key, user identifier for key creator, and digital signature(s) of those who can attest to the key being genuine. A key distribution center (KDC) can be used with this process. A KDC is a trusted node that knows all the secret keys of the parties and can easily manage changes and additions to the keys. This central design also sets the KDC up as a potential single point of attack/failure, so it is wise to consider implementing a backup/paired KDC although this increases complexity and points of attack. An alternative to a KDC is the *certificate authority* (CA). CAs are the trusted entities in a public key scenario. While each node/user has its own private key the CA has the public key; this CA has to be a trusted entity though, or there is no reliability

in the public keys stored there. To accomplish this aspect of trust, each certificate is signed by the CA. The user only needs to know the CA's public key in order to check for other public keys [6].

9.6.2.2 Authentication

Authentication is a service that assists in assuring the users that the other parties involved in the communication are who they claim to be. This can either be one-way or mutual (two-way) authentication. Having this assurance helps to undermine attempts at impersonation and spoofing that could occur. Authentication is important not only as a service on its own, but also in support of other security services, such as key exchange.

Cryptography can provide a means for supporting strong authentication and either public key or secret key algorithms can be used. A simple exchange might proceed as follows: User A wants to establish a secure connection with User B. User A sends an encrypted message (ciphertext) to B using the public key of B. User B in turn decrypts the message with its private key, and proceeds to encrypt a return message with A's public key. Now that A and B are confident of each other's identity, they may even want to exchange a secret key this way, and proceed to have an encrypted session using based on that secret key.

9.6.2.3 Access Control

Well-known access control mechanisms include passwords, or access cards to secure buildings. Access control is a means of controlling who has access to particular services or elements in a network. Access can depend on the identity of the calling party, the services requested, and/or the type of data being accessed. Assuming that information is given different levels of sensitivity, access can be granted at a given level. A traditional access policy prohibits read-up and write-down. That is, if you only have permission to read a certain level of material, or gain access to it, you can not read or access higher, more secure levels. Also, if you have material that is sensitive at a certain level, you can not provide it down to those with lesser levels of access, hence write-down is prohibited.

Access lists, closed user groups (CUGs), and even virtual private networks (VPNs) are examples of access mechanisms of a different sort. Generally speaking they all restrict access to/from users and applications and pieces of the network. Access lists can be set up for permitting or denying access to specified interfaces, traffic types, and applications/ports. CUGs, which have been defined in technologies such as ISDN as well as B-ISDN/ATM, determine access capabilities of a group of users, among themselves and with those outside the group. VPNs can help to extend private network configurations over public network facilities in a manner that is not evident to the users on the VPN.

9.6.2.4 Integrity

To be concerned with the integrity of the data stream means several things. For one, it means that you want to have the data remain unchanged, or at the very least, you want to be aware of when it is changed. Second, you want to know that the data are current, and that the sending party has in fact just sent these data. You are interested in knowing if this is the real thing, or possibly a replay of an old message. Related to this is the sequence of the data—you want to be sure that the data are in the proper order.

Digital seals and digital signatures can provide assurances for integrity and message authentication. A digital seal refers to the process in which a message digest, or hash, is computed for a message, and then the digest is encrypted using secret key encryption process. A digital signature is similar, however, a public key encryption process is used, encrypting the digest with the sender's private key. The receiver then uses the sender's public key to decrypt the digest, compute the hash, and verify that the message was from the intended party and was intact.

Without digital signatures there is no way to assure an electronic message is genuine—that it is in fact from the claimed party.

9.7 General Security Analysis Guidelines and Models

Any aspect of the network or network equipment that can be exploited for attack is a network vulnerability. To describe a network's exposure is to describe the extent of potential harm that could occur.

The goal of security analysis is to determine what requirements a security system must meet in a given context to provide adequate protection for a valued set of resources. Depending on the context, the resources can be real objects with intrinsic value, or objects whose value is basically determined by the information that they represent. The dominant focus in analyzing network security requirements is on objects that represent intellectual property or assets, and not on other types. The scope of the following discussion is limited to such information objects.

The fundamental mechanism used in secure systems to protect information is *access control.* With adequate access controls in place, both the confidentiality and the integrity of an information object can be assured to some defined confidence level. Confidentiality is at risk whenever information flows out of an object. Integrity, on the other hand, is at risk whenever information flows into an object. Information flows in and out of an object can be characterized as basic read and write operations on the object, regardless of the type of actual physical means used to access the object.

When adequate physical access controls exist, information can be stored, and moved from one place to another, without taking additional steps to hide it or to disguise it. If adequate physical path protection cannot be assured, then information needs to be transformed in some way to make it unintelligible to unauthorized parties, and able to indicate to an authorized recipient that it has not been altered in transit. To ensure the confidentiality of information flows sent via insecure paths, it is necessary to encrypt the information at the source and to decrypt it at the destination. To ensure integrity, validation methods must be used to verify that stored or transferred information is not contaminated.

To show with confidence that a system is secure is a difficult task. In the past two decades, some basic approaches have been introduced and refined that make security system analysis and synthesis tasks somewhat less daunting. No single approach, however, has emerged to displace all others, although there is a general trend toward consensus, especially in defining security policies and criteria.

A security policy establishes the scope, terms of reference, and requirements for the protection of valued resources in a given domain. The scope of a security policy can apply to a complete system, or selectively to parts of a system.

A security policy can stipulate the use of particular security methods or mechanisms, or it can leave such matters up to the manufacturer of a security product or system. It can also explicitly identify resources to be protected, and the level of protection that each requires, or just provide a set of rules that need to be satisfied. In any case, a candidate product or system must comply with all of the requirements specified in an applicable security policy to qualify.

Security criteria establish discrete reference levels for evaluating how well a candidate product or system element meets the requirements of a security policy. The criteria do not establish what protection needs to be provided in a given context: that information is provided by an applicable security policy.

Various standards for security criteria have been published. The standards reflect a general trend toward harmonization, but no universal set of security criteria has emerged to date. Newer standards tend to refine and enhance earlier work. The following list is an attempt to merge various current points of view:

- *General criteria:* Define general compliance requirements.
- *Confidentiality criteria:* Define levels of protection against authorized disclosure.
- *Integrity criteria:* Define levels of protection against unauthorized modification.

- *Accountability criteria:* Define levels of ability to correlate events and users.
- *Reliability criteria:* Establish levels to measure how well a product performs.
- *Assurance criteria:* Define overall trust levels for products or systems.

Section 9.7.4 provides additional details on the preceding criteria.

To qualify as compliant with respect to a particular criterion, a candidate product or system must satisfy all of the requirements that apply to the criterion.

Formal methods must be used to prove conclusively that a product or a system has no insecure states and transition paths. A formal proof that a complex product or system is secure may be difficult or impossible to provide without partitioning the problem into a set of suitable elements.

The next section briefly discusses the use of models to simplify security analysis problems.

9.7.1 Generic Security Models

Generic security models are abstract representations of security elements and relationships that are common to different systems. They provide a framework for analyzing security problems, synthesizing solutions, and developing system-specific models. The system blocks used in generic models can represent large or small objects in a variety of contexts.

Figure 9.1 shows a simple model of the basic relationship between user instances and protected objects. Note that there is no direct path between the protected information object in the model and the user instances outside the trusted system block. All external service requests are intercepted by the access control object in the trusted system block (TSB) and rejected unless the TSB can determine that they are legitimate. A security policy can define different sets of rules for providing services to users based on the user's identity, authentication, privilege status, information transfer path security, or other factors.

The model also illustrates that all accesses to a protected information object consist only of one or more basic read or write operations. It is convenient, however, to discuss most aspects of access control at a coarser granularity level represented by the Create, Destroy, Read, Write, and Execute blocks that comprise the access object. The purpose of these blocks is to highlight the elementary services that the TSB can perform in response to a request, but not to dwell on the unnecessary details of primitive internal operations.

The basic principles commonly used in secure systems to protect resources are:

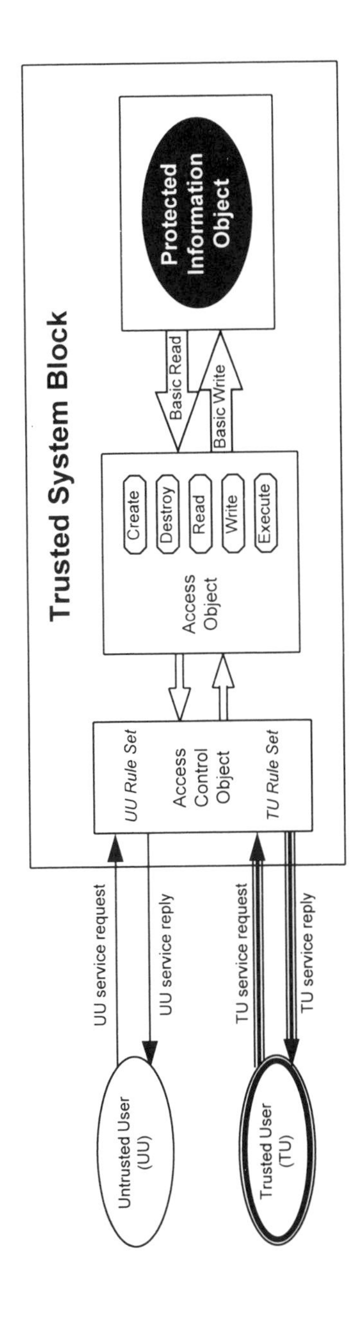

Figure 9.1 User and protected information object relationships.

- The operational state of a TSB protecting information objects is verified before the TSB can provide access services to a user.
- Users are not allowed to have direct access to information objects. All service requests are mediated by a TSB. The isolation of protected resources from direct access maintains the confidentiality and integrity of the protected resources.
- User identity, role, and privileges are verified before a TSB accedes to a service request.

9.7.2 Access Control Models

Before a TSB can agree to a service request from a user instance requesting access to a protected information object, the TSB must determine whether the privilege attributes of the user instance dominate the permission attributes the protected information object. In other words, the TSB must determine if the user has the appropriate level and defined permission to access the targeted information or process. This can be a difficult task because multiple rules expressed in a security policy may apply simultaneously.

The example in Figure 9.2 illustrates a case where three attributes, domain, role, and operation, are used to by a TSB to decide whether it will accept or reject a service request from a user instance. The example shows that the user instance privilege attributes dominate protected information permission attributes only at the intersection of Domain 1, Role C, and Operation Read. Although each object has other attribute value sets, the access represented by the intersection is the only one that the TSB will allow. In other words, the user instance can read information from the protected object under certain conditions, but it is not allowed to create, destroy, write new information to the object, nor to execute it and get the TSB to return a result.

The rules to determine when privilege attribute values dominate permission attribute values are defined in a security policy. The policy needs to define a clear set of rules, both for reasons of efficiency as well as security reasons.

The security policy for a given domain also needs to define the range of values that privilege and permission attributes are allowed to have. For example, a policy could define create authority categories to control what anonymous-user, identified-user, system operation, system administration, security administration, and supervisory role user instances could create. Similarly, there could be multiple confidentiality categories ranging from unprotected to maximally protected.

The TSB resolves the dominance issue for every protected information object designated in user's service request. If the results allow it, the TSB honors the service request, otherwise it rejects the user's service request.

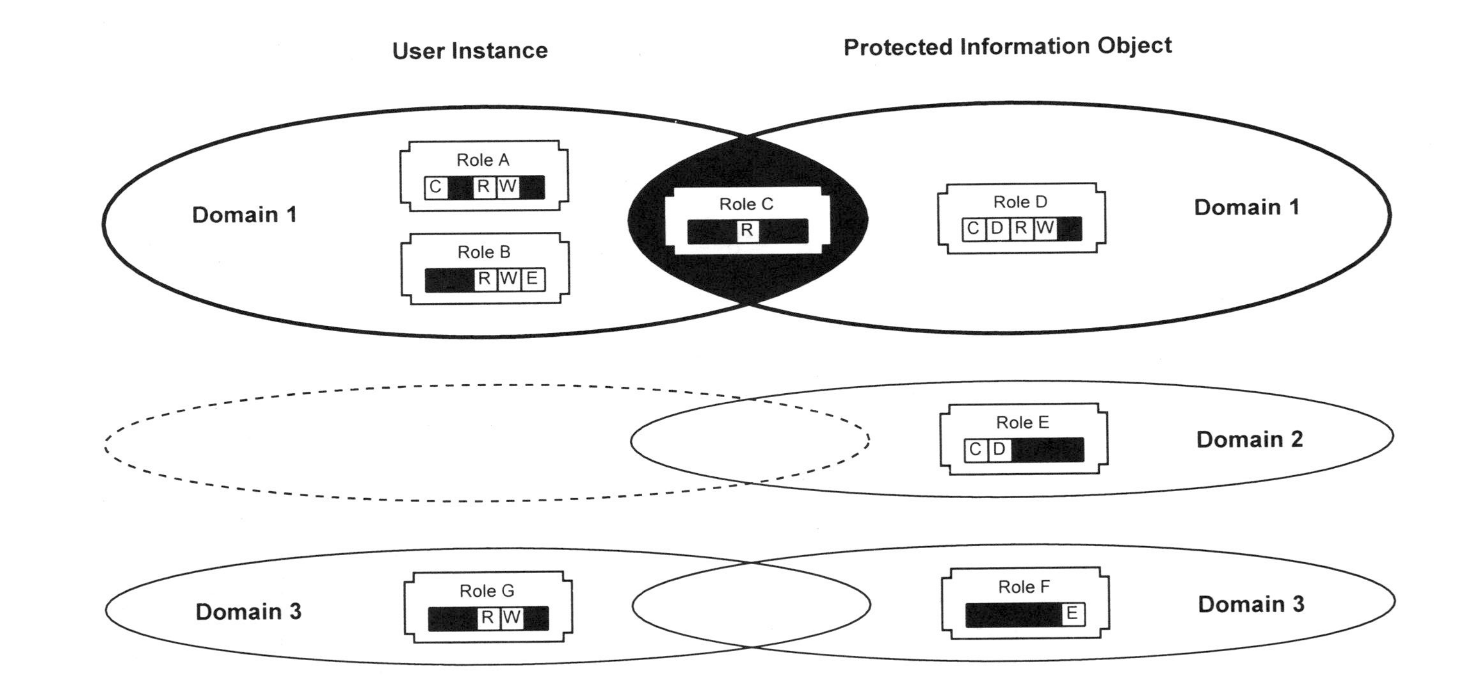

Figure 9.2 Service access relationships.

Once the basic security attribute associations are made, additional requirements can be defined to control users access to protected objects. A worksheet can help identify what levels of confidentiality, integrity, accountability, and reliability need to be associated with each protected information object or user instance. These attributes also include some choices that define the manner to be used in a product to provide protection or to control access. The worksheet can also be used to define separate create, destroy, read, write, or execute service associations with various domain and role attribute combinations.

The composite value set of the attributes (identity excepted) determines what level of protection an information object requires in a given context. The value set of a protected instance cannot be completely empty.

Aside from the necessary identification attributes, the value set of the attributes determines what security privileges a user instance has in a given context. A null value set means that the user has no access privileges to any protected information object.

The security policy needs to define how the confidentiality, integrity, accountability, and reliability security criteria attributes described later map to the privilege attribute value set.

Unless indicated otherwise in a specific requirement, the assumptions, rules, and conditions listed later apply to all system blocks. The default rules are decoupled from each other to create as much orthogonality as possible to allow class definitions to control which rules apply in a given case (see Figure 9.3).

The following list is incomplete and intended only as a sample set of items to consider when preparing a security policy document.

- Default granularity level;
- Default identity authentication rule;
- Default single thread rule;
- Default domain match rule;
- Default role match rule;
- Default confidentiality rule;
- Default privacy rule;
- Default integrity rule;
- Default audit rule.

9.7.3 Security Policies

A security policy defines the requirements for the security needs in a given domain. The scope of the policy establishes the boundaries of what constitutes

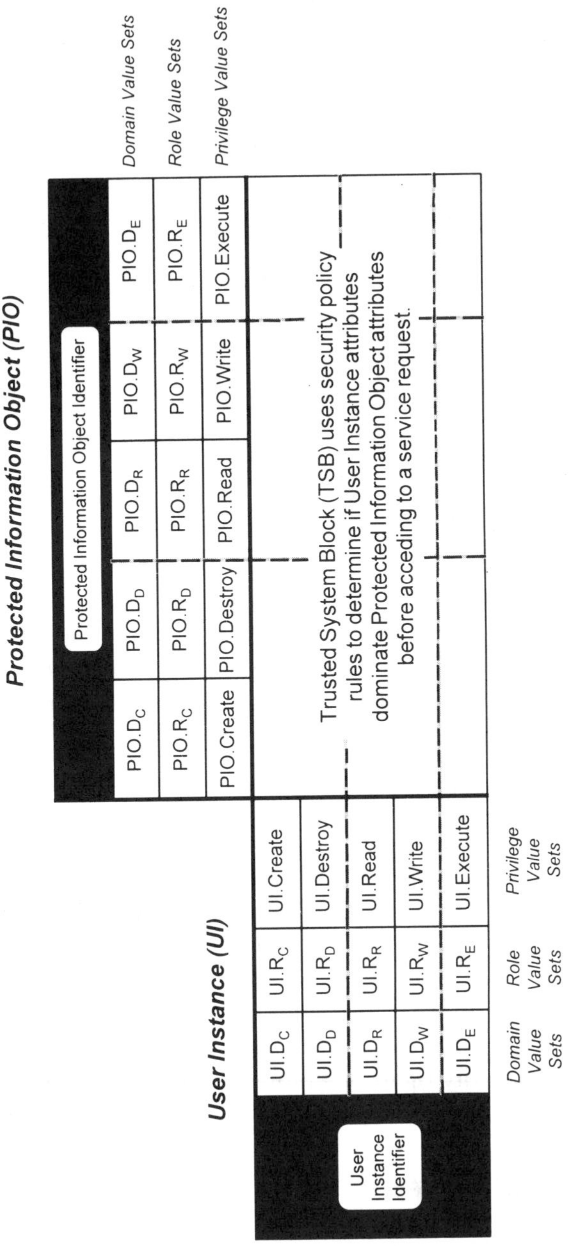

Protected Information Object (PIO)

Protected Information Object Identifier

PIO.D_C	PIO.D_D	PIO.D_R	PIO.D_W	PIO.D_E
PIO.R_C	PIO.R_D	PIO.R_R	PIO.R_W	PIO.R_E
PIO.Create	PIO.Destroy	PIO.Read	PIO.Write	PIO.Execute

Domain Value Sets

Role Value Sets

Privilege Value Sets

Trusted System Block (TSB) uses security policy rules to determine if User Instance attributes dominate Protected Information Object attributes before acceding to a service request.

User Instance (UI)

User Instance Identifier

UI.D_C	UI.R_C	UI.Create
UI.D_D	UI.R_D	UI.Destroy
UI.D_R	UI.R_R	UI.Read
UI.D_W	UI.R_W	UI.Write
UI.D_E	UI.R_E	UI.Execute

Domain Value Sets

Role Value Sets

Privilege Value Sets

Figure 9.3 Fundamental trust relationships.

the domain covered by the policy. A policy can apply to a single product, or to multiple products or subsystems within its domain.

The purpose of a security policy is to define the rules to be followed by products in its domain. The rules establish what levels of protection certain information classes need, what performance levels must be provided, and the degrees of accountability required. A policy can express the rules in general terms, or specify everything in great detail. How detailed a security policy needs to be depends on context and on the complexity of the security problem.

A security policy document needs to define the scope, terms of reference, and specific requirements that a system must meet to ensure the security of information objects contained in a defined domain. The scope of the policy includes information objects contained entirely within the system, and information objects that the system shares with other systems.

A security policy needs to address at least the following items.

Scope

Defines the scope of the system in its environment: domains, roles, related systems, and contained subsystems.

The granularity needs to be fine enough to be able to identify every user or protected object designated or implied in a service request.

The scope of the security policy is limited the protection of information objects in the joint domain, objects engaged in the transfer of protected information from one domain to another, and objects used to store and transfer information within a subdomain.

Access Control Requirements

To protect information objects, it is necessary to determine if a user has the proper authority to access a designated protected object. The authority granted to a user should not be blanket authority that permits a user unrestricted rights. If access is to be granted selectively, then certain information about users and the protected information must be available to a TSB for it to be able to make appropriate decisions.

A TSB compares user and protected object attributes to determine if user rights dominate the security rights of a protected object. The rules that a TSB uses to evaluate dominance relationships are set out in an applicable security policy.

A security policy may need to address hierarchical TSB relationships also if access to different objects in a system is provided by more than one TSB. In this case, peer, superior, and inferior trust level relationships will need to be defined explicitly to ensure product compliance with applicable confidentiality, integrity, and privacy protection requirements.

Security Attributes

To allow TSBs to determine if a user has the authority to access a protected information object, the following security attributes are defined for illustration purposes. The domain attribute category deals with the scope of a user or a protected object. Role attributes serve to partition domains on a functional basis. A security policy can stipulate, for example, that both the domain and the role attribute value sets must intersect before a TSB will consider privilege and permission attribute sets.

Each attribute has a value set that expresses the level of protection required to satisfy the physical, logical, discretionary, mandatory, and other security criteria category requirements specified for a given protected object. Attribute sets can contain subsets that are mutually exclusive, or subsets that intersect.

The following security attributes address the overall protection needs of an object from different points of view:

- *Confidentiality attribute:* The confidentiality attribute of a protected object establishes controls that limit information flows from the object to a user.
- *Integrity attribute:* The integrity attribute of a protected object establishes controls that limit information flows into the object from a user.
- *Accountability attribute:* The accountability attribute of a protected object establishes the requirements for TSBs to collect and to save access request and response information. The attribute takes into account audit, identification, information flow path, and nonrepudiation factors.
- *Reliability attribute:* The reliability attribute of a protected object establishes requirements for TSB operation under normal and under exceptional operating conditions. The attribute takes into account service availability, fault tolerance, and information recovery factors.
- *Assurance attribute:* The assurance attribute of a protected object defines the overall level of protection that a given object requires.

9.7.4 Security Criteria

Security criteria establish a finite, discrete set of terms and conditions for evaluating the ability of a product, system block, or a complete system to protect designated types of resources. The complexity of a system and the tasks it needs to perform determine how granular an appropriate set of security criteria needs to be.

A security policy specifies which security criteria must be satisfied by a qualifying product to ensure that an adequate level of protection is provided to resources. A compliant product meets or exceeds all of the requirements specified in the policy.

A product is evaluated for compliance with each criterion specified in an applicable security policy. For total compliance, a product must meet or exceed all of the requirements of the criterion.

The overall rating of a product is based on the set of criteria that the product can satisfy. If the capabilities of a product match the requirements of a security policy, then the product can be certified as compliant at a particular level.

To evaluate systems that provide more than one level of protection, different subcategories of security criteria requirements need to be defined. Security policies define which security criteria requirements apply to each defined protected information object and user class. A product is evaluated against each required security criterion subcategory for compliance one by one. If the product satisfies all applicable security critèria requirements, then it is rated as compliant with the security policy.

Each set of security criteria subcategory requirements below can apply to different objects. Informal references to objects instead of classes, made to simplify the text, should not be taken as references to unique instances.

General criteria include:

- Formal proof shall be provided that trusted system blocks are secure.
- The TSBs used to administer and enforce the security policy shall be protected against unauthorized use or tampering at all times.
- Covert channels are information flow paths that can exist in a system that are not known to the security system blocks. This criterion addresses only the issue of whether information flows into or out of a protected domain, and not the means used to achieve the information flows. All known covert channels shall be eliminated from all TSBs and from a compliant secure system. Systems shall be analyzed to determine if covert channels exist in any system blocks.
- To contain damage that may be caused by a corrupt user instance, a product should be designed to restrict instance privileges to appropriate domains, and to restrict the scope of each instance to its appropriate role in a given domain.
- Secure products must be able to verify their own integrity using only trusted resources within their own domain.

9.7.4.1 Confidentiality Criteria

Confidentiality criteria establish categories for the risk that protected information is revealed to unauthorized user instances. The criteria apply uniformly to all instances of a given system block class.

If a security policy defines a set of rules that cannot be met by an available set of TSBs, appropriate TSBs will need to be defined to provide the required capability. The effectiveness of a resulting system configuration will need to be proven.

The set of confidentiality criteria below can be modified as necessary to provide increased or decreased granularity:

- *Physical confidentiality criteria:* Protected system blocks in a product can exist in various physical forms. Different means can be used to gain access, for example, electrical, magnetic, optical, acoustic. Ultimate access, however, has to be physical. These requirements apply to TSBs and systems operating normally. Products could be highly tamper resistant, raise alarms, and/or indicate if tampering had been attempted.
- *Logical confidentiality criteria:* Even though physical processes are ultimately used to read and write information to physical media, the processes that decide to do so are logical. Logical confidentiality criteria complement the physical confidentiality requirements. TSB instances can contain permanent or transient confidential information that can exist in volatile or nonvolatile form. Copies of protected information can reside in memory or on other media. A TSB needs to clean out all storage places that were used to hold protected information, before releasing them for possible use by other processes.
- *Discretionary confidentiality criteria:* Discretionary confidentiality criteria define requirements for the control of information transfers between a TSB domain and other designated domains. In general, peer and superior-trust-level process or user instances can read protected information. However, some information can be designated private, with access limited to a restricted set of identified processes or users.
- *Mandatory confidentiality criteria:* The mandatory confidentiality criteria category applies to a class of protected system blocks that can be accessed only by privileged process or user instances. Typically, a security policy will specify the trust levels in each domain that have access privileges to a given protected system block. To protect system resources, for example, access may be restricted to privileged system or security administration processes.

9.7.4.2 Integrity Criteria

Integrity criteria are used to evaluate how well a product protects resources from unauthorized modification. Any user with write privileges is able to modify an object.

To ensure integrity, superior-trust-level user instances are not allowed to write to subordinate-trust-level objects, that is, send information "down," for two reasons: (1) a message could contain confidential information, perhaps included inadvertently, that a subordinate-trust-level process or user instance is not entitled to receive, and (2) a message could result in the modification of a protected system block.

The overall integrity of a protected object depends on the following subcriteria:

- *Physical integrity criteria:* Physical integrity criteria are analogous to the physical confidentiality criteria described in the preceding section. Whereas those criteria are to protect resources from unauthorized read access, physical integrity criteria define requirements for controlling write access to resources in a TSB domain.
- *Logical integrity criteria:* Logical integrity criteria are analogous to the logical confidentiality criteria described in the preceding section. Protected system blocks in a TSB can be susceptible to unauthorized write attempts. A product needs to provide appropriate means to prevent such security risks to qualify.
- *Discretionary integrity criteria:* Discretionary integrity criteria define requirements for the control information transfers into a TSB domain from other designated domains.
- *Mandatory integrity criteria:* Mandatory integrity criteria define requirements that ensure that only privileged process or user instances can modify protected system blocks.
- *Integrity recovery criteria:* Integrity recovery criteria define requirements for the return of a compromised protected object to a known former-integrity state.

9.7.4.3 Accountability Criteria

Accountability criteria establish reference levels for measuring a product's ability to correlate events with service users. Event and interaction histories are used to assign responsibility for actions taken by a product to a traceable service request by a user. A security policy defines which of the following criteria apply in a given context:

- *Audit criteria:* Audit criteria establish requirements for monitoring information flows between a product and its environment, as well as internal information flows. Different monitoring granularity levels can be defined by a security policy to ensure that just enough data are collected about information transfers to ensure adequate traceability of events and responsible information sources. Information flow monitoring in a product can be used to provide input to system blocks responsible for raising alarms in case of potential threat to the confidentiality or integrity of a protected system block
- *Identification and authentication criteria:* Before providing services, or using the services of another system block, a TSB needs to ascertain and to authenticate the identity of the other party. It may not be necessary for a TSB to repeat the identification and authentication process for some time, if a trusted path is established next, and then used for further communication.
- *Trusted path criteria:* Once a TSB has verified that a given information flow path can be trusted, and that trusted auditing and alarm functions are monitoring the security of the path, the TSB may be permitted by a security policy to comply with a relaxed set of identification and authentication requirements.
- *Nonrepudiation criteria:* Nonrepudiation criteria establish requirements to ensure that an originator of a signed message cannot deny the authenticity of the signature, and the document. Unlike handwritten signatures, a digital signature contains information that links in particular the contents of a document to the signature of its originator.

9.7.4.4 Reliability Criteria

Reliability criteria are used to measure a product's ability to provide services to its users under normal and under exceptional operating conditions. The following reliability criteria are defined in these guidelines:

- *Availability criteria:* A product can control what services it makes available to its users under normal conditions on the basis of quotas and privilege levels defined in a security policy. Quotas establish limits on each service user to ensure that a product provides its services on a fair basis to all users. A security policy may stipulate whether users can or cannot be denied services by a product under certain conditions.
- *Fault-tolerance criteria:* Fault-tolerance criteria establish evaluation levels for determining the capabilities of a product to continue to provide security services under exceptional operating conditions. These require-

ments focus on how well a product provides services when it is not operating normally.

- *Recovery criteria:* Recovery criteria allow a TSB to return to a know secure state after a malfunction or a detected security violation attempt.

9.7.4.5 Assurance Criteria

Assurance criteria are used to rate the overall ability of a product to pass critical review. To build sufficient confidence that a given product can be trusted, it is necessary to go beyond a suite of tests. The theoretical basis used to provide secure services needs to be known to prove that they are adequate. In addition, knowledge of a product's architecture and how it was developed and designed can reveal possible problem areas.

To ensure an appropriate level of protection and privacy, protected resources must be placed into discrete categories. Security policies define such categories, and also prescribe what level of protection needs to be provided for each category. The number of levels needed can vary with the context.

Security products that qualify at a given level can be trusted to provide the necessary protection for resources in a corresponding category. An assurance level rating for a product indicates an overall trust level that can be placed in the product.

9.8 Network Security Services

Just because multimedia communication networks may be high speed, and may be running over optical facilities, it is not safe to assume that there are no security risks. While the higher data rate and optical media may pose more challenges for would-be attackers, there are still ways in which these areas can be compromised.

Optical taps are possible, although challenging. High-speed analyzers, encryptors and decryptors provide an indication that the high-speed rates will not be much of an issue.

As identified in [2], network security requirements include accountability, alarming, archiving, assurance, audit and data analysis, authentication, authorization, availability, confidentiality, damage limitation, error recovery, integrity, logging, nonrepudiation, privacy, reporting mechanisms, robustness, short-term data backup, and survivability.

New technologies are prone to attacks since many are implemented early, with a focus on first to market, and they have not been put to the test of time and exposure that older technologies have.

9.8.1 Secure Information Transport Services

Figure 9.4 illustrates how a user can obtain various levels of security. To achieve secure source-to-destination information transfers, data and signals can be encrypted in the source block, and then decrypted at the destination block. A trusted path through intermediate system blocks can strengthen the transport system security. (The A1 and A9 reference point identifiers follow the convention defined in the DAVIC system reference model used to delimit system blocks [7].)

9.8.1.1 Intermediate Service Provider User Plane

Transparent communication through an intermediate service provider (ISP), usually a network, is shown in Figure 9.5. In this example, the ISP provides a bidirectional digital information transport service for its information provider system (IPS) and information user system (IUS) clients. The service is shown as a pipe between peer IPS and IUS transport-layer objects. All end-to-end application and content-information flows between peer objects map to the logical transport interfaces, which map to logical ATM interfaces, which finally map to their respective physical interfaces.

The logical interfaces at reference point D represent IPS and IUS information flows that are transparent to the ISP. The logical U1 service interface in Figure 9.5 represents content-information flows between peers on the DAVIC service layer 0 (SL0), and content- or control-information flows between peer SL1 application layer objects [7].

9.8.1.2 Control Plane

Figure 9.6 depicts an association between IPS and ISP control plane objects that may be used, for example, to set up, modify, or tear down ISP user plane resources. To ensure that services are provided only to authorized clients, two levels of access control are illustrated in the model. The ISP firewall object represents a first level of protection by restricting access to known clients only: The firewall will reject service requests by unauthorized users. A second level of protection is provided by the ISP gateway object. In this case, messages from known system clients are accepted and interpreted. Depending on the nature of the service request, the permission attributes associated with protected ISP objects, and the privileges of the client object, the gateway may accept or reject the service request. If the ISP system security policy requires it, the gateway may challenge the service-requesting object to authenticate its identity.

9.8.1.3 Management Plane

The management plane interface model is similar to that shown in Figure 9.6. Network implementers must be careful to protect the network elements and

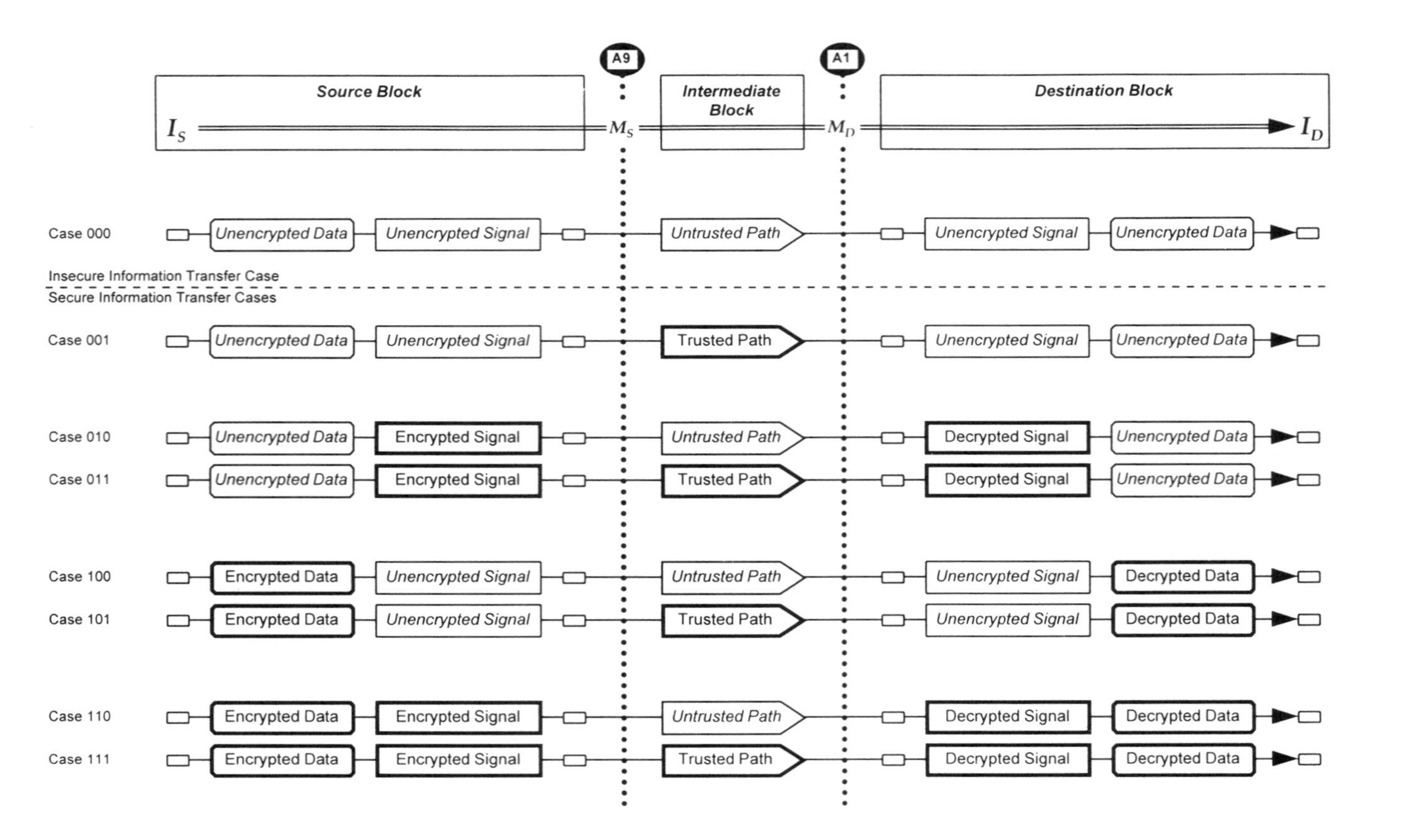

Figure 9.4 Secure information transfer options.

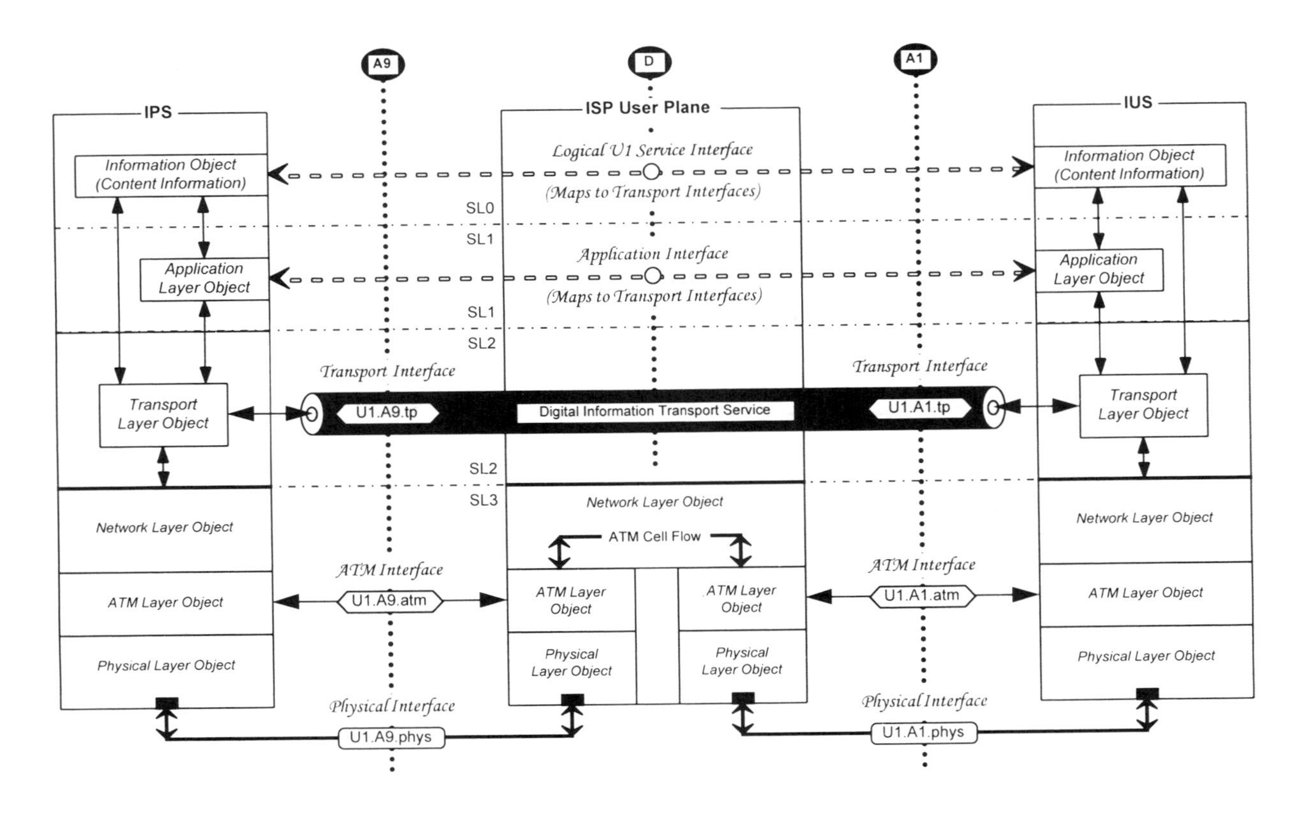

Figure 9.5 User plane digital information transport service model.

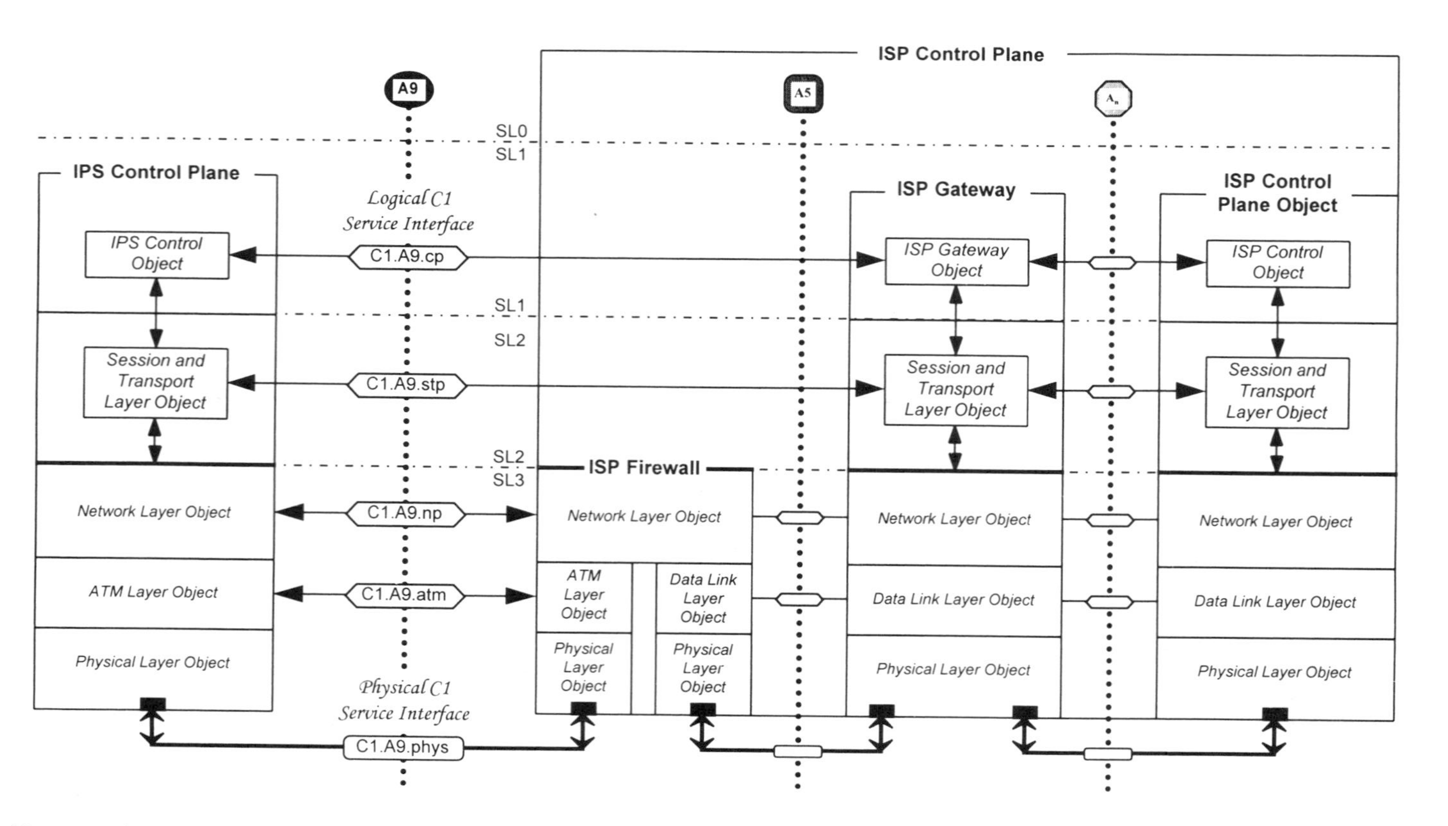

Figure 9.6 Secure control plane interface model.

systems that are used to provide multimedia services. This includes switches, routers, transmission equipment, and intelligent peripherals.

9.8.2 Configuration of Security Services

Security mechanisms or services are just one consideration for choosing a transport network protocol when planning a network, but it is an important aspect to include. There are some who will say that application level (end-to-end) is all that needs to be done. This would certainly satisfy some requirements with the use of encryption for authentication and confidentiality. But other services such as access control (endpoint-switch) and availability of service (network element protection, etc.) are still important considerations.

9.9 Security Aspects of ATM Networks

ATM technology has been under development for a number of years now, and many ATM protocol and interface standards and implementation agreements have been produced by various groups, but not very much has been done in the area of security. The ATM Forum has produced a Phase 1 Security specification that is the first industry document dedicated to security of ATM networks. Many of the ATM security services included in that specification are summarized here. Because a wide-scale, industry discussion/standards effort on the subject of ATM security was not addressed until later in the development of ATM products it is critical that these security mechanisms be incorporated into product and network design as soon as possible. Some ATM security products do exist, and some are noted in the following sections.

The challenge with security in ATM networks is to identify the type of security services required and how best to implement them; considering the weakest link guide, if the ATM network is not end to end, then it is not of much value to only implement security mechanisms at the ATM protocol layer (although support to higher layer services can still be valuable). It is likely that a combination of mechanisms, at various layers, may be most useful. Scalable solutions are needed due to the expected numbers of ATM users and the size of ATM networks. And finally security solutions must be implemented in a fashion that does not prohibit compatibility with nonsecurity aware/capable devices [8].

9.9.1 ATM Protocol

In general there is a security advantage in that ATM is connection oriented in nature as opposed to being connectionless. The greatest advantage with this condition is when the circuit is preprovisioned (PVC), and this is because the route that the cells take is preestablished and well known. This does not prevent attacks in and of itself, but it is more assured than connectionless packets traversing any route that happens to have good metrics at the time. However, the signaling associated with establishment of on-demand connections (SVC) is itself subject to attack, and should undergo security analysis to identify risks and potential threats.

In addition, on-demand connection establishment means that the determination of service type and service level is done during the connection setup process, essentially in real time. The connection will only be accepted if the network can support the level of service requested. This is an area for further analysis because any modification to the service type or service level parameters could negatively affect the user's service. If the connection cannot be accepted due to service requests and associated control mechanisms (e.g., ATM element connection admission control, or CAC), the customer can always try lowering their service requirements (provided the application/service can tolerate it) and attempt the setup again.

One of the most valuable assets of ATM is the quality of service (QoS) component it brings to the transport protocol world. Because QoS is one of the critical aspects to ATM, it may be the focus of future threats. The network needs to assure that the QoS agreed to for the customer/service can be maintained—that either unintentionally or intentionally no other user can consume so many resources as to degrade the QoS for others. Quite simply the customer should get what they paid for. If unauthorized changes to the QoS parameters could be made, then all or some parameters could either be "lowered" to damage that user's service, or they could be improved, which is the equivalent of theft of service, and could contribute toward overloading the network and degrading service for all customers.

How ATM networks manage congestion is also important. If those mechanisms can be tampered with then the network might not operate properly, buffers could overflow, and service could deteriorate.

9.9.1.1 ATM Data Confidentiality

For ATM protocol the most practical layer on which to provide data confidentiality is the ATM layer. The ATM cell payload can be encrypted easily because it is always a fixed length (48 bytes). More importantly, unless one plans to do encryption on a hop-by-hop basis with each node in the network (and establish trust with each of those nodes), the ATM cell header will have to be in the clear, or be readable by each node. This confidentiality service is likely to be useful on untrusted networks spanning between trusted networks, typically at the points where private networks attach to the public network.

An example is the Key Agile project at MCNC, funded by DARPA; this project was undertaken to study the design, fabrication, and evaluation of a key agile ATM cryptographic system in support of confidentiality service. A system design was developed that actually permitted full-duplex operation at an OC12c rate for up to 65,000 active connections, each encrypted under a different (unique) key [9]. The ATM stream was encrypted using gigabit-per-second DES chips, and the keys were generated and distributed using RSA public key encryption, certificates, and digital signatures. Normal call processing was suspended to accomplish this for SVC setup, and the results (as documented in results from work in 1995) showed establishment of secure connections in under 1 sec after certificates have been initially exchanged [10].

9.9.1.2 ATM Authentication

Authentication in ATM networks is important between communicating nodes and is a foundation or supporting service to other security services such as confidentiality and integrity. It is likely that the nodes will be security agents (SAs) associated with end-user virtual path connections (VPCs) and/or virtual channel connections (VCCs). Authentication should be considered for both the user plane (user data) and the control plane (signaling), and is typically performed once at the start of the connection.

Signaling needs to be authenticated, either bilaterally or unilaterally. Signaling must also support higher layer security exchanges, when the customer negotiates security services on an end-to-end basis.

9.9.1.3 ATM Data Integrity

This service is most useful at the ATM adaptation layer (AAL). The ATM service data unit (SDU) offered to the ATM layer can be taken as one data block for this service. Currently both the AAL5 SDU and AAL3/4 SDUs are supported in the ATM Forum security specification. And for these, the option of replay/reordering can either be used or not. If it is used, then the signature of the SDU is computed after adding a sequence number. In this way the signature will both protect and preserve the sequence number. If replay/reordering is not desired, then the signature can just be computed over the SDU by itself.

This would generally need to be done on an endpoint-to-endpoint basis because of being supported at the AAL, which is typically transparent to the intermediate switching nodes. However, it is likely that the SA will be considered the endpoint in this scenario, and due to a number of considerations the SA will not always be coincident with the end user.

9.9.2 ATM Firewalls

Some say that in an ideal world, where network security services are implemented uniformly and well, that firewalls are not needed. But this is not an ideal world, and firewalls are abundant in IP-based networks today. And with developing technologies such as ATM, where security services were not designed into the early protocols, ATM firewalls may offer some "add-on" protection for ATM users in the interim. There are ways in which ATM firewalls are expected to help, and others where they will not be helpful at all.

Firewalls have historically been based on packet-filtering approaches. For ATM, the most useful elements to filter on are the addresses (calling and/or called) and possibly the subaddresses. But this information is contained in signaling messages, and those elements are in the ATM SDU—not in the ATM cell header. To accomplish this kind of filtering, the firewall would need to intercept the signaling message (during call setup) and check policy before completing the signaling message delivery and connection establishment. And with IP packet filtering, in scenarios where IP is run over ATM, the firewall would need to identify the proper ATM cell containing the IP source and destination addresses (and any other criteria under policy). These issues bring into question the performance of these firewalls and the usefulness to high-speed networks such as ATM. With filtering on signaling packets the call setup performance may not meet user or network needs, and filtering on ATM cell payloads (for IP addresses) in theory could add significant delay.

As an example of the ability to intercept and analyze a high-speed ATM stream, there is a research product with the name of ATTILA. Developed by MCNC under funding from the DoD, ATTILA is a network analyzer that is capable of collecting cell-level traffic statistics for ATM networks; during this project MCNC found that they were also able to tap into ATM cell streams that could be filtered on specific addresses. This system has been shown to run at OC3c (ATTILA Jr.) and OC12c (ATTILA) rates [11].

9.9.3 ATM Network Elements

As with any element in a well-secured network, ATM network elements need to be protected—this includes equipment such as switches, routers, and multiplexers. For the most part general, traditional protection mechanisms can apply here (e.g., identification and authorization of users, controlling physical access to the element, establishing resource access control, and audit capabilities) if there are remote log-in capabilities on the element these need to be secured as well. Having network elements on a LAN will weaken/lessen security if access to the NE can be compromised.

Management of the ATM network is typically undertaken by the network management system, or NMS. The NMS has the ability to establish, delete, and change the ATM physical ports, logical ports, and virtual circuits and can affect other key aspects of the network element (e.g., routing tables). Clearly it is imperative that access to the NMS and privileges on the NMS be carefully managed and secured.

Across the ATM user-to-network interface a management protocol called ILMI, or integrated local management interface, can be enabled. This basically consists of SNMP over AAL type 5, so any weaknesses inherent in SNMP would need to be taken into consideration when protecting ATM resources.

An overview of ATM network management should also include some discussion of the telecommunications network management (TMN) architecture and principles. TMN is an internationally agreed-on standard that was developed to enable interoperability of management equipment/platforms and standard interfaces. The objectives are to reduce management costs, enable new services, and permit a multivendor network. Five functional layers are typically discussed in the TMN concept: fault, performance, configuration, accounting, and security management.

9.10 Security Aspects of IP Networks

IP-based networks generally have the same security concerns that other networks have. One point that should be highlighted is that IP networks are connectionless, that is they do not require a preestablished route for the data to traverse. With connectionless services integrity the focus in on the transmission from source to destination without undetected alteration; in connection-oriented networks (such as ATM-based networks) this same focus exists, but there is also concern with the data sequencing—making sure the data arrive in the proper order. This is not an issue with connectionless networks. The following sections highlight current areas of work under way at the Internet Engineering Task Force (IETF) in the area of security [12].

9.10.1 IPSEC

This group develops mechanisms that protect client protocols of IP. A network-layer security protocol will be developed to provide cryptographic security services that support combinations of authentication, integrity, access control, and confidentiality. The IPSEC focus is on two parts: IP authentication header (AH) and the IP encapsulating security payload (ESP); these will be independent of any encryption algorithm used by the client. IPSEC's preliminary goal is to pursue host-host security, followed by subnet-subnet and host-subnet topologies.

9.10.2 SNMP

SNMPv1 and SNMPv2 leave much to be desired from a security perspective. SNMPv3 is currently under development; the SNMPv3 group's goal is to "produce the necessary set of documents that will provide a single standard for the next generation of core SNMP functions" [12].

It is recognized that key security aspects for SNMP were not adopted due to differing opinions on how to implement in SNMPv2. This resulted in two versions coming to pass: V2u and V2*. The goal of this group is to converge these two implementation solutions. So the key message behind SNMPv3 is that they are not starting yet another version of SNMP, but they are focusing on how to strengthen previous versions based on material already presented. For now they are planning to address the following areas: SNMPv3 modules and interface definitions, message processing and control, security model module, local processing module, and a proxy specification. The current target date for submission of material to the IESG is April 1998 [12].

9.10.3 SSL

SSL is a security protocol that assists in providing secure communications over the Internet. The main focus is on privacy, to prevent eavesdropping, tampering, and message forgery. The two-layer protocol (SSL record protocol is at the lowest layer, layered on top of reliable transport such as TCP) does encapsulation of higher layer protocols such as the SSL handshake protocol, allowing a client and server to mutually authenticate and negotiate use of an encryption algorithm and keys. This authentication and negotiation takes place before any user data are sent by the application. Another key focus of SSL is on keeping the connection reliable, that is to do message integrity check with message transport.

9.10.4 Public Key Infrastructure

There are two efforts on public key infrastructure under way at the IETF:

1. *Simple public key infrastructure (SPKI):* The goal of this group is to develop an IETF-sponsored public key certificate format, along with associated signature and other formats, and key acquisition protocols. SPKI is intended to support a wide range of Internet applications including IPSEC protocols, encrypted e-mail and WWW documents,

payment protocols, and any other applications that need public key infrastructure (certificates and ability to access them).

2. *Public key infrastructure based on X.509 (PKIX):* The goal of PKIX is to facilitate the use of X.509 certificates in multiple applications that use the Internet for transport of information. This group is also focused on promoting interoperability between X.509 implementations.

9.10.5 DNS Security

Domain Name Systems (DNS) security will ensure enhancements to the secure DNS protocol to protect the dynamic update operation of the DNS. This is important to protect the validity of the address mappings on the DNS server. The IETF working group is focusing on the need to be able to detect the replay of update transactions and the ability to order update transactions. They also intend to address clock synchronization as well as all of the dynamic update specification. Note that this IETF group considers data in the DNS to be public—as a result they consider data confidentiality and access control proposals outside the scope of their work.

9.10.6 Pretty Good Privacy (PGP)

E-mail is far more easily intercepted and copied without the knowledge of the sender or the recipient than most users realize. The e-mail message is routinely copied as it traverses the Internet/intranet anyway, so additional copies could be made without detection. The transmission lines involved in accessing/downloading the e-mail are subject to eavesdropping via LAN or phone connections as well [13].

Pretty Good Privacy provides confidentiality and authentication security services that can be used for e-mail and file storage/retrieval applications. Through its development PGP has evolved into a general-purpose application that is platform independent, available for DOS/Windows, UNIX, and Macintosh to name a few. PGP includes RSA for public key encryption, IDEA for secret key encryption, and MD5 for hash functions. When used for authentication services, PGP offers a strong solution based on MD5 for the hash function of the original message and the RSA public-key encryption of the 128-bit hash code. For confidentiality services, PGP uses the combination of RSA (to encrypt the session key) and IDEA (to encrypt the message using the session key) to provide strong confidentiality and speed of processing (since IDEA is quicker than RSA). Note that in this process each message has its own session key (really a one-time secret key), so a session key exchange process is not required. PGP can also support both authentication and confidentiality in the same service, where the signature is applied before encryption. It is also interesting to note that the PGP process uses ZIP to compress the message after the signature is applied, and before the encryption process is applied (if confidentiality is being supported) [14].

9.10.7 Privacy Enhanced Mail (PEM)

Privacy Enhanced Mail (PEM) is most commonly used with simple mail transfer protocol (SMTP), but can be used with any e-mail scheme, including X.400. Four key areas outlined in current Internet drafts include: Message Encryption and Authentication Procedures (RFC1421), Certificate-based Key Management (RFC1422), Algorithms, Modes and Identifiers (RFC1423), and Key Certification and Related Services (RFC1424).

For PEM the prime objective with interoperability is to have mechanisms implemented at the application layer such that they are independent of the lower layer protocols or operating system. PEM is currently limited to a small set of security services: protection against unauthorized disclosure, sending party authenticity, message integrity, and nonrepudiation of the sending party when asymmetric key management is used [14].

9.10.8 IP/ATM Considerations

The combination of IP and ATM tends to follow the same issues mentioned for each technology separately. But there are also several topics of particular interest with supporting IP-based applications in an ATM network. These include the areas of addressing, routing, and member groups.

With addressing, there are traditional addressing concerns related to spoofing and masquerading, but address resolution between IP and ATM addresses adds an additional area for analysis. Two items in particular to consider relate to the trust relationship with users and other address servers. An address server should deny requests for changes to address mappings if the user requesting the change is not considered a trusted source. Likewise, address mapping updates from other address servers should not be accepted if those servers have not established their trustworthiness. An attack on addressing can redirect data from the proper destination to a fraudulent destination, or could cause service performance delay or even service loss for affected customers. While address resolution is one of the most critical functions with ATM networks, it can also be one of the most vulnerable and risky. A likely mechanism to assist with these concerns is authentication.

For the routing area, again the issues mentioned for IP and ATM independently are still of concern. But for IP/ATM implementations integrated routing techniques must be examined as well. Attacks on the routing tables or algorithms can cause the packets and cells to be misdirected or diverted to improper destinations. A key area for protection is the control plane because signaling to establish "short-cut" connections, control messages that provide updates to address resolution, and routing tables need to be analyzed for authentication services.

The grouping concept (e.g., subnets, logical IP subnets, emulated LANs) needs to be managed carefully. For group memberships/associations, any member joining that is untrusted and unauthenticated can gain access to all the information/privileges to which the group has access. This is similar to the issues with closed user groups and virtual private networks. Strict control of group access and privileges needs to be undertaken.

9.11 Intelligent Network

To understand the potential security risks associated with the intelligent network (IN), it may be beneficial to first review what an IN is and how it operates. IN has a different meaning for different people. Some may understand IN in the context of an 800-type call where routing information is stored in a database external to the switch; others may perceive IN to mean the ability to control call processing to such an extent that new services can be provided by the use of service control logic modification in network elements separate and apart from the switching fabric; and still others may view IN as an amalgamation of differing networks integrated together and stratified by the control plane and transport plane to provide whatever connection and transport capabilities are necessary to provide the service control and bandwidth demands necessary for the service needs. Multimedia falls into this later category.

In the early days of telecommunications, the control plane and transport plane traversed the same connection path; signaling information, like a needle, wove the path for the voice connection through the switching fabric. Multifrequency tones were used to signal between switches to provide the necessary routing information. This led to the opportunity for fraudulent use of the network resources. Various techniques were devised to "fool" switching systems down the line as to the actual progress status and routing of a call, which resulted in "free" calls. The fraud would often be perpetrated from a coin telephone using the circuitry in a small, self-contained box: The terms "black box," "blue box," and "red box" were used to describe the differing techniques employed.

To solve this security problem the control plane and transport plane paths were separated. The Common Channel Signaling Network (CCSN) was developed for signaling messages and it has evolved over time. IN muddied the water further by splitting apart call processing logic, allowing a high-speed processor external from the switch to provide call processing control and routing information. Specialized announcement messages were excised from the switch and placed in external network elements where they could be more quickly customized and integrated with application logic. This evolution further complicates the public switched network because the different network elements need to be provisioned and communicate synchronously to coordinate activities that provide new seamless services.

Until recently, these network elements were interconnected primarily by a closed, "trusted" environment. That also is changing. The Telecommunications of Act of 1996 requires that local exchange carriers (LECs) afford competitive local exchange carriers (CLECs) access to these network elements. The LECs' once closed CCSN must now be opened to interconnect and transport CLEC signaling information. In addition, the LECs must permit their IN service control points (SCPs) and databases to control the switches for the CLEC. The service logic and routing information once resident in the CLEC's switch may now reside in another network provider's SCP. When one considers that the LEC will, at the same time, be interconnecting and providing SCP control to competitors of the CLEC, security becomes an even more complex issue. The adage previously noted that "the network is only as strong as its weakest link" applies. For example, the CCSN is a packetized data network. It should not be possible for one CLEC to spoof the "gateway" links to the LEC and access another CLEC's network with signaling and call control commands.

The CCSN signaling protocol and the IN protocol both have very powerful capabilities. The CCSN protocol contains network management messages that can cause signaling messages to be rerouted around a particular signaling node that is in trouble. The CCSN protocol also contains messages that instruct a network element to slow down messages to a particular element that is in an overload condition. The issue is further exacerbated when IN control capabilities are layered on top. IN maintenance capabilities can instruct the switch to turn on or off specific call processing events called *triggers*. Triggers are those events in call processing that allow the switch to pause call processing and seek instructions and routing information from the SCP.

Earlier it was noted that IN functions needed to be synchronized. Consider for a moment what might happen if an IN trigger was activated and there was no service logic at the SCP to manage the service request. Service disruptions have already been experienced in a single provider environment from similar events. Or consider the havoc that could be wrought by one CLEC turning off the IN triggers on a competitor's switch. The competitors customers may become dissatisfied and change providers.

Until recently, because of the "trusted," closed network environment IN security focused on system access by maintenance personnel and system operators. Typical security included log-on and password authentication and auditing logs. However, network operators and service providers are realizing that the security measures of the closed environment are insufficient to meet the challenges of the new open network interconnection requirements. Domestically (T1) and internationally (ITU-T) standards are addressing the new security needs.

T1M1 has developed T1.233 and T.1243 defining security requirements for operational support systems; T1S1 has developed standards to secure critical aspects of the CCSN protocol, carrying credit card and other critical information; and at the time of this writing T1S1 is developing IN standards that include the use of X.500 directory services for authentication between an SCP and a remote database. Similarly, the ITU-T has adopted the X.500 recommendations for the international IN recommendations. The ITU-T is also looking at general network access and security in Study Group 11.

The following discussion provides a status of security capabilities and techniques being implemented in the realm of CCSN and IN. These techniques are not ubiquitously deployed and are frequently proprietary in nature. From an implementation perspective, vendors are beginning to offer new security tools to protect and monitor CSSN traffic.

- Fraud prevention tools have algorithms that detect fraudulent usage patterns for further investigation.
- CCSN secure gateways can inspect the messages and parameters of the total protocol stack and verify if the originator is valid for the type of message and parameter being sent, as well as their authority to address a particular network element.
- Encryption is being deployed on CCSN networks between network nodes.
- Ordering systems that link to operational support systems are using "shadow" databases, encryption techniques and authentication tools to verify that the appropriate user has access only to appropriate data.

Looking ahead to when IN will be controlling aspects of multimedia networks, security mechanisms will have be enhanced to further integrate switched broadband services, Internet services, data services, and telephony services. IN security challenges will include the need to address the spanning of multiple network types, interconnection of multiple providers' networks, and the development, delivery, and management of robust new services that give the user more access and control of service features.

9.12 Security-Related Industry Standards and Organizations

The role of standards in the security arena is very complex. The traditional benefits of technology standardization that include interoperability and lower costs are complicated by the need to have some security mechanisms protected. Security-related requirements for network elements aid in supplier design of equipment yet at the same time intellectual property rights issues have been of great concern in the industry for key areas such as encryption.

The challenges are both obvious and subtle. Detailing of specific security mechanisms and solutions is not useful when it aids in the compromise of these elements by the "bad guys." But the establishment of interoperable security platforms that enable users to implement various algorithms and modes of operation (via prior agreement or real-time "handshake" procedures) can be quite useful and flexible.

Numerous industry standards efforts are under way, many of which have been mentioned in this chapter. As a supplement to that information, the following text highlights some of the more relevant multimedia-related security efforts under way or already established.

In the ITU-T sector, a number of recommendations are being developed. While the initial focus of the work by Study Group 11, Question 29, was to focus on security related to the Intelligent Network, it is expected that this work will be expanded to include B-ISDN and universal personal telecommunications service.

ITU-T Recommendation X.509 is based on the use of public key cryptography and digital signature technology. It basically positions authentication services within the context of a directory services framework, as defined by the ITU-T X.500 series. Public key certificates could reside in the directory, with each certificate containing a user's public key and a signature (based on a hash function using secret key) from a trusted certification authority. Since certificates are not forgeable, the directory itself does not need to provide any additional protection mechanisms [14].

In the ATM Forum the Security Working Group has developed an implementation specification for ATM security. The main focus of this work is on strong encryption capabilities with real-time, in-line mechanisms for ATM networks. An interoperable security platform is defined that enables user plane authentication, integrity, confidentiality, access control, and control plane authentication. The group's focus was on specifying the application of well-known and accepted countermeasures to ATM connections. These countermeasures are primarily aimed at the ATM and AAL layer, on a per-connection (as opposed to per-link or per-node) basis. Work is also under way on control plane confidentiality and management plane security mechanisms.

The Digital Audio-Visual Council (DAVIC) industry group has published Basic Security Tools for DAVIC 1.0—DAVIC 1.2 Specification Part 10, Security System for DAVIC systems. A number of security tools (e.g., scrambling, key distribution, secure download, parental control) are identified in order to be compliant with the DAVIC security profile. Much like the ATM Forum work, the approach here is to develop a platform where users have options in the use of various algorithms and modes of operation. Two security interfaces are defined in the set-top unit (STU) that permit STU component and conditional access subsystem options.

9.13 Conclusion

Multimedia communication networks are being driven by the needs of users, the availability of new technologies, and the development of products and services from many different providers and suppliers. All communication networks require security analysis and consideration, and multimedia networks are no different. In fact, these networks reflect an increasing level of interactivity and need for interconnection, aspects which may demand that even stronger security policies and mechanisms be implemented. This is especially true in light of the use of public networks (e.g., the Internet) for transport of proprietary corporate data, electronic commerce transactions, and private personal data, as well as the delivery of critical signaling/control and network management data. In many ways multimedia has users focused on content, regardless of the location of the material or the provider of the service. As such, security issues should be seriously considered by corporate and private users alike.

If an analysis is not performed to determine what level of protection resources require, and what levels of user privileges are required, then it becomes quite difficult to properly secure and protect critical aspects of multimedia networks and content. Adequate security policies need to be described in order to determine the best security mechanisms to implement, and how and where to best implement them.

Technologies such as ATM and IP are expected to be used heavily in multimedia networks, and both have some security tools that can be used or are being developed. These tools will support security functions directly in the lower layers of multimedia networks, as well as support higher layer security applications implemented by end users. Industry standards work assists in the development of needed security mechanisms and in promoting the interoperability of various mechanisms and platforms.

9.14 Glossary of Terms

access: Unrestricted access is the ability to obtain, or make use of, an object for any purpose, for example, to create it, destroy it, observe or change its form or contents, or to get the results of some ability of the object to execute an operation.

block: A part of a system that comprises one or more blocks. A block is a recursive construct that defines a domain that provides a static (constant, consistent) interface.

create permission: An attribute of an information object (instance). *Note:* A trusted system block (TSB) can create an instance of a requested class, and populate it with default attribute values, only if the create privilege attribute value of the requesting instance dominates the create permission attribute value defined for the class.

create privilege: An attribute of a user instance (see note under *create permission*).

destroy permission: An attribute of an information object (instance). *Note:* A TSB can destroy a designated instance only if the destroy privilege attribute value of the user instance dominates the destroy permission attribute value of the designated instance.

destroy privilege: An attribute of a user instance (see note under *destroy permission*).

domain: A scope that delimits a set of objects that has at least one attribute in common. Domains can be partitioned recursively into subdomains.

dominate: When the attributes are compared in a manner specified in an applicable security policy with respect to a create, destroy, read, write, or execute operation, a user-instance privilege attribute is said to dominate a protected-object permission attribute if the privilege attribute value equals or exceeds the permission attribute value.

execute: A service request to an object to perform an operation that does not reveal how the object achieves a result.

execute permission: An attribute of an information object (instance). *Note:* A TSB can return the result of an invoked execute operation only if the execute privilege attribute value of the requesting instance dominates the execute permission value of the designated protected information object.

execute privilege: An attribute of a user instance (see note under *execute permission*).

information object: An instance of a class with defined symbolic, syntactic, and semantic attributes. (Note that information objects are ultimately represented by physical phenomena for storage and transport purposes.)

instance: The definition of a unique object with all of its properties. An instance of a class is an object created with a unique identity from a class definition. The object comprises the same set of attributes and methods, but not (except by coincidence) the same state or set of attribute values as other objects created from the same class definition.

nonce: Sometimes just called a random number, it is a value that is used only once; there are different kinds of nonces with different properties, some more secure than others. Nonces can also be time stamps or sequence numbers, in addition to random numbers.

permission attribute: An attribute of an information object used to specify access rights; separate permission attributes are used to specify create, destroy, read, write, or execute rights. These operations can be performed on a designated information object only if the corresponding privilege attribute value of a service-requesting instance dominates the permission attribute value of the protected object.

privilege attribute: An attribute of a user instance; separate attributes are used to specify create, destroy, read, write, or execute rights. A TSB will honor a service request for a designated operation only if the privilege attribute value of the requesting instance dominates the corresponding permission attribute value of the designated information object.

product: A logical or physical object that provides one or more services to users.

read permission: An attribute of an information object (instance). *Note:* A TSB will honor a read operation request only if the read privilege attribute value of the requesting instance dominates the read permission attribute value of the designated protected information object.

read privilege: An attribute of a user instance (see note under *read permission*).

role: An attribute value that identifies an object as a member of a functional subdomain (an object can have multiple roles in a domain).

security criteria: Discrete sets of requirements used to measure the ability of a product or a system to comply with the requirements of a security policy.

security policy: A set of terms and conditions that specifies the requirements for protecting resources in a defined context.

system: A collection of interacting objects that serves a useful purpose; typically, a primary subdivision of an object of any size or composition (including domains).

system block: A partition of a system, or of a system block. (Partitioning is a recursive process that divides a given domain into two or more nonintersecting subdomains.)

PSB (protected system block): A system block that requires its environment (the system that contains the block) to comply with a set of security criteria as specified in an applicable security policy.

TSB (trusted system block): A system block that is able to provide the security services required by a security policy to one or more protected system blocks contained in the TSB.

TUI (trusted user instance): An identified and authenticated user instance that initiates service requests to a TSB instances over a trusted information flow path.

user: Used for the sake of brevity in place of user instance or a trusted user instance when the context is clear that one or the other type is meant, or when differentiation is not important.

user instance: An instance that initiates service requests to a TSB instances; by default a user instance is untrusted.

write permission: An attribute of an information object (instance). *Note:* A TSB can honor a write operation request only if the write privilege value of the requesting instance dominates the write permission value of the designated protected information object.

write privilege: An attribute of a user instance (see note under *write permission*).

References

[1] Agnew, P. W., and A. S. Kellerman, *Distributed Multimedia: Technologies, Applications, and Opportunities in the Digital Information Industry: A Guide for Users and Providers,* Reading, MA: Addison Wesley, 1996.

[2] ITU-T Recommendation Q.NSEC, Q29/11, draft output from July 95 Interim Rapporteur's meeting.

[3] White House Press Release, November 15, 1996, at URL: <http://csrc.ncsl.nist.gov/keyrecovery/commerce.txt/>.

[4] Baker, S., and C. Smith, "Compliance with CALEA, The Communications Assistance For Law Enforcement Act of 1994," at URL: <http://www.steptoe.com/calea.htm>.

[5] T1 LB 605, "Draft Proposed J-STD- Lawfully Authorized Electronic Surveillance," Mar. 27, 1997, Standards Committee T1.

[6] Kaufman, C., R. Perlman, and M. Speciner, *Network Security: PRIVATE Communication in a PUBLIC World,* Englewood Cliffs, NJ: Prentice Hall, 1995.

[7] DAVIC 1.0 Specification, at URL: <http://www.davic.org/DOWN1.htm>.

[8] BTD-SECURITY-01.02 DRAFT ATM Forum Phase 1 Security Specification.

[9] Key Agile ATM Encryption Systems, at URL: <http://www.ito.darpa.mil/Summaries95/A802-MCNC.html>.

[10] MCNC ANR Projects—Enigma2, at URL: <http://www.mcnc.org/HTML/ITD/ANR/Enigma2.html>.

[11] ATTILA Network Analyzer, at URL: <http://www.mcnc.org/HTML/ITD/ANR/Attila.html>.

[12] SNMPv3 Working Group, SNMP Version 3 (snmpv3), at URL: <http://www.ietf.org/html.charters/snmpv3-charter.html>.

[13] Garfinkel, S., *PGP: Pretty Good Privacy, Encryption for Everyone,* Sebastopol, CA: O'Reilly & Associates, 1995.

[14] Stallings, W., *Network and Internetwork Security: Principles and Practice,* Englewood Cliffs, NJ: Prentice Hall, 1995.

Further Reading

ANSI T1.2xx-1996, T1M1.5/96-102R1, Application-Based Security, 6/12/96 DRAFT.

ATM Forum Security Ambassador Module, ATMSECmod1.0.

BTD-SEC-FRWK-01.00 DRAFT Security Framework for ATM Networks.

GR-1332-CORE, Issue 2, April 1996, Generic Requirements for Data Communications Network Security.

Hamilton, S., "E-Commerce for the 21st Century," *Computer,* May 1997, pp. 44–47.

ITU-T Recommendation Q.955.1 CUG Service Description, 199x.

Kimmins, J., C. Dinkel, and D. Walters, "Telecommunications Security Guidelines for Telecommunications Management Network," NIST Special Publication 800-13, U.S. Department of Commerce, Oct. 1995.

McDysan, D. E., and D. L. Spohn, *ATM: Theory and Application,* New York: McGraw-Hill, 1995.

DePrycker, M., *Asynchronous Transfer Mode: Solutions for Broadband ISDN,* 2nd ed., London: Ellis Horwood Ltd., 1993.

Schneier, B., *Applied Cryptography: Protocols, Algorithms, and Source Code in C,* 2nd ed., New York: John Wiley & Sons, 1996.

Secant Networking Technologies, at URL: <http://www.secantnet.com>.

Secant Products, at URL: <http://www.secantnet.com/product1.html#ivory>.

Tennenbaum, J. M., T. S. Chowdry, and K. Hughes, "EcoSystem: An Internet Commerce Architecture," *Computer,* May 1997, CommerceNet, pp. 48–55.

User to Network Interface 3.1 and ATM User-Network Interface Signaling Specification v4.0, can be found at URL: <http://www.atmforum.com/>.

Online References and Sources

[ATMO] ATOMIC-2 MD5 Analysis, at URL: <http://www.isi.edu/div7/atmoic2/md5.html>.

[BSA] Encryption, at URL: <http://www.bsa.org/policy/encryption/encryption.html>.

[CANA] Privacy and the Canadian Info Highway, at URL: <http://www.privacy.org/pi/countries/canada/report.html>.

[CELL1] The Cell-Relay Mailing list archive: ATM Security, at URL: <http://netlabs.ucs.indiana.edu/hypermail/cell-relay/9402/0176.html>.

[CHAR] ATM Security Page charli/levels.unisa.edu.au/~dstowww/atm_security/>.

[COMM1] CommerceNet Chairman Message, at URL: <http://www.commerce.net/about/vision.html>.

[COMM2] eTrust announcement (EFF), at URL: <http://www.commerce.net/pr/101696.html>.

[CYLI] Cylink Corp: Using the InfoGuard 100, at URL: <http://www.cylink.com/products/security/gte/>.

[DARP1] High Confidence Networking, at URL: <http://www.ito.darpa.mil/ResearchAreas/Information_Survivability/High_Confidence_Networking.html>.

[DARP2] Key Agile ATM Encryption Systems, at URL: <http://www.ito.darpa.mil/Summaries95/A802-MCNC.html>.

[EPIC] EPIC Online Guide to Privacy Resources, at URL: <http://www.epic.org/privacy/privacy_resources_faq.html>.

[EXPO] Commercial Encryption Export Controls, at URL: <http://www.bxa.doc.gov/encstart.htm>.

[FAST] FASTLANE ATM Encryptor (KG-75), at URL: <http://www.gsc.gte.com/GS/Solutions/Detail/fastlane.html>.

[FTC1] FTC: Privacy Workshop on Consumer Information Privacy, at URL: <http://www.ftc.gov/os/9703/privacy.htm>.

[FTC2] FTC Staff Report: Public Workshop on Consumer Privacy on the GII-12/96, at URL: <http://www.ftc.gov/bcp/conline/pubs/privacy/privacy1.htm>.

[GIGA] Gigabit Local ATM Testbed for MM Apps arl.wustl.edu/arl/projects/csto/intro.html>.

[HUGH] Hughes ATM Firewall paper, at URL: <http://www.network.com/~hughes/Interop-eng96-1-1.html>.

[IBM1] IBM SecureWay Home Page, at URL: <http://www.ibm.com/security>.

[IBM2] Network Security Projects (IBM), at URL: <http://www.research.ibm.com/net_security/>.

[IBM3] Security Services in High Speed Networks, at URL: <http://www.zurich.ibm.com/Technology/Security/extern/ATM/>.

[IBM4] E-commerce related, at URL: <http://www.internet.ibm/commercepoint>.

[ICRY] International Cryptography Framework, at URL: <http://hpcc920.external.hp.com/gsy/security/icf/main.html>.

[IHAC] Info Highway Advisory Council (IHAC), at URL: <http://strategis.ic.gc.ca/cgi-bin/dec/wwwfetch?/sgml/ih01094e_pr702.sgml>.

[ISO1] ISO decisions on OH&S, integration of ISO 9000 and ISO 14000, and info privacy, at URL: <http://www.iso.ch/presse/PRESSE11.html>.

[JAVA] E-commerce related, at URL: <http://www.javasoft.com/products/commerce>.

[MCNC] MCNC Project Summary, at URL: <http://www.mcnc.org:80/HTML/ITD/ANR/Projects.html>.

[MIT] The WWW Security FAQ, at URL: <http://www.genome.wi.mit.edu/WWW/faqs/www-security-faq.html>.

[NII] NII Privacy Principles June 6, 1995, at URL: <http://www.iitf.nist.gov/ipc/ipc/ipc-pubs/niiprivprin_final.html>.

[NIST] NIST Computer Security Resource Clearinghouse csrc.ncsl.nist.gov>.

[NTIA] NTIA Privacy Report 10/23/95, at URL: <http://www.ntia.doc/ntiahome/privwhitepaper.html>.

[PGP1] Pretty Good Privacy, Inc. Home Page, at URL: <http://www.pgp.com>.

[PGP2] Schlumberger Press Release—4/15/97, at URL: <http://www.slb.com/ir/news/et-pgp0497.html>.

[PGP3] BAL's PGP Public Key Server www-swiss.ai.mit.edu/~bal/pks-toplev.html>.

[PRIN] Secure Internet Programming: Java Security: From HotJava to Netscape and Beyond, at URL: <http://www.cs.princeton.edu/sip/pub/secure96.html>.

[PSU] Portland State Univ. Secure Mobile Networking Project, at URL: <http://www.cs.pdx.edu/research/SMN/>.

[PURD] COAST Hotlist (Computer Operations, Audit, and Security Technology), at URL: <http://www.cs.purdue.edu/homes/spaf/hotlists/csec-top.html>.

[RENN] ATM Security, at URL: <http://www.rennes.enst-bretagne.fr/~rolin/Anglais/ATM-securite.html>.

[RSA1] RSA Home Page, at URL: <http://www.rsa.com>.

[SMIM] S/MIME Central, at URL: <http://www.rsa.com/rsa/S-MIME/>.

[SRIC] SRI-CSL-Security-Research, at URL: <http://www.csl.sri.com/sri-csl-security/recent-projects.html>.

[STOR] StorageTek Network Systems Group, at URL: <http://www.network.com>.

[TELS] Telstra Corp: Computer and Network Security Reference Index, at URL: <http://www.telstra.com.au/info/security.html>.

[TET] TET Encryption Technology, at URL: <http://www.ifnsec.com/EncrypTech.html>.

[TIS1] Policy-based Crypto Key Release System, at URL: <http://www.tis.com/docs/research/crypto/ckrpolicy.html>.

[TIS2] TIS Worldwide Survey of Cryptographic Products, at URL: <http://www.tis.com/docs/research/crypto/survey/index.html>.

10

Software Architecture for Multimedia Communication and Management

Distributed multimedia systems have received a great deal of attention in the past few years. Several experimental video-on-demand and news-on-demand systems have been on trial. However, the development of large-scale distributed applications is not fully supported by existing distributed platforms. In this chapter, we present the current trends for incorporating multimedia computing and communications in distributed platforms. We initially discuss the need for introducing special mechanisms into the operating system for the manipulation of multimedia data (Section 10.1). We then review the open distributed processing reference models (Section 10.2) and the CORBA platform (Section 10.3). We proceed by showing some extensions for supporting multimedia on distributed platforms (Section 10.4) and some proposals for multimedia on CORBA (Section 10.5). Multimedia in Java is briefly commented on (Section 10.6). Finally, we illustrate how ODP can be used to specify large-scale multimedia systems (Section 10.7).

10.1 Multimedia Operating Systems

Operating systems (OSs) hide all the hardware resource management details from user applications. Multimedia applications have different requirements than traditional data applications, which makes conventional operating systems inappropriate for multimedia computing and communications. In this section, we briefly discuss the need for incorporating special mechanisms into the operating system for the manipulation of multimedia data.

Although a multimedia operating system deals with time-critical data, it differs from a traditional real-time system in several aspects [1]. In a real-time system, speed and efficiency are the main requirements. In multimedia systems, we also need to consider timing and logical dependencies among related tasks, for instance, the synchronization of audio and video streams during a movie exhibition. Contrary to strict fault-tolerant requirements of real-time applications, a short-duration failure of a multimedia system may not result in serious consequences such as a threat to life. Missing a deadline in multimedia systems may not be as serious as in real-time systems. Actually, it may even go unnoticed. For example, the loss of a voice packet in a telephone conversation may not impact the perceived quality of service (QoS). Another difference between the requirements of multimedia and real-time systems is that multimedia data are periodic, which makes scheduling easier.

Multimedia applications have strict delay and jitter requirements, which makes scheduling a critical issue in multimedia OS. An operating system that supports both time-critical and non-time-critical data has to deal with conflicting goals: providing service to time-critical tasks and not unnecessarily delaying non-time-critical tasks. In addition, time-critical tasks should never suffer from priority inversion in order to avoid starvation of non-time-critical tasks. Due to the periodic nature of continuous media, scheduling may deteriorate an application QoS if scheduling overhead is not minimized. Moreover, in multimedia OSs, a scheduling mechanism needs to be designed not only to maximize throughput but also to achieve deadlines.

Most of the scheduling algorithms in multimedia OSs are variations of the earliest deadline first (EDF) and the rate monotonic algorithm. In the EDF algorithm, at every arrival time, a new scheduling order is defined. In the rate monotonic algorithm, static priorities are assigned at the connection setup time according to the required rate. In the ARTS real-time OS (Carnegie Mellon University) a variation of the EDF algorithm, the time-driven scheduler (TDS), was adopted [2]. In the DASH project (University of California at Berkeley) an EDF algorithm is used for scheduling time-critical tasks. At Digital Equipment Corporation, a scheduler based on the rate monotonic algorithm and on weighted round-robin was designed for a video-on-demand server. Three priority classes were defined according to their throughput, delay, and jitter requirements [3]. IBM developed a metascheduler based on the rate monotonic algorithm for the AIX and OS/2 operating systems in order to support real-time continuous media [4]. Experiments with UNIX SVR4 show that it is inappropriate for continuous multimedia data. Only by trial-and-error is the SVR4 scheduler capable of finding a set of priorities to meet a set of requirements [5].

Storage and retrieval of continuous media requires real-time read and write operations. These operations differ from their traditional file system counterparts with regard to timing constraints. The retrieval delay together with the computation and presentations latencies should not exceed a certain deadline. Read and write latencies should also support jitter requirements. Thus, additional buffer may be necessary to smooth a data stream. Moreover, video and audio files require much larger space than data files, therefore, leading to the need for optimized use of limited storage space. Another characteristic of a multimedia file system is that it has to support various media at once. There are two approaches for dealing with multimedia file systems design requirements. In the first approach, we try to minimize the seek time, and consequently maximize the throughput by optimizing the placement of data on several volumes (disk arrays). Striping and data grouping are used in this approach. In the second scheme, real-time guarantees are achieved by disk scheduling and by providing enough buffer to meet jitter requirements.

Multimedia file systems can use either CD-ROMs or magnetic disks. CD-ROM file systems are specified by the ISO 9660 standard with very few variations allowed. On the other hand, several disk scheduling algorithms have been proposed to use magnetic disks efficiently. The EDF CPU scheduling policy may also be used for disk scheduling. Since EDF is a preemptive discipline, it leads to poor throughput and excessive seek times. In the SCAN-EDF discipline, EDF is applied and requests with the same deadline are served according to the scan direction. Requests with the same deadline can be grouped together. The group sweeping strategy (GSS) divides the flows in groups and serves groups in round-robin fashion. SCAN is used within a group, and consequently buffering is required to guarantee jitter requirements caused by consecutive seek time differences. GSS is a trade-off between buffer space and arm movement.

An important issue for multimedia distributed systems is the OS support to end-to-end QoS of a multimedia flow. A multimedia flow should maintain the same QoS values along the production–communication–consumption path. Microkernel mechanisms for enforcing end-to-end QoS requirements have been developed at Lancaster University [6]. Two types of flow were defined as primitive service: QoS controlled message and QoS controlled stream. The former is oriented to real-time control whereas the latter is used for continuous media transmission. These services are simplex and can be used as building blocks for higher level compound services, such as invocation (QoS controlled remote procedure call) and pipeline. The pipeline service is the process of transforming a software signal by successive hardware/software devices and it can be implemented as the concatenation of QoS controlled stream flows. Some principles guided the implementation of QoS controlled flows: (1) the availability of resource consumption estimates at flow establishment time, (2) reports about the maintenance of QoS from a QoS mechanism layer to a policy layer in a two-layer architecture, and (3) the integration of communication and scheduling by running the user code and the protocol in a single thread. The QoS controlled flow service is implemented on top of Chorus. Although Chorus has real-time features, its connectionless communication and the lack of mechanisms to specify individual thread requirements make it improper for distributed multimedia support. Therefore, the Chorus application program interface was extended by the following mechanisms:

- *rtport:* a communication endpoint with associated QoS and internal buffer which can be directly accessed by users;
- *Device:* a producer, consumer, or filter of real-time data that supports the creation of rtports;
- *Handler:* user-specified procedures which may (optionally) be attached to rtports for data manipulation. (Handlers are upcalled by their associated rtports and may perform two distinct functions: event notification and data transfer. The separation of these two functions allows users to choose whether they want to actively process the flow or just monitor it. Real-time programming can be simplified by using handlers to structure applications to react to events and delegate the system the responsibility for initiating events.);
- *QoS controlled connections:* simplex connection-oriented communications with associated QoS;
- *QoS handlers:* upcalled similarly to handlers but they are used to notify the application layer whenever a violation of a connection QoS contract happens.

Implementation details as well as examples of how these mechanisms are used can be found in [7].

10.2 Middleware and RM-ODP

The *middleware* is a layer between the operating system and the applications that provides open distributed processing support, allowing for application development, use, and maintenance. The middleware layer is a shell over the operating system, adding new functionalities to support the distribution of applications.

ISO and ITU-T defined a specification to deal with open distributed environments, the reference model for open distributed processing (RM-ODP) [8–10]. The environment is open in the sense that, in addition to the systems being open, the services are also open. In an open distributed environment, exporters offer services and importers search for services, independent of the identity and location of the exporter.

An open service environment is also characterized by the unlimited access to accept a new user. On the other hand, in this environment, an exporter is autonomous when deciding how to perform the service. The RM-ODP specification deals with this dichotomy: openness and privacy. The RM-ODP specification is object oriented and has four parts: overview, foundations, architecture, and architectural semantics. The first part describes the concept of an open distributed environment and presents an outline of the ODP architecture. The second part contains the definitions of the main concepts used in these environments following an object-oriented modeling. The third part prescribes the building of these environments, considering the required open characteristics. The last part shows how the ODP concepts can be formalized using an FDT (formal description technique).

The first contribution of the RM-ODP is the definition of viewpoints. Viewpoints make easier the specification of complex open distributed environments. The RM-ODP proposes the following viewpoints: enterprise, information, computational, engineering, and technology.

The *enterprise viewpoint* considers the environment's requirements and the scope and policies of a system. It defines the agents (objects that begin actions), artifacts (objects that do not begin actions), communities (composition of objects with a common goal), and activities of the system. Policies are expressed in terms of permissions (what can be done), prohibitions (what must not be done), and obligations (what should be done). Activities are associated with changes of policies.

The *information viewpoint* describes the information structures in an ODP system. It defines invariant schemata (sets of predicates on one or more information objects which are true in any point in time), static schemata (specifications of the state and the structure of one or more information objects at some point in time), and dynamic schemata (action templates which specify the allowable state changes of one or more information objects). Static and dynamic schemata are constructed with reference (subject to the constraints) to invariant schemata.

The *computational viewpoint* deals with the functional decomposition of an ODP system into distributed objects. A computational specification defines objects, their interfaces, their activities, and the interactions. There are three kinds of interfaces:

1. *Signal interface*, in which the interactions are signals;
2. *Stream interface*, in which the interactions are flows;
3. *Operation interface*, in which the interactions are operations.

A signal is an atomic shared interaction from an initiating object to a responding object. A flow is an information stream (representing an abstraction of a sequence of interactions) from a producer object to a consumer object. An operation is an interaction between a client object and a server object (client–server model). It may be either an interrogation or an announcement. In an announcement (one-way communication), the client object invokes an operation of the server object interface and no result is sent. In an interrogation (two-way communication), a result is sent in response. In the server side, subtypes of an interface can be substituted by the father type (or any supertype).

The RM-ODP defines a contract as an agreement that governs the cooperation among objects. In an environment contract an object specifies its constraints to the environment, and vice-versa. These constraints are usually QoS attributes [11]. Although the RM-ODP defines environment contract, it does not provide a notation to specify QoS yet.

Interactions among objects happen through bindings among interfaces. A binding establishes a contract between two or more interfaces (and consequently between the supported objects). In ODP systems an object can have multiple interfaces. Figure 10.1 presents examples of bindings in the computational viewpoint. In an explicit binding, an object called a *binder* can control this binding offering a control interface that enables a QoS negotiation.

Besides failure, persistence, replication, and transaction transparencies, the RM-ODP defines some transparencies that an open distributed environment may need:

- *Access:* Masks differences of heterogeneous computer systems with different operating systems and the use of different languages;
- *Location:* Hides the location of an interface (object). With this transparency the service environment becomes open, because a client can access a server interface without using the interface location;
- *Relocation:* Hides the change of location of an interface;
- *Migration:* Masks, from an object, the change of location of that object.

Considering multimedia communication, new transparencies may be added to these, such as:

- *Stream synchronization:* for intramedia synchronization;
- *Synchronization for streams:* for intermedia synchronization.

The *engineering viewpoint* considers the infrastructure of the distributed system. It defines:

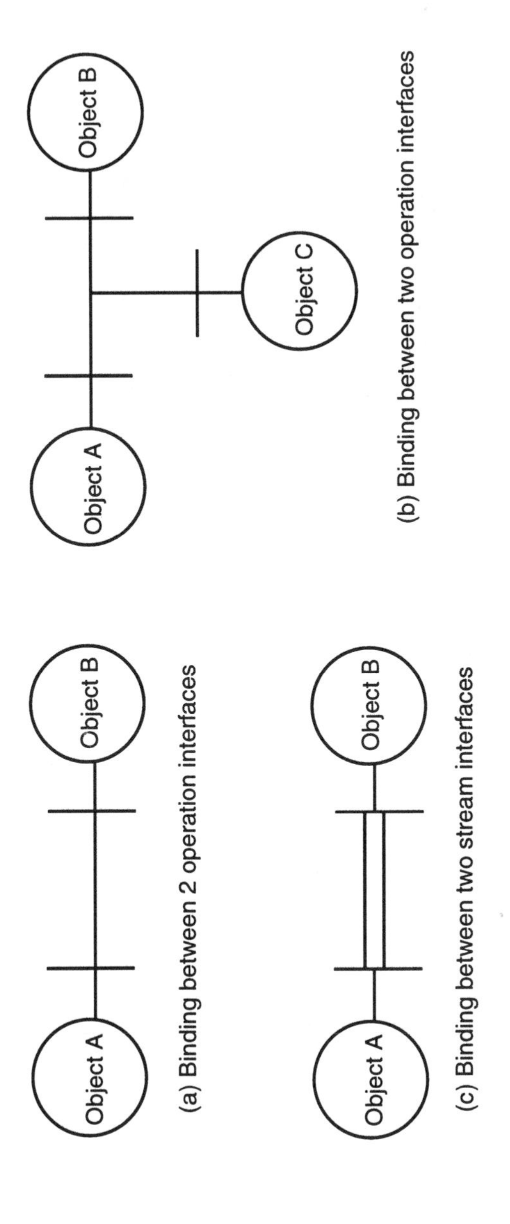

Figure 10.1 Bindings in the computational viewpoint.

- How to structure a distributed application, using BEOs (basic engineering objects), clusters, capsules, and nodes;
- How to support bindings, which have selective transparencies, using channels with stubs, binders, and protocol objects.

BEOs are application-distributed objects. A computational object corresponds to one or more BEOs. The engineering objects can be BEOs or infrastructure objects.

A node represents the unit of resource independence that defines a resource management domain. It has a nucleus and can support one or more capsules. The capsule is the unit of resource allocation, whereas the cluster is the unit of deactivation, reactivation, and migration. A capsule can have one or more clusters. A cluster is composed of BEOs that intercommunicate directly or communicate with other BEOs, outside this cluster, via channels. A nucleus is supported by a *trader* [12], possibly at a different node. Figure 10.2 shows an example of a node with three capsules, four clusters, and eight BEOs.

Associated with each cluster there is a cluster manager that controls the BEOs in that cluster, and each capsule has a capsule manager that controls the engineering objects in the capsule.

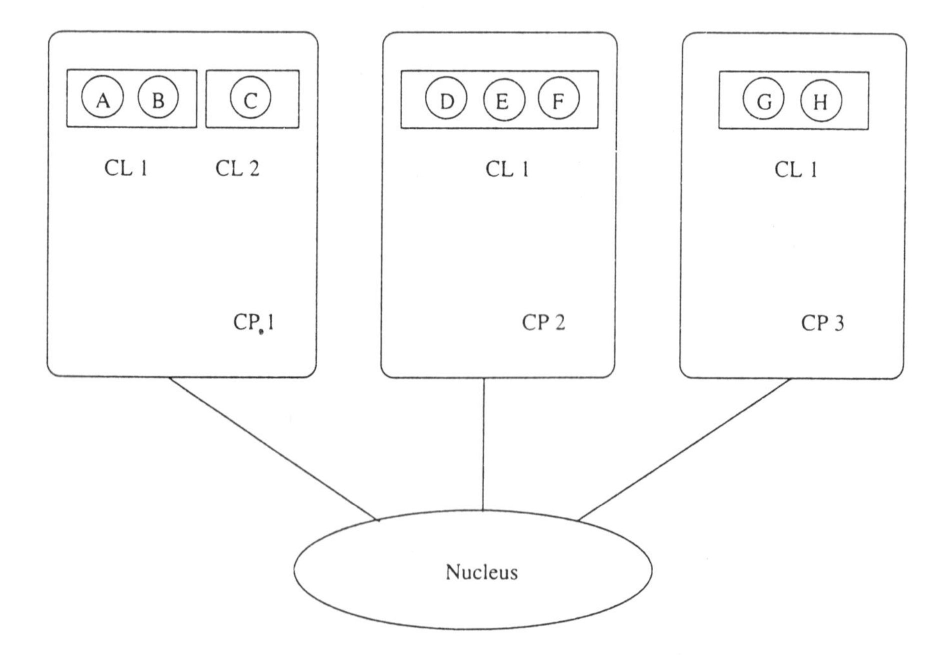

Figure 10.2 Example of a node with capsules, clusters, and BEOs.

Considering units of resource allocation, deactivation, reactivation, and migration, a designer can distribute his or her application in different nodes, capsules, clusters, and BEOs.

A channel corresponds to the binding in the computational viewpoint and it is composed by stubs, binders, protocol objects, and an interceptor (Figure 10.3).

A stub and a binder are infrastructure objects that support the distribution transparency. Stubs may transform the information conveyed by a channel in the interaction (e.g., performing the marshalling of parameters) and/or monitor this information (e.g., for audit purposes). A binder is responsible for the maintenance of the distributed binding between engineering objects. A protocol object communicates with a peer protocol object in another node in order to achieve an interaction between the nuclei within a channel. Protocol objects may be interconnected by an interceptor to mask technology and/or administrative boundaries crossed by the interaction.

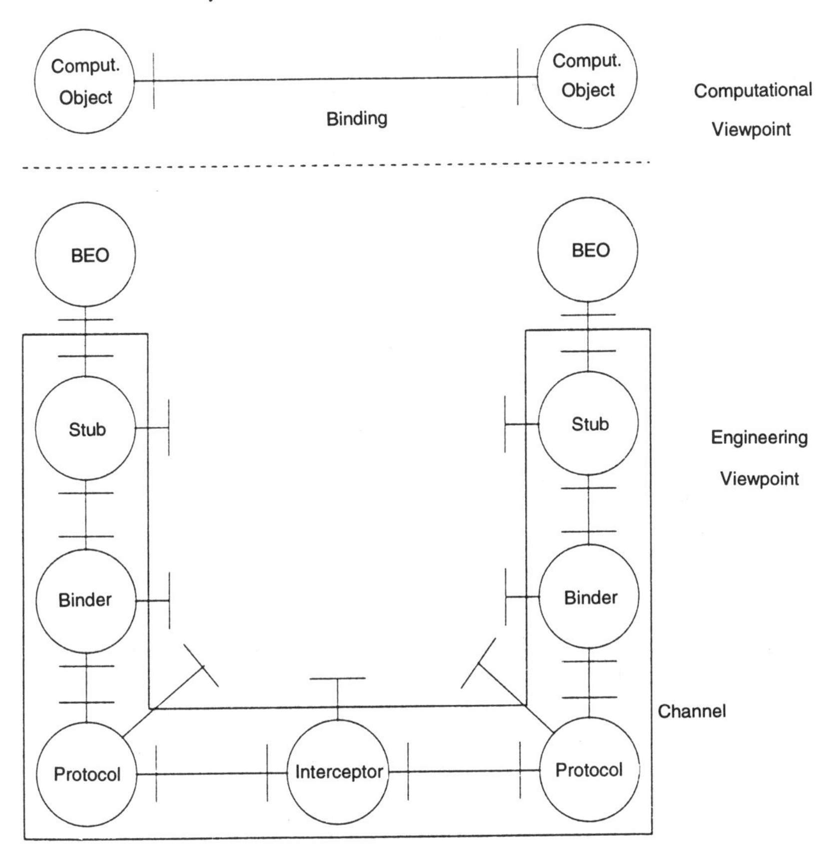

Figure 10.3 Correspondence between computational binding and channel.

Stubs, binders, protocol objects, and interceptors have control interfaces. For example, a binder may be supported by a relocator object. RM-ODP also defines multiple-endpoint channels to bind three or more BEOs.

The RM-ODP specification also defines the engineering interface identifier and the engineering interface reference. The former is the interface identifier of a BEO in the capsule's name context, used for interaction purposes, and the latter is the interface identifier of a BEO in the context of the interface reference management, used for binding purposes.

The *technology viewpoint* considers the choice of equipment, operating systems, networks, and platforms, as well as tests and their maintenance.

Several platforms were developed to create an open service environment. Among them, we mention distributed computing environment (DCE) from Open Software Foundation (OSF) and object request broker (ORB) from Object Management Group (OMG). DCE is process oriented whereas ORB is object oriented.

10.3 CORBA: Common Object Request Broker Architecture

The CORBA architecture was developed to support the integration of different object systems, and it was defined by OMG to be the base of an open distributed platform. OMG is a consortium of almost 700 entities among corporations, universities, and research centers, and it continues to grow. OMG specified the object management architecture (OMA) as an architecture to support the distribution of objects. OMA has five components (Figure 10.4):

1. *ORB* is responsible for these mechanisms: (i) to locate the remote server object (object implementation), in a request of the client object; (ii) to prepare the object implementation to receive a request; (iii) to convey the data performing the request; and (iv) to deliver the possible results to the client object, in a transparent manner. ORB offers the distribution transparencies of access and location, and allows "to put" the client and server objects over it, becoming an *object bus* [13].
2. *Object services* (OSs) offer basic functions to use and implement distributed objects [14]. Some object services defined by OMG are life cycle, persistence, naming, event notification, concurrency control, transaction, relationship, security, and trader. Some functionalities are:
   - *Life cycle* defines methods to create, copy, move, and destroy objects.
   - *Persistence* offers the persistent storage of objects independent of the kind of the storage server.

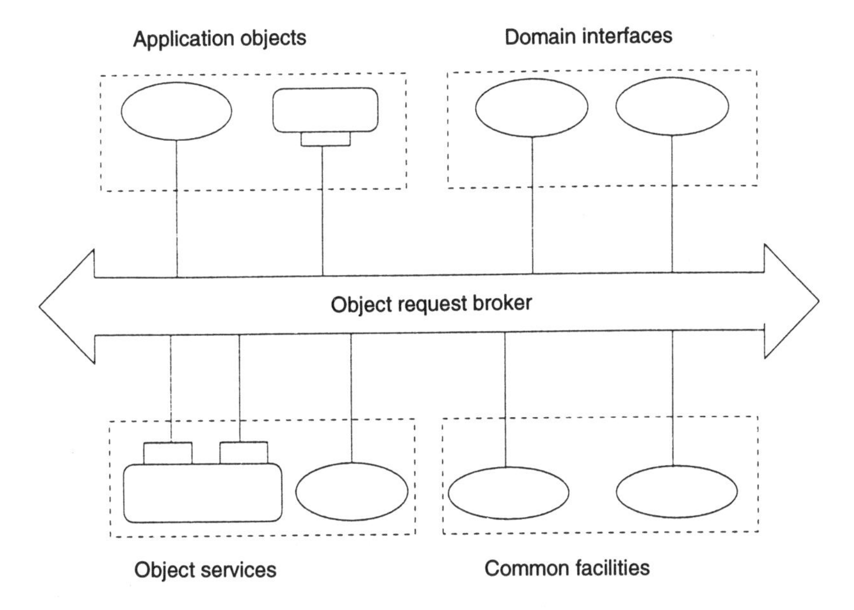

Figure 10.4 Object management architecture.

- *Naming* supports the definition of names and contexts of names.
- *Event notification* allows the definition of events and their notification to the interested objects.

3. *Common facilities (CFs)* offer general use functions to several distributed applications [15]. OMG defines common facilities related to user interface, information management, system management, and task management. Considering the definitions, the boundary between object services and common facilities is flexible in the sense that a common facility can become an object service if its functionalities are considered as basic.
4. *Domain interfaces* are related to specific application domains, such as telecommunications, distributed simulation, and manufacturing.
5. *Application objects* are the user's application distributed objects. They are not standardized.

Figure 10.5 presents the logical relationship among application objects, domain interfaces, OSs, CFs, and ORB. An application object can directly access the ORB, or can use OSs, CFs, and domain interfaces. A domain interface can directly access the ORB or can use OSs and CFs; whereas a CF can directly access the ORB or can use OSs.

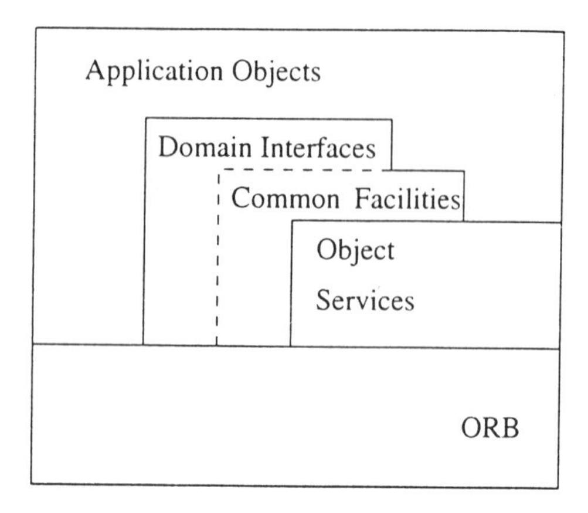

Figure 10.5 OMA logical relationships.

The CORBA 2.0 document specifies the architecture of the ORB (Figure 10.6) [13]. The access transparency is obtained through an IDL (interface definition language) used to write all interfaces and through some identical interfaces for all ORB implementations. The location transparency is obtained through an implementation repository that stores for each interface information about which objects have this interface and the locations of these interfaces. CORBA only allows one interface per object, but several objects can have the same interface type.

The client object can invoke a remote method of the object implementation in two ways:

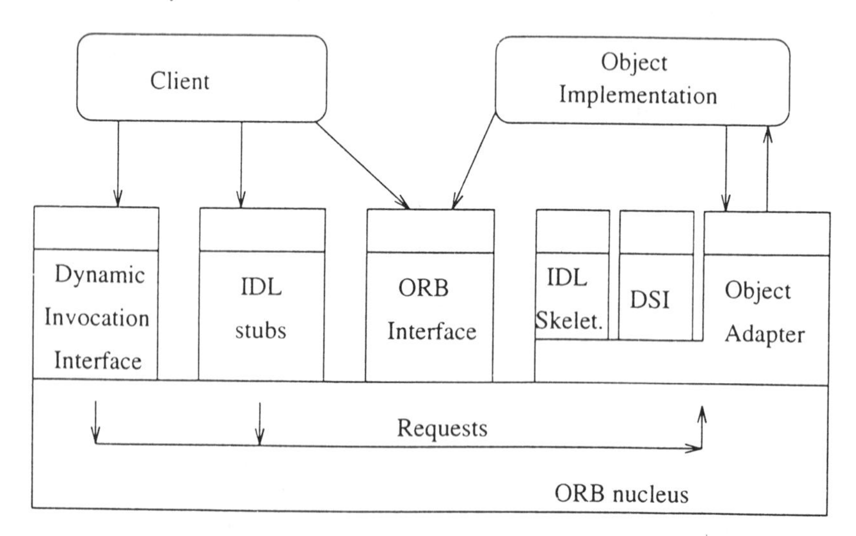

Figure 10.6 CORBA 2.0.

- *IDL stubs:* For use when the client knows the interface, method, and parameters in compilation time. For each method of each object implementation a stub is generated. The client should link its code with the stubs that will be used.
- *Dynamic invocation interface (DII):* For use when the client does not know the interface, method, and parameters in compilation time. Either the client will only know the interface, method, and parameters in execution time or the client knows partially what it wants to invoke (e.g., only the interface, or the interface and the method). In the first case, the DII offers primitives for the client to build the request in execution time, and in the second, to query the desirable information in an interface repository.

The interface repository contains the interfaces, methods, and parameters of the environment, which are structured in modules to facilitate searching. CORBA has two repositories: interface repository and implementation repository. The implementation repository contains the necessary information for the ORB to locate and activate implementation objects.

At the CORBA client side there are the IDL stubs, DII, an API for the interface repository, and an ORB interface that offers few local services (like to translate an object reference to a string). The API for the interface repository allows us to get and modify the registered interfaces and their methods and their parameters. In a CORBA environment, an object has a unique identifier called object reference (*objref*).

Besides the implementation repository and the ORB interface, at the CORBA server side we have the following:

- *Skeletons* are the IDL stubs of the server side. A skeleton provides the static interface for the services exported by the implementation object. An implementation object can have one or more objects. There is a skeleton for each object.
- *Object adapter* receives the request from the ORB and activates the implementation object (if it is not in the main memory). It uses information from the implementation repository and invokes the appropriate skeleton. It provides the main interface to an implementation object to access the services offered by the ORB. It is also responsible for the generation and interpretation of *objrefs*.
- *Dynamic skeleton interface* (DSI) is invoked when the client request is for a method of an implementation object that does not have a compiled skeleton. It is similar to the DII in the client side. It is useful when

a client invokes a method of an implementation object that is at another ORB (B) linked to the client's ORB (A). The DSI of the ORB A can receive the client request and deliver it to a bridge (or two half bridges), which should invoke an ORB B implementation object, as shown in Figure 10.7. In this case, these two ORBs are called *interoperables.*

To allow the communication between ORBs, OMG defined the general inter-ORB protocol (GIOP), which specified a set of messages (and their formats) and common data representations. OMG also defined the internet inter-ORB protocol (IIOP), which specified how GIOP messages are exchanged in a TCP/IP network. In these cases, the involved ORBs are also called *interoperables.* To obtain interoperability with other environments, OMG created the environment-specific inter-ORB protocols (ESIOPs) and defined the DCE-ESIOP for the DCE.

Some constraints of ORB platforms are as follows:

1. Performance, but its use allows programming in a higher level of abstraction (interfaces, methods, and parameters) to occur.
2. The security OS was recently approved.
3. The products that follow the CORBA specification do not incorporate the majority of the OSs and CFs yet. Only the naming and event notification OSs are products.
4. Object passing is done by reference, not by value.
5. An object can only have one interface.
6. ORB does not support nonblocked operations.
7. ORB does not support bulked data transfer.
8. ORB does not support multimedia transmission.

OMG created working groups to propose solutions for items 4 to 8. Specifically to resolve the multimedia transmission support, OMG issued an RFI (request for information) [16] asking for works on real-time object-oriented technologies, such as real-time ORBs, and service and facility extensions, such as time-sensitive delivery of messages and real-time schedules. The solution should also consider QoS aspects. The issues related to a real-time ORB suggested by OMG are clock synchronization, bounded delays, priority-based scheduling and priority inheritance, resource monitoring, multithreading and concurrency, and support for multiple protocols.

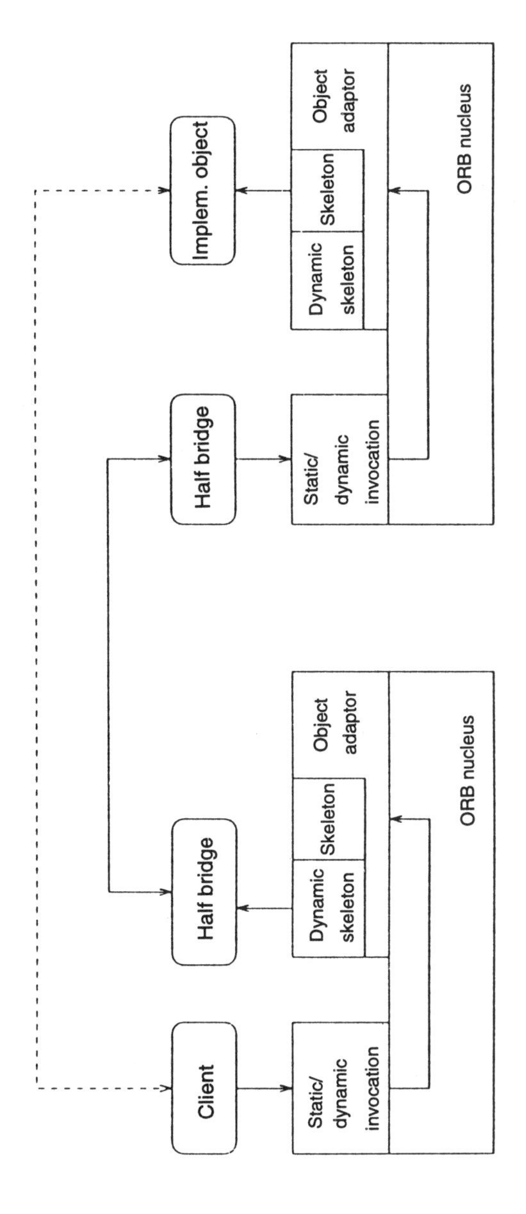

Figure 10.7 An example of interoperability between ORBs.

Besides facilitating the programming in a higher level of abstraction, some advantages of using the ORB platform follow:

- It is a specification and commercial products are available for implementing this specification.
- Several entities are involved in the process of specification.
- It offers ODP functionalities and supports ODP transparencies.
- It allows the incremental addition of new OSs.
- It is an integrated solution for a distributed platform together with OSs and CFs.

10.4 Extensions for Supporting Multimedia on Distributed Platforms

In this section, we show three different approaches for supporting multimedia computing and communication. The first approach extends the concept of stream to incorporate multimedia flow. The second scheme shows how multimedia objects can be incorporated in distributed platforms. Finally, we show some proposed architectures for multimedia processing.

10.4.1 An Extended Stream Channel

In [17] the composition of a stream channel is presented (Figure 10.8). A multimedia object is associated with source and sink devices, which use the stream channels. These channels are composed of a stream stub, stream binder, and stream protocol.

As in RM-ODP, the role of a stub is to provide data conversion. Here, a stub performs the compression and decompression of the data. As in RM-ODP, a binder is responsible for the channel maintenance. Here, the binder allows the configuration of the channel. In a certain way, a stub may control the intramedia synchronization and a binder may control the intermedia synchronization. As in RM-ODP, the stream protocol offers communication primitives. Here, this protocol should guarantee certain QoS.

In this model, an explicit connection control interface is offered, which can negotiate QoS constraints.

10.4.2 PREMO

The ISO/IEC 14478-3 specification [18] defines the multimedia system services (MSS) component of the PREMO (Presentation Environment for Multimedia

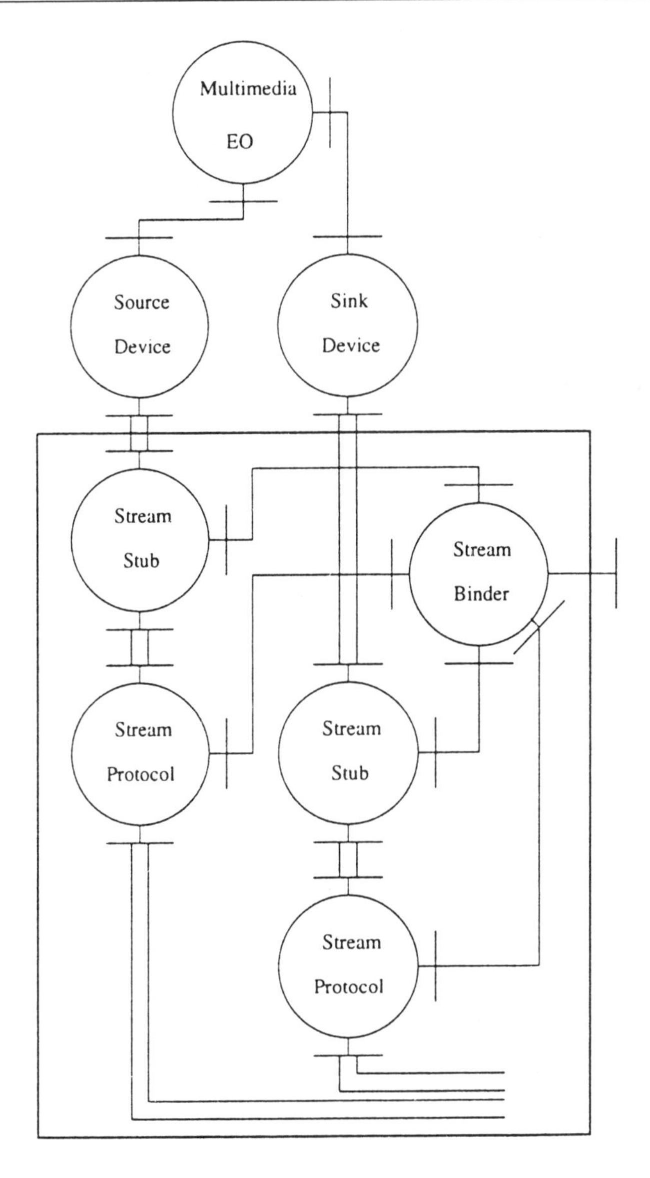

Figure 10.8 Stream channel.

Objects) to provide an infrastructure for building multimedia computing platforms in heterogeneous and distributed environments. MSS is a framework for middleware layers that enables multimedia applications in these environments. This framework encompasses a set of interrelated abstract classes that offer the same application programming interface (API) in different operating system platforms. When the framework is specialized, concrete classes are introduced to implement the API in the specific platform. Some concepts used follow:

- *Virtual resource* is an abstraction of a physical resource. There are three basic kinds of virtual resources: virtual devices, virtual connections, and groups.
- *Virtual device* is a subtype of a virtual resource and an abstraction of a media device, such as a capture card or a compressor.
- *Virtual connection* is an abstraction of media transport between virtual devices in the application view. A virtual connection has QoS parameters and synchronization capabilities. A virtual connection can be unicast or multicast.
- *Group* allows several resources to be handled as a group.

Figure 10.9 shows a partial subtyping diagram, with the main types needed to implement MSS, such as format, QoS, stream, video, audio, and virtual device types. Note that the video device is a subtype of video and of virtual device.

Figure 10.10 presents a typical MSS client interaction. The client is communicating with two virtual devices and a virtual connection. The stream object provides an abstraction to control a media stream, whereas the format object provides abstraction of the details of media formatting. Ports describe input and output mechanisms for the virtual device. The virtual connection provides operation to create a connection between an output port of a virtual device and an input port of another virtual device, hiding low-level transport semantics. Multicast connections are also supported by virtual connections. The group object assists the client for atomic resource allocation and QoS specification. The Factory is used to instantiate MSS objects. Factories can create certain types of objects. A FactoryFinder allows us to select the appropriate Factory. EventHandler and Callback are related to the event model provided by the PREMO.

One way to develop a middleware layer, which supports multimedia communication, would be to consider objects defined in PREMO as CORBA objects, and use a real-time ORB or a stream channel outside the ORB for the communication.

10.4.3 Examples of Multimedia Architectures

In this section we show two architectures for multimedia distributed systems, which are based on adding new layers over the distributed platform layer, which is above an operating system layer.

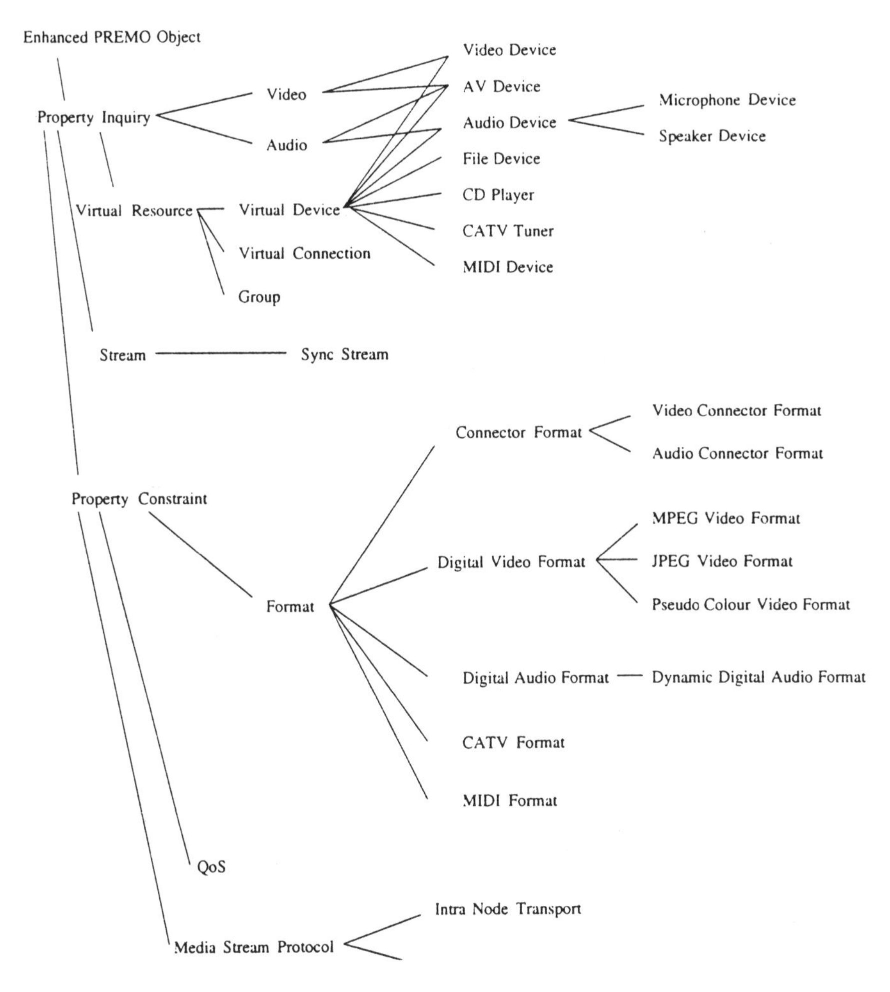

Figure 10.9 Partial subtyping diagram of PREMO.

In [19] a service platform to be used in a service open market was defined. The distributed object system platform and the high level object runtime service (e.g., CORBA, OSs, and CFs) layers are enhanced by a dedicated object runtime services layer (e.g., IMA multimedia system services) as shown in Figure 10.11.

An application platform can be added in parallel with the dedicated object runtime services layer, allowing the construction of applications from generic building blocks, such as packages of communication, conferencing, and cooperation services. The application platform can use the dedicated Object Runtime Services layer, for example, to establish a media channel.

In [20] a software architecture for broadband multimedia based on the Orlando Full Service Network (FSN) was presented [21]. This network is a

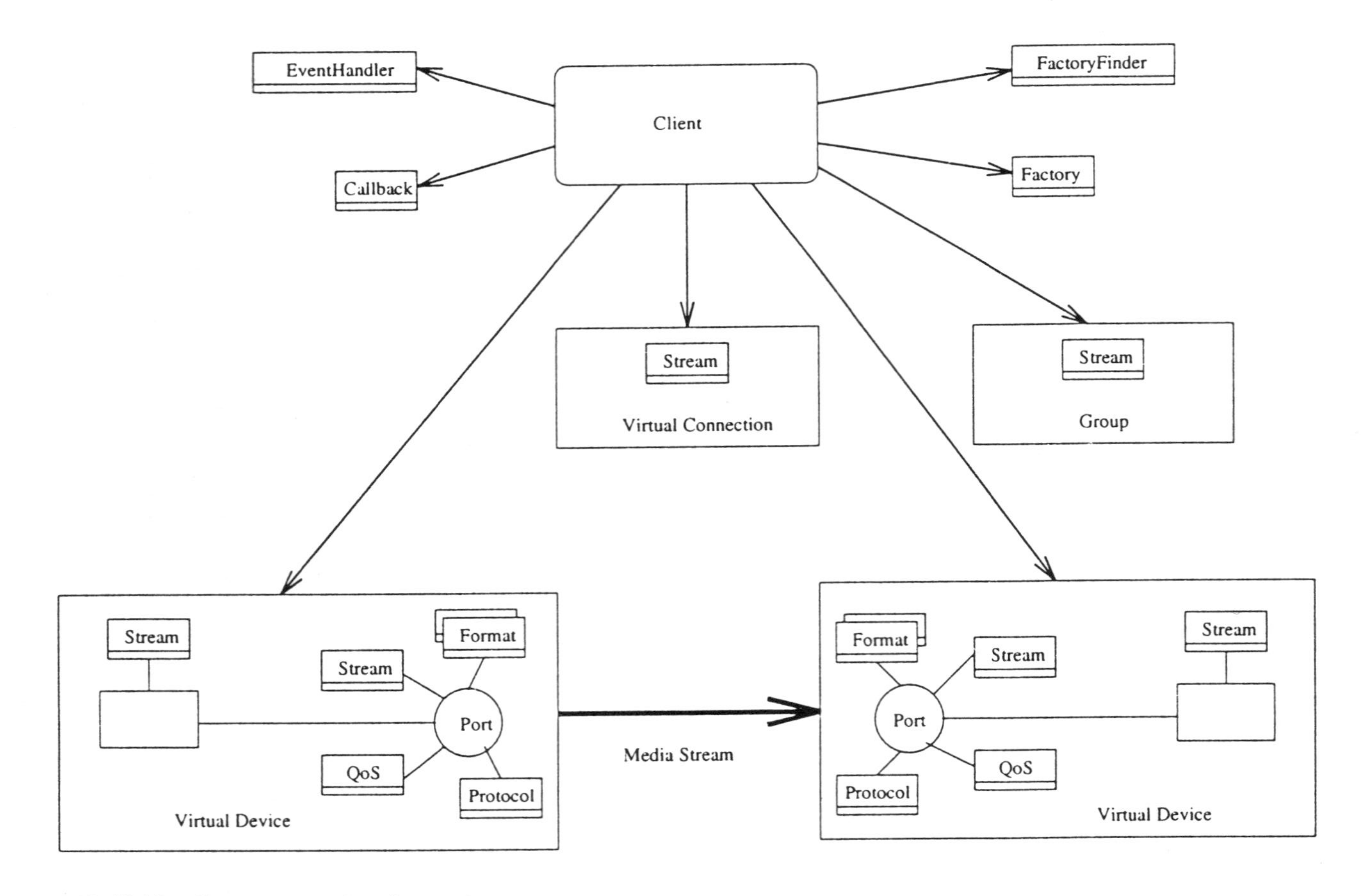

Figure 10.10 Multimedia system services interaction.

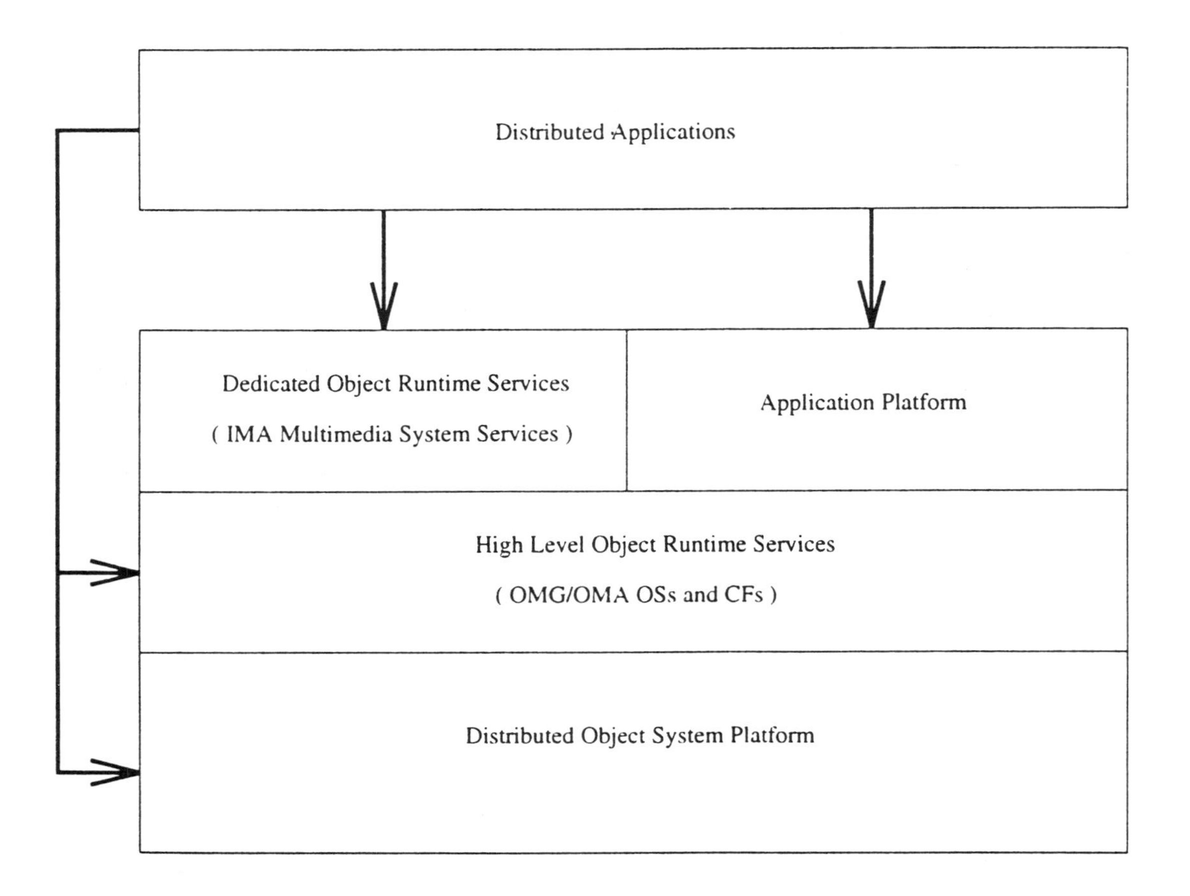

Figure 10.11 Service and application platform.

distributed architecture where audio and video streams produced in a server complex are switched to home terminals at the customer's residence, allowing video-on-demand, news-on-demand, and other services.

The Orlando Full Service Network's service is based on CORBA and IDL, and is shown in Figure 10.12. Basic services are shown in the lower levels, whereas interface processes of the client are shown in the higher levels.

The server system runs on the IRX operating system environment (UNIX with real-time extensions), which contains a media delivery service to convey real-time MPEG video data. The client system runs on a developed microkernel, which offers run-time and service libraries, multimedia libraries, a multimedia scripting environment, and an application manager.

10.5 Multimedia in CORBA

CORBA does not currently support multimedia transmission. Two approaches can be applied to provide this support:

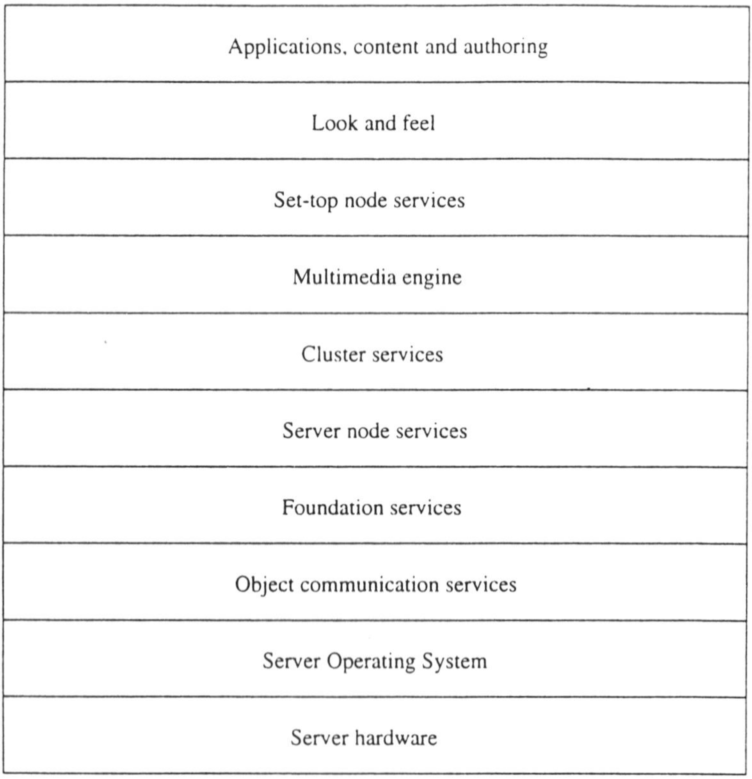

Applications, content and authoring
Look and feel
Set-top node services
Multimedia engine
Cluster services
Server node services
Foundation services
Object communication services
Server Operating System
Server hardware

Figure 10.12 FSN system software architecture.

- Using the ORB only to negotiate the QoS of a multimedia channel, creating the multimedia channel outside the ORB;
- Extending the ORB's concepts, creating a multimedia ORB that conveys multimedia information.

The next two sections illustrate how the ORB can be used to negotiate the desired QoS and a CORBA extension for supporting multimedia QoS. Finally, we discuss CORBA limitations on a high-speed environment, and its implication for broadband multimedia communications.

10.5.1 Multiware Platform

This platform is being developed at UNICAMP, State University of Campinas, in Brazil. It considers the ODP concepts and is based on CORBA [22]. Figure 10.13 shows the architecture of the multiware platform. The multiware platform has three layers: software/hardware, middleware, and groupware. The multiware platform environment is composed of SUN, IBM, and PS stations.

The first layer is composed of operating systems, protocols, and microkernels of these machines. It does not provide distribution support. The second layer adds this support for the distribution of applications and contains three parts:

1. An ORB sublayer;
2. An object service sublayer, over the ORB, which contains some OSs from OMG and other objects developed in the project;
3. A multimedia processing sublayer, in parallel with the other two sublayers.

The third layer offers support to different kinds of CSCW applications, such as co-editing and teleconferencing.

Some of the objects developed in the object service sublayer are listed here:

- *Trader* manages the knowledge of the currently available services and finds service offers that match the client requirements [23].
- *Group support* provides the creation and handling of groups independent of the application type [24].
- *QoS negotiation* allows the QoS negotiation for a media channel [25].

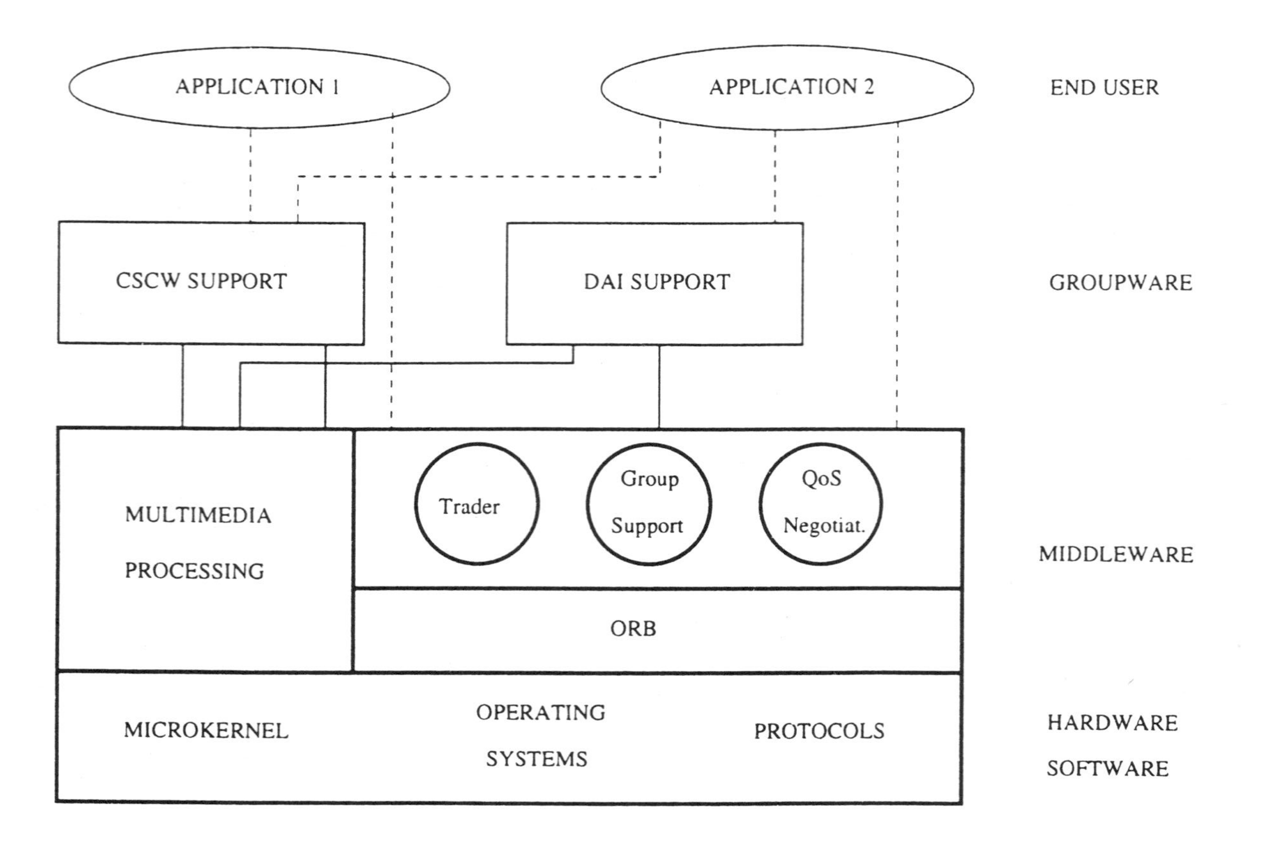

Figure 10.13 Multiware platform.

- *Distributed system management* offers monitoring management of CORBA objects, including the collection of management data, its logging, processing, and presentation [26].
- *Bridge* allows interoperability with other ORBs from different vendors [27].

An application object that establishes a media channel negotiates the QoS with the QoS negotiation object. The negotiated requirements are synchronism, interactivity, priority, capacity, error handling, fault tolerance, confidentiality, authenticity, integrity, availability, and cost.

A negotiation involves two phases: a query to the trader about the requirements related to the target object (the last eight requirements presented earlier) and a negotiation with the transport protocol object on the requirements associated with the interaction (the first three requirements presented earlier). For example, if an application object wants a channel to transmit voice with authentication, the authentication requirement is negotiated with the trader, and the requirements to transmit the voice (synchronism, interactivity and priority) are negotiated with the transport protocol object.

A QoS negotiation returns a channel identifier related to a channel established through the multimedia processing sublayer. Then, the application object can either produce or consume streams via this channel. The stream channel considers the ideas presented in Section 10.4.1. The multimedia processing sublayer uses a microkernel, which offers thread capabilities, because media channels require performance constraints.

The group support object decreases the gap between a specific groupware support and a generic middleware. This object allows group creation, group termination, group joining, member departure, policy change, authorization change, and member role change, and also sends messages to a member, all group members, and all members that have the same role in the group.

10.5.2 Extending CORBA to Support QoS of Continuous Media Applications

In this section, we present an extension of CORBA developed at Lancaster University that tries to overcome the limitation of lack of support for both the "streaming" interaction style and for the QoS required by continuous media applications [7]. The proposed architecture goes beyond multimedia-in-CORBA platforms by integrating a continuous media data type as first class types in the application computational model, therefore, allowing programmers to address sources, sinks, and the process real-time stream in a manner similar to that used for conventional static data types in CORBA. The new design makes the definition of new interfaces and associated binding types easier. This type of feature is obtained by extending the CORBA computational model in two ways:

1. *Augmented interface definition:* In addition to IDL control interface, programmers may declare interfaces containing interaction points called *events.*
2. *Binding objects:* Binding objects are used for the specification and the realization of communications with a given QoS when run-time bindings are established. QoS specifications are translated into specific QoS support mechanisms.

This design contrasts with proposed platforms that adopt the "off-line plumbing" approach in which programmers connect multimedia objects and then monitor and control the flow of media inside these objects. In line with that, programmers are not expected to require access to the internals of the objects. This approach lacks the flexibility of performing application-specific processing on a continuous media stream. Consequently, programmers must drop the level of the generic computational model into a system-specific OS and network protocol that is highly undesirable.

The extended computational model differs from the CORBA model by adding an event interface. Event interfaces contain event interactions rather than operation interactions. Events are unidirectional and may carry typed data objects as arguments. Events were introduced to capture the concept of information flow over time and also to avoid the representation of media flows as repeated operation invocation. In line with that, we overcome the limitation of the bidirectional request/reply style of operation, which cannot represent the concept of a connection governed by an overall QoS because each invocation is a separate, isolated event. An extended IDL language named Event Definition Language (EDL) was created to support events. In EDL, we need to define the directionality of an event (keywords in or out). Both IDL and EDL interfaces are supported by the extended object.

Binding objects were created for connecting two or more event interface references so that communication may happen in a way such that the underlying communication infrastructure is encapsulated. Binding objects may also be used as control points in which applications may specify, monitor, and adjust ongoing communication QoS requirements. All binding objects support a standard operational interface through which we request generic control functions. This generic interface is inherited by new binding objects. In contrast to standard CORBA, binding objects support third-party binding, which allows a remote application setup for the binding of two interfaces. Binding objects are created with a given end-to-end QoS specification that governs the flow of events. Binding objects can also be used to bind standard IDL operational interfaces. Binding object implementations deduce the necessary components and instantiate and bind these components to form the end-to-end binding. Binding object implementation may need to perform an admission test in order to guarantee the required QoS.

The object support environment for the extended model is basically the CORBA environment and aspects such as writing and registering servers are unchanged except that the extended model does not maintain a clear separation between client and server roles. However, some differences between the two environments are essential for the extended model. Among the differences, we mention a new object called an *event object adapter* which allows programmers to implement classes corresponding to EDL interfaces. The EDL compiler generates procedures, called *QoS monitors*, that evaluate actual QoS by recording the occurrence of events at the event interface and comparing these records with the QoS requirements. Discrepancies (violations) between the measured and specified QoS values are signaled to the binding object. QoS monitors are generated from the EDL interface specifications and their associated QoS specifications. Finally, we can have either active or passive binding. Active binding allows the delegation of as much as possible of the responsibility for the QoS management to the binding itself, whereas in passive binding the application is responsible for initiating events.

In addition to the ORB demon, three other run-time services should be available on each node to support the extended CORBA platform. They are QoS mapper, event interface type checking service, and node QoS manager. The QoS mapper translates end-to-end QoS specifications into a set of component-level QoS specifications. The event interface type checking service determines whether or not two event interfaces are type compatible and consequently eligible to be bound by a binding object. The node QoS manager is a per-binding, per-node service component that manages resources and the coordination of components associated with a particular stream through a particular node.

QoS is managed in three hierarchical levels. The component-level QoS management tries to maintain QoS according to their agreed-on specification. QoS monitoring codes generated from a user's QoS specifications are responsible for the maintenance task. The QoS manager receives indications from low-level QoS monitors and decides the appropriate action to be taken. Finally, the client QoS control is responsible for QoS maintenance in case all other mechanisms fail to do so.

10.5.3 High-Speed CORBA

Experience with CORBA has shown that it is suitable for low-speed networks such as Ethernet. However, recent studies have pointed out that current CORBA implementation leads to significant overhead, which makes CORBA inappropriate for performance-sensitive applications over high-speed networks. As network technology moves to gigabit data rates, users will be forced to use low-level transfer mechanisms such as RPC, which increases development efforts and reduces system reliability and flexibility.

Reference [28] describes an experiment used to evaluate the adequacy of CORBA over high-speed networks. In this experiment, a 16-port OC3-155 Mbps/port ATM switch and two SPARC workstations were used. Each workstation has a 70-MHz CPU with 1 Mbyte of CPU cache, 128 Mbytes of RAM, and an ATM adapter card, which supports 155-Mbps SONET multimode fiber. To run the experiment at speeds higher than 155 Mbps, the I/O backplane of a dual-CPU SPARC workstation was used to emulate a "high-speed network." The user-level memory-to-memory bandwidth was approximately 1.4 Gbps, which is comparable to an OC24 gigabit ATM network. Traffic was generated and measured by an extended version of the TTCP protocol benchmarking tool. The TTCP tool measures end-to-end data transfer throughput. Several parameters such as the size of the socket transmitting and receiving queue, the size of data buffers, and the type of transferred data were varied in the experiment. Extensive experiment results are available in [28] and major findings are transcribed next.

Based on measurements of transferring different data types at different transfer rates, the current CORBA implementation (Orbix 2.0 and ORBeline 2.0) implies high processing overheads due to [29] the following:

- Excessive presentation layer conversion and data copying—by far, the major bottleneck;
- Inefficient server demultiplexing techniques; for a large interface, demultiplexing based on linear algorithm is a significant bottleneck;
- Long chains of intra-ORB function calls;
- Nonoptimized buffering used for network read and write.

One way to guarantee the requested end-to-end QoS for CORBA application over high-speed networks is to integrate the network, transport protocol, operating system, and middleware. In this sense, some challenges that might be addressed follow [30]:

- *Real-time OS and network scheduling:* The operating system and the network should provide scheduling mechanisms that guarantee CORBA middleware and applications the target QoS. For instance, high-priority tasks should run to completion at the OS level and should not be blocked by low-priority applications at the network level.
- *Lightweight transport mechanisms:* If, on one hand, reliable transport mechanism are implemented by heavy-weighted protocols, on the other hand, unreliable protocols lack the mechanisms to avoid network congestion and, consequently, QoS degradation. One possible solution is the OS providing lightweight real-time implementation of transport protocols that can be customized by the application.
- *Efficient and predictable demultiplexing:* CORBA requests must be demultiplexed to the appropriate method of the target object implementation. Layered demultiplexing as in current CORBA implementations can be inappropriate for latency-sensitive application. OS mechanisms should minimize demultiplexing overhead. Demultiplexing mechanisms should provide consistent QoS performance regardless of the number of protocols, application-level target object implementations, and operations defined by the IDL interface of these objects.
- *Reduced data copying:* OS device drivers, protocol stacks, and CORBA middleware should collaborate to minimize data copying. Memory management used by OS and ORB must behave predictably, irrespective of user buffer sizes and workload.
- *Efficient presentation layer conversion:* Compiled marshalling code is efficient, but may require excessive amounts of memory, whereas interpreted code is compact but slower. CORBA should offer the flexibility of selecting compiled marshalling for heavily used data types, and interpreted marshalling for infrequently used data types.

10.6 Multimedia in Java

Existing distributed platforms do not provide support to multimedia communication yet. However, there are some solutions for specific environments. The Java development kit (JDK) allows object communication, defines classes to facilitate the graphical interface creation, and offers audio and video handling through media players. The Java media framework (JMF) API incorporates media data types into Java applications and applets [31]. This framework provides APIs for media players, allowing the delivery of statically stored, synchronized media types in an integrated way with the Java environment, and in the near future will provide APIs for media capture and conferencing. The media player considers reliable protocols, such as hypertext transfer protocol (HTTP), and streaming protocols, such as real-time protocol (RTP). A typical application of media player is to present an MPEG movie from within a Java applet.

10.7 Using ODP for Designing Distributed Multimedia Applications

In this section, we illustrate how ODP can be used to design large-scale distributed multimedia systems. We show a full ODP specification of a distributed home theater system.

In a distributed home theater (DHT) application, a movie is simultaneously displayed on the screen of a group of users [32,33]. Users can issue VCR operations and debate about specific scenes. DHT can be seen as an extension of teleconferencing in which a movie is displayed. Interaction among users can be either through voice or through voice and still image as in current teleconferencing systems. DHT has a potentially large use in education and in professional teleconferencing.

DHT can be implemented with different degrees of user control (operational modes). In a system with centralized control whenever a user wants to issue a VCR operation or engage in a conversation, he or she needs to request authorization from a coordinator. DHT with centralized control can be used in classroom where a central authority controls both movie session duration and users' interventions. In DHT with decentralized control, there is no hierarchical relationship among users and anyone can temporarily control a debate. In addition, a coordinator with superuser privileges is necessary in order to create/terminate a movie session as well as to solve unforeseen situations. DHT with decentralized control is more appropriate to professional teleconferencing. We proceed by showing the specification of a DHT system with different modes of operation (Figure 10.14).

1) *Enterprise Viewpoint*

1.1) DHT with Centralized Control

i) Artifacts

- *Video server:* The video server displays a movie on request. It receives and executes VCR operations on a movie being displayed.
- *Teleconference server:* The teleconference server receives voice (+ still image) flows and distributes these flows among users of a DHT session.

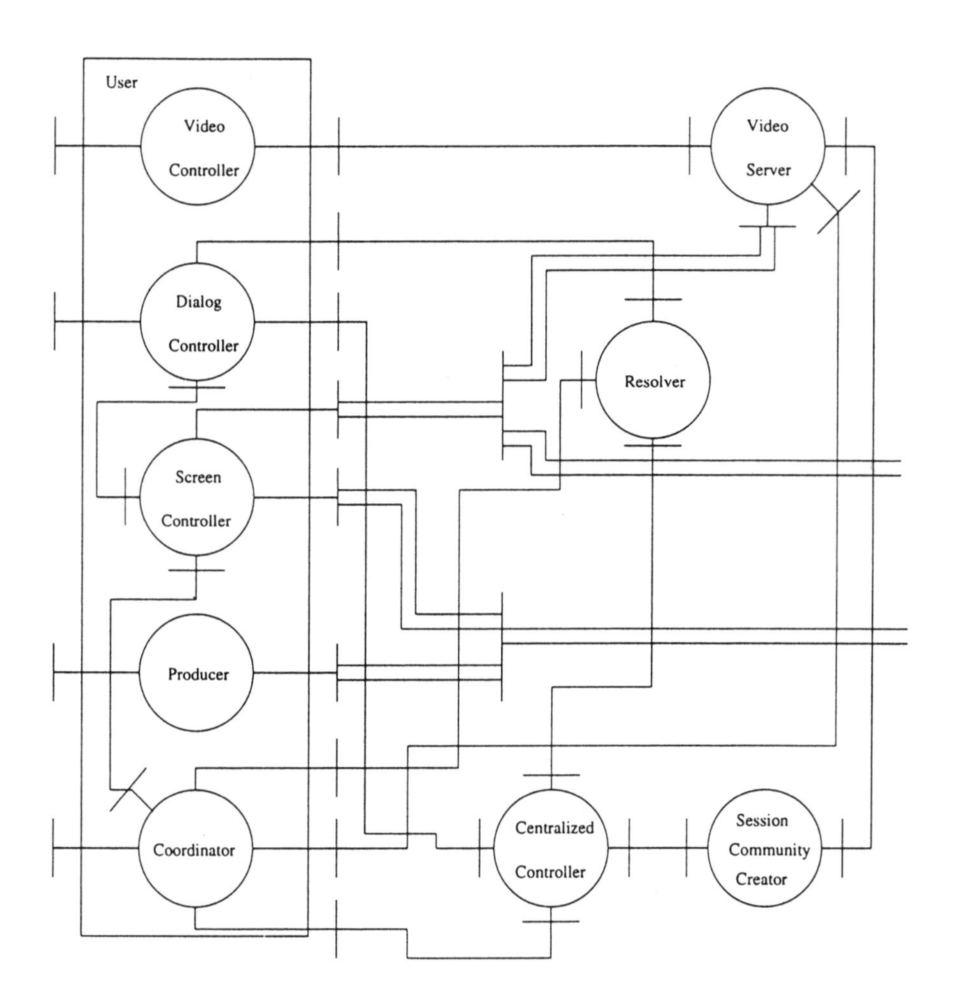

Figure 10.14 Distributed home theater system.

ii) Agents

- Subscribers;
- Users;
- Coordinator.

The time interval between the establishment of a group of users and the end of the debate of a film is called a DHT session. Every DHT session has one and only one coordinator. The coordinator is a user of a DHT session with special privileges, such as selecting and starting the exhibition of a movie. Subscribers to a DHT service should request inclusion in a specific session to the coordinator. If accepted they become users of that session. During a DHT session, whenever a user wants to either issue a VCR operation such as pause and resume or engage in a debate he or she needs to ask permission (a token) to the coordinator. On the other hand, the coordinator may remove a token from a user at any time, canceling the current operation. In the beginning, there is neither a DHT session nor a group of users. Subscribers with special privileges (such as instructors) may request the creation of a session community. If it is created, the subscriber who requested its creation is assigned to be the coordinator of that community.

iii) Communities

- *Session community:* composed of users, coordinator, and teleconference and video servers;
- *Service community:* composed of subscribers, users, coordinator, and teleconference and video servers.

iv) Policies
Permissions
Users are allowed to:

- Request a video token while the film is being displayed;
- Request a voice token;
- Request to be excluded from a session community;
- Talk when voice token is granted;
- Consult the video server directory.

The subscriber is allowed to:

- Request to be included in a session community;
- Create a session community.

The coordinator is allowed to:

- Include a subscriber into a session community;
- Exclude a user from a session community;
- Release a video/voice token to a user;
- Remove a video/voice token from a user;
- Use a video/voice token any time;
- Consult the video server directory;
- Elect a new coordinator.

Prohibitions
Users cannot:

- Perform a video/voice operation without a token;
- Cancel another user's token;
- Elect a new coordinator;
- Exclude himself or herself from the community;
- Select a film to watch.

Subscribers cannot:

- Include themselves into a session community.

The coordinator cannot:

- Leave the community without electing a new coordinator.

Obligations
Users must:

- Use granted video token;
- Release a video token after using it.

The coordinator must:

- Notify user of a granted token;
- Notify user of its exclusion from the session community;
- Notify subscriber of its inclusion in the session community;
- Notify user of a token cancellation;
- Select a film to watch;
- Elect new coordinator before leaving the session community;
- Notify session community of new elected coordinator;
- Resume the movie exhibition after a user finishes using a video token;
- Initiate film exhibition;
- Extinguish the session community.

1.2) DHT with Distributed Control

i) Agents

- Subscribers;
- Users;
- Coordinator;
- Temporary coordinator.

In the distributed control operational mode, a temporary coordinator controls a debate. Different debates may have different coordinators. A debate should be extinguished as soon as a conversation about a specific scene or topic is done. During the exhibition of a movie, there is no predefined temporal coordinator. Whenever any user requests to be the temporary coordinator, a decision is made based on the number of requests within a time window. The coordinator has special privileges as in the centralized control mode in order to manage the session. A user with coordinator privileges should also compete with other users to be the temporary coordinator of a debate except when he or she wants to use his or her privileges to control the session.

ii) Artifacts

- Video server;
- Teleconference server;
- Resolver.

The resolver is a device that decides which user will be the temporary coordinator based on the request arrival times and possibly on other informations such as user priority.

iii) Communities

- *Session community:* composed of users, coordinator, temporary coordinator, resolver, and teleconference and video servers;
- *Service community:* composed of subscribers, users, coordinator, temporary coordinator, resolver, and teleconference and video servers;

iv) Policies
Permissions
Subscribers are allowed to:

- Request to be included in a session community;
- Request the creation of a session community.

Users are allowed to:

- Request the temporary coordinator a video/voice token;
- Request the temporary coordinator to replace him or her (be the new temporary coordinator);
- Submit a request to the resolver to be the temporary coordinator;
- Talk when voice token is granted;
- Notify coordinator of his or her exclusion from the session community;
- Consult video server directory.

A temporary coordinator is allowed to:

- Grant video/voice token;
- Cancel video/voice token;
- Elect a new temporary coordinator.

The coordinator is allowed to:

- Remove privileges from the temporary coordinator;
- Include subscriber into the session community;

- Elect new temporary coordinator;
- Elect new coordinator;
- Select a film to be displayed.

Prohibitions:
Users cannot:

- Select a movie to be displayed;
- Request a video/voice out of a debate interval;
- Elect neither temporary coordinator nor coordinator;
- Remove another user's token.

The temporary coordinator cannot:

- Elect a new coordinator.

The coordinator cannot:

- Leave the session community without electing a new coordinator.

Resolver cannot:

- Elect a new temporary coordinator during a debate.

Obligations:
The temporary coordinator must:

- Notify user of granted token;
- Notify user of token removal;
- Notify end of debate.

The coordinator must:

- Notify inclusion/exclusion of a user from the session community;
- Select a movie to be displayed;
- Extinguish the session community;

- Elect a new coordinator before leaving the session community;
- Notify temporary coordinator and users of the removal of the current temporary coordinator privileges.

The resolver must:

- Choose a temporary coordinator if there are pending requests;
- Notify the selection of the new temporary coordinator.

The user must:

- Use video token when granted;
- Notify end of video token use.

2) *Information Viewpoint*

i) Invariant schemata

- Each group has one coordinator;
- In the distributed control case, each group has at most one temporary coordinator.

ii) Static schemata

- Each service community has a group list;
- Each group has a record with the field member list and coordinator (and temporary coordinator in the distributed control case).

iii) Dynamic schemata

- Group creation increases by one the group list;
- Member joining increases by one the member list;
- Exchange of coordinator alters the coordinator field;
- Exchange of temporary coordinator alters the temporary coordinator field;
- Group extinguish decreases by one the group list;
- Member departure decreases by one the member list.

3) *Computational Viewpoint*

i) Objects

From the client side:

- *Video controller* deals with video control operations;
- *Dialog controller* deals with the dialog control during a debate;
- *Screen controller* is responsible for information presentation at the user screen;
- *Coordinator* is responsible for all coordinator attributes;
- *Producer* produces video/voice flows.

From the server side:

- Video server.

System infrastructure objects:

- Session community creator;
- Centralized controller;
- Resolver.

ii) User Interfaces

ii.a) Video Controller

An operational interface with the following operations:

- FF, Rew, Play, Pause, Stop, and Consult movie directory.

ii.b) Dialog Controller

An operational interface with the following operations:

- Request video token;
- Release video token;
- Request voice token;
- Release voice token;
- Request to leave the session community;
- Notify when leaving the session community.

ii.c) Producer
A stream interface with the following operations:

- Voice stream;
- Video stream.

ii.d) Coordinator
An operational interface with the following operations:

- Grant voice token;
- Grant video token;
- Allow subscriber to be part of the session community;
- Allow user to leave community;
- Notify new coordinator;
- Extinguish session;
- Cancel video token;
- Cancel voice token;
- Elect new temporary coordinator;
- Cancel temporary coordinator privileges;
- Release temporary coordinator privileges;
- Select film.

iii) System internal interfaces
iii.a) Video server
An operational interface with the following operations:

- FF, Rew, Play, Pause, Stop, and Consult movie directory.

A stream interface with the following operations:

- Voice stream;
- Video stream.

iii.b) Central Controller
An operational interface with session community creator with the following operations:

- Create a session community;
- Extinguish session community;
- Include member in the community.

An operational interface with the resolver with an operation to:

- Establish new temporary coordinator.

An operational interface with the dialog controller with the following operations:

- Request video token;
- Request voice token;
- Notify when leaving the session community;
- Request to be excluded.

An operational interface with the coordinator with the following operations:

- Grant video token;
- Grant voice token;
- Allow subscriber to be part of the session community;
- Allow user to leave community;
- Notify new coordinator;
- Extinguish session;
- Cancel video token;
- Cancel voice token;
- Elect new temporary coordinator;
- Cancel temporary coordinator privileges;
- Release temporary coordinator privileges;
- Select film.

iii.c) Resolver

An operational interface with the dialog controller with the following operation:

- Request to be the temporary coordinator.

An operational interface with central controller with the following operations:

- Assign new coordinator;
- Assign new temporary coordinator;
- Cancel temporary coordinator privileges.

iii.d) Video Controller

An operational interface with the central controller with the following operations:

- Get video token;
- Notify loss of video token.

iii.e) Screen controller

A stream interface with the video server with the following operations (multiple binding):

- Video stream;
- Voice stream.

A stream interface with the producer with the following operations (multiple binding):

- Video stream;
- Voice stream.

An operational interface with the dialog controller with the following operations:

- Show video token owner;
- Show pending video token request;
- Show voice token owner;
- Show pending voice token request;
- Show user exclusion;
- Show user inclusion;
- Notify new temporary coordinator.

An operational interface with the coordinator with the following interface:

- Show request for voice token;
- Show request for video token;
- Show subscriber inclusion in the session community;
- Show user request for leaving the community;
- Show request to be the new temporary coordinator;
- Confirm film selection.

iii.f) Subscriber Interface

An operational interface with session community creator with the following operations:

- Create session community;
- Request to be included in the session community;
- Consult the video server directory.

4) *Engineering Viewpoint*

Each object presented in Figure 10.14 is a BEO. The structure of BEOs can be a cluster composed of video controller, dialog controller, screen controller, producer and coordinator BEOs for each user, and a cluster composed of resolver and centralized controller BEOs for each group. In each node, a capsule should have all clusters related with the same group. The session community creator BEO should be in a different capsule because it can be used by subscribers associated with different groups. The video server BEO may be in the same capsule of the session community creator BEO or in another capsule (for instance, if the video server BEO is called by users that are not subscribers of the DHT system).

Support for the following transparencies are required:

- *Access, location:* as basic distribution transparencies;
- *Failure (partial):* allowing failure tolerance;
- *Persistence (partial):* to store the group status;
- *Transaction (partial):* allowing event (message) ordering.

Support for the following transparency is not required:

- *Replication:* the users need to know other users of the same group.

5) *Technology Viewpoint*

The DHT system can be implemented over a multimedia distributed platform, such as the platforms described in this chapter.

10.8 Conclusion

The deployment of large-scale distributed multimedia applications demands not only the support of strict QoS requirements but also the support of access and location transparencies. Existing distributed platforms furnish access and location transparencies. Nevertheless, providing multimedia QoS is still an open problem. In this chapter, we showed current extensions of distributed platforms to support multimedia applications at different function levels. These extensions are as follows:

- An extension of the concept of channel to stream channel;
- A framework (set of abstract classes) to compose a middleware that supports multimedia communication;
- Architectures for multimedia distributed platforms;
- Multimedia distributed platforms based on CORBA, using the ORB platform only to control QoS and the ORB platform to control QoS and to convey multimedia information.

The impact of multimedia on how we communicate and live depends on our ability to develop distributed multimedia applications, which in turn relies on how effectively multimedia is incorporated in distributed platforms.

References

[1] Steinmetz, R., "Analyzing the Multimedia Operating System," *IEEE Multimedia*, Spring 1995, pp. 68–84.

[2] Mercer, C. W., and H. Tokuda, "The ARTS Real-Time Object Model," *Proc. IEEE Real-Time Systems Symp.*, pp. 2–10, New York: IEEE Press, 1990.

[3] Ramakrishnan, K. K., et al., "Operating Systems Support for a Video-on-Demand File Service," *Proc. 4th Int. Workshop on Network and Operating System Support for Digital Audio and Video*, pp. 225–236, Berlin: Springer-Verlag, 1993.

[4] Mauthe, A., W. Schulz, and R. Steinmetz, "Inside the Heidelberg Multimedia Operating System Support: Real-Time Processing of Continuous Media in OS/2," IBM Technical Report 43.9214, 1992.

[5] Nieh, J., et al., "SVR4Unix Scheduler Unacceptable for Multimedia Applications," *Proc. 4th Int. Workshop on Network and Operating Systems Support for Digital Audio and Video*, 1993.

[6] Coulson, G., Blair, G. S., and Robin, P., "Micro-kernel Support for Continuous Media in Distributed Systems," *Computer Networks & ISDN Syst.*, Vol. 26, 1994, pp. 1323–1341.

[7] Coulson, G., and Waddington, D. G., "A CORBA Compliant Real-time Multimedia Platform for Broadband Networks," *Proc. Workshop Trends in Distributed Systems*, Aachen, Germany, Oct. 1996.

[8] ITU-T X901—ISO/IEC 10746-1, "ODP Reference Model—Part 1: Overview," 1996.

[9] ITU-T X902—ISO/IEC 10746-2, "ODP Reference Model—Part 2: Foundations," 1995.

[10] ITU-T X903—ISO/IEC 10746-3, "ODP Reference Model—Part 3: Architecture," 1995.

[11] Vogel, A., B. Kerherve, and G. V. Bochmann, "Distributed Multimedia Applications and Quality of Service—A Survey," *Proc. CASCON'94*, Toronto, 1994.

[12] ITU-T X9tr—ISO/IEC 13235, "ODP Trading Functions," Jan. 1997.

[13] OMG, "The Common Object Request Broker: Architecture and Specification," rev. 2.0, July 1995.

[14] OMG, "CORBAservices: Common Object Services Specification," June 1997.

[15] OMG, "CORBAfacilities: Common Facilities Architecture," rev. 4.0, June 1997.

[16] OMG, "Realtime Technologies," Draft: Request for Information, Doc. 96-08-02, 1996.

[17] OMG-TelSIG, "Stream Model and Object Multiple Interfaces," Issue 1.7, 1995.

[18] ISO/IEC 14478-3, "Presentation Environments for Multimedia Objects (PREMO)—Part 3: Multimedia Systems Services," 1996.

[19] Tschammer, V., and K. -P. Eckert, "A Platform Architecture for Future Telecommunication Services and Open Distributed Applications," *Proc. 5th IEEE Workship on Future Trends in Distributed Computing Systems, FTDCS'95*, Korea, Aug. 1995.

[20] Vecchi, M. P., "Broadband Networks and Services: Architecture and Control," *IEEE Commun. Mag.*, Aug. 1995, pp. 24–32.

[21] Williamson, L. D., "FSN Technology," *SCTE'95 Conf. Emerging Technologies, Society of Cable Television Engineering*, Orlando, FL, 1995, pp. 27–35.

[22] Loyolla, W. P. C., et al., "Multiware Platform: An Open Distributed Environment for Multimedia Cooperative Applications," *Proc. IEEE Computer Software & Applications Conf., COMPSAC'94*, Taipei, Taiwan, Nov. 1994.

[23] Lima Jr., L. A. P., and E. R. M. Madeira, "A Model for a Federative Trader," pp. 173–184 in *Open Distributed Processing: Experiences with Distributed Environment*, London: Chapman & Hall, 1995.

[24] Costa, F. M., and Madeira, E. R. M., "An Object Group Model and Its Implementation to Support Cooperative Applications on CORBA," pp. 213–229 in *Distributed Platforms*, London: Chapman & Hall, 1996.

[25] Lima, F. H. S., and E. R. M. Madeira, "ODP-based QoS Specification for the Multiware Platform," *Proc. IFIP IWQoS'96—4th Int. Workshop on Quality of Service*, Paris, Mar. 1996, pp. 45–54.

[26] Queiroz, J. A. G., and E. R. M. Madeira, "Management of CORBA Objects Monitoring for the Multiware Platform," *Open Distributed Processing and Distributed Platforms*, London: Chapman & Hall, 1997, pp. 122–133.

[27] Zuquello, N. M. S., and E. R. M. Madeira, "A Mechanism to Provide Interoperability Between ORBs with Relocation Transparency," *Proc. IEEE Third Int. Symp. Autonomous Decentralized Systems, ISADS'97*, Berlin, Germany, Apr. 1997, pp. 195–202.

[28] Gokhale, A., and D. Schmidt, "Measuring the Performance of Communication Middleware on High-Speed Networks," *ACM SIGCOMM Conf.*, 1996.

[29] Gokhale, A., and D. Schmidt, "The Performance of the CORBA Dynamic Invocation Interface and Dynamic Skeleton Interface over High-Speed ATM Networks," *IEEE GLOBECOM'96 Conf.*, London, Nov. 1996.

[30] Gokhale, A., et al., "Operating System Support for High-Performance for Real-time CORBA," *Proc. 5th Int. Workshop on Object-Orientation in Operating Systems, IWOOOS'96*, Seattle, WA, Oct. 1996.

[31] Sun Microsystems Inc., Silicon Graphics Inc., and Intel Corporation, "Java Media Players," http://java.sun.com/products/java-media/jmf/index.html, May 1997.

[32] Fonseca, N. L. S., C. M. R. Franco, and F. Schaffa, "Network Design for the Provision of Distributed Home Theatre Services," *Proc. IEEE ICC'97*, pp. 534–539.

[33] Fonseca, N. L. S., C. M. R. Franco, and F. Schaffa, "Comparing Network Design for the Provision of Distributed Home Theatre Services," *Proc. IEEE Globecom'97.*

11

Standards for Multimedia Communications

11.1 Introduction

Standards for telecommunications were formerly and primarily developed by the International Telecommunication Union (ITU). More specifically, the CCITT (Consultative Committee International Telegraph et Telephone, now the ITU-T) and the CCIR (Consultative Committee International Radiocommunication, now the ITU-R) played the preeminent roles in telecommunication and radio communication standardization.

As a result of remarkable progress in digital and information technology, since the 1960s telecommunication networks have been digitized with the goal of realizing fully digital networks in the form of the integrated services digital network (ISDN). Standards for telecommunications in these new digital environments should make full use of the relevant digital and information technology. Therefore, the ITU-T maintains strong liaisons with external standardization organizations, such as ISO/IEC JTC 1, to establish harmonized telecommunication standards that take into account the progress of information technology.

The environment for telecommunications is changing drastically and quickly, as well as both politically and technically. Deregulation of the telecommunication market is proceeding in several countries and is being scheduled or planned for in many other countries. This deregulation will affect the

---

We express our sincere thanks to Mr. Hisashi Kasahara, Mr. Kouzou Sakae, and Mr. Naotaka Morita of NTT for their help in compiling this information.

process of standardization among international organizations. The emerging multimedia technologies and global information infrastructure (GII) require broader collaboration among not only de jure standardization organizations but also with the de facto industry standardization bodies.

This chapter presents the activities of telecommunication and information standardization organizations, forums, and consortia, described from the viewpoint of the current status of standards related to multimedia communications, such as ATM/B-ISDN, UPT, FPLMTS, and TMN. Multimedia communication is one of the most important features of the GII, and this chapter also describes the standardization activities related to achieving a GII. It explains the structures and recent activities of telecommunication and information standardization organizations, such as the ITU-T, ISO/IEC JTC 1, and the ISO/IEC, and forums and consortia, such as the ATM Forum, the IETF, DAVIC, and IMA, from the viewpoint of the current status of standards for telecommunications.

11.2 Telecommunication and Information Standardization Organizations

In the past, telecommunication standardization was the province of international organizations, such as the ITU, the ISO, the IEC, and ISO/IEC JTC 1. Now these activities are also being addressed by regional and domestic standardization bodies, such as the European Telecommunications Standards Institute (ETSI) in Europe, the T1 Committee in the United States, and the Telecommunication Technology Committee (TTC) in Japan. Coordination among these bodies is important for ensuring that resources are not wasted in reaching the final goal: global standards. Such coordination activities were informally started with the formation of the Interregional Telecommunication Standards Conference (ITSC) in 1989. The members of the ITSC included the ETSI, the T1 Committee, the TTC, the TSACC, and the TTA. The CCITT was invited as an observer because it was the place where members of the ITSC came together to work on global standards. The ITSC identified several targets for collaboration:

- Synchronous digital hierarchy (SDH);
- Broadband ISDN/asynchronous transfer mode (B-ISDN/ATM);
- Telecommunication management network (TMN);
- Intelligent networks (IN);
- Future Public Land Mobile Telecommunication System/Universal Personal Telecommunications (FPLMTS/UPT);
- Audio-visual multimedia service (AVMMS).

The ITSC was reorganized as the Global Standardization Collaboration (GSC) Group in 1994, and its activities continue. The forums and consortia established for the targeted areas achieved most of their objectives very quickly. Several international standardization organizations, including the ITU, the ISO, the IEC, and JTC 1, are thus reengineering themselves to enable them to act more quickly. They have also established rather formal relationships with forums and consortia as well as other regional standardization bodies, such as the ETSI, the T1 Committee, and the TTC. The GSC Group and its members also recognized the usefulness of liaisons with forums and consortia in the targeted areas.

The global framework for telecommunications standardization is shown in Figure 11.1 [1]. The entities shown have formal or informal liaisons with each other in the form of submitting contributions from lower entities, sending liaison statements, and exchanging information.

11.2.1 ITU-T

11.2.1.1 General

The International Telecommunication Union-Telecommunication Standardization Sector (the ITU-T; formerly the CCITT) is the major player among the global standardization organizations studying technical, operating, and tariff issues related to telecommunications. The members of the ITU-T include administrations, recognized operating agencies (ROAs), scientific or industrial organizations (SIOs) from about 180 countries, and international organizations and agencies, such as INTELSAT and the EBU. Members are classified into two categories, Member States, or *Members,* and sector members, or *members.* Member States represent countries. Sector members include network operators, equipment vendors, and others who are not Member States.

The ITU-T consists of 14 study groups (SGs) and the Telecommunication Standardization Advisory Group (TSAG) for the 1997–2000 study period (Table 11.1) [2]. Each SG has assigned questions. Some SGs are assigned to be the lead SGs for designated areas. For examples, SGs 4, 11, and 16 are the lead SGs on TMN, IMT 2000 (formerly FPLMTS), and multimedia services, respectively. Study Group 13 is the lead SG on B-ISDN, GII, and general network aspects.

The ITU-T's standardization activities have resulted in several series of ITU-T recommendations, with each series designated by a letter. For example, the recommendations on ISDN and B-ISDN form the I-series, and on TMN the M-series. The Q-series recommendations are dedicated to signaling specifications, and the G-series recommendations are for digital transmission systems and equipment.

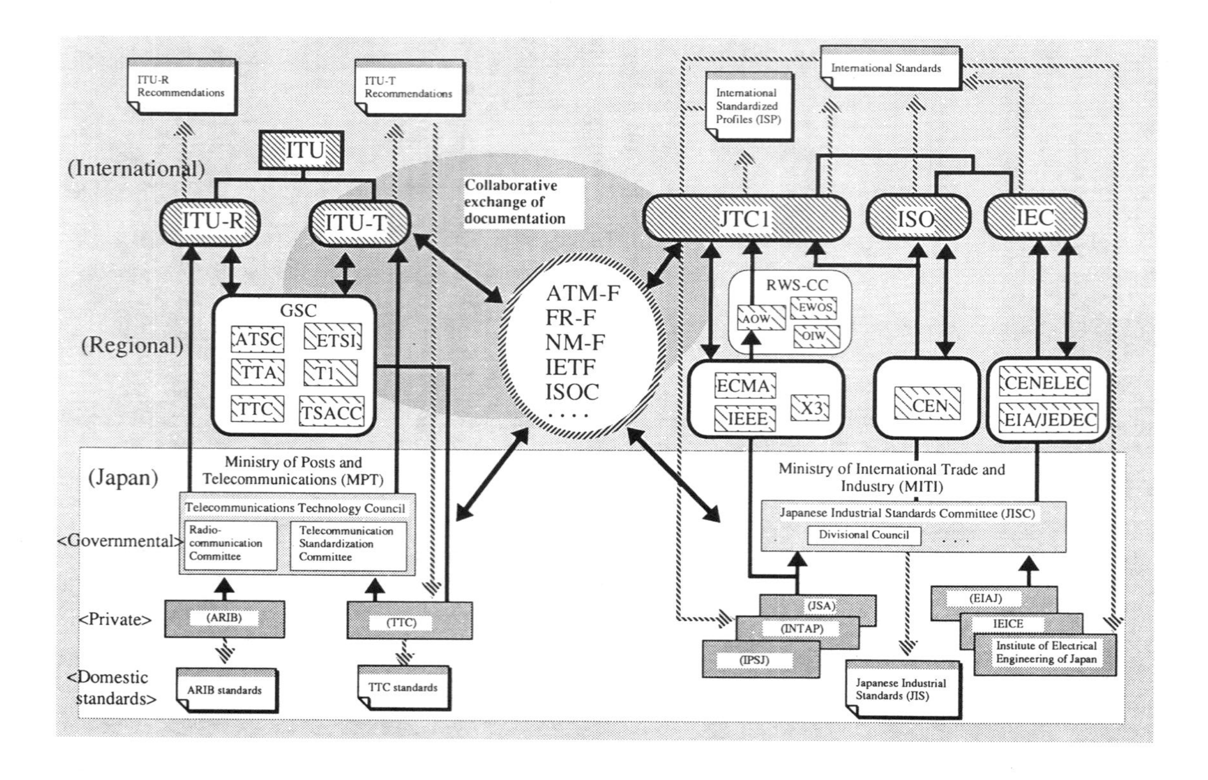

Figure 11.1 Global framework of telecommunications standardization.

Table 11.1
Structure of ITU-T Study Groups (1997–2000)

SG 2:	Network and service operation
SG 3:	Tariff and accounting principles including related telecommunications, economic, and policy issues
SG 4:	TMN and network maintenance
SG 5:	Protection against electromagnetic environmental effects
SG 6:	Outside plant
SG 7:	Data networks and open system communications
SG 8:	Characteristics of telematic systems
SG 9:	Television and sound transmission
SG 10:	Languages and general software aspects for telecommunication systems
SG 11:	Signaling requirements and protocols
SG 12:	End-to-end transmission performance of networks and terminals
SG 13:	General network aspects
SG 15:	Transport networks, systems, and equipment
SG 16:	Multimedia services and systems

At the World Telecommunication Standardization Conference held in October 1996 (WTSC96), two topics in particular were discussed: (1) the structure of study groups and (2) work methods suitable for the multimedia age and rapidly changing environments.

Structure of Study Groups

The structure of the 14 SGs for the 1997–2000 study period (Table 11.1) is one of the key results of restructuring the ITU-T.

Study Group 1 (service definition) and SG 14 (modems and transmission techniques for data, telegraph, and telematic services) for the 1993–1996 study period were terminated. Many of the questions assigned to these two SGs were transferred and rearranged among other SGs. Most of the questions assigned to SG 1 were transferred to current SG 2 in order to reduce the interaction between SGs.

Study Group 16 was formed to cover multimedia telecommunications; it took over the relevant questions from the former SGs 8, 14, and 15 (multimedia telematic services, multimedia modems, and audio/video coding, respectively). It was defined as a leading SG for multimedia.

Questions on TMN were studied by SGs 4, 7, 11, and 15 in the previous study period. The questions assigned to SGs 7, 11, and 15 were primarily transferred to SG 4 to reduce SG interactions.

Questions on switching and transmissions including coding were previously assigned to SGs 11 and 15, respectively. The questions on ATM switching equipment assigned to SG 11 were transferred to SG 15. Study

Group 11 is now specifically responsible for software-oriented questions regarding signaling requirements and protocols, whereas SG 15 now handles hardware-oriented questions on transport systems and equipment, including ATM transmission and switching equipment.

Work Methods

Work methods was the second major issue. Two key results were reached related to multimedia communications: (1) establishment of focus groups and their work method and (2) identification of lead SGs.

Focus Groups. The ITU-T has an official communication process for cooperating with the many forums and consortia that have been established in the telecommunication area [3]. These forums and consortia generally achieve quicker results than the ITU-T because their targets are more limited and because they use majority-based decision making instead of the unanimous decision requirements of the ITU-T. Although this majority-based decision making results in quicker decisions, the specifications developed are likely to be less stable. Unanimous decision making may be slow, but the recommendations are more stable.

Some of the specifications produced by the forums and consortia overlap standards covered by ITU-T recommendations. Because most members of these forums and consortia are also members of the ITU-T, harmonization between the ITU-T and the forums and consortia is required to avoid inefficient utilization of resources due to the overlapping of tasks posed to the ITU-T and the forums and consortia. The TSAG decided to introduce a formal process for communicating with forums and consortia in January 1995. Currently, the ITU-T has official liaisons with the ATM Forum, DAVIC, and the IETF in several areas.

To keep up with the speed of technical innovation, specific questions are now treated in a "project" manner. A *focus group* is assigned to each question and is expected to produce specifications for that question within one year. The specifications are turned into a recommendations if they are approved by the relevant SG. Example candidates for specific questions include video-on-demand (VOD), signaling systems for multimedia services, and architectures and interfaces for advanced networks.

These focus groups are also expected to interface with other relevant standardization organizations, forums, and consortia on their assigned topics.

Lead SG. Even after the restructuring of the SGs, some big areas are still not covered by a single SG, such as B-ISDN, IMT 2000/UPT, and TMN. Lead SGs were thus introduced to lead and coordinate the studies done by the SGs in these areas. Formerly, joint coordination groups (JCGs) were formed for these large areas; their activities ran parallel to those of the SGs. This JCG approach was therefore not effective for coordination. Table 11.2 shows the list of lead SGs now responsible for coordination.

11.2.1.2 ITU-T Recommendations Related to Multimedia Communications

B-ISDN/ATM

The CCITT SG XVIII (currently ITU-T SG 13) began standardization of B-ISDN in 1985. They have expanded their activities to cover implementation-related subjects and customer-premises equipment in both private and public networks.

Detailed protocol specifications are being released in three steps. Requirements for a user-network interface (UNI) to support various service media were discussed and led to a three-step approach to achieving full-service capability. The Release 1 recommendations, the first step, are for an ATM adaptation layer (AAL) applicable to data transfer, operations, and maintenance (OA&M), and traffic management for basic functions to support point-to-point and constant bit rate (CBR) services. Constant bit rate services have been renamed deterministic bit rate (DBR) services to take into account the typical features of this service class.

The Release 2 recommendations cover variable bit rate (VBR) capability for voice and video, point-to-multipoint connections, and multiple levels of quality of service (QoS). The Release 3 recommendations will cover enhanced connection configurations and broadcast-type communications.

The structure of the B-ISDN recommendations is shown in Figure 11.2 [4]. The various topics are covered by the following SGs:

- B-ISDN service principles and definitions are covered by the SG2 part of the former SG1.

Table 11.2
ITU-T Lead Study Groups

SG 2	Service definitions, numbering, routing, and global mobility
SG 4	TMN
SG 7	Open distributed processing, frame relay, and security of communication systems
SG 8	Facsimile
SG 11	Intelligent networks and IMT 2000/FPLMTS
SG 13	General network aspects, global information infrastructure (GII), and B-ISDN
SG 15	Access network transport
SG 16	Multimedia services and systems

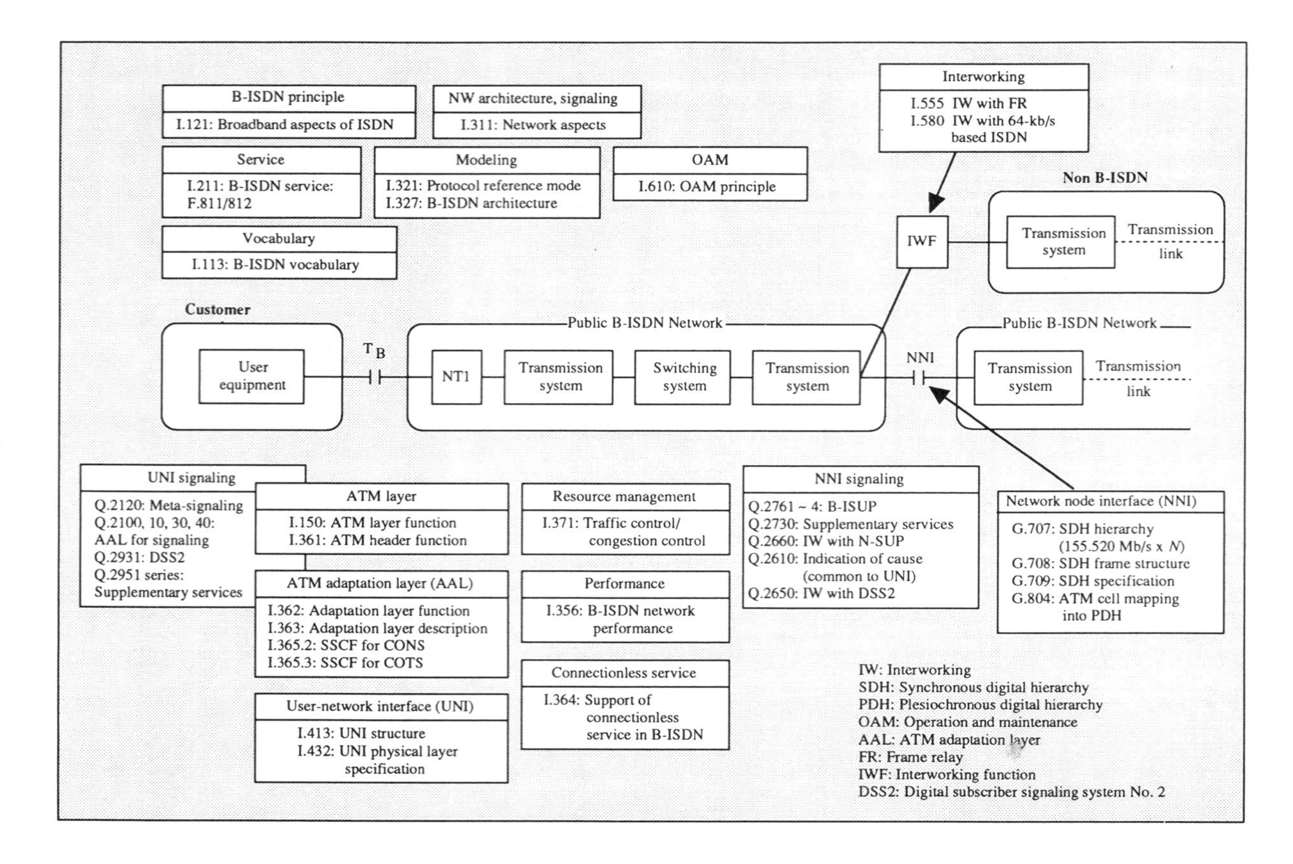

Figure 11.2 Structure of B-ISDN/ATM recommendations.

- UNI specifications, including the physical layer, ATM layer, and AAL, are covered by SG13.
- The signaling system is covered by SG 11.
- Network node interface (NNI) specifications are covered by SG13 and SG15.
- Interworking with other networks, such as frame relay networks and N-ISDN, are covered by SG 13.

Detailed technical descriptions of B-ISDN and ATM can be found in various references [5].

Audio-Visual Multimedia Services (AVMMS)

Multimedia will play a major role in the application of many communication services. It will improve the performance and quality of these services. Multimedia owes much of its growth to high-performance computers and high-speed digital transmission technologies. Multimedia is an emerging technology based on telecommunication, computing, and consumer electronics technologies. Therefore, various standardization organizations are involved in setting standards for multimedia, including the ITU-T, the ITU-R, ISO/IEC JTC 1, the ISO, and the IEC, as well as various forums and consortia.

The ITU-T set up the Joint Coordination Group on Audio-Visual Multimedia Services (JCG-AVMMS) in 1993. The activities of this JCG have been supported by the relevant SGs of the ITU-T (1, 2, 7, 8, 9, 11, 12, 13, 14, and 15) working together to advance AVMMS harmoniously.

As the lead SG for the JCG-AVMMS, SG 15 has been developing the H-series and T-series recommendations for audiovisual and teleconferencing services, respectively as shown in Table 11.3. The H-series recommendations include those for video telephony, video conferencing, and AVMMS in general switched telephone network (GSTN), N-ISDN, B-ISDN, and local-area network (LAN) environments.

The MPEG2 standards on multiplexing, video coding, and audio coding are referred to as the MPEG2 system, MPEG2 video, and MPEG2 audio standards. The MPEG2 system standard (ITU-T Recommendation H.222) and video standard (ITU-T Recommendation H.262) were jointly developed by JTC 1 SC29 and ITU-T SG15. The MPEG2 standards comprise not only the standards already mentioned, but also such other standards as digital storage media command and control extensions, and real-time interface. Activities to improve the MPEG2 standards continue.

After the completion of the MPEG2 standards, study of MPEG4 began for very-low-bit-rate audiovisual coding. The goals of this MPEG4 study are

Table 11.3
ITU-T Recommendations Related to AVMMS and Multimedia Communications

GSTN-Based Coding and Multiplexing Protocol	
H.263	The low-bit rate video coding method
G.723.1	Speech coding method at 5.6 and 6.3 Kbps (previously G.723)
H.223	Multiplexing protocol for low-bit-rate multimedia communication
N-ISDN-Based Audiovisual Services and Protocols	
H.261	Video codec for audio-visual service at p × 64 Kbps with applications to video teleconferencing
H.320	Umbrella recommendation on N-ISDN audio-visual terminals and systems, including multimedia multiplexing, the procedure and protocol, video and audio coding methods, and audio-visual terminals and systems
H.321	Adaptation to the ATM environment of H.320 terminals
H.322	Adaptation to a guaranteed QoS LAN or H.320 terminals
H.324	Terminal configuration for audio-visual services over GSTN
H.310	Terminal in ATM environments
H.323	Terminal in a nonguaranteed QoS LAN like Ethernet
N-ISDN-Based Teleconferencing and for Transmission Protocol of Multimedia Data	
T.123	Protocol stacks for audio-visual teleconferencing
T.124	Generic conference control
T.126	Multipoint applications of still image
T.127	Multipoint applications of file transfer
H.244	Channel aggregation method for audio-visual communications; this enables several B-channels to behave as a single higher bit rate channel
H.230	Control and indication signals for audiovisual systems
H.231	Multiunit control for audiovisual systems using channels up to 2 Mbps
H.243	Procedures for establishing communication between three or more audiovisual terminals using channels up to 2 Mbps
H.242	Procedures for establishing communication between audiovisual terminals using channels up to 2 Mbps
H.233	Confidentiality system for audio-visual services
ATM-Based Broadband Communication	
H.262	Video codec for moving pictures, known as MPEG2
H.222.0	General form of elementary stream multiplexing as the MPEG2 system part
H.222.1	Specifies the parameters in H.222.0 for communications use
H.245	Multimedia system control, defining negotiation procedure and signaling for the B-ISDN and GSTN

to improve coding efficiency, enable content-based manipulation, make bit-stream editing more robust in an error-prone environment, and develop content-based multimedia data access tools. The experts group for MPEG4 has constructed a verification model and used it to estimate potential methods as a core experiment.

In the ATM Forum, MPEG over ATM has been studied for expanding multimedia applications; the results were reflected in the relevant ITU-T recommendations.

Universal Personal Telecommunications

Universal personal telecommunications (UPT) enables users to access various services from virtually anywhere. It enables each user to participate in a user-defined set of subscribed services. It also enables the user to originate and receive calls on the basis of a personal, network-transparent UPT number through multiple access networks, using any fixed or mobile terminal, irrespective of geographical location. This service is limited only by terminal and network capabilities, plus any restrictions imposed by the network operator. Telecommunication services featuring user mobility, for example, mobile telephony, are rapidly increasing, but maintaining user satisfaction will require global network-transparent telephone numbers and access capability from any terminal.

The concept of UPT has been the subject of lively discussion in several ITU-T study groups since 1988, and recommendations for UPT are being developed in the ITU-T.

The UPT service definition was studied in SG 1 (currently SG 2). UPT is expected to expand over time; it will take an evolutionary path heavily affected by evolving market needs and advances in technology. Standardization of UPT by the ITU-T has been separated in to the following three phases:

1. *UPT Service Set 1* (restricted short-term UPT service scenario): Provides only for telephone service over public switched telephone networks (PSTNs), N-ISDNs, and the public land mobile network (PLMN);
2. *UPT Service Set 2* (basic UPT service scenario): Will incorporate more services and networks and provide various data services;
3. *UPT Service Set 3* (long-term enhanced UPT service scenario): Future technological and market developments will give rise to advances in UPT services that are unknown today.

Recommendation F.851 specifying UPT Service Set 1 was approved in 1995. This service set covers both essential and optional UPT features. The essential UPT features are related to the basic operation of UPT and are therefore essential for UPT implementation. Four essential UPT features are defined:

1. UPT user-identity authentication;
2. In-call registration;
3. Outgoing UPT calling;
4. In-call delivery.

The optional UPT features enhance the operation of the basic UPT services.

Recommendation F.852, which defines UPT Service Set 2, and Recommendation F. 853, which defines UPT supplementary services, were defined by SG 2 (formerly SG 1). The supplementary services are classified by their relationship to the N-ISDN I-series recommendations, as shown in Figure 11.3. Category a covers the supplementary services specified for UPT; they are applicable only within UPT. Category b covers the supplementary services defined in the N-ISDN I-series recommendations; they can be used in various UPT environments. Services in this category require modifications in their definitions. Category c covers the supplementary services for UPT. Recommendation F.853 also defines three types of interactions between the environments and the supplementary services and features:

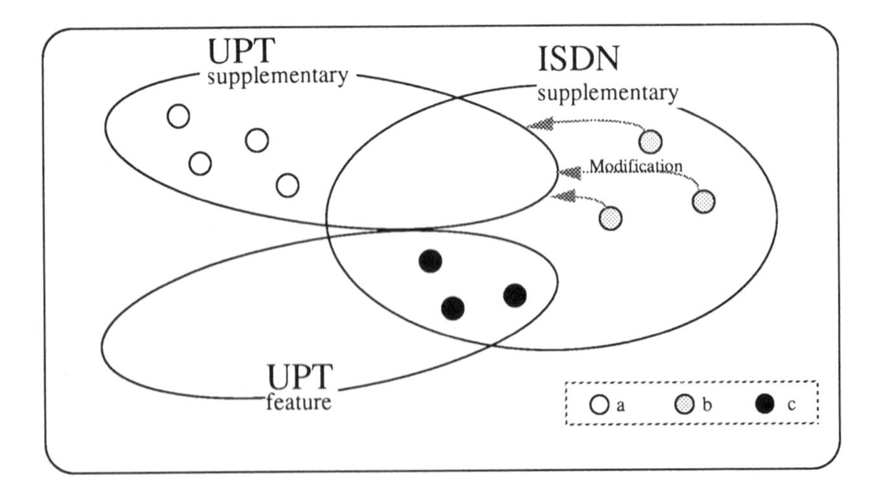

Figure 11.3 UPT supplementary services.

1. UPT feature;
2. Supplementary services outside the UPT environments (UPT supplementary services subscribed to at a particular access);
3. Features in the UPT environments (subscribed to at a particular access).

The numbering plans are the responsibility of SG 2. The numbering plan for UPT is specified in Recommendation E.168. It provides the framework for numbering *in-calls* (i.e., calls incoming to the user) and *in-call registration* (i.e., transactions between the user and service profile). Recommendation E.168 originally only described the UPT numbering scenarios. It was enhanced by adding a description of the UPT numbering structure, the UPT access code (UPTAC), the UPT access number (UPTAN), and the personal user identity (PUI) principles.

The PUI identifies the user to the UPT service provider. Two methods are specified in both UPTAC and UPTAN to access a UPT service profile, one for national access and one for international access. UPTAC puts higher demands on the interconnection of international IN than UPTAN. It identifies the international call-routing principles and provides guidance in the user plane for a restricted short-term UPT service scenario. It also identifies the start point of the enhanced UPT service scenario applicable to in-calls, UPT out-calls, and UPT-to-UPT calls. Recommendation E.174 specifies the UPT routing. Traffic engineering for UPT is specified in Recommendations E.755 (Reference Connection), E.775 (UPT Grade of Services), E.776 (Network Grade of Service Parameters), and E.785.

Recommendation D.280 covers UPT charging and accounting; it was developed by SG 3. It specifies the principles for charging for UPT calls and for defining UPT service profile management functions. It also covers traffic accounting principles, principles for billing UPT calls, and reimbursement between service providers for various types of UPT calls.

The study by SG 11 on UPT signaling and switching standardization is based on the signaling network defined for the intelligent network (IN). Provisioning of service access to multiple networks should be done at the initiation of UPT services. Recommendation Q.76 (Service Procedure for UPT) defines the functional entities and information flows for UPT. It also specifies the UPT signaling based on UPT Service Set 1, defined in Recommendation F.851, for a UPT user using dual-tone multi-frequency (DTMF) access on PSTN. The functional entities and information flow defined for IN Capability Set 1 (IN CS-1) are the base for UPT modeling. DTMF in-band user access is the basis for interaction between a user and a UPT service provider, so a voice announcement is used for the interaction. Information flow diagrams are classified into the four categories:

1. UPT elementary procedures and common sequences;
2. Procedures for personal mobility (generic names);
3. UPT call handling;
4. Procedures for UPT services profile management.

Recommendation Q.1551 covers protocol specification for UPT; it is based on Recommendation Q.76. It defines the data model shown in Figure 11.4, which is based on the X.500-series used for IN modeling.

Studies supporting various UPT Service Set 1 functions not provided by IN CS-1 will be covered by IN CS-2; UPT signaling is being studied using the out-channel user interaction defined for the ISDN environment.

International Mobile Telecommunications (IMT 2000)/Future Public Land Mobile Telecommunication Systems (FPLMTS)

IMT 2000 requires international standards for services, numbering, signaling, network management, voice coding, security algorithms, radio transmission, and so on. ITU-T SGs 1, 2, 4, 11, 13, and 15 are working on IMT 2000-related standardization. In particular, SG 11 has been working since 1996 to standardize signaling. Therefore, here we focus on the current status of IMT 2000/FPLMTS standardization activities in SG 11.

The targeted mobility and portability in FPLMTS are shown in Figure 11.5. The objectives are global terminal mobility and transmission at bit rates up to 2 Mbps, enabling multiple calls to be connected simultaneously to an IMT 2000 mobile terminal. FPLMTS supports UPT and FPLMTS user mobility. FPLMTS user mobility is available only within the FPLMTS, whereas UPT can be used across various telecommunication networks. In FPLMTS, the same procedure for connecting mobile terminals and networks is applied to both UPT and FPLMTS. Therefore, a user can communicate by using his or her ID number and an FPLMTS mobile terminal.

For terminal mobility and FPLMTS user mobility, a standardized international mobile user identity (IMUI) is required. It is currently thought that Recommendation E.212 (Identification Plan for Land Mobile Stations) can be applied to IMUI. FPLMTS provides user identity module (UIM) portability, which enables the FPLMTS user identity to be independent of the physical FPLMTS terminal. In addition, FPLMTS provides portability for a standard terminal enabling it to be connected to an FPLMTS mobile terminal.

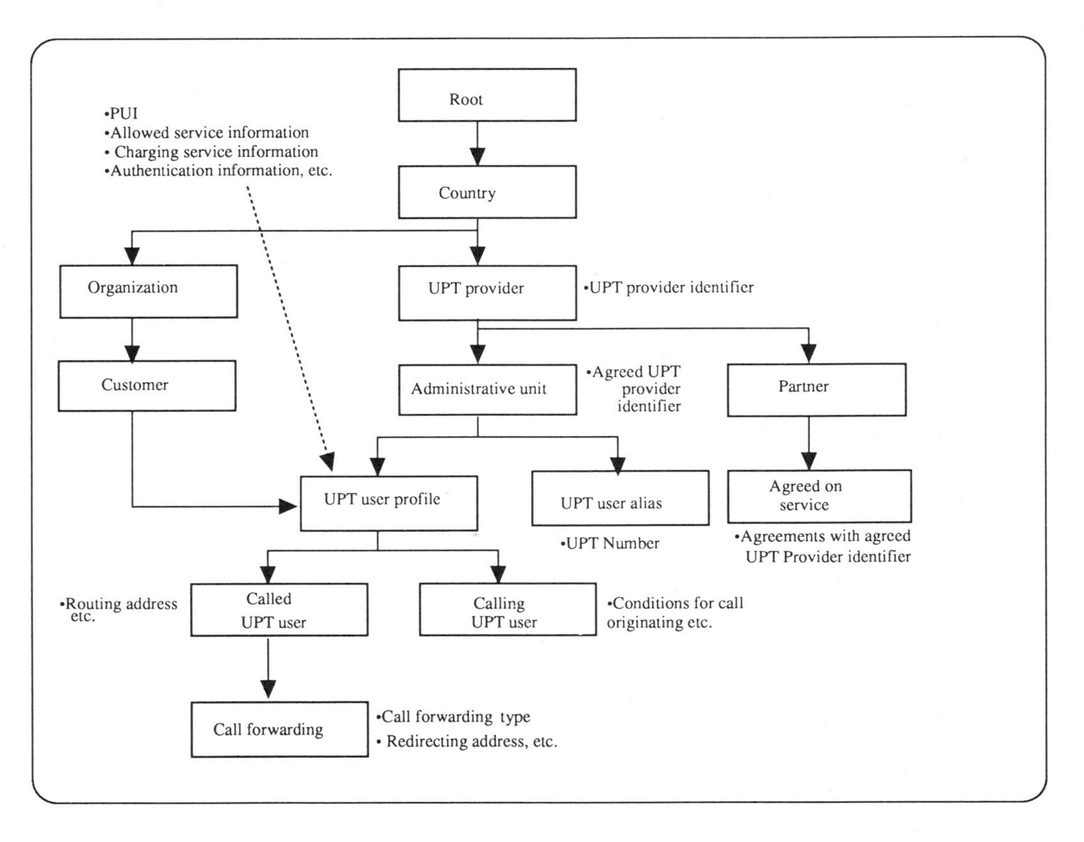

Figure 11.4 UPT data model.

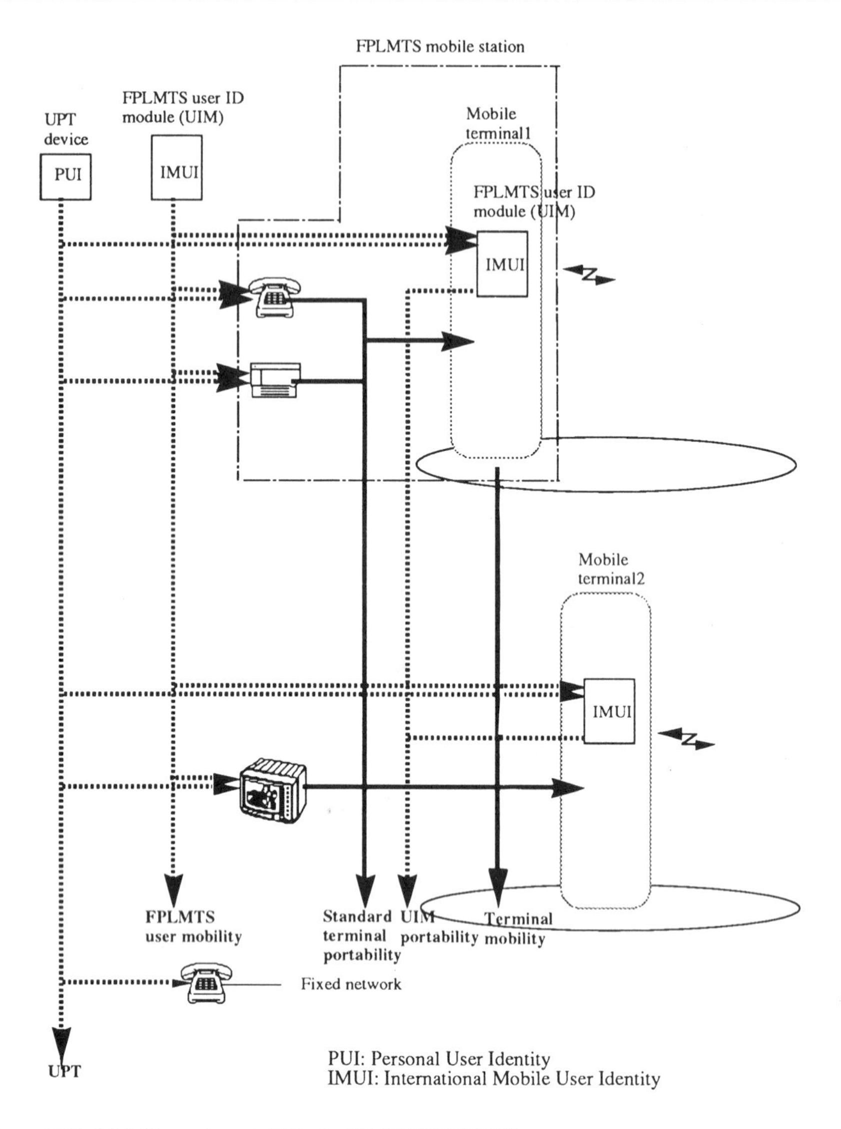

Figure 11.5 Mobility and portability in FPLMTS/IMT 2000.

Draft Recommendation Q.FNA specifies the FPLMTS functional network architecture (Figure 11.6). The FPLMTS network architecture uses the IN distributed functional plane (DFP) model. A clear distinction has been made between the functions on the mobile terminal side and those on the network side, so the functional relations spanning the radio interface are clear. The FPLMTS network architecture forms two separate planes, one for communication control (CC) and one for radio resource control (RRC). Overall call-

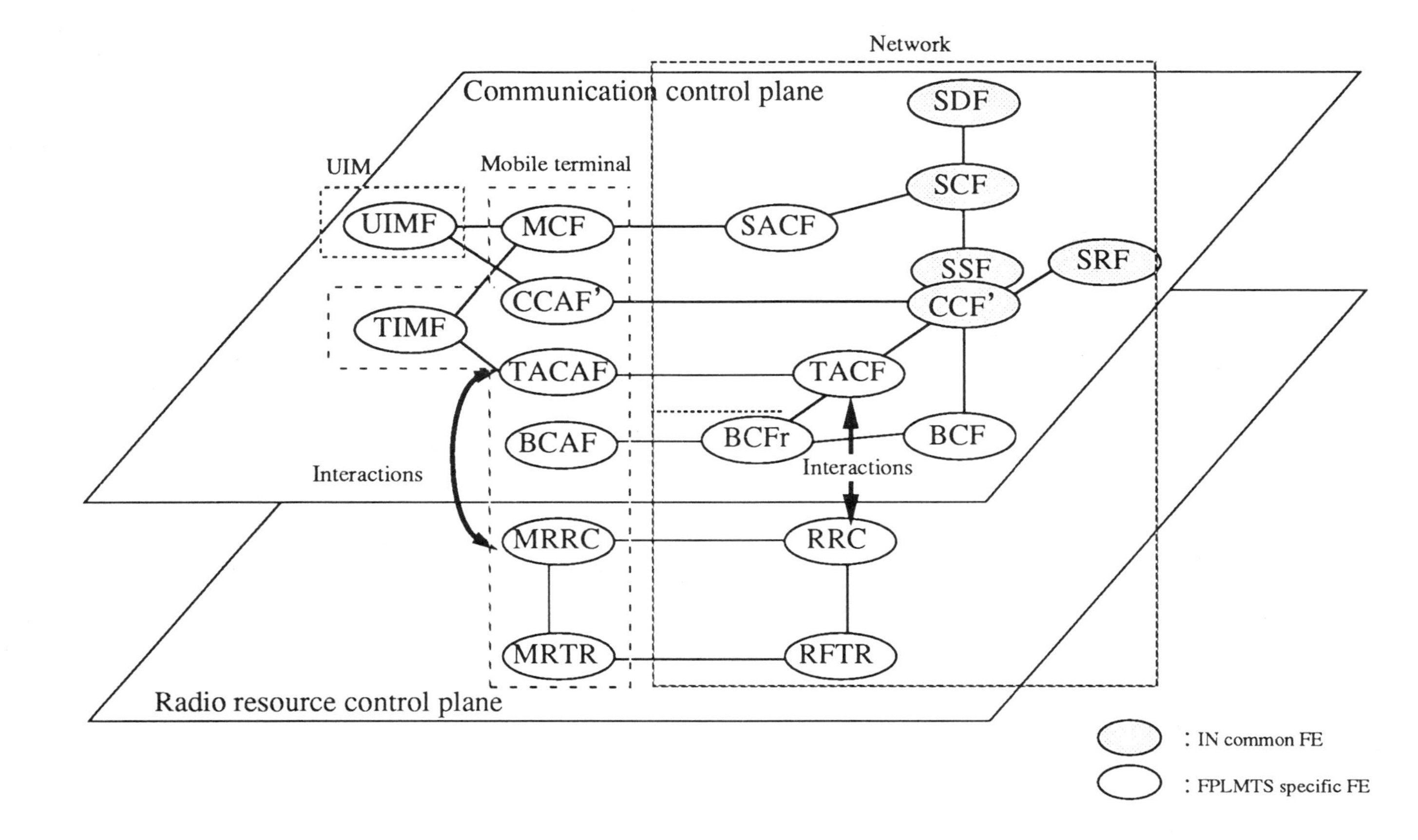

Figure 11.6 FPLMTS/IMT 2000 functional network architecture.

related control, call-unrelated control (e.g., location registration), and connection control (e.g., handover and paging) are carried out in the CC plane. Basic radio resource management (e.g., selection and reservation of radio resources, radio channel and radio environment supervision, and handover initiation) are carried out in the RRC plane.

Interaction between the two planes is needed to establish and maintain communication. The IN DFP model is being enhanced to incorporate this and other FPLMTS requirements. It includes state models of the functional entities and of the information flow; these state models are in the annex of Recommendation Q.1224 (IN CS-2). An IN signaling protocol, INAP for FPLMTS, will be provided in IN CS-3.

Draft Recommendation Q.FIF specifies the information flows over the FPLMTS functional network architecture; it includes such basic procedures as call origination/termination, handover, and terminal/user location registration. A framework based on the information flow is being developed by SG 11 for the signaling protocol over a radio interface. The B-ISDN UNI protocol, Digital Subscriber Signaling System 2 (DSS2), in Q.2931 is expected to be applied to the FPLMTS call control protocol to enable the provision of mobile multimedia services.

These activities are expected to be completed by the end of 1998.

Telecommunications Management Network

When studies on the telecommunications management network (TMN) started, the target was to standardize the interfaces between the management systems and network elements. The fundamental TMN framework was developed first; it included generic TMN functional and physical models, a protocol profile, and a management information model. The studies on the protocol profiles for the TMN Q3 and Qx interfaces were carried out based on the existing system management protocols. The Q3 interface is located between network elements and the TMN operation system function (OSF). The Qx interface accommodates non-TMN-standard network elements; it is located between the Q-adaptation function (QAF) for such network elements and the TMN mediation function (MF, adaptation of Qx to Q3). The TMN MF connects to the TMN OSF through the Q3 interface. The structure of the TMN-related recommendations is given in Figure 11.7. An overview of the TMN recommendations is contained in Recommendation M.3000.

The TMN studies have focused on specifying the tools needed for management/managed system interoperability, taking into account the need for user-machine cooperation in the management environment.

The TMN studies were related to areas covered by several study groups. Therefore, the Joint Coordination Group (JCG) on TMN was formed to facilitate collaboration among these SGs (2, 4, 7, 10, 11, 13, and 15).

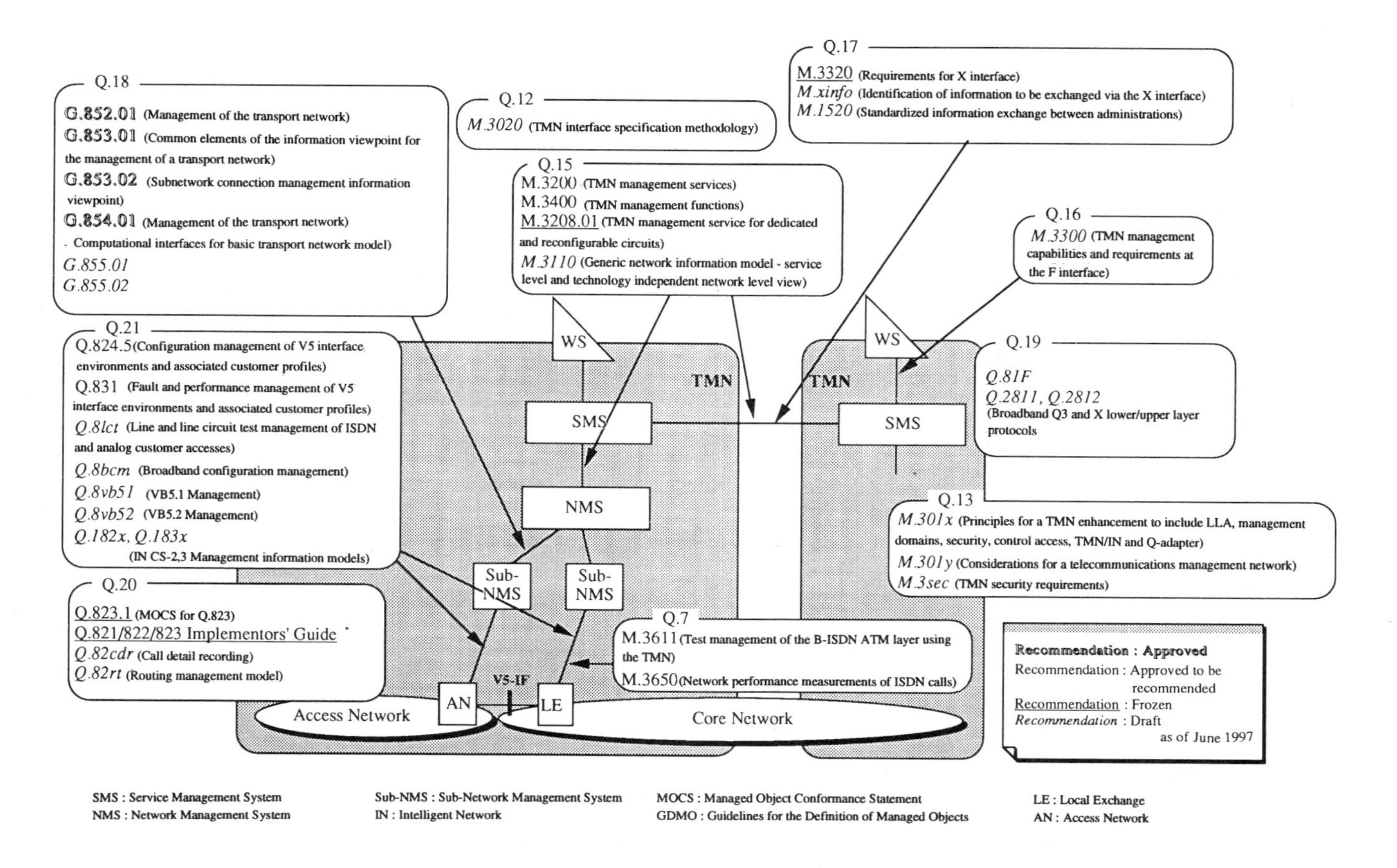

Figure 11.7 Structure of TMN-related recommendations.

Study Group 4 covered TMN principles, architecture, management services, functions, and generic network information models. It also covered improving the tools, methodology for TMN study, and the conformance requirements (e.g., managed object conformance statements).

Study Group 7 developed OSI systems management models for enhancing the basic management architecture and functions, in collaboration with the ISO/IEC.

Study Group 11 was responsible for the protocols used in TMN, including the TMN Q3 interface based on Recommendations DSS1 and Signaling System No. 7 (SS No. 7). It also developed specific application-layer protocols (messages) and associated support (managed) objects.

Study Group 15 was responsible for management information models (network-element level) for transmission equipment related to synchronous digital hierarchy (SDH), ATM, plesiochronous digital hierarchy (PDH), digital circuit multiplication equipment (DCME), echo cancellers, video coders/decoders (codecs), low-bit-rate voice codecs, and optical fiber transmission systems. It also covered the network level of the management information model for managing a transmission network as a single entity.

As a result of reengineering of the ITU-T, initiated at WTSC 96, most of the related questions covered by these study groups were transferred to SG4, to achieve a more unified and concentrated study on TMN. (See Section 7.5.1 for TMN technical issues on management.)

11.2.2 ISO/IEC JTC 1

11.2.2.1 General

In 1987, Joint Technical Committee 1 (JTC 1) was formed by amalgamating technical committees of the ISO and the IEC in order to reduce the overlap of standardization activities between these two organizations. JTC 1 works on standardization in the industrial application of information technology. Each member is a national body representing a member country. JTC 1 has 18 subcommittees. To better meet the needs of the multimedia age, JTC 1 has identified the areas that need standardization. The *Fast Track* procedure is one in which existing industry standards are directly voted on and agreed to as Draft International Standards to speed up the standardization process. It has introduced the *International Standardized Profile* (ISP), a technical standards document, to facilitate implementation. It has also introduced the *Publicly Available Specification* (PAS) process for adopting de facto industry standards that have the open characteristics needed for wide use in the information area. The aim of the PAS process is to speed up the development of international standards.

11.2.2.2 Major Achievements of JTC 1

As explained in an earlier section, the MPEG2 standards on multiplexing, video coding, and audio coding are called the MPEG2 system, MPEG2 video, and MPEG2 audio standards. The MPEG2 system and video standards were jointly developed by JTC 1 SC29 and ITU-T SG 15. After the completion of MPEG2, the MPEG4 study started for very-low-bit-rate audiovisual coding. (Details on MPEG2 and MPEG4 were given earlier.)

In 1989, the Multimedia and Hypermedia Information Coding Experts Group (MHEG) was established in ISO/IEC JTC 1 SC29/WG12 to develop international standards for describing the coding of hypermedia information. These international standards, originally called "Coded Representation of Multimedia and Hypermedia Information Objects," provide generic multimedia information structures that are suited to real-time multimedia applications/synchronization and real-time interchange of applications [6].

Table 11.4 shows the JTC 1 standards related to MPEG and MHEG.

11.2.3 The ISO and the IEC

The International Organization for Standardization (ISO) is a nongovernmental organization that performs standardization in industrial fields other than electricity and electronics. The ISO has 167 technical committees (TCs) and has member bodies from about 80 countries.

The International Electrotechnical Commission (IEC) performs standardization in the industrial fields of electricity and electronics. The IEC is also a nongovernmental organization and has about 50 national committees consisting of one member from each country. The IEC consists of about 80 TCs, including a General Policy Committee (GPC) and an International Special Committee on Radio Interface (CISPR). The council held in Dresden, Germany, in September 1996 approved a large-scale restructuring to enhance operations. The objectives follow:

1. To speed up the decision-making process to enable the IEC to keep up with the speed of technical innovation;
2. To simplify the management system concerning technology as necessary;
3. To implement a decision-making organization that can execute high-level judgments related to conformity assessment systems.

The ISO and IEC have adopted two procedures for promoting an international standardization process that complements the examination of standards.

Table 11.4
ISO/IEC JTC 1 Standards Related to MPEG and MHEG

	Standards	Fields	Contents	Approved
MPEG1	ISO/IEC 11172-1	Multimedia multiplexing	Multimedia multiplexing/synchronization for digital storage media at up to about 1.5 Mbps	1992
	ISO/IEC 11172-2	Video	Coding of moving pictures for digital storage media at up to about 1.5 Mbps	
	ISO/IEC 11172-3	Audio	Coding of 20-KHz stereo audio signals	
JPEG	ISO/IEC 10918-1 ITU-T T.81	Still images	Digital compression and coding of continuous-tone still images	1992
MPEG2	ISO/IEC 13818-1 ITU-T H.222.0	Multimedia multiplexing	Generic multimedia multiplexing/synchronization	1994 (ISO/IEC) 1995 (ITU-T)
	ISO/IEC 13818-2 ITU-T H.262	Video	Generic coding of high-quality moving pictures	1994 (ISO/IEC) 1995 (ITU-T)
	ISO/IEC 13818-3	Audio	Application of 20-KHz stereo audio signals to low-sampling frequency and multichannel audio signals	1994
MHEG	ISO/IEC 13522-1	Multimedia/ hypermedia	Coded representation of multimedia and hypermedia information objects	1995

The first one, mentioned earlier, is the Fast Track procedure, which omits the examination of standards that have already achieved positive results at the TC or subcommittee level in order to speed up the development of international standards. The second one is the Technical Report procedure, in which a report is issued to provide early disclosure of important technical information. Such a report is issued in the following cases:

- The draft for an international standard was not accepted during the approval stage.
- It is still technically in the developing stage, so it is too early for international standardization.
- Data collected are different from that in the standard.

11.3 Forums and Consortia

11.3.1 The ATM Forum

The ATM Forum was established in October 1991 to achieve two objectives [7]:

1. To accelerate the deployment of ATM products and to quickly prepare implementation specifications to enable ATM interconnection;
2. To promote industry cooperation and general acceptance of ATM technology.

The ITU-T is promoting the standardization of the public B-ISDN, so the ATM Forum initially targeted the utilization of ATM in private networks. It has promoted the development of common implementation specifications for ATM LANs.

The number of participating members had reached 953 as of March 1997 (246 principal members, 505 auditing members, and 202 user members).

The ATM Forum has close relations not only with the ITU-T, but also with other standards organizations and forums, including ANSI, the ETSI, the IETF, the FRF, and DAVIC. The ATM Forum contributes to standardization by adopting existing standards or specifications and through upstream activities. Its work in developing basic implementation specifications reached a peak in 1996. The ATM Forum is shifting its focus to developing implementation specifications and enlightenment activities concerning such specifications.

The ATM Forum has accomplished several important tasks (Table 11.5). It developed UNI specification version 4.0 for improved signaling specifications prior to the ITU-T. It includes ABR specifications for traffic management (TM) 4.0 and leaf-initiated multipoint connections. It completed broadband intercarrier interface (B-ICI) versions 2.0 and 2.1, including NSAP support, NCCI, and best effort type. The interim local management interface (ILMI)

Table 11.5
The ATM Forum Specifications

(Working groups/Title of specifications)
B-ICI: Defines interswitch communications in public networks.
ATM B-ICI v1.1, v2.0, v2.1 Specification
Signaling: Defines procedures for call setup and negotiation of QoS for an end-user connection.
ATM UNI v2.0, 3.0, 3.1 ILMI MIB for UNI v3.0, UNI v3.1
Physical layer: Defines the physical and electro/optical characteristics for interfaces and signals.
ATM DS 1 PHY v3.1 UTOPIA Specification Level 1, v2.01
6,312 kbps UNI v1.0 Specification ATM 52 Mb/s Category 3 UTP
ATM 155 Mb/s Category 5 UTP v1.0 Specification
Testing: Establishes test suites to evaluate conformance, interoperability, and performance of implementations of the ATM specifications.
Introduction to ATM Forum Test Specifications v1.0
PICS Proforma for the DS3 Physical Layer Interface v1.0 Specifications
PICS Proforma for the SONET STS-3c Physical Layer Interface v1.0 Specifications
PICS Proforma for the 100 Mbps Multimode Physical Layer Interface v1.0 Specifications
PICS Proforma for the UNI 3.0 ATM Layer
PICS Proforma for the DSI Physical Layer Interface
Interoperability Abstract Test Suite for the ATM Layer
Interoperability Abstract Test Suite for the Physical Layer
Frame-Based User-to-Network Interface (FUNI): Defines frame-based interface for ATM services.
Frame Based User-to-Network Interface (FUNI) Specifications
Traffic Management: Provides specifications for traffic control to limit congestion and maximize bandwidth use.
Traffic Management v4.0
Service Aspects and Applications (SAA): Specifies services, such as application programming interfaces, and interworking with frame relay, SMDS, and circuit emulation.
FUNI CES Interoperability Specifications
Private Network-to-Node Interface (P-NNI): Specifies the protocol by which ATM switches communicate within a private ATM network.
Interim Interswitch Signaling Protocol P-NNI v1.0

for UNI was improved. Private network node interface (P-NNI) 1.0 was developed to define the routing protocols required for on-demand connection setups. Local-area network emulation (LANE) was developed to enable customers to put their upper layer protocols on an ATM network without any changes. A method for encapsulating CBR-coded MPEG2 signals into AAL 5 was defined for VOD service. Various kinds of physical interfaces, including ATM 25M, were specified.

In the ATM Forum, MPEG over ATM was studied for expanding multimedia applications, and the results were reflected in the relevant ITU-T recommendations. The ATM Forum also published the Phase 1 documents, which describe ATM terminal specifications for VOD services. (See Section 7.5.3 for management issues.)

11.3.2 The Network Management Forum (NMF)

The NMF was established primarily by network operators in July 1988. Its objectives follow:

1. To promote and accelerate implementation standards for network management and system management based on OSI and related de facto standards;
2. To promote product development by equipment vendors through cooperation among carriers, users, and vendors;
3. To contribute to ITU-T standardization and regional standardization from the carrier and user viewpoints;
4. To provide guidelines for selecting options and integrating systems with the cooperation of industry and partner organizations.

To respond to rapidly changing business requirements, a consensus is needed among network operators, network managers, and network technology suppliers on the following:

- Detailed requirements for automating end-to-end business processes and reengineering;
- Interfaces between customers, service providers, and network technology suppliers;
- A framework for service management that satisfies the business purpose and technology and supports interoperability for management;
- A common computing platform on which service management and network management applications can be built.

The NMF has been pursuing these objectives through three work plans: SMART, OMNI Point, and SPIRITS [8]. The OMNI Point program was established under the umbrella of the NMF. This program is aimed at achieving a unified approach to management and a decisions-making process that uses the help and guidance of experts from many areas of industry. The objective is to minimize the investment outlays and risks of all concerned: commercial users, service providers, vendors/suppliers, and standards organizations. It links the activities and results of 16 partner organizations with those of the de jure standardization organizations, such as the ITU-T, the ISO, and the TTC, and with those of *de facto* industry standardization organizations, such as the IETF. The partner organizations include the European Workshop for Open Systems (EWOS), the Asia-Oceania Workshop (AOW) on OSI Network Management Special Interest Group (NM-SIG), the Object Management Group (OMG), X/Open, the Corporation for Open Systems International (COS), and Standards Promotion and Application for Information Processing (INTAP).

The Information Infrastructure Standards Panel (IISP) was established in the United States in June 1994 with the primary objective of coordinating the activities of related standardization organizations, such as the T1 Committee and the IEEE for standardizing a national information infrastructure (NII) in the United States [9]. This panel discusses user requirements mainly those of governmental and large corporate business users. They expanded NMF Release 1, referred to as OMNI Point 0, to include Internet management, application programming interfaces (APIs), integration technologies, and object libraries.

The specifications to be used by service providers for procuring general-purpose computers are being studied by the Service Providers' Integrated Requirements for Information Technology (SPIRIT) action team, under the umbrella of the NMF. The objective of SPIRIT is to establish common requirements for the computing platforms to be used by service providers. The SPIRIT team has achieved this objective and has published three issues, which are the core of OMNI Point.

The NMF has been focusing on developing two management information models for managing customer services. For automation of service management and reengineering, the SMART team started in October 1994. The subjects being addressed are trouble management (between service providers and between customers and service providers), performance reporting, service ordering, and billing. (See Section 7.5.2 for management issues.)

11.3.3 The Internet Engineering Task Force (IETF)

The Internet Society (ISOC) was established in 1992 to promote international cooperation in the development of Internet-related technologies and applications. The IETF covers the standardization activities regarding these areas. It verifies and examines the standardization of Internet protocols. Final approval of standards is done by the Internet Engineering Steering Group (IESG), which consists of 14 members elected from the IETF.

The IETF has 72 working groups (WGs) covering 10 areas [10]:

1. Applications;
2. Next generation IP;
3. The Internet;
4. Network management;
5. Operational requirements;
6. Routing;
7. Security;
8. Transport;
9. User service;
10. Miscellaneous.

A Request for Comments (RFC) is a document describing the protocol specification of an IETF standard. They are classified into four classes:

1. *Standard track:* for proposed standards, draft standards, and standard protocols;
2. *Experimental:* protocols being tested or not yet mature;
3. *Informational:* protocols specified by bodies other than the IETF, such as the OSI, or by specific vendors, or specifications for implementation;
4. *Historic:* protocols that have been replaced by newer versions.

Five statuses are defined for each protocol, according to the degree of mandate:

1. *Required protocols:* Implementation is mandated in all systems.
2. *Recommended protocols:* Implementation is recommended.
3. *Effective protocols:* Implementation is optional.
4. *Limited use protocols:* Limited implementation is recommended.
5. *Nonrecommended protocols:* Implementation is not recommended.

The standardization process in the IETF is based on the following policies:

- Working groups are open to anyone.
- E-mail is the primary tool for discussions.
- Decisions are made on a consensus basis, that is, no voting for decisions or approval.
- Implemented protocols are the only candidates for standardization.
- Members at any level represent individuals, not organizations, and have equal footing.

Table 11.6 lists the standard Internet protocols. It includes 33 draft standard protocols, 152 proposed standard protocols, and 22 historic standard protocols.

The IETF is currently discussing the following topics: IP version 6 (IPv6) is being discussed as a next-generation Internet protocol on the network layer. The aim of this protocol is to increase the IP address capacity and to reduce the cost of calculating the routing control table in the core router. It has an address field of 128 bits, compared to the original IP address field of 32 bits.

The next hop resolution protocol (NHRP) is being discussed for determining the address (internetworking layer address and nonbroadcast multiple access (NBMA) subnetwork address) of the next hop along the route from the sender (host or router) to the destination.

The goal of the Inter-Domain Multicast Routing (IDMR) project is to develop a scalable routing protocol for multicasting and to propose a standard. The project members are discussing several promising draft methods, including Protocol Independent Multicast–Sparse Mode (PIM-SM), Protocol Independent Multicast–Dense Mode (PIM-DM), and Core Based Tree (CBT).

The IP security working group is discussing all aspects of IP-level security, including combinations of authentication, integrity, access control, and confidentiality. The protocols they are discussing include secure IP headers and encapsulation and key exchange. The protocols allow different encryption algorithms. IP encryption can take place between hosts or between routers.

The resource reservation setup protocol (RSVP) is used when the host requires the network QoS to specify the data stream by reserving resources on the network.

The IETF is collaborating with other standardization organizations in order to promote interoperability between Internet protocols and other protocols. The closest relationship is with the ITU-T. The IETF, when it was the ISOC, became a member of the ITU-T in 1995 to cement formal relationships in common areas of concern. The Internet protocols are considered to be candidates for the IETF's shortcut voting system (PAS). IETF also has close relationships with other standardization organizations and forums/consortia,

Table 11.6
Internet Protocol Standards

Standard Protocol	RFC Number	Status
0001 Internet Official Protocol Standards	RFC1880	REQ
0002 Assigned Numbers	RFC1700	REQ
0003 Host Requirements	RFC1122,1123	REQ
0005 Internet Protocol (IP)	RFC0791,0950, 0919,0922,792	REQ
	RFC1112	REC
0006 User Datagram Protocol (UDP)	RFC0768	REQ
0007 Transmission Control Protocol (TCP)	RFC0793	REC
0008 Telnet Protocol	RFC0854,0855	REC
0009 File Transfer Protocol (FTP)	RFC0959	REC
0010 Simple Mail Transfer Protocol (SMTP)	RFC0821,1870, 1869	REC
0011 Format for Electronic Mail Messages	RFC0822,1049	REC
0012 Network Time Protocol	RFC1119	REC
0013 Domain Name System (DNS)	RFC1034,1035	REC
0014 Mailing Routing and the Domain System	RFC0974	REC
0015 Simple Network Management Protocol (SNMP)	RFC1157	REC
0016 Structure of Management Information	RFC1155,1212	REC
0017 Management Information Base	RFC1213	REC
0019 NetBIOS Service Protocol	RFC1001,1002	REC
0020 Echo Protocol	RFC0862	REC
0021 Discard Protocol	RFC0863	ELE
0022 Character Generator Protocol	RFC0864	ELE
0023 Quote of the Day Protocol	RFC0865	ELE
0024 Active Users Protocol	RFC0866	ELE
0025 Daytime Protocol	RFC0867	ELE
0026 Time Server Protocol	RFC0868	ELE
0033 Trivial File Transfer Protocol (TFTP)	RFC1350	ELE
0034 Routing Information Protocol (RIP)	RFC1058	ELE
0035 ISO Transport Service on top of the TCP	RFC1006	ELE
0050 Ethernet MIB	RFC1643	ELE
0051 PPP in HDLC-like Framing	RFC1662	ELE
0051 The Point-to-Point Protocol (PPP)	RFC1661	ELE
0052 IP Datagrams over the SMDS Services	RFC1209	ELE

Status:
REQ: Required Protocol
REC: Recommended Protocol
ELE: Elective Protocol

such as ISO/IEC JTC 1 SC6 WG2 on CNLP, the ISODE Consortium on X.500, RARE WG-MSG on X.400, W3C (MIT World Wide Web Consortium) on WWW, and the ATM Forum on ATM technology.

11.3.4 Digital Audio-Visual Council (DAVIC)

The first meeting of the Digital Audio-Visual Council (DAVIC) was held in March 1994 in Geneva, Switzerland. The objective of DAVIC is to provide specifications for implementing system architectures, interfaces between systems, and communication protocols for multimedia services and applications, such as VOD. DAVIC is a forum for addressing general issues, such as storage, broadcasting, and telecommunication. The work of DAVIC is based on the MPEG2 standard for moving picture compression, as formalized in ITU-T Recommendation 262.

DAVIC is composed of seven technical committees (TCs): Technology, System Integration, Server, Set-Top Unit, Delivery System, Applications, and Security. DAVIC specifications are defined as follows:

- Identify standards or specifications that are being widely used or circulated in the market.
- Select the options of those that are suitable for the digital video services and applications that are in the scope of DAVIC.
- Identify the issues to be reconsidered in the standards or specifications and feed them back to the bodies that developed them.
- Define original specifications as needed.

The contents of the first DAVIC specification, DAVIC 1.0, are summarized in Table 11.7. DAVIC 1.1 is an enhanced version, and the features added are as follows:

1. Fundamental specifications for access from a set-top unit (STU) to the Internet;
2. Specification of JAVA API for DAVIC so that STU can operate as a virtual machine by downloading software;
3. Transmission specifications for terrestrial broadcasting by microwave.

Further enhancements were made in DAVIC 1.2. It added the following:

Table 11.7
DAVIC Specifications

Name	Contents	Date of Finalization
DAVIC 1.0	This first version aimed to define the interoperable specifications in retrieval and distribution systems like VOD.	December 1995
DAVIC 1.1	Following points were added: • Fundamental specification for access from a set-top terminal to the Internet • JAVA API for DAVIC, which is the specification that changes "set-top box" into "virtual machine" • Transmission standards for terrestrial broadcast by microwave	September 1996
DAVIC 1.2	Following points were added: • Basic security for DAVIC 1.0 systems • High-quality audio and video standards • Full Internet access by high-speed video network • ADSL ATM mapping.	December 1996
DAVIC 1.3	Following points are included: • DAVIC service and system management • Communication services (telephony, conferencing, and multiplayer games) • Multiple server and services • Still picture display control API for set-top unit	September 1997

1. Basic security for DAVIC 1.0 systems;
2. Support of high-quality audio and video;
3. Full Internet access through high-speed video networks;
4. ADSL ATM mapping.

DAVIC 1.3 will include the following [11]:

1. Service and system management;
2. Support of telecommunication services (telephony, conferencing, and multi-player games);
3. Support of multiple servers and services;
4. Still-picture display control API for STUs.

DAVIC has formal liaisons with international and regional standardization organizations, forums and consortia, including the ITU-T, the ISO, the

IEC, the ETSI, the IEEE, ANSI, The ATM Forum, and the IETF. (See Section 7.5.5 for system reference model.)

11.3.5 The IMA and Others

The Interactive Multimedia Association (IMA) was established in 1988 to promote the development of interactive multimedia applications and to expand the use of multimedia technology. More than 240 companies are participating in this forum [12].

The Intellectual Property Project in the IMA published guidelines concerning IPR, and the Compatibility Project has developed recommendations on digital audio and on interactive video systems (level 3).

The IMA held the IMA Publishers Forum with the American Booksellers Association in 1993 and holds the National Association of Broadcasters Multimedia World Conference & Exhibition jointly with the NAB every year. It has a close relationship with the COSE and Kaleida forums.

Other forums and consortia are also developing implementation specifications, educating the public about the standards in their specifications, and promoting markets for their products. Table 11.8 shows some of the active forums and consortia for multimedia communications. A comprehensive survey of these organizations is given in [3].

11.4 GII-Related Activities [13]

Many activities have been directed toward achieving a global information infrastructure (GII) since it was first advocated in March 1994. Based on the growing consensus on GII, standardization organizations have committed themselves to GII by establishing special committees or groups to identify tasks for achieving GII and by coordinating these tasks within their organizations and with other organizations.

In some regional and domestic standardization areas, there has also been some movement to develop a GII or NII.

The Information Infrastructure Standards Panel (IISP) was established in June 1994 with the primary objective of coordinating the activities of ANSI, the T1 Committee, and the IEEE for standardizing NII in the United States [9]. It established a working group for international liaisons to discuss GII issues and extended membership to organizations and to telecommunication and information industries outside the United States.

The T1 Committee established an ad hoc project group in its Technical Subcommittee T1P1. This ad hoc group issued a report identifying key items

Table 11.8
Forums and Consortia Related to Multimedia Communications

Name	Purpose	Objective Fields	Date of Establishment
The APPI Forum	Prestandards/implementation specifications/interconnection To combine the elaborate interconnection properties of the TCP/IP network	Telecommunications multimedia	1992
TINA-C (Telecommunications Information Networking Architecture Consortium)	Prestandards Joint research project for construction of common architecture specifications for carriers concerned with communication software	Telecommunications multimedia	March 1993
The ATM Forum	To take measures that promote technical enlightenment and diffusion by developing technical specifications while placing importance on ATM interconnection and on positive participation by users	Telecommunications multimedia	October 1991
DAVIC	De facto directions/prestandards International standardization of digital AV and multimedia service systems	Multimedia	June 1994
EMF (European Multimedia Forum)	Exchange of neutral opinions and experiments Exploration of the multimedia industry Establishment of business plans Study of the legality of regulations Promotion of cooperation Planning/information exchange/diffusion	Multimedia	June 1994
IMA (Interactive Multimedia Association)	Implementation specifications, interoperability	Multimedia	1988

Table 11.8
Forums and Consortia Related to Multimedia Communications (continued)

Name	Purpose	Objective Fields	Date of Establishment
IMTC (Intenational Multimedia Teleconferencing Consortium)	Jointly organized by CATS, MCCOI, and PCWG to ensure the interoperability of multimedia desktop conferencing Promotes diffusion of multimedia conferencing systems based on ITU- T standards H.320 and T.120 Creates implementation standards needed for realization Supports testing for confirmation of interoperability	Multimedia desktop conferencing systems	October 1994
IETF (Internet Engineering Task Force)	Development and access of Internet, and expansion of relevant technologies and applications	Telecommunications/ multimedia/ information processing	January 1992
MDG (Multimedia Development Group)	Information exchange/diffusion To develop the multimedia market in order to promote information exchange between developers, service companies, infrastructure providers, government agencies, etc., for multimedia diffusion	Multimedia	February 1993
MMCF (Multi-Media Communications Forum)	Information exchange with vendors, network providers, multimedia communication users To define the functions that stretch over the platform by completing the architecture and standard, collect requirements for multimedia communication, and put them in order To educate users about recommendations for service quality To recommend standards for various technologies and interfaces To promote publicizing of interoperability between multiple vendors To provide data concerning products introduced to the market and end-user applications	Multimedia	June 1993

for NII/GII, such as the required information and telecommunication functions and functional models for ensuring interoperability, interworking, integration, and operation of various media [14].

In Europe, Strategic Review Committee 6 (SRC 6) was established in the ETSI to define a coordinated work program on standards for the European Information Infrastructure (EII). Their report, which was approved at the ETSI General Assembly, identifies key issues for EII, such as the characterization of EII, conceptual models, fields to be standardized, and management of standardization by the ETSI [15]. The High Level Strategy Group (HLSG) was organized by four European information and communication technology (ICT) industry organizations: the European Association of Consumer Electronics Manufacturers (EACEM), the European Telecommunications and Professional Electronics Industry (ECTEL), the European Public Telecommunications Network Operations Association (ETNO), and the European Association of Business Machines Manufacturers and Information Technology Industry (EUROBIT). The mission of HLSG is to provide total coordination in developing ICT standards. The HLSG is expected to play a role similar to that of the IISP in the United States. One of the objectives of the HLSG is to identify obstacles blocking the implementation of GII and to give advice to the relevant organizations as to how they can be overcome. Other objectives include identifying areas requiring immediate standardization and helping the standardization, thereby ensuring that European industries play leading roles in the future global ICT market. It also identified the following key projects to facilitate these objectives:

- Interoperability of broadband networks;
- City information services;
- Electronic commerce for small- and medium-sized enterprises (SMEs).

In Japan, the TTC organized the Information Infrastructure Task Group (IITG) in March 1995, under the umbrella of the Strategic Research and Planning Committee. A report issued by the IITG clarifies the role of the TTC in promoting standardization for GII and NII in Japan. The report identifies the basic principles to be incorporated into TTC standards for GII and NII in Japan [16]. A two-layered approach for standards is proposed. Low-layer standards (i.e., those for the network infrastructure) should be based on *de jure* standards, and high-layer standards (i.e., those related to applications and services) should primarily adopt the specifications produced by forums and consortia as pre-TTC standards (TTC technical documents), which may or may not be approved as regular TTC standards after a few years of experience.

To track global GII-related activities and to coordinate with other standardization organizations, a Special Working Group on Information Infrastructure (SRPC SWG-II) was established in April 1996, replacing the IITG.

To coordinate among the international, regional, and domestic standardization organizations, the Global Standards Collaboration Group (GSC) was organized in 1994, originally as the Inter-Regional Telecommunication Standardization Conference (ITSC). Members include the ITU, the T1 committee, the ETSI, the TTC, the Telecommunication Technology Association (TTA) of Korea, and other standardization organizations. In the second GSC meeting held in Ottawa in June 1995, aspects of the GII were discussed, as well as other key areas, such as B-ISDN/ATM, UPT/IMT 2000, and TMN. It was agreed that the GII standardization should be developed by the ITU-T. The ITU-T then initiated a study on standardization for GII by ITU-T SG 13 in July 1995, based on the agreement made at GSC2. The Joint Rapporteurs Group on GII (JRG GII) was established under SG 13 leadership; it consists of rapporteurs and experts from various ITU-T study groups who have an interest in GII standardization. This establishment of the JRG GII was agreed on by the TSAG of the ITU-T in September 1995.

The objectives of the JRG GII are to further develop the ITU-T work plan for GII standardization, draft the required recommendations, and participate in the development of joint work programs with other ITU sectors and standardization organizations. The first meeting of the JRG GII was held in January 1996, and experts from interested ITU-T study groups were invited. This meeting produced a first draft clarifying the direction and scope of GII, evolution scenarios, GII modeling and items to be standardized, and standardization organizations relevant to GII [17]. It was noted that joint activities could be useful in developing areas of common interest, such as GII scenarios. The second meeting was held jointly with experts from the ISO/IEC JTC 1 Special Group on GII in June 1996. The third meeting was held in September 1996. The progress of their activities was reported at WTSC96, where future work plans and working methodologies were also decided. Through these activities, GII-related recommendations have been identified and drafted, as is shown in Table 11.9 [18].

The following ITU-T draft recommendations are awaiting approval:

- *I.GII—Overview:* Developed by ITU-T SG 13 and ISO/IEC JTC 1 jointly. It provides an overview of the GII standards for use in developing GII standards by the ISO/IEC/ITU and other standardization organizations.
- *I.GII—PFA:* Gives principles and framework architecture for GII. It describes an enterprise model for identifying interfaces that are likely to be of general commercial importance.

Table 11.9
GII-Overview of ITU-T Recommendations and ISO/IEC JTC 1 Standards

ITU-T	ISO/IEC JTC1
GII Overview	GII Overview
GII—Principles Framework Architecture	GII Roadmap
GII—Scenario Methodology	SWG—GII has adopted no specific methodology (but supports scenarios as a mean to identify standards requirements)
GII—Scenario Methodology—Examples of Use	See above
GII—Terminology & Definitions	JTC 1/SC1
GII—Performance	Performance is a generic requirement included in the attributes and fundamental building blocks of the GII Roadmap for all SCs to note

- *I.GII—Scenario Methodology:* Defines the fundamental tools for describing scenarios needed to achieve GII. Practical applications of this recommendation are given in the following recommendations.
- *I.GII—Scenarios:* Describes major scenarios for identifying the key interface locations in a scenario, the services supported by the interfaces, and the profiles of protocols across the interfaces.
- *I.GII—Terms:* Provides a list of the GII terms and definitions used in ITU-T and JTC 1 recommendations/standards.

ISO/IEC JTC 1 formally established a Special Working Group on GII (JTC 1 SWG-GII) at the 9th JTC 1 Sydney (Australia) Plenary in March 1996. The tasks of SWG-GII are similar to those of the JRG GII of the ITU, such as coordinating GII-related activities within ISO/IEC JTC 1 and with outside organizations. JTC 1 SWG GII produced the GII Roadmap and GII Standards Roadmap [19]. The GII Roadmap is used as a guideline for evolution, management, and development of GII standards. The GII Roadmap describes a strategic, high-level overview of the core services and concepts that serve as the foundation necessary to enable GII as a whole. The GII Standards Roadmap advances the required attributes and fundamental building blocks within the scope, mandate, and expertise of JTC 1. It is used for planning and tracking the development of global standards. The objectives of the Roadmap are (1) to establish a common understanding of GII language and concepts across JTC 1 groups, (2) to identify group-related standards and use existing standards where possible in new work programs, (3) to identify current gaps, areas, and requirements for future standards, and (4) to identify and track related work items and more effectively share workloads among the different standards development bodies.

Cooperation among the ITU, the ISO, and the IEC is essential for GII. The ISO/IEC/ITU Joint Seminar on GII was held in Geneva in January 1996, the week following the first ITU JRG GII meeting described earlier. The key issue of this joint seminar was the need to encourage the three standardization organizations to cooperate in developing GII standards. As the first step, the ITU-T JRG GII and ISO/IEC JTC 1 Special Working Group on GII held a joint meeting in June 1996 to develop a work program and scenarios as described earlier. The ITU-T JRG-GII invites representative from ISO/IEC JTC 1 SWG-GII and vice versa.

11.5 Conclusion

In this chapter, the structures and activities of telecommunication and information standardization organizations, forums, and consortia were described from the viewpoint of the current status of standards related to multimedia communications. In specific fields of multimedia communications, the activities of forums and consortia will become more lively. Broader and stronger harmonization among standardization organizations, forums, and consortia will thus be needed to make the most of each group's efforts.

References

[1] Global Framework for Telecommunication Standardization.

[2] ITU-T WTSC Report of WTSC'96.

[3] The Telecommunication Technology Committee, Surveys on Forums and Consortia in Telecommunication and Information, Apr. 1996.

[4] Asatani, K., Y. Maeda, and N. Morita, "Research Trend of Multimedia Communication Technology in ITU-T," *Proc. 1997 IEICE Annual Conf.*, paper TB-2-1, pp. 869–872, Mar. 1997 (in Japanese).

[5] See, for example, Asatani, K., et al., *Introduction to ATM Networks and B-ISDN*, New York: John Wiley and Sons, 1997.

[6] JTC 1: http://www.iso.ch/meme/JTC 1.html.

[7] ATM Forum: http//www/atmforum.com/.

[8] NMF: http://www.nmf.org/.

[9] For example, IISP/95-0076, Aug. 1996.

[10] IETF: http://www.ietf.org/.

[11] Kasahara, H., "Multimedia Services and Technologies at DAVIC," *Proc. 1997 IEICE Annual Conf.,* paper TB-2-3, pp. 873–874, Mar. 1997 (in Japanese).

[12] IMA: http://www.ima.org/.

[13] Asatani, K., "Standardization of Telecommunications for Global Information Infrastructure," *Encyclopedia Telecommunications,* Vol. 16, Marcel Dekkers, (to be published in 1998).

[14] 1995 Annual Report of Committee T1-Telecommunications, "Alliance for Telecommunications Industry Solutions," Mar. 1996.

[15] Report of the Sixth Strategic Review Committee on European Information Infrastructure, European Telecommunications Standards Institute, Sophia Antipolis, June 1995.

[16] "Preparations for Promotion of NII of Japan and GII - Proposal for TTC Standardization Activities," Strategic Research and Planning Committee Information Infrastructure Task Group, The Telecommunication Technology Committee, Nov. 1995.

[17] "Draft Report Preparatory Steering Group meeting 29.02.1996," Mar. 1996.

[18] ITU-T JRG GII, TD 33-4 (GII), June 1997.

[19] ISO/IEC JTC 1/SWG-GII, "ISO/IEC JTC 1 GII Roadmap: Guidelines for Evolution, Management and Development of GII Standards," June 1997.

List of Acronyms and Abbreviations

AAL	ATM Adaptation Layer
AB	Address Broadcaster
ABR	Available Bit Rate
ACK	ACKnowledgment
ADM	Add/Drop Multiplexer
ADSL	Asymmetrical Digital Subscriber Line
AF	Address Filter
AH	Authentication Header
AIR	Additive Increment Rate
ANSI	American National Standards Institute
AOW	Asia-Oceania Workshop
AP	Access Point
API	Application Programming Interface
APS	Automatic Protection Switching
AQ	Address Queue
AR(1)	1st Order Auto Regressive Process
ARPA	Advanced Research Project Agency
ASN.1	Abstract Syntax Notation 1
ATC	ATM Transfer Capability
ATD	Asynchronous Time Division
ATM	Asynchronous Transfer Mode
ATMF	ATM Forum
ATMOS	ATM Transport Networks Management and Operations System

AU	Administrative Unit
AU-n	Administrative Unit-n
AVMMS	Audio-Visual Multi-Media Services
AVR	Asymptotic Variance Rate
BBN	Buffered Banyan Network
BCN	Broadcast Channel Number
BEO	Basic Engineering Object
B-ICI	Broadband Inter-Carrier Interface
Bi-CMOS	Bipolar Complementary Metal Oxide Semiconductor
B-ISDN	Broadband-Integrated Services Digital Network
BLSR	Bi-directional Line Switched Ring
BMAP	Batch Markovian Arrival Process
BML	Business Management Layer
BS	Base Station or Base Station Controller (BSC)
BSS	Basic Service Set or Broadband Switching System
CA	Certificate Authority
CAC	Connection Admission Control
CALEA	Communications Assistance for Law Enforcement Act
CAM	Content Addressable Memory
CATV	Community Antenna Television or Cable Television
CBR	Constant Bit Rate
CBT	Core Based Tree
CC	Communication Control
CCIR	Consultative Committee of International Radiocommunication
CCITT	Comite Consultatif Internationale de Telegraphique et Telephonique
CCR	Current Cell Rate
CCSN	Common Channel Signaling Network
CDMA	Code-Division Multiple Access
CDPD	Cellular Digital Packet Data
CDV	Cell Delay Variations
CDVT	Cell Delay Variation Tolerance
CEPT	Conference of European Post and Telecommunications Administrations

CF	Common Facility
CFP	Call For Proposal
CH	Correspondent Host
CI	Congestion Indicator
CISPR	Comite International Special des Perturbations Radioelectriques
CLEC	Competitive Local Exchange Carrier
CLP	Cell Loss Priority
CLR	Cell Loss Ratio
CMIP	Common Management Information Protocol
CMIS	Common Management Information Service
CMISE	Common Management Information Service Element
CMOS	Complementary Metal Oxide Semiconductor
CMTP	Continuous Media Transport Protocol
CNM	Customer Network Management
CNU	Coaxial Network Unit
CO	Connection Oriented
CORBA	Common Object Request Broker Architecture
COS	Corporation for Open Systems International
CP	Content Provider
CPE	Customer Premise Equipment
CPS	Content Provider System
CPU	Central Processing Unit
CRAN	Cyclic Running Address Network
CS	Circuit Switching
CSCW	Computer Supported Cooperative Work
CSMA/CA	Carrier Sense Multiple Access with Collision Avoidance
CSMA/CD	Carrier Sense Multiple Access with Collision Detection
CTD	Cell Transfer Delay
CTP	Connection Termination Point
CTS	Clear To Send
CUG	Closed User Group
DAC	Dual Attachment Concentrator
DAE	Dummy Address Encoder
DAN	Desk Area Network
DAS	Dual Attachment Station

DAVIC	Digital Audio-Visual Council
DB	Data Base
D-BMAP	Discrete Time Batch Markovian Arrival Process
DBMS	Data Base Management System
DBR	Deterministic Bit Rate
DBS	Direct Broadcast Satellite
DCC	Data Communication Channel
DCE	Distributed Computing Environment
DCF	Data Communication Function or Distributed Coordination Function
DCME	Digital Circuit Multiplication Equipment
DCN	Data Communication Network
DCS	Digital Cross-Connect System
DES	Data Encryption Standard
DFP	Distributed Functional Plane
DFWMAC	Distributed Foundation Wireless Media Access Control
DHT	Distributed Home Theater
DIFS	Distributed Coordination Function Inter-Frame Space
DII	Dynamic Invocation Interface
DNS	Domain Name Service
DP	Demand Priority
DPE	Distributed Processing Environment
DS	Delivery System or Distribution System
DSA	Directory Service Agent
DSI	Dynamic Skeleton Interface
DSL	Digital Subscriber Line
DSM-CC UN	Digital Storage Media Command and Control User-Network
DSM-CC UU	Digital Storage Media Command and Control User-to-User
DSRM	DAVIC System Reference Model
DSS2	Digital Subscriber Signaling System 2
DTL	Destination Transit List
DTMF	Dual Tone Multiple Frequency
EACEM	European Association of Consumer Electronics Manufacturers

EBB	Exponentially Bounded Burstiness
EC	End Customer
ECTEL	European Telecommunications and Professional Electronics Industry
EDF	Earliest Deadline First
EDFA	Erbium-Doped-Fiber Amplifier
EDL	Event Definition Language
EFCI	Explicit FCI
EII	European Information Infrastructure
EML	Element Management Layer
EPRCA	Enhanced PRCA
EPRCA	Enhanced Proportional Rate Control Algorithm
ER	Explicit Rate
ESIOP	Environment Specific Inter-ORB Protocol
ESP	Encapsulating Security Payload
ESS	Extended Service Set
ETNO	European Public Telecommunications Network Operations Association
ETSI	European Telecommunications Standards Institute
EUROBIT	European Association of Business Machines Manufacturers and Information Technology Industry
EWMA	Exponentially Weighted Moving Average
EWOS	European Workshop for Open Systems
FA	Foreign Agent
FBM	Fractional Brownian Motion
FCI	Forward Congestion Indicator
FDDI	Fiber Distributed Data Interface
FDMA	Frequency Division Multiple Access
FDT	Formal Description Technique
FIFD	First in First Drop
FP	Feedback Priority
FPLMTS	Future Public Land Mobile Telecommunication Systems
FQ	Fair Queueing
FRF	Frame Relay Forum or Frequency Reuse Factor
FSN	Full Service Network
FTTC	Fiber To The Curb

FTTH	Fiber To The Home
FTP	File Transfer Protocol
Gbps	Gigabits per second
GCRA	Generic Cell Rate Algorithm
GDMO	Guideline for Definition of Managed Object
GII	Global Information Infrastructure
GIOP	General Inter-ORB Protocol
GIS	Global Information Society
GOP	Group Of Picture
GPC	General Policy Committee
GPS	Generalized Processor Sharing or Global Positioning System
GSC	Global Standardization Collaboration
GSM	Global System for Mobility
GSS	Group Sweeping Strategy
GSTN	General Switched Telephone Network
GUI	Graphical User Interface
GUI-WS	GUI-Workstation
HA	Home Agent
HDT	Host Digital Terminal
HE	Head End
HEC	Header Error Control
HFC	Hybrid Fiber Coax
HLSG	High Level Strategy Group
HOL	Head Of Line
HTTP	HyperText Transfer Protocol
IAFB	Idle Address FIFO Buffer
IBT	Intrinsic Burst Tolerance
IC	International Carrier
ICT	Information and Communication Technology
IDC	Index of Dispersion for Counts
IDL	Interface Definition Language
IDMR	Inter-Domain Multicast Routing
IDSP	Intelligent Dynamic Service Provisioning

IEC	International Electrotechnical Commission
IEEE	Institute of Electrical and Electronics Engineers
IEICE	Institute of Electronics, Information and Communication Engineers
IESG	Internet Engineering Steering Group
IETF	Internet Engineering Task Force
IFS	Inter-Frame Space
IIOP	Internet Inter-ORB Protocol
IISP	Information Infrastructure Standards Panel
IITG	Information Infrastructure Task Group
ILMI	Interim Local Management Interface
IMA	Interactive Multimedia Association
IMT 2000	International Mobile Telecommunications 2000
IMUI	International Mobile User Identity
IN	Intelligent Network
IN CS-1	IN Capability Set 1
INTAP	Standards Promotion and Application for Information Processing
int-serv	Integrated Services
I/O	Input/Output
IP	Information Provider or Internet Protocol
IPC	Input Port Controller
IPP	Input Port Processor
IPR	Intellectual Property Right
IPS	Information Provider System
IPSEC	Internet Protocol SECurity Protocol
IPv6	Internet Protocol Version 6
IS	Interim Standard
ISDN	Integrated Services Digital Network
ISM	Industrial Scientific and Medical
ISO	International Organization for Standardization
ISOC	Internet Society
ISP	Internet Service Provider or Intermediate Service Provider
ITM	Intelligent Trouble Management
ITMS	Intelligent Trouble Management System
ITSC	Inter-regional Telecommunication Standardization Conference

ITU	International Telecommunication Union
ITU-T	ITU-Telecommunication Standardization Sector
IUS	Information User System
JCG	Joint Coordination Group
JDK	Java Development Kit
J-EDD	Jitter Early Due Date
JMF	Java Media Framework
JRG	Joint Rapporteur Group
JTC1	Joint Technical Committee 1
JW	Jumping Window
Kbps	Kilobits per second
KDC	Key Distribution Center
LAES	Lawful Authorized Electronic Surveillance
LAN	Local-Area Network
LANE	LAN Emulation
LB	Leaky Bucket
LC	Late Counter
LCD	Loss of Cell Delineation
LEC	Local Exchange Carrier
LGN	Logical Group Node
LIFD	Last In First Drop
LIS	Logical IP Subnet
LMDS	Local Multipoint Distribution Service
LP	Local Priority
LR	Latency-Rate
MAC	Media Access Control
MAN	Metropolitan-Area Network
MAU	Medium Attachment Unit
Mbps	Megabits per second
MBS	Maximum Burst Size
MCBP	Multiple Class Buffer Priority Algorithm
MCC	Multicast Cell Counter
MCN	Multimedia Communication Network

MCR	Minimum Cell Rate
MD	Mediation Device
MF	Mediation Function
MGN	Multicast Grouping Network
MH	Mobile Host
MHEG	Multimedia and Hypermedia Information Coding Experts Group
MIB	Management Information Base
MII	Media Independent Interface
MIT	Management Information Tree
MLLB	Multi-Level Leaky Bucket
MMBP	Markov Modulated Bernoulli Process
MMDS	Multichannel Multipoint Distribution Service
MMPP	Markov Modulated Poisson Process
MO	Managed Object
MP	Multicast Pattern
MPEG	Moving Picture Experts Group
MRDB	Memory Resident DB
MS	Multiplex Section
MSC	Mobile Switching Center
MSOH	Multiplex Section OverHead
MSS	Multimedia System Services
MTT	Multicast Translation Table
MUX	Mutiplexer
MW	Moving Window
NAC	Null Attachment Concentrator
NAV	Network Allocation Vector
NBMA	Non-Broadcast Multi-Access
NCCE	Native Computing and Communications Environment
NE	Network Element
NEF	Network Element Function
NHRP	Next Hop Resolution Protocol
NIC	Network Interface Card
NII	National Information Infrastructure
NMF	Network Managment Forum
NML	Network Management Layer

NMS	Network Management System
NM-SIG	Network Management Special Interest Group
NNI	Network Node Interface
NP	Network Provider or Navigation Provider or Nondeterministic Polynomial
NPC	Network Parameter Control
NS	Next Station
NT	Network Terminal
NW	Network
OA	Office Automation
OAM or OA&M	Operation, Administration, and Maintenance (or Management)
OAM&P	Operation, Administration, Maintenance, and Provisioning
ODN	Optical Distribution Network
ODP	Open Distributed Processing
OMA	Object Management Architecture
OMG	Object Management Group
ONU	Optical Network Unit
OODB	Object Oriented DB
OPC	Output Port Controller
OPP	Output Port Processor
OpS	Operating System
ORB	Object Request Broker
OS	Object Service or Operating System
OSF	Operation Systems Function or Open Software Foundation
OSI	Open Systems Interconnection
OSPF	Open Shortest Path First
PAS	Publicity Available Specification
PCF	Point Coordination Function
PC/NT	Personal Computer/Network Terminal
PCR	Peak Cell Rate
PCS	Personal Communications Systems
PDA	Personal Digital Assistant
PDH	Plesiochronous Digital Hierarchy
PEM	Privacy Enhanced Mail

PG	Peer Group
PGID	Peer Group IDentifier
PGL	Peer Group Leader
PGP	Pretty Good Privacy
PGPS	Packet-By-Packet Generalized Processor Sharing
PIFS	Point Coordination Function Inter-Frame Space
PIM	Personal Information Management
PIM-DM	Protocol Independent Multicast-Dense Mode
PIM-SM	Protocol Independent Multicast-Sparse Mode
PIO	Protected Information Object
PLC	Planar Lightwave Circuit
PLL	Phase Lock Loop
PLMN	Public Land Mobile Network
PMS	Path Management System
P-NNI	Private Network-to-Network Interface
POH	Path OverHead
POT	Push Out with Threshold
POTS	Plain Old Telephone Service
PR	Packet Radio
PRCA	Proportional Rate Control Algorithm
PREMO	PResentation Environment for Multimedia Objects
PRM	Protocol Reference Model
PRMA	Packet Reservation Multiple Access
PS	Previous Station
PSTN	Public Switched Telephone Network
PTSPs	P-NNI Topology State Packets
PUI	Personal User Identity
PVC	Permanent Virtual Channel or Circuit
PVP	Permanent Virtual Path
QA	Q Adaptation
QAF	Q-Adaptation Function
QM	Quality of Service Manager
QoS	Quality of Service
QoS-A	Quality of Service Architecture
RAM	Random Access Memory

RAND	RANDomly Selected
RAR	Read Address Register
RAT	Resource Administration Technique
RCAP	Real-Time Channel Administration Protocol
RDB	Relational DB
RED	Random Early Detection
REM	Rate Envelope Multiplexing
RF	Radio Frequency
RFC	Request For Comments
RFI	Request For Information
RIP	Routing Information Protocol
RM	Routing Module or Resource Management
RM-ODP	Reference Model for Open Distributed Programming (or Environment)
RTT	Round Trip Time
ROA	Recognized Operating Agency
RPC	Remote Procedure Call
RRC	Radio Resource Control
RS	Regenerator Section
RSA	An algorithm named after Rivest, Shamir and Adleman
RSOH	Regenerator Section OverHead
RSVP	Resource Reservation Setup Protocol
RTIP	Real Time Internet Protocol
RTP	Real Time Protocol
RTS	Request To Send
RTU	Real Time Upcall
SA	Security Agent
SAC	Single Attachment Concentrator
SAP	Service Access Point
SAS	Single Attachment Station
SBM	Shared Buffer Memory
SCFQ	Self-Clocked Fair Queueing
SCR	Sustainable Cell Rate
SDH	Synchronous Digital Hierarchy
SDLC	Synchronous Data Link Control
SDS	Space Division Switching

SHD	Super High Definition
SHS	Scenario Handling System
SIFS	Short Inter-Frame Space
SIO	Scientific or Industrial Organization
SIR	Signal to Interference Ratio
SL0	Service Layer Zero
SLA	Service Level Agreement
SLT	Subscriber Line Terminal
SMDS	Switched Multi-Megabit Data Service
SME	Small and Medium-Sized Enterprise
SMK	Shared Management Knowledge
SML	Service Management Layer
SMS	Service Management System
SNMP	Simple Network Management Protocol
SO	Serivce Order
SOH	Section OverHead
SONET	Synchronous Optical NETwork
SP/IP	Service Provider/Information Provider
SPIRIT	Service Provider Integrated Requirements for Information Technology
SPS	Service Provider System
SS	Standardization Sector
SS7	Signaling System No. 7
SSL	Secure Sockets Layer
SSM	Small Switch Modules
STB	Set-Top Box
STM	Synchronous Transfer Mode
STM-N	Synchronous Transport Module Level-N
STP	Shielded Twisted Pair
STS	Space-Time-Space
STU	Set Top Unit
SUS	Service User System
SVC	Switched Virtual Circuit or Connection
SW	Sliding Window
SWE	SWitch Element
SWG	Sub-Working Group
T1	Committee T1

TC	Technical Committee
TCP/IP	Transmission Control Protocol/Internet Protocol
TD	Tail Dropping
TDM	Time-Division Multiplexing
TDMA	Time-Division Multiple Access
TDPE	TINA Distributed Processing Environment
THT	Token Hold Time
TINA	Telecommunications Information Networking Architecture
TJW	Triggered Jumping Window
TM	Traffic Management
TMN	Telecommunication Management Network
TNMS	Telecommunication Network Management System
TRT	Token Rotation Timer
TSAG	Telecommunication Standardization Advisory Group
TSB	Trusted System Block
TSP	Telecommunications Service Provider
TTA	Telecommunication Technology Association
TTC	Telecommunication Technology Committee
TTP	Trial Termination Point
TTRT	Target Token Rotation Time
TU	Trusted User
TV	Television
UBR	Unspecified Bit Rate
UDP	User Datagram Protocol
UI	User Instance
UIM	User Identity Module
UNI	User Network Interface
UPC	Usage Parameter Control
UPSR	Unidirectional Path Switched Ring
UPT	Universal Personal Telecommunications
UPTAC	UPT Access Node
UPTAN	UPT Access Number
UTP	Unshielded Twisted Pair
UU	Untrusted User
VAR	Variance

VBR	Variable Bit Rate
VC	Virtual Container or Virtual Clock or Virtual Channel
VCC	Virtual Channel Connection
VCI	Virtual Channel Identifier
VOD	Video On Demand
VP	Virtual Path
VP-AIS	Virtual Path-Alarm Indication Signal
VPC	Virtual Path Connection
VPI	Virtual Path Identifier
VPN	Virtual Private Network
VPT	VP Termination Point
VT	Virtual Tributary
VWP	Virtual Wavelength Path
WF^2Q	Worst-Case Fair Queueing
WAN	Wide-Area Network
WDM	Wavelength-Division Multiplexing
WP	Wavelength Path
WRR	Weighted Round Robin
WSF	Workstation Function
WTSC	World Telecommunication Standardization Conference
WUGS	Washington University Gigabit Switch
WWW	World Wide Web
XC	Cross-Connect system
XRM	eXtended Integrated Reference Model

About the Authors

Bhumip Khasnabish (Editor) is a principal member of the technical staff at GTE Labs. in Waltham, Massachusetts, where he works on various multimedia, wireless, and enterprise networking projects. He earned the Ph.D. degree (1992) in electrical engineering from UW, Canada, and received the Roy Saheb Memorial Gold Medal (1975) and Commonwealth Scholarship (1984–1988). From 1992 to 1995, Dr. Khasnabish worked on and led various trunking and traffic management software development projects of Northern Telecom's award (in 1995 NetWorld+Interop in Atlanta) winning Magellan Passport Frame-Cell-Switching (FCS) product. Prior to that he worked as a research and teaching assistant at the University of Windsor, University of Waterloo, and McMaster University, Hamilton, all in Ontario, Canada. From 1982 to 1984, he worked as a lecturer in the Department of Electrical and Electronic Engineering of the Bangladesh University of Engineering and Technology (BUET), Dhaka, Bangladesh. During that time he also worked as an instructor for courses on computer and microprocessor programming, and as a member of the Electricity Licensing Board (ELB) of the Bangladesh Power Development Board (BPDB). He is the author of more than 75 articles published in various IEEE and other international archival journals/magazines and refereed conference proceedings. He also contributes to the WATM, TM, and SAA-AMS working groups of the ATM forum. He is the information director of ACM SIG-Mobile and the general chair of IEEE ComSoc's EnterNet TC sponsored 1998 conference on Enterprise Networking and Computing. Dr. Khasnabish's home page is at http://www1.acm.org:82/~bhumip/; his e-mail addresses are: bhumip@gte.com or bhumip@acm.org or b.khasnabish@ieee.org. Bhumip is a senior member of the IEEE.

Mallikarjun Tatipamula (Editor) is a senior systems engineer in Ground Systems Division, Cellular Networks and Space Sector, Motorola, Phoenix, AZ. He is involved in Gateway Systems Design for Satellite/Terrestrial Telecommunication Networks. Prior to joining Motorola in July, 1997, he was a senior member of the scientific staff and team leader at Northern Telecom (Nortel), Ottawa, Canada. While with Nortel, he was involved in various projects of the Local Access and Transport Networks Design and Development Group, Broadband Networks Division, and CDMA Development Group, Wireless Networks Division. At Nortel, he was also a prime technical consultant to leading telecommunications service providers in Canada and the United States. From 1990 to 1992, he was a senior project officer in the Electrical Engineering Department at the Indian Institute of Technology, Madras, India. He was a group leader and a systems designer, responsible for design and development of high speed lightwave systems. From March 1990 to August 1990, he was with Indian Telephone Industries, Bangalore, India, as an assistant executive engineer, in R&D Transmission Laboratories, involved in design and development of high speed optical receivers. He received his Master of Technology in electrical engineering with a specialization in communications systems and high frequency technologies, from the Indian Institute of Technology, Madras, India, in January 1990 and Bachelor of Technology in electronics and communications engineering, from the Regional Engineering College, Warangal, India, in May 1988. He has organized and chaired technical sessions in the area of broadband networks and also has served as a member of technical and organizing committees of several international conferences.

Koichi Asatani (Chapter 11) received B.E., M.E., and Ph.D. degrees from Kyoto University in 1969, 1971, and 1974, respectively. He joined Nippon Telegraph and Telephone (NTT) Public Corporation in 1974. He has been engaged in R&D on fiber optic transmission systems including fiber optic subscriber loop systems, high-definition television transmission systems, ISDN, B-ISDN, and R&D strategic planning on these technologies. He was executive manager, Technical Standards and Strategy, R&D Headquarters, and was responsible for corporate strategy on technical standardizations during 1993–1996. He has been vice-chair of ITU-T Study Group 13 (formerly CCITT SG XVIII) since 1989. He is currently professor in the Department of Electronic Engineering, Faculty of Engineering, Kogakuin University, Tokyo, Japan. He was vice-chair of Asian ISDN Council (AIC) during 1988–1993. He is also currently chair of the Strategic Research and Planning Committee, The Telecommunication Technology Committee (TTC), in Japan. He is a co-author of *Optical Fiber Communications* (Shokodo Publishing Co.) and *Introduction* *to B-ISDN* (John Wiley & Sons). He is a member of IEEE and IEICEJ. His email address is asatani@sin.cc.kogakuin.ac.jp.

V. Michael Bove, Jr. (Chapter 1) holds an S.B.E.E., an S.M. in visual studies, and a Ph.D. in media technology, all from the Massachusetts Institute of Technology, where he is currently head of the object-based media group at the Media Laboratory. He is the author or co-author of more than 40 journal or conference papers on digital television systems, video processing hardware/software design, multimedia, scene modeling, and optics. He holds patents on inventions relating to video recording and hardcopy, and has been a member of several professional and government committees. In December 1995, *Boston Magazine* named him one of the "People Shaping Boston's High-Tech Future." He is on the Board of Editors of the *Journal of the Society of Motion Picture and Television Engineers*, and served as general chair of the 1996 ACM multimedia conference.

Andrew T. Campbell (Chapter 3) is an assistant professor in the Department of Electrical Engineering at Columbia University and a member of the COMET Group at the Center for Telecommunications Research. He is currently leading a new research effort in wireless media systems focusing on the development of QoS-aware middleware for mobile networks that comprise ad hoc, broadband, and Internet technologies. Dr. Campbell is currently a co-chair of the IFIP Fifth International Workshop on Quality of Service.

H. Jonathan Chao (Chapter 5) received his B.S.E.E. and M.S.E.E. degrees from National Chiao Tung University, Taiwan, in 1977 and 1980, respectively, and the Ph.D. degree in electrical engineering from Ohio State University, Columbus, in 1985. He is professor of electrical engineering at Polytechnic University, New York, which he joined in January 1992. His areas of research include large-scale multicast ATM switches, gigabit switch routers, multimedia services over IP/ATM networks, and congestion/flow control in ATM networks. He holds 16 patents and has published more than 60 journal and conference papers in his subject areas. From 1985 to 1991, he was a member of the technical staff at Bellcore, New Jersey, where he was involved in prototyping an experimental broadband local access system. He proposed various architectures for SONET/ATM networks and implemented them in high-speed ASIC chips. He received the Bellcore Excellence Award in 1987. From 1977 to 1981, he worked at Taiwan Telecommunication Laboratories, where he was engaged in the development of a digital switching system. He served as a guest editor for the *IEEE Journal on Selected Areas in Communications* with special topic on Advances in ATM Switching Systems for B-ISDN, June 1997. He is currently serving as editor for the *IEEE/ACM Transactions on Networking*. He has been giving short courses to industry on IP/ATM Networks: Switching, Routing, and Internetworking.

Reuven Cohen (Chapter 4) received the B.Sc., M.Sc., and Ph.D. degrees in computer science from the Technion–Israel Institute of Technology in 1986, 1988, and 1991, respectively. From 1991 to 1993, he was with IBM T. J. Watson Research Center working on protocols for ATM and high-speed LANs. Since 1993, he has been with the Department of Computer Science, at the Technion–Israel Institute of Technology. His most recent work focuses on the design of protocols for routing, multicasting, and transport. In the last three years he has also been working with HP Labs on residential broadband networking.

Patrick Dowd (Chapter 2) was born in Buffalo, NY and received the B.A. (1983) in electrical engineering and computer science from the State University of New York at Buffalo; and the M.S. (1985) and Ph.D. (1988) degrees in electrical engineering from Syracuse University.

Dowd was with IBM Corporation between 1983 and 1989 as a staff engineer with System Design in Processor Development at the IBM Endicott Laboratory. His principal design effort was in the area of communication subsystems for future computer systems, involving token ring, FDDI, and ISDN. Early assignments at IBM were in the areas of processor design: microcode, fault detection, isolation, and diagnosability. Dowd returned to Buffalo in 1989 and joined the Department of Electrical and Computer Engineering at the State University of New York at Buffalo. In 1996 Dowd joined the Department of Electrical Engineering and UMIACS at the University of Maryland at College Park and also holds the position of senior research scientist at the Advanced Networking Research Department with the U.S. Department of Defense in Ft. Meade, MD.

Dowd has served as conference/general chair, program chair, and program committee member of many conferences during the last 10 years in the area of telecommunications, high-speed networking, simulation and performance analysis, and computer architecture. He has served as a guest editor of special issues related to both high-speed networking and also network security. Patrick serves on the editorial board of the *International Journal in Computer Simulation* and is associate editor of the SCS Transactions on Simulation. He currently serves on the executive committee of ACM SIGCOMM. Patrick is a member of the IEEE and ACM with research interests in cluster-based computing, distributed shared memory architectures, WDM and ATM networks, and optical communication.

Nelson L. S. Fonseca (Chapters 4 and 10) received an electrical engineer degree (1984) and an M.Sc. in computer science (1987) degree from Pontificial Catholic University at Rio de Janeiro, Brazil, and the M.Sc. (1993) and Ph.D. (1994) degrees in computer engineering from the University of Southern California, Los Angeles. He held lecturer positions at Pontificial Catholic University (1985–1987) and worked in the Computer Communications group of IBM Rio Scientific Center (1989). He is the recipient of the USC International Book award and of the Brazilian Computing Society First Thesis and Dissertations award. Dr. Fonseca was recently listed in *Marqui's Who Is Who in the World*, 1997 edition. In 1995, he joined the Institute of Computing of The State University of Campinas, Brazil, where he has been as an assistant professor. He has served as a member of technical and organizing committees of several national and IEEE conferences. Dr. Fonseca's main interests are ATM traffic control, queueing network models, and video on demand.

Mario Gerla (Chapter 8) was born in Milan, Italy. He received a graduate degree in engineering from the Politecnico di Milano, in 1966, and the M.S. and Ph.D. degrees in engineering from UCLA in 1970 and 1973, respectively. He joined the faculty of the UCLA Computer Science Department in 1977. His research interests cover the performance evaluation, design, and control of distributed computer communication systems and high-speed computer networks (ATM, optical networks and wireless). He is currently involved in the design and evaluation of network protocols and strategies for wireless, mobile, and multimedia communications systems under sponsorship of the government and industry.

Dale A. Harris (Chapter 1) is currently executive director of the Center for Telecommunications at Stanford University and professor of electrical engineering (consulting). The center is composed of faculty from the Departments of Computer Science, Electrical Engineering, and Engineering Economic-Systems and Operations Research. Professor Harris's own research interest is in the area of high-speed networking applications, particularly distance learning. Prior to joining Stanford, Dr. Harris held management positions with Pacific Bell, Bank of America, and the Department of the Army. He has served on the faculty of Harvard University and the visiting faculty of the California Institute of Technology. Dr. Harris received his B.S. degree in engineering science from the University of Texas at Austin and his M.S. and Ph.D. degrees in electrical engineering and computer science from the University of California at Berkeley. He is a member of the New York Academy of Sciences and a senior member of the IEEE.

Masaharu Kaihara (Chapter 7), senior research engineer, supervisor, received the B.S. and M.S. degrees in precision engineering from Hiroshima University in 1975 and in 1977, respectively. He joined the Customer Equipment Development Laboratory, NTT Yokosuka Electrical Communication Laboratories, in 1977, where he was involved in research and development of the magnetic card public telephone from 1977 to 1987. From 1988 to 1990, he was engaged in the research planning in NTT Telecommunication Networks Laboratories. From 1990 to 1994, he was involved in research on operation systems engineering and operation architecture in NTT Network Information Systems Laboratories and has been involved in research and development of multimedia service operations system since 1994 at the NTT Information and Communication Systems Laboratories. He is a member of the Institute of Electronics, Information and Communication Engineers (IEICE) of Japan.

Randy H. Katz (Chapter 8) received the A.B. degree from Cornell University in 1976, and an M.S. and Ph.D. degree from the University of California, Berkeley, in 1978 and 1980, respectively. After a year in industry, he taught at the University of Wisconsin–Madison, joining the Berkeley faculty in 1983. He now holds the title of United Microelectronics Corporation Distinguished Professor of Electrical Engineering and Computer Science and chairman of the department. He is a fellow of the ACM and the IEEE. Professor Katz has published more than 140 papers on various aspects of computing system design, including databases, computer-aided design, multiprocessor systems, disk arrays, multimedia, wireless communications, and mobile computing. During 1993–1994, he was on assignment at the Defense Advanced Research Projects Agency, where one of his duties was to wire the White House to the Internet.

Eugene Lubchenko (Chapter 9) has an international background in research, development, design, and production of electronic systems and products. He began his career in Canada after graduating with a B.S.E.E. degree from the University of Miami, Florida. Since then he has managed projects and engineering organizations in both consumer and industrial sectors in the United States, Canada, and the Netherlands. His present responsibilities at Bell Atlantic in White Plains, New York, include system analysis, specification, and model development. The scope of the work includes architectural as well as security issues.

Edmundo R. M. Madeira (Chapter 10) is an assistant professor in the Institute of Computing at State University of Campinas (UNICAMP), Brazil. He received his Ph.D. degree in electrical engineering at UNICAMP in 1991. He currently is a coordinator member of the multiware platform and head of the

Computer Systems Department. His interest areas include open distributed processing, multimedia communication, and distributed system management.

Masahiko Matsushita (Chapter 7) is a professor of the Division of Information Engineering, Maebashi Institute of Technology, Maebashi-shi, Japan. Prior to that, at NTT's Laboratories, he was engaged in the R&D of digital networks. He has spent most of the last 10 years in the R&D of network operation and management. He was active in the international and domestic standards arenas. He was a vice-chairman of ITU-T SG 4 from 1989 to 1996. He was in the TMN Experts Group of ITU-T for 10 years. He was also active at the technical team of NM forum. He received the B.E., M.E., and D.E. degrees from Tohoku University, Sendai, Japan, in 1968, 1970, and 1995, respectively. He is a member of IEEE, IEICE of Japan, and the Information Processing Society of Japan.

Osamu Miyagishi (Chapter 7) received the B.E. degree in electronics engineering from the University of Hokkaido in 1973. In 1973 he joined Musashino Electrical Communication Laboratories, NTT. From 1973 to 1987, he was engaged in research and development of teleprinter and telephone sets. Since 1987 he has been in charge of research of network architecture and operation architecture for multimedia services. He is now in charge of promotion of R&D products in Hokkaido prefecture. He is a member of the Institute of Electronics, Information and Communications Engineers (IEICE) of Japan and the IEEE Communication Society.

Noriharu Miyaho (Chapter 1) received the B.E. and Ph.D. degrees in electrical engineering from the University of Electro-Communications, Tokyo, Japan, in 1974 and 1997, respectively. He is currently a director in the Global R&D Strategic Planning and Promoting Office, NTT Research and Development Headquarters. He is responsible for the R&D strategic aspects of developing enterprise networks and multimedia applications. Since joining the NTT Communication Laboratory in 1974, he has been engaged in research on time-division data switch design for the digital data switching system (DDX), VLSI processor design, the packet/circuit hybrid switching systems, and, more recently, the research and development of ATM switching systems and ATM network management and operation systems for B-ISDN as the Research Group Leader in the Network Service Systems Laboratories. His Ph.D. work focused on digital switching system architecture and design principles for handling multimedia traffic. He has been the guest speaker and chairman at many international conferences and workshops. He received the NTT President Award for his contribution to the development of an ATM switching system in 1994 and 1996. Dr. Miyaho is a member of IEEE, IEICE (The Institute of Electronics, Information and Communication Engineers), and IPS (Information Processing Society) of Japan.

Takahiko Mori (Chapter 7) received B.E. and M.E. degrees in electronics engineering from Nagoya University in 1977 and 1979, respectively. He joined the Yokosuka Electrical Communication Laboratory, NTT, Japan, in 1979. From 1979 to 1988, he was engaged in research and development of remote maintenance systems for computer hardware and software. Since 1988 he has been in charge of research and development of network and/or service operation and management systems for multimedia services. He is a member of the Institute of Electronics, Information and Communications Engineers (IEICE) of Japan and of the Information Processing Society of Japan.

Shinya Nogami (Chapter 11) received B.E., M.E., and Dr.Eng. degrees in communication engineering from Tohoku University in 1979, 1981, and 1984, respectively. In 1984, he joined Nippon Telegraph and Telephone (NTT) Public Corporation Laboratories, where he was engaged in research on traffic evaluation, traffic design, and system architecture. He was responsible for corporate strategy on technical standardizations in Technical Standards and Strategy, R&D Headquarters during 1994–1996. He is currently senior research engineer, Communication Assessment Laboratory, Multimedia Networks Laboratories, NTT, Tokyo, Japan. He is a member of IEICEJ and Information Processing Society of Japan. His e-mail address is nogami@hashi.tnl.ntt.co.jp.

Hody Salloum (Chapter 6) has been with Bell Communications Research (Bellcore) since 1986. He is currently director, Broadband Access Department, holding management and business development responsibilities with a focus on international markets and projects in the areas of subscriber local loop access, specifically full-service networks, interactive multimedia services, systems integration, wireless cable (LMDS and MMDS), asymmetric digital subscriber line (ADSL), xDSL, fiber-to-the-home (FTTH), fiber-to-the-curb (FTTC), and hybrid fiber coax (HFC). Prior to his current position, he worked on multiple teams providing consulting services focusing on the network platforms that offer broadband services to residential and business end users. Areas of work included end-to-end systems engineering, powering, traffic engineering, multiple access protocols, and voice frequency analog transmission. He also led the team that planned and executed the fiber-to-the-curb/home requirements document (TR-909) to include requirements that enable the delivery of broadband video services. He also authored many chapters of TR-909, the first system-level requirements documents for fiber-in-the-loop (FITL) systems. He was also responsible for Bellcore's technical involvement in all national and international standards deliberations on FITL system issues. He authored many key technical contributions for ITU-T and ANSI T1 Committees. Also, he served as editor for ITU-TFITL Recommendation (ITU-T SG 15/24). Prior to Bellcore, Dr. Salloum worked as a power supply design engineer at Preferred Electronics, leading a small team to design custom switch-mode and ferro-resonant power supplies and magnetics. Dr. Salloum holds a B.S., M.S., and Ph.D. in electrical engineering from Stevens Institute of Technology. He is a member of Tau Beta Pi, senior member of the IEEE, *IEEE Communications Magazine* technical editor since 1994, *IEEE Communications Magazine* feature editor since 1996, and has served IEEE in many capacities such as chairing local chapters, serving on technical committees, holding treasurer responsibility, and editing various papers and publications. He was also guest editor for the April 1995 issue of the *IEEE Communications Magazine.* He has published more than 75 papers, technical memoranda, and documents in his areas of specialty. He also lectures at universities, conferences, and seminars in the area of broadband access.

Ken-ichi Sato (Chapter 6) is currently a senior research engineer and leads the Photonic Network Systems Research Group and Optical Network Systems Laboratories at NTT. His main research activity includes information transport network architectures, variable network architectures with virtual paths, broadband transport technology, ATM network performance analysis, and photonic network technologies that include architectural and OA&M issues and hardware development. He authored/co-authored more than 100 research publications in international journal and conferences. He received the B.S., M.S., and Ph.D. degrees in electronics engineering from the University of Tokyo, Tokyo, Japan, in 1976, 1978, and 1986, respectively. In 1978, he joined Yokosuka Electrical Communication Laboratories, NTT. During 1978–1984, he was engaged in the research and development of optical fiber transmission technologies. His R&D experiences cover fiber optic video transmission systems for CATV distribution systems and subscriber loop systems. Since 1985, he has been active in the development of broadband ISDN based on ATM techniques. Dr. Sato was awarded the NTT Research and Development Award in 1991 and the NTT President Award in 1995 for his contributions to ATM network architecture and system technology development. He received the Young Engineer Award in 1984 and the Excellent Paper Award in 1991 both from the Institute of Electronics, Information and Communication Engineers (IEICE) of Japan. He is a member of the IEICE and a senior member of the IEEE Communications Society.

Kathryn A. Szot (Chapter 9) has been involved with ATM technology planning for Bell Atlantic since 1990. Ms. Szot is currently in the Business Services Technology district where technical support pertaining to ATM, frame relay, SMDS, and IP is provided to sales and product management. Her primary role is to provide ATM expertise in support of Bell Atlantic's Cell Relay Service. Implementation-oriented experience has included providing technical and planning assistance to customers in the education, medical, and government arenas, for applications such as distance learning, telemedicine, and LAN connectivity. This experience also spans a wide variety of supplier ATM wide area and local area networking equipment.

For three and a half years, Ms. Szot had responsibility for the coordination of Bell Atlantic's BISDN/ATM industry standards strategy, activities, and positions. During this time, she was elected to the ATM Forum Board of Directors for two terms. From March 1995 to April 1997, she actively participated in many board activities and held the psotion of VP Worldwide Operations and Management. She has been a regular participant in the ATM Forum Technical Committee since March 1992. Areas of participation have included the Service Aspects and Applications Working Group—Audiovisual Multimedia Services (SAA-AMS); and access signalling work contributing toward User-to-Network Interface (UNI) specification development. She has also been monitoring the work of the ATM Forum Security Working Group since its formation. In 1997 Ms. Szot received the ATM Forum Spotlight Award for her involvement with ATM Forum activities.

Ms. Szot's ATM responsibilities have included technical planning for both residential video services and broadband business data services. Prior to her work in BISDN/ATM, Ms. Szot was involved in architecture planning and protocol work for Bell Atlantic in the area of integrated services digital network (ISDN).

Ms. Szot received her Bachelor of Science degree in mathematics, with a concentration in computer science, from the University of Richmond in 1984.

Shohei Takeuchi (Chapter 7) received B.E. and M.E. degrees in instrumentation engineering from Keio University in 1981 and 1983, respectively. He joined the Yokosuka Electrical Communication Laboratory, NTT, Japan in 1983. From 1983 to 1990, he was engaged in research and development of OSI-based communication control software. Since 1990, he has been in charge of research and development of network and/or service operation and management systems for multimedia services. He is a member of the Information Processing Society of Japan.

Terence D. Todd (Chapter 2) received the B.A.Sc., M.A.Sc., and Ph.D. degrees in electrical engineering from the University of Waterloo, Ontario, Canada, in 1978, 1980, and 1984, respectively. During that time he spent three years as a research associate with the Computer Communications Networks Group at the University of Waterloo. He is currently a professor of electrical and computer engineering at McMaster University in Hamilton, Ontario, and a member of the Communications Research Laboratory (CRL). Dr. Todd spent 1991 on research leave in the Distributed Systems Research Department at AT&T Bell Laboratories in Murray Hill, New Jersey, and is currently head of the Wireless Network Protocols Project in the Telecommunications Research Institute of Ontario (TRIO). Dr. Todd's current research interests include metropolitan/local-area networks, wireless communications, and the performance analysis of computer communication networks and systems. Dr. Todd is a Professional Engineer in the province of Ontario and is currently an editor of the *IEEE/ACM Transactions on Networking.*

Makoto Yoshida (Chapter 7) is executive manager, Network and Systems Operations, Networking Divisions Group, NTT-AT (Advanced Technology) Corporation, currently responsible for the management of computing network/ systems integration and operations services. He holds a B.E. in electronics engineering and a Ph.D. in electrical engineering from the University of Tokyo. Before joining NTT-AT, Dr. Yoshida worked for NTT with considerable experience in the research and developmental fields of teletraffic engineering, network architecture/design, operations and management architecture/systems, including the international standardization arena (CCITT). He has also been involved in the activities of the Network Management Forum since 1988 and is currently chairman of the board of trustees. He is a member of IEEE, the Institute of Electronics, Information and Communication Engineers (IEICE), and the Information Processing Society of Japan. Dr. Yoshida has served as a committee member to international symposiums such as NOMS, DSOM, and IM (formerly ISINM).

Tatsuhiko Yoshida (Chapter 7) received the B.E. degree in electronics engineering from University of Tokyo in 1977. In 1977 he joined the Yokosuka Electrical Communication Laboratories, NTT. During 1977–1981 he was engaged in research and development of digital subscriber transmission systems. During 1982–1986 he was engaged in the development of an N-ISDN prototyping system in the Tokyo area. After having engaged in the development of the operating system of the high-speed digital leased circuit system, he is now active in development of the ATM operation systems. He is also a member of SG4 and SG15 of ITU-T and is working for standardization of the ATM network management.

Moshe Zukerman (Chapter 4) received his B.Sc. in industrial engineering and management and an M.Sc. in operation research from The Technion–Israel Institute of Technology and a Ph.D. degree in electrical engineering from the University of California at Los Angeles (UCLA) in 1985. Dr. Zukerman was an independent consultant with IRI Corporation and a postdoctoral fellow at UCLA during 1985–1986. He joined Telstra Research Laboratories (TRL) in 1986 where he served until March 1997. At TRL he managed a team of researchers conducting research and providing advice to Telstra on design and traffic management of its modern networks, and on traffic aspects of evolving telecommunications standards. He is the recipient of the TRL Outstanding Achievement Award in 1990. During 1990–1997, he also taught and supervised graduate students at Monash University. In March 1997 he joined the Electrical and Electronic Engineering Department at the University of Melbourne where he is responsible for expanding telecommunications research, teaching, and industry collaboration. He served as editor of the *Australian Telecommunications Research Journal* during 1991–1996. Currently he serves as an IEEE JSAC guest editor of the upcoming special issue on Future Voice Technologies, and on the Editorial Board of the *International Journal of Communication Systems.* He is also a senior member of the IEEE. Dr. Zukerman has published more than 100 papers in scientific journals and conference proceedings and has been awarded nine national and international patents. He has served as a session chair and member of technical and organizing committees of numerous national and international conferences.

Index

The Artech House Telecommunications Library

Vinton G. Cerf, Series Editor

Access Networks: Technology and V5 Interfacing, Alex Gillespie

Advanced High-Frequency Radio Communications, Eric E. Johnson, Robert I. Desourdis, Jr., et al.

Advanced Technology for Road Transport: IVHS and ATT, Ian Catling, editor

Advances in Computer Systems Security, Vol. 3, Rein Turn, editor

Advances in Telecommunications Networks, William S. Lee and Derrick C. Brown

Advances in Transport Network Technologies: Photonics Networks, ATM, and SDH, Ken-ichi Sato

An Introduction to International Telecommunications Law, Charles H. Kennedy and M. Veronica Pastor

Asynchronous Transfer Mode Networks: Performance Issues, Second Edition, Raif O. Onvural

ATM Switches, Edwin R. Coover

ATM Switching Systems, Thomas M. Chen and Stephen S. Liu

Broadband: Business Services, Technologies, and Strategic Impact, David Wright

Broadband Network Analysis and Design, Daniel Minoli

Broadband Telecommunications Technology, Byeong Lee, Minho Kang and Jonghee Lee

Cellular Mobile Systems Engineering, Saleh Faruque

Cellular Radio: Analog and Digital Systems, Asha Mehrotra

Cellular Radio: Performance Engineering, Asha Mehrotra

Cellular Radio Systems, D. M. Balston and R. C. V. Macario, editors

CDMA for Wireless Personal Communications, Ramjee Prasad

Client/Server Computing: Architecture, Applications, and Distributed Systems Management, Bruce Elbert and Bobby Martyna

Communication and Computing for Distributed Multimedia Systems, Guojun Lu

Communications Technology Guide for Business, Richard Downey, et al.

Community Networks: Lessons from Blacksburg, Virginia, Andrew Cohill and Andrea Kavanaugh, editors

Computer Networks: Architecture, Protocols, and Software, John Y. Hsu

Computer Mediated Communications: Multimedia Applications, Rob Walters

Computer Telephone Integration, Rob Walters

Convolutional Coding: Fundamentals and Applications, Charles Lee

Corporate Networks: The Strategic Use of Telecommunications, Thomas Valovic

The Definitive Guide to Business Resumption Planning, Leo A. Wrobel

Digital Beamforming in Wireless Communications, John Litva and Titus Kwok-Yeung Lo

Digital Cellular Radio, George Calhoun

Digital Hardware Testing: Transistor-Level Fault Modeling and Testing, Rochit Rajsuman, editor

Digital Switching Control Architectures, Giuseppe Fantauzzi

Digital Video Communications, Martyn J. Riley and Iain E. G. Richardson

Distributed Multimedia Through Broadband Communications Services, Daniel Minoli and Robert Keinath

Distance Learning Technology and Applications, Daniel Minoli

EDI Security, Control, and Audit, Albert J. Marcella and Sally Chen

Electronic Mail, Jacob Palme

Enterprise Networking: Fractional T1 to SONET, Frame Relay to BISDN, Daniel Minoli

Expert Systems Applications in Integrated Network Management, E. C. Ericson, L. T. Ericson, and D. Minoli, editors

FAX: Digital Facsimile Technology and Applications, Second Edition, Dennis Bodson, Kenneth McConnell, and Richard Schaphorst

FDDI and FDDI-II: Architecture, Protocols, and Performance, Bernhard Albert and Anura P. Jayasumana

Fiber Network Service Survivability, Tsong-Ho Wu

Future Codes: Essays in Advanced Computer Technology and the Law, Curtis E. A. Karnow

Guide to Telecommunications Transmission Systems, Anton A. Huurdeman

A Guide to the TCP/IP Protocol Suite, Floyd Wilder

Implementing EDI, Mike Hendry

Implementing X.400 and X.500: The PP and QUIPU Systems, Steve Kille

Inbound Call Centers: Design, Implementation, and Management, Robert A. Gable

Information Superhighways Revisited: The Economics of Multimedia, Bruce Egan

Integrated Broadband Networks, Amit Bhargava

International Telecommunications Management, Bruce R. Elbert

International Telecommunication Standards Organizations, Andrew Macpherson

Internetworking LANs: Operation, Design, and Management, Robert Davidson and Nathan Muller

Introduction to Document Image Processing Techniques, Ronald G. Matteson

Introduction to Error-Correcting Codes, Michael Purser

An Introduction to GSM, Siegmund Redl, Matthias K. Weber and Malcom W. Oliphant

Introduction to Radio Propagation for Fixed and Mobile Communications, John Doble

Introduction to Satellite Communication, Bruce R. Elbert

Introduction to T1/T3 Networking, Regis J. (Bud) Bates

Introduction to Telephones and Telephone Systems, Second Edition, A. Michael Noll

Introduction to X.400, Cemil Betanov

LAN, ATM, and LAN Emulation Technologies, Daniel Minoli and Anthony Alles

Land-Mobile Radio System Engineering, Garry C. Hess

LAN/WAN Optimization Techniques, Harrell Van Norman

LANs to WANs: Network Management in the 1990s, Nathan J. Muller and Robert P. Davidson

Minimum Risk Strategy for Acquiring Communications Equipment and Services, Nathan J. Muller

Mobile Antenna Systems Handbook, Kyohei Fujimoto and J. R. James, editors

Mobile Communications in the U.S. and Europe: Regulation, Technology, and Markets, Michael Paetsch

Mobile Data Communications Systems, Peter Wong and David Britland

Mobile Information Systems, John Walker

Multimedia Communications Networks: Technologies and Services, Mallikarjun Tatipamula and Bhumip Khasnabish

Networking Strategies for Information Technology, Bruce Elbert

Packet Switching Evolution from Narrowband to Broadband ISDN, M. Smouts

Packet Video: Modeling and Signal Processing, Naohisa Ohta

Performance Evaluation of Communication Networks, Gary N. Higginbottom

Personal Communication Networks: Practical Implementation, Alan Hadden

Personal Communication Systems and Technologies, John Gardiner and Barry West, editors

Practical Computer Network Security, Mike Hendry

Principles of Secure Communication Systems, Second Edition, Don J. Torrieri

Principles of Signaling for Cell Relay and Frame Relay, Daniel Minoli and George Dobrowski

Principles of Signals and Systems: Deterministic Signals, B. Picinbono

Private Telecommunication Networks, Bruce Elbert

Radio-Relay Systems, Anton A. Huurdeman

RF and Microwave Circuit Design for Wireless Communications, Lawrence E. Larson

The Satellite Communication Applications Handbook, Bruce R. Elbert

Secure Data Networking, Michael Purser

Service Management in Computing and Telecommunications, Richard Hallows

Signaling in ATM Networks, Raif O. Onvural, Rao Cherukuri

Smart Cards, José Manuel Otón and José Luis Zoreda

Smart Card Security and Applications, Mike Hendry

Smart Highways, Smart Cars, Richard Whelan

Successful Business Strategies Using Telecommunications Services, Martin F. Bartholomew

Super-High-Definition Images: Beyond HDTV, Naohisa Ohta, et al.

Telecommunications Deregulation, James Shaw

Television Technology: Fundamentals and Future Prospects, A. Michael Noll

Telecommunications Technology Handbook, Daniel Minoli

Telecommuting, Osman Eldib and Daniel Minoli

Telemetry Systems Design, Frank Carden

Teletraffic Technologies in ATM Networks, Hiroshi Saito

Toll-Free Services: A Complete Guide to Design, Implementation, and Management, Robert A. Gable

Transmission Networking: SONET and the SDH, Mike Sexton and Andy Reid

Troposcatter Radio Links, G. Roda

Understanding Emerging Network Services, Pricing, and Regulation, Leo A. Wrobel and Eddie M. Pope

Understanding GPS: Principles and Applications, Elliot D. Kaplan, editor

Understanding Networking Technology: Concepts, Terms and Trends, Mark Norris

UNIX Internetworking, Second Edition, Uday O. Pabrai

Videoconferencing and Videotelephony: Technology and Standards, Richard Schaphorst

Voice Recognition, Richard L. Klevans and Robert D. Rodman

Wireless Access and the Local Telephone Network, George Calhoun

Wireless Communications in Developing Countries: Cellular and Satellite Systems, Rachael E. Schwartz

Wireless Communications for Intelligent Transportation Systems, Scott D. Elliot and Daniel J. Dailey

Wireless Data Networking, Nathan J. Muller

Wireless LAN Systems, A. Santamaría and F. J. López-Hernández

Wireless: The Revolution in Personal Telecommunications, Ira Brodsky

Writing Disaster Recovery Plans for Telecommunications Networks and LANs, Leo A. Wrobel

X Window System User's Guide, Uday O. Pabrai

For further information on these and other Artech House titles, including previously considered out-of-print books now available through our In-Print-Forever™ (IPF™) program, contact:

Artech House
685 Canton Street
Norwood, MA 02062
781-769-9750
Fax: 781-769-6334
Telex: 951-659
email: artech@artech-house.com

Artech House
Portland House, Stag Place
London SW1E 5XA England
+44 (0) 171-973-8077
Fax: +44 (0) 171-630-0166
Telex: 951-659
email: artech-uk@artech-house.com

Find us on the World Wide Web at: www.artech-house.com